*NUCLEAR AND
PARTICLE PHYSICS*

NUCLEAR AND PARTICLE PHYSICS

E. B. PAUL

1969

NORTH-HOLLAND PUBLISHING COMPANY
AMSTERDAM · LONDON

WILEY INTERSCIENCE DIVISION
JOHN WILEY & SONS, INC.
NEW YORK

Library of Congress Catalog Card Number: 69–18377

SBN 7204 0146 1

Printed in The Netherlands

Publishers:

NORTH-HOLLAND PUBLISHING COMPANY – AMSTERDAM
NORTH-HOLLAND PUBLISHING COMPANY, LTD – LONDON

Sole Distributors for the Western Hemisphere:

WILEY INTERSCIENCE DIVISION
JOHN WILEY & SONS, INC. – NEW YORK

Foreword

When a student of physics reaches an advanced stage in his undergraduate course and decides to specialise in nuclear physics research he finds himself faced with the formidable task of absorbing a vast number of new concepts, new facts and new techniques, particularly the quantum-mechanical tools specific to modern nuclear theory. This task is best begun with the help of good texts written by scientists who are actively pursuing research in the field and who are able through clarity of mind and ease of expression to overcome many of the students' difficulties and inhibitions. There are not many textbooks of recent date satisfying these requirements. The present volume fills this need.

The broad scope of Professor Paul's text is clear from a perusal of his chapter headings and subsections. The inclusion in the book of the results and interpretation of current research in high energy physics is timely. At the moment the pure nuclear physicist has little chance to grasp the relevance of discoveries made with the help of modern high energy accelerators and he finds it difficult to appreciate the bold ideas put forward by the theoretical physicists. It is a great pity for him to be deprived of the excitement of these developments. I feel that this book will help the student to bridge the gap between nuclear and elementary particle physics, technically and conceptually so very different, but nevertheless closely related through their common interest in the origin of nuclear forces.

In normal circumstances a Preface to this volume would have been written by the author. Professor Paul's sudden and untimely death prevented him from putting the final touches to the book, though the text was complete when he died.

The arduous task of proof-reading has been shouldered by Dr. B. E. Bonner, a physicist at the Rice University, Texas while Professor Paul was there, and presently a Research Fellow of the Rutherford Laboratory, England. Dr. Bonner's willing efforts in preparing the book for publication, and Mrs. Vivien Paul's continued support, deserve the grateful thanks of nuclear physicists.

To complete this foreword I feel it appropriate to indicate in a few lines some of the landmarks in Professor Paul's career:

Born in Canada in 1919, he studied physics at Queen's University, Kingston, Ontario. After working at the National Research Council laboratories in Ottawa during the war years, he went to the Cavendish Laboratory, Cambridge, England, obtaining his Ph.D. degree there in 1948. Research in nuclear physics was Paul's great interest which he felt could best be pursued at one of the large National Laboratories set up after the war. He went first to the Chalk River Laboratory of the Atomic Energy of Canada Ltd., and subsequently joined the Atomic Energy Research Establishment, Harwell. It was during the Chalk River period (1948–56) that Paul made the important observation that the energy levels of fluorine could be simply and elegantly interpreted in terms of the collective model of the nucleus. This discovery in particular established Paul's international reputation. Joining the Nuclear Physics Division of AERE in 1956, he soon became Group Leader of the High Voltage Laboratory, and later Deputy Head of the Division. He left AERE in 1961 for a period at universities, occupying a Chair first at Manchester University and then at Rice University. He returned to Britain again in 1966, accepting a senior appointment at the National Physical Laboratory, Teddington, which position he was holding at the time of his death.

E. Bretscher

List of Contents

Interaction of Radiation with Matter — Chapter 3

Detection and Measurement of Radiation — Chapter 4

Accelerators *Chapter 5*

Nuclear Reactions　*Chapter 6*

Reaction Mechanisms　*Chapter 7*

Nuclear Models: Shell Model *Chapter 8*

Nuclear Models: Collective Model *Chapter 9*

1

Nuclear Properties

1 Introduction

We shall first make a brief survey of our knowledge of the physical world so as to place the atomic nucleus in its correct perspective. We believe that our physical world is controlled by four kinds of force of widely different strengths, as listed in table 1.1.

TABLE 1.1

	Force	Relative strength	Range
1)	Strong or nuclear	15	short: 10^{-13} cm
2)	Electromagnetic	$\frac{1}{137}$	r^{-2}
3)	Weak	10^{-15}	short: 10^{-14} cm or less
4)	Gravitational	10^{-40}	r^{-2}

It is clear at once that forces 1) and 3) act only in subatomic systems while 2) and 4) act also in macroscopic systems.

In quantum electrodynamics a fruitful and elegant connection is made between the electromagnetic force and the quanta of electromagnetic radiation, the photons. We can by analogy consider that the other force fields also are associated with field quanta. Thus the mesons (pions, kaons, etc.) can be considered as field quanta of the strong force, the photon can be considered the field quantum of the electromagnetic force, the as yet unobserved 'vector boson' would be the field quantum of the

1

weak force and the as yet unobserved 'graviton' would be the field quantum of the gravitational force.

Field particles can become observable when a high momentum is given the 'source' of the field. The field is then, as it were, left behind. The sources of the four fields are as follows:

1) nuclear force: nucleons or hyperons or mesons,
2) electromagnetic force: electric charge,
3) weak force: leptons or baryons,
4) gravitational force: mass.

The structure of atoms and molecules depends primarily on force 2), the electromagnetic. The structure of nuclei depends primarily on force 1) although forces 2) and 3) are of significance when we consider the decay of nuclei on a long time scale.

The lines of force of the field terminate in the sources. However a line has two ends. At the opposite end will be a particle with properties opposite to that of the source particle. When a field line is broken, one particle of each kind results: i.e. a pair is produced. For example annihilation of a photon results in an electron–positron pair, one with negative charge, the other with positive charge. There are similar effects connected with the strong and weak force fields when a nucleon pair of proton and antiproton are produced, or a lepton–antilepton pair is produced. Thus our fundamental particles are often associated with the appropriate anti-particles.

We are concerned with the nucleus of the atom. Atoms are made up of a small central nucleus where essentially all the mass resides and carrying net positive charge Ze. In the neutral atom this nucleus is surrounded by a cloud of Z electrons each carrying negative charge e. These electrons are arranged in different energy states and a rather oversimplified view is that they are orbiting the nucleus at various distances. The more tightly bound electrons are said to occupy 'inner' shells while the more loosely bound electrons are said to occupy 'outer' shells. The number and binding energies of these outer electrons give the atom its chemical character, in particular its valency.

These outer electrons are only bound rather loosely and when an atom loses one or more electrons or attaches an extra electron, it has a net positive (or negative) charge and is called an *ion*. In a gas or liquid an ion is free to migrate under an applied electric or magnetic field. In a solid or crystal lattice however it cannot usually move but instead contributes to the polarization of the solid in an applied field.

The electrons themselves can move freely also in gases and liquids and in solid conductors. In the latter case a certain number can divorce themselves from the crystal lattice and act as conduction electrons.

The atom is an object of the order of 10^{-8} cm in diameter. The nucleus is very much smaller: of the order of 10^{-12} cm diameter. Even though it carries most of the atomic mass it is 10^4 times smaller. The nucleus is composed of nucleons of two kinds, positively charged *protons* and *neutrons* of similar mass but carrying no electric charge. Hence there are Z protons and $N = A - Z$ neutrons, where A represents the total number of nucleons present and Z the nuclear charge. Since A represents the total number of nuclear masses it is called the *mass number*. Z represents the nuclear charge or the number of protons in the nucleus or the number of electrons in a neutral atom, and is called the *atomic number*.

The nucleons are held together in the nucleus by the strong force which acts over the short distances involved within a nucleus. This strong force acts between all nucleons, no matter whether they are electrically charged or not. That is, the strong force is charge independent. Of course the electromagnetic force also acts within the nucleus causing the protons to repel each other and try to fly apart. But since it is so much weaker than the strong force attraction, the nucleus stays together provided it is not too large. We do not see nuclei with $A > 250$ because of the coulomb repulsion.

The subject of nuclear physics therefore has to do with the structure of the atomic nucleus and the manner in which it can decay or change its energy. From this study we try to infer something about the nature of the forces involved in the nucleus. Closely connected with this is the study of the kinds of fundamental particles which we find associated with the forces.

2   *Historical development*

Although we shall discuss nuclear physics in a logical manner and therefore not directly related to the historical sequence, we should briefly sketch the early history of the subject. It is a most exciting period in the history of knowledge encompassing only two and a half generations. In 1868 Mendeleev had devised the periodic table of the elements based on their chemical properties, and in 1895 Röntgen had discovered continuous X-rays. This latter discovery had excited extreme interest among physicists and the ground was prepared for Becquerel's observation in 1896 of the radioactivity of uranium. He noted the similarity of the radiation to Röntgen's rays. In 1897 J. J. Thomson identified the electron as a fundamental particle. In 1898 Pierre and Marie Curie separated polonium and radium as new radioactive elements. In 1899 Ernest Rutherford at McGill University in Canada observed the exponential decay of thoron gas. In 1900 Planck proposed his quantum hypothesis. In 1903 Rutherford and Soddy proposed the transformation theory of α- and β-decay, that is, radioactivity involved the change of one element into another. In the period 1906–1911 Soddy, Russell and Fajans proposed the displacement laws in which an α-decay decreases Z by 2 units and decreases A by 4 units while β-decay increases Z by 1 unit and leaves A unchanged. From this Soddy inferred the existence of isotopes.

In 1908 Rutherford (now at Manchester) and Geiger measured the charge of the α-particle to be twice that of the electron. In 1909 Rutherford and Royds identified the α-particle as a helium ion by detecting the helium gas evolved from radon. In 1911 the α-particle scattering experiments of Marsden and Geiger led Rutherford to postulate the nuclear atom. In 1913 Niels Bohr came to Manchester to work out the theory of the nuclear-atom model which at one stroke explained so much of atomic spectroscopy. Moseley in the same year derived the atomic numbers Z from the characteristic X-ray spectra. In 1914 Rutherford and Robinson measured the e/m ratio for the α-particle and found its mass to be four times the proton mass. In the next few years Thomson and Aston observed the isotopes of neon, and Rutherford, Marsden and others observed the first nuclear reaction initiated by an α-particle and resulting in a proton.

The next decade was a period of great theoretical advance while the experimentalists consolidated the position, filled in details and attempted to devise new techniques. De Broglie proposed his matter waves in 1924, Schrödinger his wave equation in 1926. In 1927 diffraction of electrons was observed by Davisson, Germer and G. P. Thomson. In the same year Heisenberg proposed the uncertainty principle and in 1928 the theory of potential barrier penetration was published by Gamow, Gurney and Condon.

The realization that quantum mechanics allowed a barrier to be penetrated led to a remarkable period of progress in devising artificial methods of accelerating nuclear particles. In the period 1930–1932, the cyclotron of Lawrence, the electrostatic accelerator of Van de Graaff and the observation of the first artificially induced nuclear reaction by Cockcroft and Walton, all took place. Also in 1932 Chadwick identified the neutron. All the experimental advances were spurred on by theoretical predictions but were made possible by the recent availability of mechanical vacuum pumps, electronic scalers and new detecting devices.

Artificial radioactivity was discovered by Irene Curie and Frédéric Joliot in 1934 and neutron capture by Fermi in the same year. Nuclear β-decay theory was put on a sound basis by Fermi and Pauli in 1933 and 1934. The discovery of nuclear fission in 1939 by Hahn and Strassmann started then the enormous efforts in which physicists are still engaged to utilize the nuclear forces as a source of energy. The artificial production of mesons came in 1948 by Gardner and Lattes, and ushered in the era of high energy physics in which the substructure of the nucleons themselves and of the nuclear fields is being probed. Such particles had been observed previously in cosmic radiation; μ-mesons by Anderson and Neddermeyer in 1936 and π-mesons by Powell in 1946.

3 *Nuclear mass*

The mass of an atom can be expressed in several different ways. It is rarely convenient to express the mass absolutely in grams as this tends to conceal the physical
significance. We have noted that the atomic mass is rather close to being an integral multiple of the nucleon mass. We called this A, the mass number. The physical
mass scale is then related to this. The unit of mass is taken to be one twelfth of
the mass of the atom ^{12}C. This unit is termed a mass unit (mu). It should be noted
that before 1961 the mass unit was based on ^{16}O which is larger by about three
parts in ten thousand. Finally, since the interchangeability of mass and energy is
commonplace in nuclear physics, masses may also be expressed in energy units.
In other words the rest mass $m_0 c^2$ is quoted in MeV or sometimes more precisely
as MeV/c^2. Our atomic mass unit therefore has the value

$$1 \text{ mu} = 931.478 \text{ MeV}/c^2 = 1.66042 \times 10^{-24} \text{ gram.}$$

Note that conventionally the mass of the neutral atom including its full complement of electrons is used and the actual mass of the nucleus is not used, even though
we may refer to the 'mass' of such and such a nucleus. Since the mass of an electron
is only 1/1836 times the mass of a nucleon this would not seem to be a very serious
error. However when we are dealing with small differences in mass this is of great
importance.

We find that the masses of nuclei defined on the physical scale are close to integral numbers but are not exact multiples of the proton or neutron masses. The
difference between the exact atomic mass of an isotope $M(A, Z)$ and its mass
number A is called the *mass excess* or the *mass defect* $\Delta = M(A, Z) - A$. The mass
excess per unit mass number is called the *packing fraction*

$$P = \frac{\Delta}{A} = \frac{M(A, Z) - A}{A}. \tag{1.1}$$

If we assume that a nucleus is made up of protons and neutrons we can calculate
the energy needed to break it up into its constituent nucleons. This energy is called
the *total binding energy*.

$$B(A, Z) = ZM_H + (A - Z)M_n - M(A, Z). \tag{1.2}$$

Thus B measures the same thing as the mass excess Δ except that now we use the
exact masses of the hydrogen atom M_H and of the neutron M_n. In analogy with the
packing fraction we define the average binding energy per nucleon as B/A.

We see that the deviations of the mass values from exact multiples of the nuclear
masses just represents the conversion of mass into binding energy.

We may also define the minimum energy required to remove a specific particle

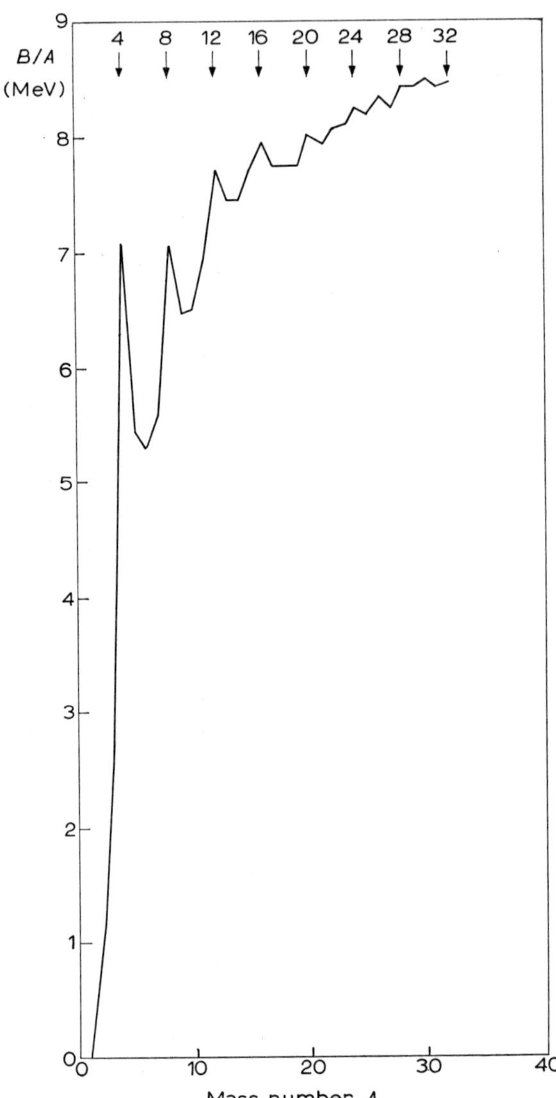

Fig. 1.1a.   The binding energy per nucleon for the light nuclei

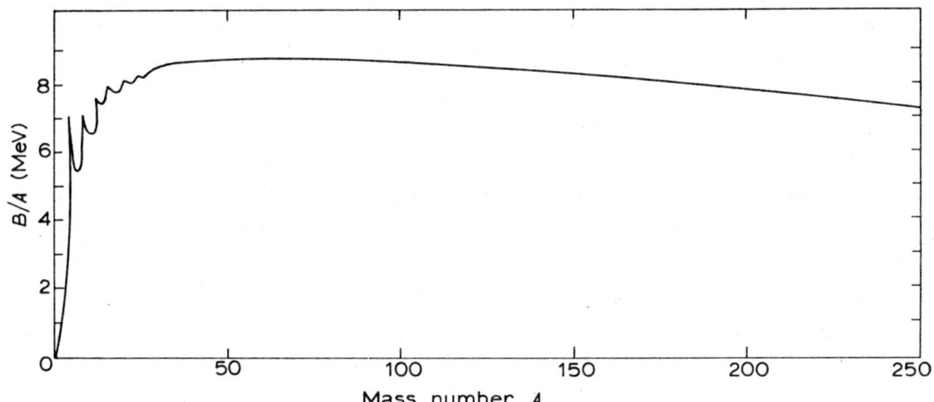

R. D. Evans, The Atomic Nucleus (McGraw-Hill, New York 1955) p. 299

Fig. 1.1b.   The binding energy per nucleon for all nuclei

such as a proton, neutron or α-particle from a nucleus. We call this the *separation
energy* for that particle;

for a neutron
$$\begin{aligned}
S_\mathrm{n} &= B(A, Z) - B(A-1, Z) \\
&= M(A-1, Z) - M(A, Z) + M_\mathrm{n},
\end{aligned} \tag{1.3}$$

for a proton
$$\begin{aligned}
S_\mathrm{p} &= B(A, Z) - B(A-1, Z-1) \\
&= M(A-1, Z-1) - M(A, Z) + M_\mathrm{H},
\end{aligned} \tag{1.4}$$

for an α-particle
$$\begin{aligned}
S_\alpha &= B(A, Z) - B(A-4, Z-2) \\
&= M(A-4, Z-2) - M(A, Z) + M_\mathrm{He}.
\end{aligned} \tag{1.5}$$

The separation energy is also termed the binding energy for that particle in the
nucleus (A, Z). In general this is not the same as B/A, the average binding energy
per nucleon.

It is instructive to present the average binding energy per nucleon as a function
of mass number A (fig. 1.1). This curve illustrates many properties of nuclear
binding:

1)   The binding energy is *positive* for all nuclei; that is, the nucleus is more stable
than the sum of its constituents. This means that the nuclear force between nucleons is *attractive*. On the other hand nuclei do not collapse to very small dimensions so the force must become somehow repulsive at very small distances.

2)   The peaks at low values of A occur at values of A which are multiples of 4
and for values of $Z = \frac{1}{2}A$. These are therefore multiples of the two proton–two
neutron structure like the α-particle and must be unusually stable.

3)   For A greater than about 20 the value of B/A does not vary very much but lies
between 7.5 and 8.5 MeV/nucleon. That is, for most nuclei B must be approxi-

mately proportional to the number of nucleons A. However if each nucleon interacted with every other nucleon we should expect B to be proportional to A^2. Hence each nucleon must interact with only a limited number of neighbouring nucleons. We call this property *saturation* of the nuclear force.

4)   The slow decrease in binding energy between $A=60$ and the heaviest nuclei can be attributed to the increased electrical repulsion of the protons, which tends to disrupt the nucleus. This effect is compensated somewhat by adding more neutrons than protons in the heavier nuclei and maintaining stability by the increased nuclear force between the neutrons. However there are limits to this as we shall see later and the coulomb force eventually limits the number of stable elements.

5)   If we look closely at individual binding energy values we also observe that there is a tendency for nuclei with even numbers of protons (even Z) or even numbers of neutrons (even N) to be more stable than those with odd Z or N. This we shall see is the result of a *pairing force* between like nucleons with opposite spins.

Table 1.2 lists some masses and binding energies and illustrates some of these

TABLE 1.2

	Mass (mu)	$B(A, Z)$ (MeV)	B/A (MeV)
n	1.008665	–	–
^{1}H	1.007825	–	–
^{2}H	2.014102	2.225	1.113
^{3}H	3.016050	8.482	2.827
^{3}He	3.016030	7.717	2.572
^{4}He	4.002603	28.295	7.074
^{5}He	5.012296	27.338	5.468
^{10}B	10.012939	64.749	6.475
^{13}C	13.003354	97.107	7.470
^{13}N	13.005739	105.284	8.099
^{14}N	14.003074	104.657	7.476

features. We note here the large binding energy for ^{4}He. The binding energy of the extra neutron in ^{5}He is in fact negative and this nucleus at once breaks up into ^{4}He and a neutron.

Using some of the other masses in the table we can calculate the separation energies for various particles in ^{14}N

$$M(^{13}\text{N}+\text{n}) = 14.014404$$
$$M(^{14}\text{N}) = 14.003074$$

$$S_\text{n} = 0.011330 \text{ mu} = 10.553 \text{ MeV.}$$

$$M(^{13}\text{C}+\text{p}) = 14.011179$$
$$M(^{14}\text{N}) = 14.003074$$

$$S_\text{p} = 0.008105 \text{ mu} = 7.549 \text{ MeV.}$$

$$M(^{10}\text{B}+^{4}\text{He}) = 14.015542$$
$$M(^{14}\text{N}) = 14.003074$$

$$S_\alpha = 0.012468 \text{ mu} = 11.613 \text{ MeV.}$$

The effect of pairing can most easily be illustrated by looking at the effect of adding neutrons to nuclei with fixed Z (table 1.3). We define the pairing energy $P_\text{n}(A, Z) = S_\text{n}(A, Z) - S_\text{n}(A-1, Z)$.

TABLE 1.3

	$S_\text{n}(A, Z)$ (MeV)	$P_\text{n}(A, Z)$ (MeV)
^{40}Ca	15.73	2.74
^{41}Ca	8.36	-7.37
^{42}Ca	11.47	3.11
^{43}Ca	7.93	-3.54
^{44}Ca	11.14	3.21
^{45}Ca	7.42	-3.72
^{46}Ca	10.40	2.98
^{47}Ca	7.30	-3.10
^{48}Ca	9.93	2.63
^{49}Ca	5.14	-4.79

This table shows clearly that about 3 MeV of extra binding can be attributed to the pairing of two neutrons. The rather larger binding observed when 20 protons are combined with 20 or 28 neutrons is a more subtle effect which we shall discuss in ch. 8.

A more compact method of listing atomic masses is to list the *mass excess* rather than the mass itself and modern mass tables normally quote this quantity.

The first three conclusions drawn from the binding energy curve suggest that nuclear matter behaves somewhat similarly to a liquid drop wherein the short range attractive intermolecular forces lead also to saturation and a fixed potential energy per molecule. Using this analogy the nuclear masses are often described by a formula which is designed to represent analytically the binding energy curve but containing empirically determined constants chosen to give the best fit to the measured masses. Such a semi-empirical mass formula was first devised by von Weizsäcker and has been refined and improved by Fermi and others to include more details.

Let us assume that the neutron–proton liquid drop is of constant density. The volume is then proportional to the number of nucleons A. The radius is then proportional to $A^{\frac{1}{3}}$, that is $R=r_0 A^{\frac{1}{3}}$ where r_0 is an empirically determined constant and has a value 1.2×10^{-13} cm. A formula for the total binding energy $B(A, Z)$ will then include the following terms.

1)   The main binding term is the result of the short range attractive forces between nucleons. It was seen to be roughly proportional to the number of nucleons A since B/A was approximately constant. This term is called the *volume term*, $+a_v A$ where a_v is the empirical constant.

2)   The nucleons at the surface of the nucleus will not contribute so much to the binding as those inside the nucleus. The binding will be *reduced* by a factor proportional to the surface area which is proportional to $A^{\frac{2}{3}}$. So the *surface term* is $-a_s A^{\frac{2}{3}}$.

3)   In the light nuclei there is a tendency for nuclei with the same number of neutrons and protons to be most stable. We introduce a term which represents the departure of our nucleus from this symmetrical situation and is thus proportional to $(\frac{1}{2}A-Z)^2$, squared so that excess protons or neutrons have the same effect in *reducing* the binding. This effect however decreases in importance for heavier nuclei so we also include a factor A^{-1}. Thus the *asymmetry* term is $-a_a(A-2Z)^2/A$.

4)   The nuclear electric charge also has a disruptive effect and reduces the binding energy. If the charge Ze is confined to a sphere of radius $R=r_0 A^{\frac{1}{3}}$ the potential energy of such a uniformly charged sphere is $\frac{3}{5}(Ze)^2/R \propto Z^2/A^{\frac{1}{3}}$. So the coulomb term is $-a_c Z^2/A^{\frac{1}{3}}$.

5)   We also noted that for nuclei with even A, those with even Z, even N are more stable than those with odd Z, odd N. The term to take account of this pairing effect is written $+\delta(A, Z)$ where

$$\begin{aligned}
\delta &> 0 \quad \text{for } A \text{ even, } Z, N \text{ both even,} \\
\delta &< 0 \quad \text{for } A \text{ even, } Z, N \text{ both odd,} \\
\delta &= 0 \quad \text{for } A \text{ odd.}
\end{aligned}$$

The complete semi-empirical formula is then

$$B(A, Z) = a_v A - a_s A^{\frac{2}{3}} - a_a \frac{(A-2Z)^2}{A} - a_c \frac{Z^2}{A^{\frac{1}{3}}} + \delta(A, Z). \tag{1.6}$$

A reasonable fit to the binding energy curve is obtained with the following values of the constants

$$a_v = 15.6 \text{ MeV} \qquad a_c = 0.70 \text{ MeV}$$
$$a_s = 17.3 \text{ MeV} \qquad \delta = 33.5A^{-\frac{3}{4}} \text{ MeV}$$
$$a_a = 23.3 \text{ MeV}$$

4   Measurement of nuclear masses

There are two distinct methods for determining masses. One involves direct observation of the deflection of charged ions in electric and magnetic fields. Such instruments are called mass spectrographs or spectrometers. The other method involves a study of the energy change in nuclear reactions. Since this is related to the mass change the difference in nuclear masses can be derived.

The earliest type of mass spectrograph was Thomson's parabola method in 1913. This was primarily a qualitative device. The first instrument suitable for precise determinations was Aston's mass spectrograph.

The ions to be studied are produced in the discharge tube and emerge with charge ze through the collimation slits with a range of velocities. The electric field acts as an energy filter since the radius of curvature R_E is given by

$$zeV = \frac{mv^2}{R_E} , \qquad R_E = \frac{mv^2}{zeV} . \tag{1.7}$$

R_E is defined by the two sets of apertures. The magnetic field acts as a momentum filter since

$$Hzev = \frac{mv^2 c}{R_M} , \qquad R_M = \frac{mvc}{zeH} . \tag{1.8}$$

Thus for ions having a given ratio of charge to mass

$$\frac{ze}{m} = \frac{R_E Vc^2}{R_M^2 H^2} . \tag{1.9}$$

Then if V, R_E and H are kept constant the position where the ions register on the photographic plate corresponds to a particular value of R_M^2 which is proportional to ze/m and independent of the ion velocity. Since we record our mass spectrum photographically we term this instrument a *mass spectrograph*. Aston's instrument could measure masses to about 1 part in 10^3 (fig. 1.2). Modern versions achieve accuracies of about 1 part in 10^5.

At about the same time Dempster developed another type of mass measuring instrument (cf. fig. 1.3). The ions are initially accelerated through a potential

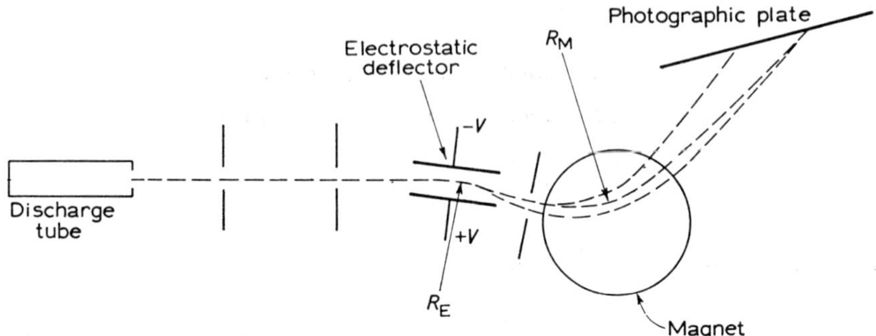

Fig. 1.2.   The mass spectrograph

$V \approx 1000$ V so that they pass into the magnetic field with energy zeV. Their velocity is therefore given by

$$\tfrac{1}{2}mv^2 = zeV, \qquad v = (2zeV/m)^{\frac{1}{2}}. \tag{1.10}$$

In the magnetic field

$$HR = mvc/ze = (2mVc^2/ze)^{\frac{1}{2}}. \tag{1.11}$$

Thus for fixed values of H and V ions having a given value of ze/m will traverse the magnetic field with a definite radius of curvature R and be detected in the detector. We can then vary V and successive values of ze/m will be detected. Since the

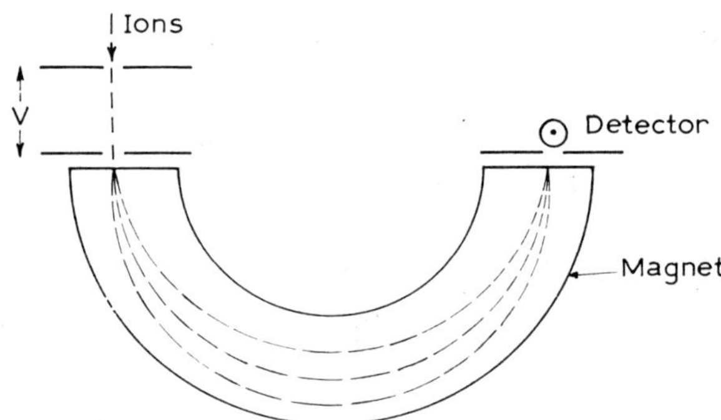

Fig. 1.3.   The mass spectrometer

ions are detected electrically this device is called a *mass spectrometer*. Modern versions can give masses accurate to 1 part in 10^6. For precise measurements two mass lines are compared which lie close together, one known and the other unknown.

Other types of mass spectrometer have been devised depending for example on time-of-flight of ions of specified energy. Modern mass spectrometers and spectrographs often contain a second set of energy and momentum filters following the first set to achieve better resolution.

In the second general method of mass determination the energy changes in a nuclear reaction are carefully measured and a relationship between the masses of nuclei involved is then determined. In many cases a series of such reactions can be used so that a complete cycle of masses is determined. Where this can be done the accuracy is generally somewhat better than can be achieved with mass spectrometers.

There is a very large body of such data available at present both from nuclear reactions and from mass spectrometers. A best set of masses is extracted from these by an elaborate procedure of least square fitting to the weighted experimental values.

5   Nuclear stability

We now consider further details of nuclear stability, making use of the semi-empirical mass formula (1.6) to clarify them. In fig. 1.4 we plot the known stable nuclides on a graph of neutron number N against proton number Z. This is often referred to as a Segre chart. The unstable radioactive nuclides lie in a fringe on both sides of the band of stable nuclides. The deviation from the $Z=N$ line for the heavier elements is just a result of requiring more neutrons to be added to offset the progressively disruptive effect of the coulomb repulsion of the Z protons. If we take the semi-empirical binding energy formula for odd-A nuclei ($\delta=0$) and differentiate with respect to Z we can determine the charge Z of the nucleus of maximum stability for given A. The result is

$$Z = \frac{A}{2+0.015A^{\frac{2}{3}}} \tag{1.12}$$

which shows how the proton number falls short of $\frac{1}{2}A$ for the stable nuclei as A increases.

If in addition to plotting the nuclides as a function of Z and N we also plot the values of $M(A, Z)$ in the z-direction, we obtain a mass surface in which the stable nuclides lie in the bottom of a valley (smallest mass, largest binding) while the less stable ones lie on the sides of the valley. If we take a section through this mass surface for constant A we get a parabolic curve. This can be seen from the form of the semi-empirical formula for the mass

$$M(A, Z) = ZM_{\mathrm{H}}+(A-Z)M_{\mathrm{n}}-B(A, Z). \tag{1.13}$$

This is

$$M(A, Z) = aZ^2+bZ+c\pm\delta(A, Z)$$

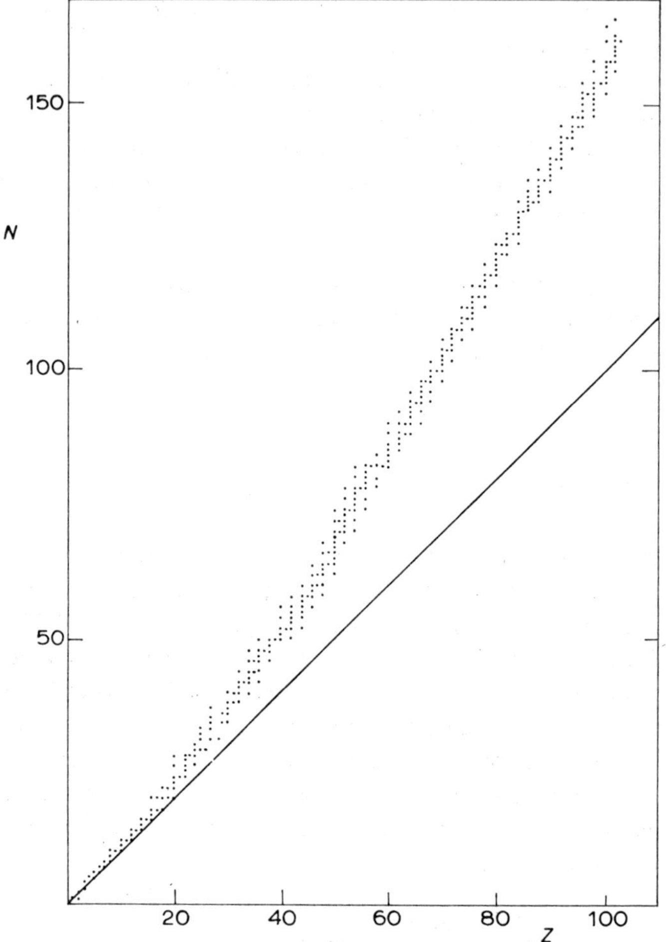

Fig. 1.4.   The Segre chart for the stable nuclei. The solid line is for $Z=N$

where a, b, c are constants. Approximately a is inversely proportional to A, b is independent of A and c is proportional to A. This demonstrates that the mass parabola slowly flattens out for increasing A being steepest for low A. For odd-A nuclei $\delta=0$ and we have one parabola. For even A we have two parabolae, the lower for even-Z, even-N nuclei and the upper at 2δ higher for odd–odd nuclei. Thus for odd-A nuclei there will be one stable nuclide only, whilst for even A there may be two or more stable isobars. In general we should expect no odd–odd nuclei to be stable, but in fact four of the lightest odd–odd nuclei ^{2}H, ^{6}Li, ^{10}B and ^{14}N are stable. This is because the asymmetry term has a large effect for light nuclei. However heavier odd–odd nuclei are all unstable. Transitions between isobaric nuclei take place by negative electron (β^-) emission or by positive electron (β^+) emission or by electron capture. The transitions result in a more stable nucleus being formed.

The β^--transitions result in a change of Z of one unit to the right while β^+ or electron capture transitions result in a change of Z of one unit to the left. The energy changes involved are typically a few hundred keV to a few MeV. Thus we must take care to account for the electron masses involved since the electron rest mass is equivalent to 511 keV.

Suppose we designate the mass of the neutral atom as $M_a(A, Z)$ and the mass of the nucleus without electrons as $M_N(A, Z)$. We have not hitherto used the latter mass. Then in β^--decay we consider the transition

$$Z^A \rightarrow (Z+1)^A + \beta^-.$$

The energy available for the transition is then

$$
\begin{aligned}
E_{\beta^-} &= M_N(Z, A) - M_N(Z+1, A) - m_{\beta^-} \\
&= [M_a(Z, A) - Zm_e] - [M_a(Z+1, A) - (Z+1)m_e] - m_e \\
&= M_a(Z, A) - M_a(Z+1, A).
\end{aligned}
\tag{1.14}
$$

In β^+-decay we have the transition

$$Z^A \rightarrow (Z-1)^A + \beta^+,$$

$$
\begin{aligned}
E_{\beta^+} &= M_N(Z, A) - M_N(Z-1, A) - m_{\beta^+} \\
&= [M_a(Z, A) - Zm_e] - [M_a(Z-1, A) - (Z-1)m_e] - m_e \\
&= M_a(Z, A) - M_a(Z-1, A) - 2m_e.
\end{aligned}
\tag{1.15}
$$

In electron capture an atomic electron is captured from an inner atomic shell, typically the K-shell in which it is bound with energy E_K

$$Z^A + e_K^- \rightarrow (Z-1)^A,$$

$$
\begin{aligned}
E_{EC} &= M_N(Z, A) - M_N(Z-1, A) - E_K + m_e \\
&= [M_a(Z, A) - Zm_e] - [M_a(Z-1, A) - (Z-1)m_e] + m_e - E_K \\
&= M_a(Z, A) - M_a(Z-1, A) - E_K.
\end{aligned}
\tag{1.16}
$$

The final line in each case enables us to calculate the energy available in terms of the atomic masses which are normally used. This energy is carried off as kinetic energy by the β-particle, a neutrino (in the case of electron capture, a neutrino only), and by recoil of the resulting nucleus. Since the neutrino has zero rest mass it does not enter into the energy balance.

We can see from the mass surface sections that as we move further from the stable valley the mass changes become larger, that is, the energy change involved in the transition becomes larger. This results in a faster decay or shorter lifetime. Thus as we observe nuclides further from the stability line we may expect them to have shorter lifetimes on average. We note also that the most stable odd–odd nucleus can frequently decay to either of two even–even neighbours and show both β^- and β^+ decay or electron capture decay. The relative probability of the two decays will again be governed mainly by the energy available for each.

6 Nuclear size

6.1 Definitions. Since the nucleus has a net charge of $+Ze$ an electrostatic force will be exerted between it and another charged object. If we plot the potential energy against radius we must include both the strong force of short and definite range which is attractive $(-V)$ and the electromagnetic force of long range with potential falling off as $1/r$. When these are combined we get a potential as shown in fig. 1.5. We define the barrier height at distance R from the centre of the

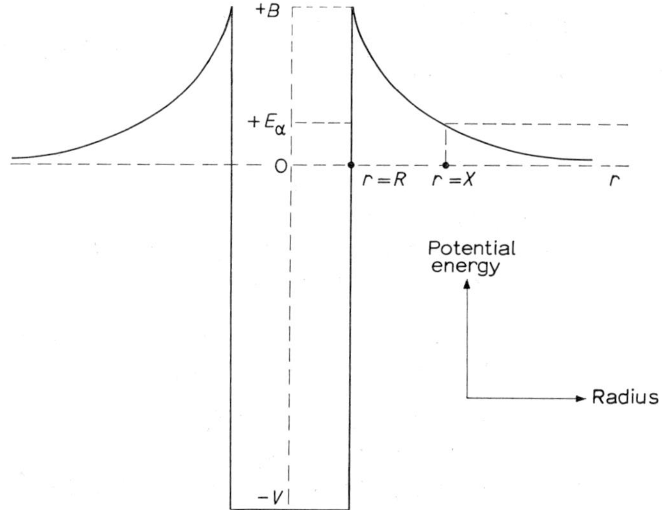

Fig. 1.5.   Nuclear potential energy vs nuclear radius

nucleus where the attractive force is just zero. The barrier height is then $B = zZe^2/R$ for an incident charged particle of charge ze. For example for uranium $R = 8 \times 10^{-13}$ cm and $B = 17$ MeV for protons $(z = 1)$ and $B = 34$ MeV for α-particles $(z = 2)$.

Classically only particles of this energy or greater could penetrate or escape from the nucleus but quantum mechanically the position of a particle of lower energy inside the nucleus is not well defined because of the uncertainty principle and there is a small but finite chance of finding it outside the nucleus. If it finds itself sometimes at the outside surface of the barrier at point X it is then repelled by the coulomb force of the rest of the protons in the nucleus and attains its original energy

E_α at a distance from the nucleus. It is as if the charged particle had 'tunnelled' through the barrier. In fact uranium emits α-particles of energy 4.2 MeV even though the barrier is 34 MeV high. It is only *very rarely* that such α-particles are emitted since the lifetime of uranium is 4.5×10^{10} years while in the absence of the barrier the lifetime would be 10^{-20} sec. This illustrates the small probability of finding the α-particle outside.

The radius R we defined above is actually the radius of the *potential*. This will be slightly larger than the *mass radius* or the *charge radius* because of the range of the nuclear force. On the other hand because of the strong nuclear forces we expect that the *charge radius*, that is the radius of the proton distribution, will be very close to the *mass radius*, that is the radius of neutrons plus protons. The charge radius is the most directly measurable of these two. The potential radius must be measured separately as we do not know a priori the range of the nuclear force. Nuclear radii are measured in units of 10^{-13} cm called by international convention femtometers (10^{-15} m), abbreviated fm. These units are sometimes referred to as fermis, an earlier unofficial designation. The mean square radius of the charge distribution is defined as

$$\overline{r^2} = \frac{\displaystyle\int_0^\infty r^2\, 4\pi r^2 \varrho(r)\,\mathrm{d}r}{\displaystyle\int_0^\infty 4\pi r^2 \varrho(r)\,\mathrm{d}r} \tag{1.17}$$

where $\varrho(r)$ is the charge density. The radius of a uniformly charged object is then $R = (\tfrac{5}{3}\overline{r^2})^{\frac{1}{2}}$.

6.2   Measurement of charge radii. The most direct method is to study the interaction of electrons with nuclei. Since electrons have no strong force interaction but only electromagnetic interaction with the nucleus it is as if pure electric charges were being used. The de Broglie wavelength of 200 MeV electrons is about 10^{-13} cm so they are suitably 'small' objects to use as probes. The *elastic scattering of electrons* is thus one of the most used methods for charge radius determination. An electron linear accelerator (fig. 1.6) is used to produce well collimated beams of electrons of energy up to 1 GeV, and the angular distribution of the elastically scattered electrons is observed. We shall see later that the differential cross section is related to the number of scattered electrons per unit solid angle and has the form

$$\sigma(\theta) = \left[\left(\frac{Ze^2}{2E}\right)^2 \frac{\cos^2 \tfrac{1}{2}\theta}{\sin^4 \tfrac{1}{2}\theta}\right] [F(q)]^2 \tag{1.18}$$

where Z is the target nuclear charge, E is the electron energy and θ is the scattering angle. The first bracket is the expression for scattering from a *point* charge. It is called the Mott scattering and is valid for $2Ze^2/\hbar c \ll 1$. $F(q)$ (≤ 1) is called a form

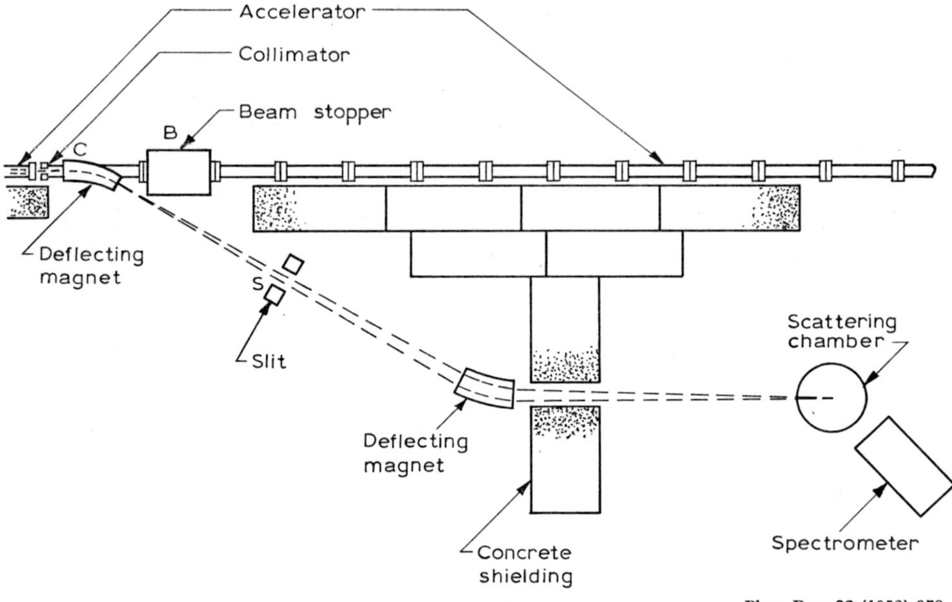

Phys. Rev. **92** (1953) 978

Fig. 1.6.   Experimental arrangement for electron scattering measurements

factor, which expresses how much the scattering is *reduced* if the charge is spread out over a volume. It is less than that for a point charge because of destructive interference between waves scattered from different parts of the object. We can write $F(q)$ as

$$F(q) = \frac{1}{Ze} \int \varrho(r) \exp i[(\mathbf{k} - \mathbf{k}') \cdot \mathbf{r}] d\tau \qquad (1.19)$$

where $\varrho(r)$ is the charge density and the other factor is a phase factor over the volume. This is in terms of

$$q = \frac{|\mathbf{p} - \mathbf{p}'|}{\hbar} = |\mathbf{k} - \mathbf{k}'| = \frac{2p}{\hbar} \sin \tfrac{1}{2}\theta, \qquad (1.20)$$

the momentum transfer in the elastic scattering. The form factor is thus just the Fourier transform of the charge density. Crudely speaking $F(q)$ corresponds to the charge effective in a given collision or the amount of charge in a sphere of radius $1/q$.

When the angle θ reaches a point where the destructive interference is a maximum there will be a marked reduction over Mott scattering and there will be a minimum in the cross section. The first minimum corresponds roughly to $\theta_{\min} \approx \lambdabar/R$ where $\lambdabar$ is the electron wavelength and R is the charge radius. If $F(q)$ is different from unity however the minimum will not be sharp. We then have to choose a charge density distribution $\varrho(r)$ which will fit the observed curve. The density which fits is

$$\varrho(r) = \frac{\varrho_0(r)}{1+\exp\left[(r-R)/a\right]} \tag{1.21}$$

where $R \propto A^{\frac{1}{3}}$ and $a \approx 0.5 \times 10^{-13}$ cm. This corresponds to a fuzzy edge to the nucleus with the charge distribution falling to half value at $r = R$ and with a surface thickness independent of A falling from 90% to 10% in about 2.5×10^{-13} cm (fig. 1.7).

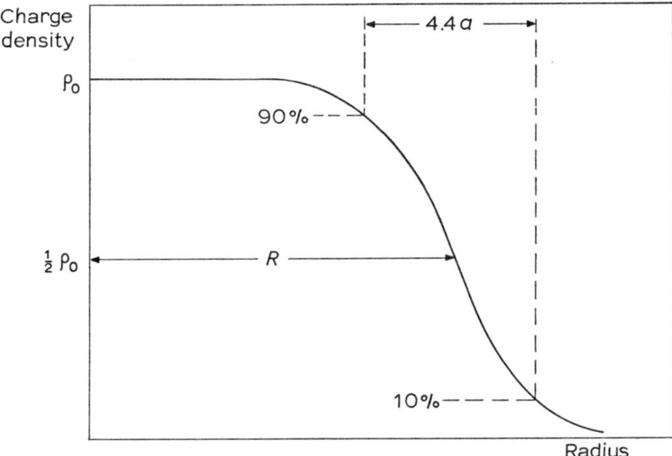

Fig. 1.7.   The nuclear charge density distribution

If we approximate this distribution with a fuzzy edge by a uniform charge distribution we can write $R = r_0 A^{\frac{1}{3}}$ and the constant r_0 then has a value 1.32×10^{-13} cm ($A < 50$) and 1.21×10^{-13} cm ($A > 50$). This confirms that nuclear matter is very nearly of constant density if we assume that $R_{\text{mass}} \approx R_{\text{charge}}$.

The potential energy of the inner atomic electrons is also sensitive to the charge radius. We can use this to detect small differences in radius of isotopes of the same element. These have of course the same electronic structure. Such *isotope shifts* have been observed in optical and in X-ray spectra and lead to a value of $r_0 = 1.2 \times 10^{-13}$ cm. In some nuclei excited states have fairly long lifetimes and these are called *isomers*. An optical isomer shift can be observed and the difference, if any, of the radii of excited and ground state derived.

The fundamental particle called the muon (or μ-meson) is observed as the decay product of heavier fundamental particles. We shall see that it has properties essentially identical to those of the electron except that it is 207 times heavier. They could be used in scattering experiments instead of electrons except that muon beams are not readily available. Very elaborate special accelerators are planned to achieve this (meson factories). However negative mesons are readily captured into electron-like orbits around nuclei in place of one of the electrons. The muon orbits are however smaller than the corresponding electron orbits by a factor $m_e/m_\mu = \frac{1}{207}$. Hence in its bound state it lies very close to or even inside the nucleus. When it is captured it passes from the loosely bound optical levels to the more tightly bound X-ray levels and we can observe the muonic X-ray. Measurement of the X-ray energy then tells us the binding energy in the K muonic orbit for example. This binding will be greatly *reduced* if part of the wavefunction lies inside the charge distribution of the nucleus. Since the binding energy is 207 times that for an electron the X-rays will be in the MeV energy range (fig. 1.8). The reduction of the energy from

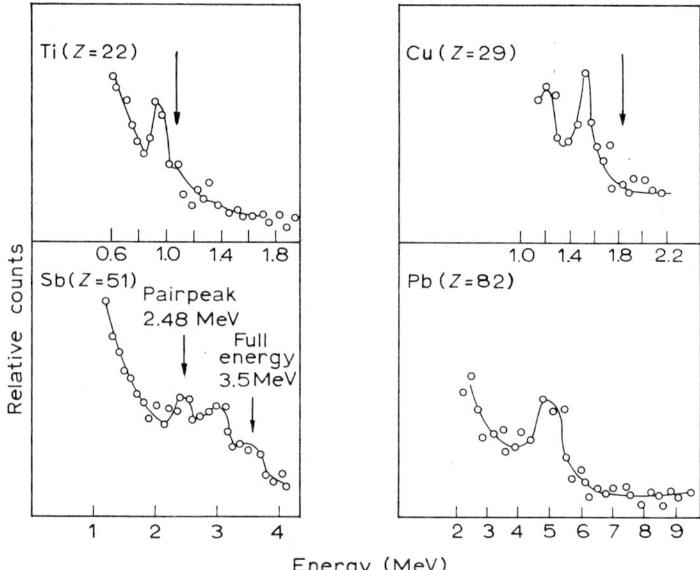

Prog. Nucl. Phys. **6** (1957) 108

Fig. 1.8. Muonic X-ray spectra. For Ti and Cu the energy expected for a point nuclear charge is indicated by the arrows. For Sb and Pb the point nucleus energies are 5.80 and 16.2 MeV respectively.

that expected for a point charge can be related to the mean square radius of the charge distribution. The results give a value of $r_0 = 1.15 \pm 0.03 \times 10^{-13}$ cm.

In principle since the semi-empirical mass formula contains a term depending on the coulomb repulsion of the protons we could estimate charge radius from nuclear binding energies. However this turns out to be much too inaccurate. But if we compare isobaric nuclei which differ only in the replacement of a neutron by a proton and make the assumption of *charge symmetry* of nuclear forces i.e. that n–n bonds are equivalent to p–p bonds, then the only difference in binding of the two must be the coulomb energy resulting from the extra proton and the neutron–proton mass difference. Such nuclei are called *mirror nuclei*. Examples are

	^{11}C	^{11}B	^{13}N	^{13}C	^{15}O	^{15}N
Neutrons	5	6	6	7	7	8
Protons	6	5	7	6	8	7

In each case the coulomb energy difference should be (assuming a uniformly charged sphere)

$$\Delta E_c = \frac{3e^2}{5r_0 A^{\frac{1}{3}}} [Z^2 - (Z-1)^2] = \frac{3(2Z-1)e^2}{5r_0 A^{\frac{1}{3}}}. \tag{1.22}$$

The nucleus of greater charge is unstable to β^+-decay to the other nucleus and so we can find the mass difference by measuring the energy release in the positron decay, T_0

$$\begin{aligned}
T_0 &= \Delta E_c - (M_n - M_p) - 2m_e c^2 \\
&= \Delta E_c - 1.80 \text{ MeV} \\
&= \tfrac{3}{5}e^2 A^{\frac{1}{3}}/r_0 - 1.80 \text{ MeV}
\end{aligned} \tag{1.23}$$

since $A = 2Z - 1$ for the nucleus of higher Z. Thus we can solve for r_0 and when corrections are made to our formula for ΔE_c to take the charge distribution with a 'fuzzy' edge into account, we get $r_0 = 1.28 \pm 0.05$ fm.

Thus all the methods for measuring charge radius combine to give a radius parameter (slightly sensitive to A) $r_0(\text{charge}) = (1.19 + 0.1A^{-\frac{2}{3}}) \times 10^{-13}$ cm.

6.3   Measurement of potential radii. In this case we want to detect the edge of the nuclear potential well so we must use strong force interactions. The most straightforward way is to study the interaction of neutrons with nuclei since these exhibit only strong interactions and have negligible electromagnetic interaction. We should expect the *total cross section* for fast neutrons to be related directly to the geometrical area i.e. $\sigma_T = 2\pi R^2$ or more precisely $\sigma_T = 2\pi(R + \lambda)^2$ where λ is the de Broglie wavelength of the neutron. Also at high energies the *absorption cross section* approaches πR^2 if we assume a perfectly absorbing nucleus which is not quite true. Such measurements lead to a radius parameter for the potential $r_0 = 1.25 + 0.6A^{-\frac{2}{3}}$ fm. However neutron measurements are somewhat difficult and so methods using charged strongly interacting particles are also used. The elastic scattering of α-particles by nuclei has been studied. 20–30 MeV α-particles are used. The results appear as shown in fig. 1.9.

The scattering is confined to the edge of the nucleus since an α-particle which penetrates the nucleus will lose energy before it re-emerges. The curves show that

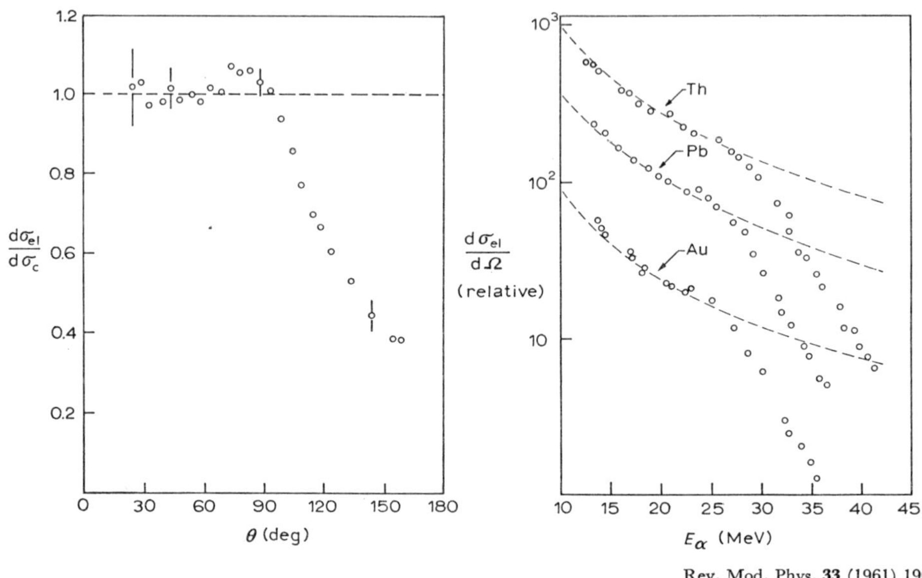

Rev. Mod. Phys. **33** (1961) 190

Fig. 1.9.   Alpha-particle scattering cross sections as a function of angle (for Pb) and as a function of energy (for Th, Pb and Au). The dashed curves are the coulomb scattering cross sections

the scattering falls below the charge or coulomb scattering for a certain critical distance of approach. This is just where the strong nuclear absorption sets in, and the shape of the deviation can be simply related to the potential radius. We get $r_0 = 1.35 + 1.0A^{-\frac{1}{3}}$ fm.

If we study the elastic scattering of protons of energy 5–300 MeV we observe diffraction patterns due to the extent of the potential. To fit these results the following radial dependence is required where R_0 and a have a similar significance to the charge distribution we met previously.

$$f(r) = \frac{1}{1 + \exp\left[(r - R_0)/a\right]}.$$ (1.24)

A value of $r_0 = 1.33$ fm is derived. The potential radius is then expressed in terms of a radius parameter $r_0 = 1.25 + 0.6A^{-\frac{1}{3}}$ fm.

For both the charge radius and the potential radius the surface thickness measured from the 90% point to the 10% point $(4.4a)$ is 2.5 fm. The potential radius is thus about 0.7 fm greater than the charge radius and this is interpreted as the range of the nuclear force. Typical radius values are listed in table 1.4 in units of 10^{-13} cm (fm).

TABLE 1.4

Nuclear radii

A	R_{charge}	R_{pot}	$R_{\text{pot}} - R_{\text{charge}}$
10	2.67	3.29	0.62
50	4.49	5.25	0.76
100	5.64	6.45	0.81
200	7.05	7.91	0.86
250	7.62	8.50	0.88

7   Angular momentum

Nuclei and nuclear particles may possess angular momentum. This property is conserved in nuclear reactions of all kinds.

The observable feature of angular momentum is the projection of the angular momentum vector on a spatial axis (fig. 1.10). Quantum theory requires that this projection be quantised in units of $I\hbar$ where I is integral or half-integral. The absolute value of the angular momentum vector is thus $\hbar[I(I+1)]^{\frac{1}{2}}$ and $m\hbar$, the projection of this vector on the space axis, has values of

$$m = I, I-1, I-2, \ldots -(I-2), -(I-1), -I.$$

The angular momentum is defined as the largest value of $m\hbar$, that is, $I\hbar$ or simply I.

Individual particles may possess intrinsic angular momentum about some axis of the particle. This is generally called spin or intrinsic spin and designated by s. Nucleons and electrons all have $s=\frac{1}{2}$.

Particles rotating about some centre of force may possess orbital angular momentum. This is similarly quantised, is designated as l and is always integral. The lowest values of l have acquired special designation for historical reasons. Thus a wave function having $l=0$ is termed an s-wave, that having $l=1$ a p-wave, that having $l=2$ a d-wave, and that having $l=3$ an f-wave, that having $l=4$ a g-wave and alphabetically thereafter.

Finally a nuclear system may possess a *total angular momentum* being the sum of intrinsic spins s and orbital angular momenta l. This total angular momentum is designated by J (or I). Thus J is the vector sum of the intrinsic spins S added vectorially to the vector sum of the orbital angular momenta L. It is this total angular momentum which must be conserved in a nuclear reaction.

The nuclear angular momentum is often referred to as the nuclear spin where confusion with intrinsic spin of particles is unlikely. It can be inferred from nuclear

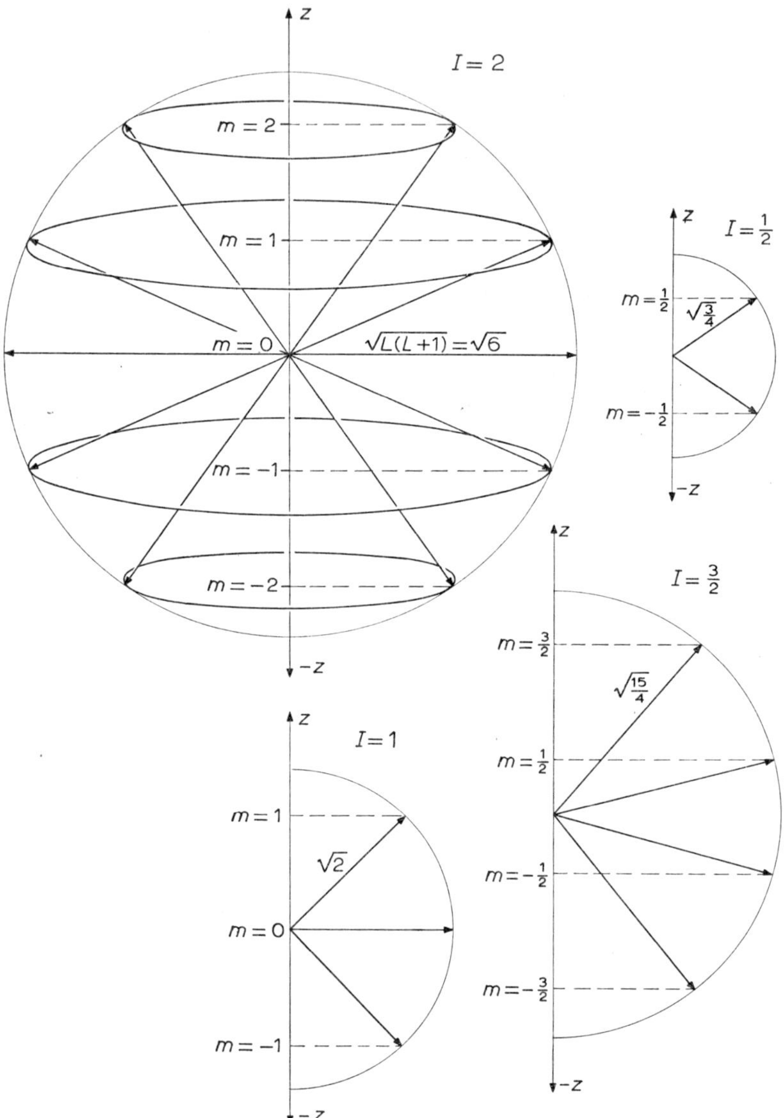

Fig. 1.10.   Vector representations of angular momenta

reactions or decays since it must be conserved. It can also be inferred from a study of angular correlations. In addition there are a number of 'atomic' methods from which spin can be inferred. These will be discussed when we consider nuclear moments.

8 Parity

Nuclei and fundamental particles in their normal and excited states have another definite property called parity which defines their symmetry under reflection of all axes, that is space inversion. This can be demonstrated for a classical system by considering a vibrating string. When in fig. 1.11a the string is struck at the origin

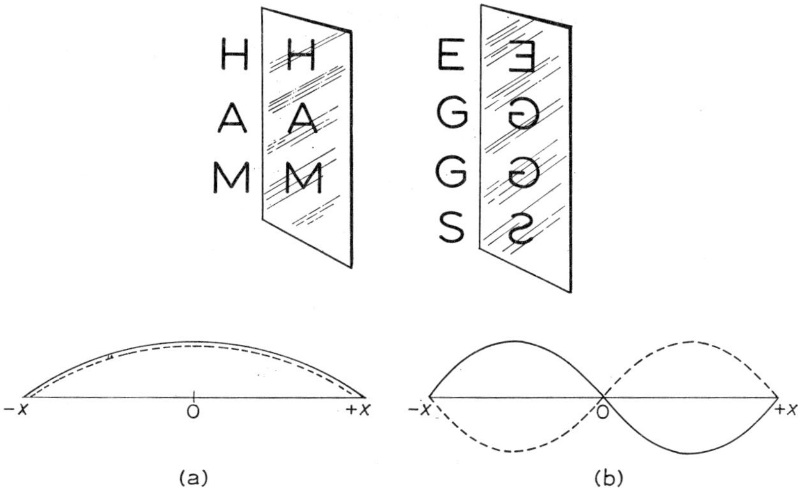

(a)   (b)

Fig. 1.11.   Parity or reflection symmetry

it will initially assume the shape shown. If we invert the x-axis (x becomes $-x$) there is no change in sign of the amplitude. When struck away from the origin (fig. 1.11b) it will initially assume the shape shown by the solid line. If we invert the x-axis there is a change in sign of the amplitude (dotted curve).

The first situation is called a symmetric state and is said to have even parity (an even harmonic). The second is called an antisymmetric state and is said to have odd parity (an odd harmonic).

For a quantum system we define the parity to be even or odd ($+1$ or -1) depending on whether the wavefunction (probability amplitude) does not or does change sign upon space inversion, that is $(x, y, z) \rightarrow (-x, -y, -z)$.

As in the case of angular momentum, fundamental particles may possess intrinsic parity. This is related to inversion of some internal axes of the particle. It can only be defined relatively however and by convention nucleons have even parity. The parity of a nucleon in a potential is fixed by its orbital angular momentum so that $P_{\text{orb}} = (-1)^l$. In other words systems with odd values of orbital angular momentum have odd parity and systems with even values of orbital angular momentum have even parity. Finally we can define the total parity of a system as $P_{\text{total}} = P_{\text{orbital}} \times P_{\text{intrinsic}}$. The total parity is conserved in nuclear and electromagnetic interactions but not in the weak interactions such as β-decay. Because of this we cannot define an intrinsic parity of electrons or muons relative to the nucleon. In general the parity of particle and antiparticle is opposite.

Parity can be inferred because of its conservation in strong nuclear reactions. There are also reactions from which the change in orbital angular momentum can be inferred, and this gives at once the parity change.

Let us look at some examples of how angular momentum and parity conservation operate. Consider first the reaction

$$^{7}\text{Li} + \text{p} \rightarrow {}^{8}\text{Be}$$

^{7}Li has angular momentum $\frac{3}{2}$ and negative parity, p has intrinsic spin $\frac{1}{2}$ and positive parity. The ^{7}Li + p system has negative parity. Thus we consider the parity of states in ^{8}Be for different orbital angular momenta for the ^{7}Li + p system:

Orbital	P_{o}	P_{i}	^{8}Be parity
s	$+$	$-$	$-$
p	$-$	$-$	$+$
d	$+$	$-$	$-$

Now let us consider possible angular momenta for the ^{8}Be states:

$$
\begin{array}{llll}
^{7}\text{Li} + \text{p} + & \text{orbital} & \rightarrow & ^{8}\text{Be} \\
\tfrac{3}{2}^{-} \quad \tfrac{1}{2}^{+} & \text{s } (l=0) & & 1^{-}, 2^{-} \\
& \text{p } (l=1) & & 0^{+}, 1^{+}, 2^{+}, 3^{+} \\
& \text{d } (l=2) & & 0^{-}, 1^{-}, 2^{-}, 3^{-}, 4^{-} \\
& \text{f } (l=3) & & 1^{+}, 2^{+}, 3^{+}, 4^{+}, 5^{+}
\end{array}
$$

Another example will show how some states cannot be formed because of parity violation. This is the reaction $^{19}\text{F}(\text{p}, \alpha)^{16}\text{O}$ leading to the ground state of ^{16}O.

$$
\begin{array}{llllll}
^{19}\text{F} + \text{p} + \text{orbital} & \rightarrow {}^{20}\text{Ne} & \rightarrow & {}^{16}\text{O} + \alpha + \text{orbital} \\
\tfrac{1}{2}^{+} \quad \tfrac{1}{2}^{+} \quad \text{s} & 0^{+} & & 0^{+} \quad 0^{+} \text{ s} \\
& 1^{+} & & \text{parity violated} \\
\quad\quad\quad\quad\quad \text{p} & 1^{-} & & 0^{+} \quad 0^{+} \text{ p} \\
& 2^{-} & & \text{parity violated} \\
\quad\quad\quad\quad\quad \text{d} & 2^{+} & & 0^{+} \quad 0^{+} \text{ d} \\
& 3^{+} & & \text{parity violated}
\end{array}
$$

Thus, in this case, only so-called 'natural' parity states of ^{20}Ne ($J = 0^{+}, 1^{-}, 2^{+}, 3^{-}$ etc.) can break up into the ground state of ^{16}O and an α-particle. Note that it is convenient to add the angular momenta of the initial particles and then combine this with the orbital angular momentum. The vector sum of the target angular momentum and bombarding particle spin is called the channel spin. In the first example the channel spin is 1^- or 2^-, in the second example the incoming channel spin is 0^+ or 1^+ and the outgoing channel spin is 0^+.

9   Statistics

There is a further symmetry property of systems of particles which defines whether or not the sign of the total wave function changes when the coordinates of two *identical* particles are interchanged. If the sign is unchanged we say the exchanged particles obey Bose–Einstein statistics and the Pauli exclusion principle does not apply. If the sign of the wavefunction does change we say the particles obey Fermi–Dirac statistics and the Pauli exclusion principle does apply. This is clear since if the wavefunction did not change the exchanged particles could have been in the same state of motion which is not allowed. It is observed that particles and nuclei with half-integral spin are *fermions* while particles and nuclei with integral spin are *bosons*. The reason for this is not clearly understood. This exchange symmetry is also conserved in nuclear reactions. However since angular momentum is also conserved it need not be considered separately. The two go together.

10   Nuclear moments

10.1   Measurement. Nuclei may have electric moments due to non-spherical charge distributions and magnetic moments due to currents resulting from the charge distribution. We shall discuss later nuclear models which consider how these arise.

We discuss now some of the atomic methods by which such moments can be measured. Many of these methods also lead to a determination of the nuclear angular momentum. Basically we study the alteration in the atomic states through the interaction of the magnetic moment of the nucleus of spin I, μ_I. This interaction results in a splitting of the normal atomic spectral line into a multiplet of lines called the *hyperfine structure* (hfs). This is to distinguish it from the *fine* structure which results from the electron spin combining in different ways with the electron orbital angular momentum.

We characterize the total angular momentum of the system as

$$F = I+L+S = I+J \qquad (1.25)$$

where I is the nuclear angular momentum, L is the electronic orbital angular momentum, S is the electronic spin and J is the total electronic angular momentum. The energy values for a group of hfs levels can be written

$$E_{HF} = E_0 + \tfrac{1}{2}h\Delta v_0[F(F+1)-I(I+1)-J(J+1)]. \qquad (1.26)$$

The hyperfine separation constant Δv_0 is then a function of the *nuclear magnetic moment,* and the electronic wavefunction at the nucleus. If the hfs can be resolved completely and $I<J$ then merely by counting the $2I+1$ lines of a level J we get a value for the *nuclear spin* directly. Where there are deviations from the spacings indicated above, this can be attributed to a *nuclear electric quadrupole moment* and a value for this can be deduced as well.

The spectrum can be further complicated by using an external magnetic field which splits the hfs into Zeeman components but this sometimes aids in interpretation.

Another technique used is that of microwave spectroscopy. The absorption spectra in molecular gases are observed. Again we see a splitting of lines which is now not due to the nuclear moment directly but due to internal molecular fields. We can then see effects of the *nuclear quadrupole moment* interacting with the molecular electric field. This in turn can lead to a determination of the *nuclear spin.* By study of the Zeeman splitting of the absorption spectra the *nuclear magnetic moment* can also be determined.

By far the largest number of measurements of nuclear moments use magnetic resonance effects. A particle with a magnetic moment will, if free, orient itself in a magnetic field. Its orientation is quantized according to m_I which is the component of the angular momentum along the magnetic axis. In the simplest form, nuclei with spin I will have $2I+1$ orientations β with respect to the magnetic field H_0. H_0 interacts with the dipole magnetic moment μ_I and exerts a couple but since the angular momentum is collinear with the magnetic moment there will be a precession of the spin vector about the field direction. This precession will have angular velocity

$$\omega_0 = \frac{\text{couple}}{\text{angular momentum}} = \frac{|\mu_I|\sin\beta H_0}{|I|\sin\beta} = \gamma_I H_0 = \frac{g_I\mu_N H_0}{\hbar}. \qquad (1.27)$$

$\gamma_I = |\mu_I|/|I|$ is called the gyromagnetic ratio, g is called the g-factor, μ_N is the nuclear magneton that is, the classical magnetic moment of the proton $e\hbar/2m_p c$. Thus g is the gyromagnetic ratio in units of the nuclear magneton. Another unit is used for the electron magnetic moment, the Bohr magneton. This is equal to $e\hbar/2m_e c$ and is larger by the ratio $m_p/m_e = 1836$. For free protons $\omega_0 = 2.6753 \times 10^4 H_0$ radians/ sec where H_0 is in gauss or the frequency

$$v_0 = \omega_0/2\pi = 4.258 \times 10^3 H_0 \text{ Hz.}$$

Thus for magnetic fields of a few thousand gauss v_0 is some tens of MHz.

The energy associated with the precession is the potential energy of the dipole in the field

$$W = -(\boldsymbol{\mu}_I \cdot \boldsymbol{H}_0) = -|I|\omega_0 \cos \beta$$
$$= -m_I \hbar \omega_0$$
$$= m_I h \nu_0 \,. \tag{1.28}$$

Thus the $2I+1$ values of the orientation lead to a set of energy levels of spacing $h\nu_0 = g_I \mu_0 H_0 = \mu_I H_0/I$. These states are not usually equally populated because of thermal contact with the surroundings and the Boltzmann factor $\exp(-W/kT)$ increases the population of the lower energy state. At a given value of H_0, transitions between particular states may be induced, that is energy is resonantly absorbed or emitted and this energy change may be detected electronically.

Thus in the technique of *nuclear magnetic resonance absorption* (NMR) the

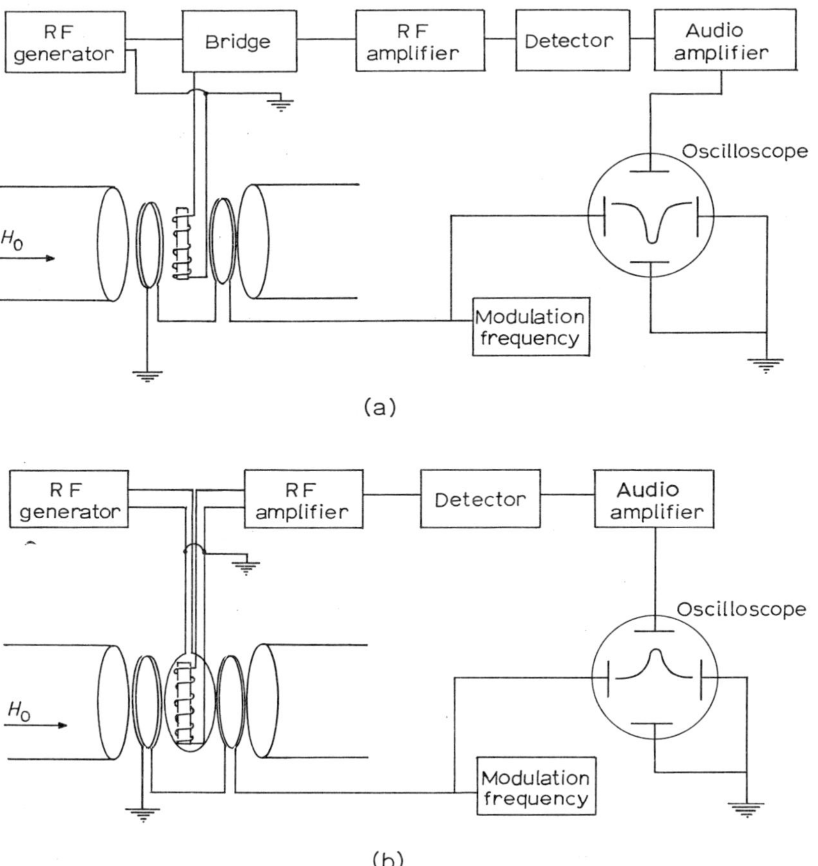

Fig. 1.12.   (a) Nuclear magnetic resonance absorption. (b) Nuclear magnetic induction

sample forms part of a resonant circuit in the steady magnetic field H_0 (fig. 1.12). When $H_0 = \omega/\gamma_I$ the absorption of power by the precessing nuclei can be detected in a coil around the sample at right angles to H_0. The Q of the resonant circuit is decreased. If H_0 is modulated by auxiliary coils over a small region of field strength the sharp dip in signal level as H_0 passes through resonance can be detected.

In the technique of *nuclear magnetic induction* the sample is in a strong field H_z and an alternating field is applied at right angles, $H_x = H \cos \omega t$. A pickup coil at right angles to both H_z and H_x will then detect the reorientation of the magnetic vector when $H_z = \omega/\gamma_I$ and an increase in signal is seen at this point. H_z is modulated by auxiliary coils as before, and we see an increase in the signal at resonance.

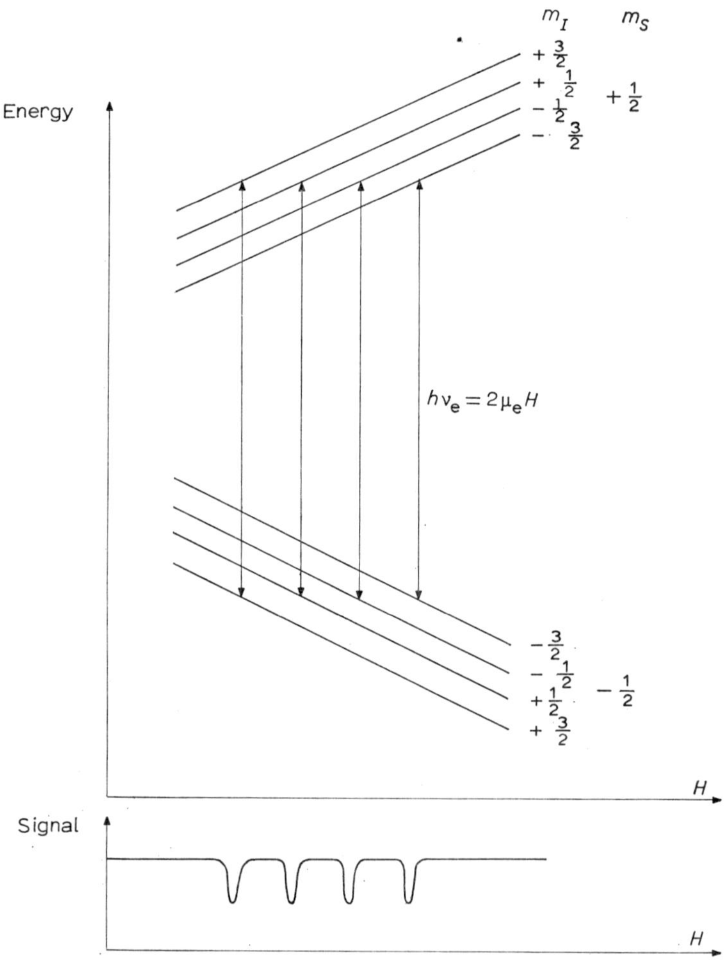

Fig. 1.13.   Electron paramagnetic resonance

Again weak modulation of the main field enables us to see when the field passes through resonance.

The technique of *electron paramagnetic resonance* (EPR) is closely similar. The value of g for electrons is much larger than for nuclei, and thus the frequencies involved are much greater being thousands of MHz rather than MHz. Paramagnetic ions in a solid often behave as though their magnetic effect were due to a single electron so the state is characterised by spin $\frac{1}{2}$ and substates $m_s = +\frac{1}{2}, -\frac{1}{2}$ (fig. 1.13). In an external magnetic field the separation of these two substates is seen but in addition each substate is split up due to the nuclear spin interaction into $2I+1$ hyperfine states.

When the spacing between substates with the same M_I is just such that $h\nu_e = 2\mu_e H$ the electron spin direction can reverse with respect to the field, a transition occurs and absorption of microwave power of frequency ν_e takes place. Thus simply by counting the dips as H is varied we can determine the nuclear spin I.

All of the methods so far described apply to solid or liquid samples. We can also study free atoms or molecules in the form of a *beam* (fig. 1.14a). Originally the Stern–Gerlach experiment determined the magnetic moment of atoms. The force on a magnetic moment μ in a non-uniform magnetic field with gradient $\partial H/\partial z$ is

$$F_z = \mu \cos \theta \; \partial H/\partial z. \tag{1.29}$$

If instead of an atomic beam we use a molecular beam in which the molecular magnetic moment is zero, then any deflection observed must be due to the nuclear moment. This deflection will be very small since $\mu_{\text{nucl}} \approx \frac{1}{1836}\mu_{\text{atomic}}$.

An important development of this system is due to Rabi and his group who incorporated a magnetic resonance element as well (fig. 1.14b). A and B are two inhomogeneous Stein–Gerlach type magnets with opposite gradients. C produces a homogeneous field H and contains a loop to produce an oscillating field H_{osc} at right angles to H. Thus H_A defocusses the beam and H_B refocusses it onto the detector on the axis. This occurs independent of velocity provided no magnetic change occurs at the centre. However if the frequency of H_{osc} is such that

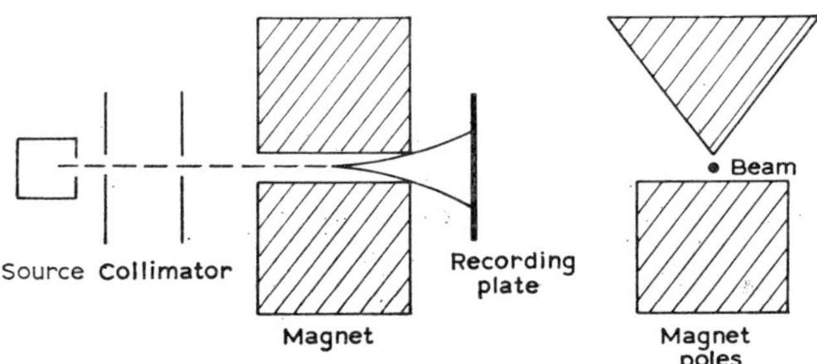

Fig. 1.14a.   The Stern–Gerlach molecular beam method for nuclear moments, side and end view.

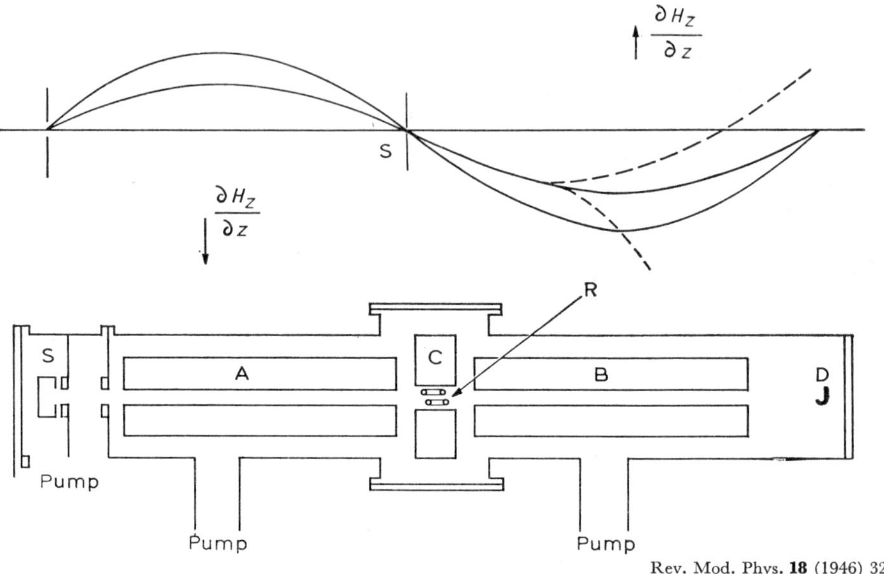

Rev. Mod. Phys. **18** (1946) 323

Fig. 1.14b.   The Rabi molecular beam apparatus. S is the source of molecules, A and B are identical inhomogeneous field magnets with opposite field gradients, C is a homogeneous field magnet, R is a coil producing an rf field at right angles to H, and D is the detector

$hv = \mu H/I$ there will be a change. The atom or molecule will gain or lose energy due to the resonant transition. It will then be deflected differently by H_B and fall off the axis at the detector.

Thus if v is fixed and H is varied until a dip in the detector counts is noted we can then determine $\mu/I = g$ and if we know I, μ can be determined.

If the nucleus has an electric quadrupole moment this will interact with the electric field of the molecule and affect the spacing of the resonant dips. From this we can derive a value of the quadrupole moment.

Magnetic moments can sometimes be measured by studying the change produced in the angular distribution of a nuclear reaction when an external magnetic field is applied.

In a few cases the nuclear Zeeman effect can be observed directly where recoilless resonant absorption of γ-rays occurs (the Mössbauer effect; ch. 15 § 5).

Nuclear quadrupole moments can be deduced also from a study of nuclear rotational states which occur when the nucleus has a permanent non-spherical shape (ch. 9).

10.2   Values of nuclear magnetic moments. We consider first the magnetic moments of the nucleons. The proton has spin $\frac{1}{2}$ and charge e exactly as the electron does. Thus we should expect the magnetic moment due to the spinning charge to be $e\hbar/2m_{\rm p}c$, that is just one nuclear magneton. The neutron having no charge should have zero magnetic moment. In fact we find the proton has $\mu_{\rm p}=2.793$ n.m and the neutron has $\mu_{\rm n}=-1.913$ n.m, the minus sign indicating that the magnetic moment vector is directed opposite to the spin vector. Qualitatively we can consider the extra anomalous moment of 1.793 n.m for the proton and -1.913 n.m for the neutron to be the effect of the cloud of virtual charged mesons which surround the nucleon. If we consider the dissociation into the bare Dirac nucleon and a meson

$$\rm p \rightarrow n+\pi^{+}$$
$$\rm n \rightarrow p+\pi^{-}$$

then the orbital motion of the positive meson gives rise to $+\mu_{\pi}$ and that of the negative meson to $-\mu_{\pi}$. So if the nucleon is dissociated for a fraction of the time, f

$$\begin{aligned}
\mu_{\rm p} &= +\mu_{\pi}f+(1-f)\times 1 = (\mu_{\pi}-1)f+1,\\
\mu_{\rm n} &= -\mu_{\pi}f+f\times 1 = -(\mu_{\pi}-1)f.
\end{aligned} \tag{1.30}$$

This $(\mu_{\pi}-1)f$ is 1.793 n.m for the proton and 1.913 n.m for the neutron which agreement is surprisingly good considering the crudeness of our argument. We refer later in ch. 15 § 7 to attempts to calculate this with greater precision.

The deuteron has a magnetic moment almost exactly equal to the sum of the magnetic moments of neutron and proton. We consider them further in ch. 11.

It is found that the spins of odd-A nuclei frequently appear to correspond to the angular momentum of the last added nucleon moving in the field of the core made up of the remaining nucleons. Let us consider the moments to be expected if this is the case. A single nucleon moving in a potential well with total angular momentum $j=l+s$ where l and s are orbital and spin angular momentum gives rise to a magnetic moment $\mu_j=g_j\mu_{\rm N}I$ where

$$g_j = \frac{j}{2}\left[(g_l+g_s)+(g_l-g_s)\frac{l(l+1)-\frac{3}{4}}{j(j+1)}\right] \tag{1.31}$$

where $\mu_{\rm N}$ is the nuclear magneton $e\hbar/2Mc$, g_l is the nuclear g-factor due to the orbital motion and g_s is the nuclear g-factor due to the intrinsic spin of the nucleon.

If the magnetic moment μ_I of an odd-A nucleus of spin I is due to an odd proton (Z odd, N even) then $g_l=1$, $g_s=\mu_{\rm p}/s=5.585$ and

$$\begin{aligned}
\mu_I &= (j-\tfrac{1}{2})\mu_{\rm N}+\mu_{\rm p} && \text{for } I = j = l+\tfrac{1}{2},\\
\mu_I &= \frac{j}{j+1}\left[(j+\tfrac{3}{2})\mu_{\rm N}-\mu_{\rm p}\right] && \text{for } I = j = l-\tfrac{1}{2}.
\end{aligned} \tag{1.32}$$

For the case of an odd neutron (Z even, N odd) $g_l=0$ since the neutron is uncharged and $g_s=-3.826$ then

$$\mu_I = \tfrac{1}{2}g_s\mu_N = \mu_n \qquad\qquad \text{for } I = j = l+\tfrac{1}{2},$$
$$\mu_I = -\frac{j}{j+1}\tfrac{1}{2}g_s\mu_N = -\frac{j}{j+1}\mu_n \qquad \text{for } I = j = l-\tfrac{1}{2}. \tag{1.33}$$

In ch. 8 we shall consider in detail the measured magnetic moments of odd-A nuclei. They differ significantly from the single particle estimates above although the overall trends are predicted correctly.

The groundstates of even–even nuclei invariably have zero angular momentum and thus have no magnetic moment. In the case of odd–odd nuclei the measured magnetic moments can sometimes be considered as the combination of contributions from the odd neutron and the odd proton, as in the deuteron case.

10.3 Values of nuclear electric quadrupole moments. Nuclei do not have any static electric dipole moment since this operation corresponds simply to a translation of the centre of mass of the nucleus. However nuclei with non-spherical charge distributions may possess electric quadrupole (and higher) moments. The quadrupole moment is defined as

$$Q = \frac{1}{r}\int r^2(3\cos^2\theta - 1)\varrho_{m=j}(r)\mathrm{d}z \tag{1.34}$$

where θ is the angle between r and the z-axis and $\varrho(r)$ is the charge distribution for maximum value of m. Thus if the charge distribution is symmetric $Q=0$. Thus all even–even nuclear ground states with $I=0$ have zero quadrupole moment. The charge distribution for a nucleus with spin $\tfrac{1}{2}$, $m=\pm\tfrac{1}{2}$ is also symmetric and $Q=0$.

For a single odd-proton nucleus we expect

$$Q^p_{sp} = -\overline{r^2}\,\frac{2j-1}{2(j+1)}. \tag{1.35}$$

The negative value indicates a charge distribution such that the shape is an oblate spheroid. A single hole leads to a prolate shape, that is, Q^p_{sn} is positive (fig. 1.15). A single neutron, although it cannot contribute directly to the quadrupole moment,

Fig. 1.15. Single particle and single hole nuclei showing the associated change in shape

affects the proton distribution by shifting the centre of mass and

$$Q_{sp}^{n} = \frac{Z}{A^2} Q_{sp}^{p}. \tag{1.36}$$

The measured values of Q for odd-A nuclei are frequently much larger than the above estimates would indicate and show that the nucleus as a whole sometimes acquires a permanent deformation. This is considered further in ch. 9.

11   Conserved quantities in nuclear interactions

We have described a number of nuclear properties and have mentioned how these are sometimes conserved in nuclear reactions. Conservation laws are related in general to invariance of physical laws under some symmetry operation. If we want to tell whether an object is spherical we look at it while rotating it in every direction. If it looks the same we say it is spherical. 'Roundness' is conserved under rotations. A rectangular box similarly tested would look different after some rotations (cf. fig. 1.16). In this case 'squareness' is not conserved under rotations. In other words we ask of our system, are the physical laws changed or not when we make a certain specified change in the space, time or other coordinates of our system?

a) *Conservation of energy*. This is valid for all known types of force. It is related to invariance under translation along the time axis. That is, the laws of the interaction do not depend on the time they are measured.

b) *Conservation of linear momentum*. This is valid for all known types of force. It is related to invariance under translation in position in space. That is the laws of the interaction do not depend on the place they are measured. Space is homogeneous.

c) *Conservation of angular momentum*. This is valid for all known forces. It is related to invariance under rotations in space. That is the laws of the interaction do not depend on the orientation in space. Space is isotropic. Note that what is conserved is the *total* angular momentum of a system. Intrinsic spin and orbital angular momentum may not be conserved separately.

d) *Conservation of parity*. This is valid for the strong nuclear force and for the electromagnetic force but is completely violated for the weak force. It is related to invariance under inversion of the space axis or equivalently to a combined reflection and rotation. That is the interaction laws do not depend on right or left handedness.

e) *Conservation of C*. This is related to invariance under 'charge conjugation' that is invariance when all particles are changed to their antiparticles. If C is conserved the interaction laws are the same in a 'world' and an 'anti-world'. This is

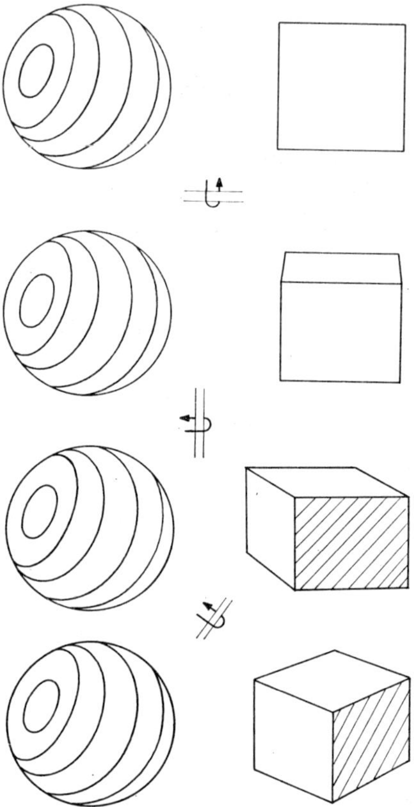

Fig. 1.16.   Testing symmetry by rotation operations

conserved for the strong nuclear force and for the electromagnetic force. It is completely violated for the weak force.

f)   *Conservation of CP*. The product of space inversion and charge conjugation is conserved in strong and electromagnetic interactions and approximately in weak interactions. Since a particle in a given state of motion is formally identical to its antiparticle with opposite linear momentum, opposite spin direction and negative energy, then it can be seen that if we perform space inversion and charge conjugation everything comes out the same, except that *time* is reversed.

g)   *Conservation of T* is thus equivalent to conservation of *CP*. It is related to invariance under reversal of the time axis.

h)   *Conservation of baryons and leptons*. This accounts for the fact that baryons and leptons appear and disappear only in pairs of particle plus antiparticle. Hence the total number is invariant. If we break a field line we get a pair of source + sink. If we bring a pair together both disappear. We count $B = +1$ for baryons and

$B = -1$ for antibaryons; $L = +1$ for leptons and $L = -1$ for antileptons. It has the effect of preventing stray sources or sinks from floating about or 'preventing the cancerous infection of our universe by antiparticles'.

This conservation law is related to the quantum mechanical property of *gauge invariance* of strong and weak fields.

i)   *Conservation of charge* is similarly related to the gauge invariance of the electromagnetic field.

We shall refer again to these and other symmetry properties of nuclear systems.

2

Nuclear Instability
and Decay

1 Radioactivity

It was first noted by Becquerel that certain minerals containing uranium seemed to emit rays rather like the newly discovered X-rays which could affect a photographic plate, discharge an electroscope and to some extent be deflected by electric and magnetic fields (unlike X-rays). In a magnetic field there were three components; one bent like a positively charged particle, one bent easily like a negatively charged particle, and one was unaffected (fig. 2.1). These were labeled α-, β- and γ-rays. In a magnetic field a charged particle is deflected in radius ρ as $H\rho = mv/e$. Thus the magnetic rigidity $H\rho$ is inversely proportional to the e/m ratio of

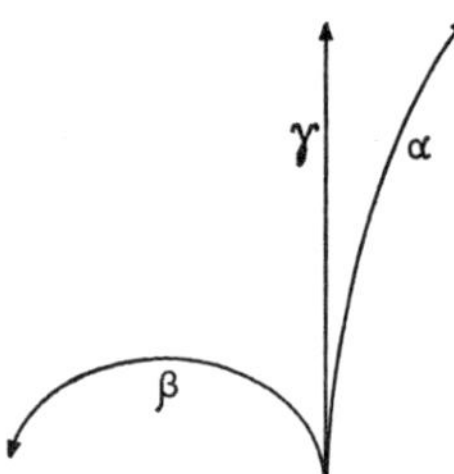

Fig. 2.1. Radiation types

the particle for given velocity. Alpha-rays showed definite energies, emitted as lines or discrete groups, β-rays showed a continuous spectrum up to a definite upper limit (fig. 2.2), and γ-rays were unaffected by electric or magnetic fields, and

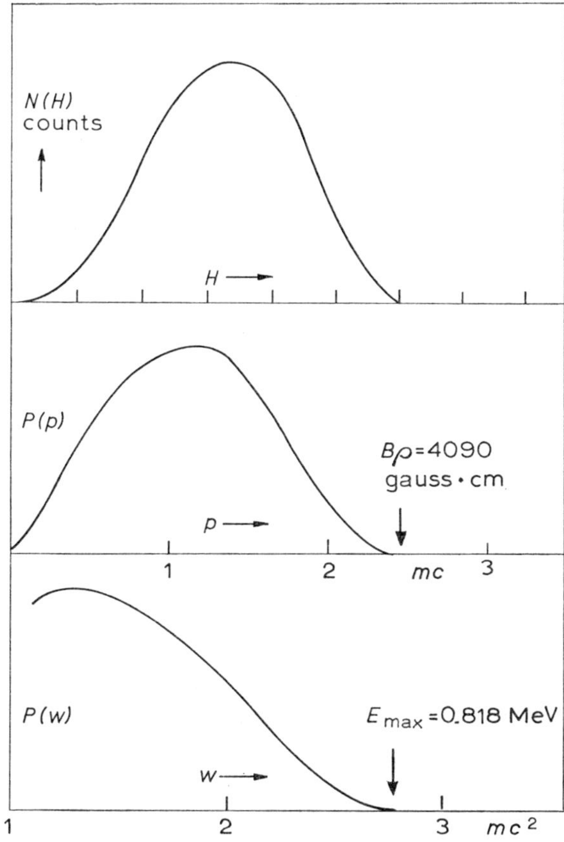

E. Segre (ed.), Experimental Nuclear Physics **3** (Wiley, New York 1959) p. 456

Fig. 2.2.   The β-spectrum as a function of magnetic field H in a magnetic spectrometer, as a function of momentum p and as a function of energy w

were found to be electromagnetic radiation of high frequency. Early experiments using crystal diffraction showed them to be emitted as line spectra in all probability.

Rutherford studied the radioactive decay of thoron, a radioactive gas. He swept a quantity of thoron into an electroscope by blowing an air current over a sample of thorium oxide. When the air current was shut off, the conductivity of the chamber decreased with time – falling to half value in about a minute. Further the decay was observed to follow an *exponential law* with great accuracy.

This follows if we assume an individual atom-by-atom process. We can write the activity (disintegrations/unit time) dN_t/dt. Then the observed exponential requires

$$dN_t/dt = -\lambda N_t$$

where λ is a constant, or

$$N_t = N_0 e^{-\lambda t} \tag{2.1}$$

where N_0 is the number of radioactive atoms at time zero. λ is called the radioactive *decay constant* and is just the probability/sec of an atom decaying.

The mean life is defined as

$$\tau = \frac{\int_{N_0}^0 t\,dN_t}{\int_{N_0}^0 dN_t} = \frac{\int_0^\infty t(dN_t/dt)\,dt}{\int_0^\infty (dN_t/dt)\,dt} = \frac{1}{\lambda}. \tag{2.2}$$

The time for half the radioactive atoms to decay is termed the half-value period or half-life $t_{\frac{1}{2}}$

$$\tfrac{1}{2}N_0 = N_0 \exp(-\lambda t_{\frac{1}{2}})$$

$$t_{\frac{1}{2}} = \frac{\log_e 2}{\lambda} = \frac{0.693}{\lambda} = 0.693\tau. \tag{2.3}$$

It was shown at an early stage that radioactive decay is a completely random process. A study of the fluctuations in the number of particles emitted in a given time interval showed that they followed a Poisson distribution. The probability of observing n particles if an average of N is expected is

$$W(n) = \frac{N^n e^{-N}}{n!} \tag{2.4}$$

if the process is random and the probability of decay is small and constant. Rutherford and Geiger showed this to be true.

We have already mentioned the transformation theory of Rutherford which states that radioactivity is really the change of one kind of nuclide to another by emission of an α- or β-particle. Successive changes lead to radioactive series in which the nuclides successively occurring are governed by the displacement laws. These state that emission of an α-particle changes nucleus (A, Z) to nucleus $(A-4, Z-2)$ and emission of a β-particle changes nucleus (A, Z) to nucleus $(A, Z+1)$.

For a long time there were three such series known whose ultimate parents were the three radioactive nuclei with lifetimes comparable to the age of the earth so that an appreciable quantity still can be found on earth. These are

1) the thorium series with parent ^{232}Th, $t_{\frac{1}{2}}=1.39\times10^{10}$ y; also called the $4n$ series since $A=4n$;

2) the uranium series with parent ^{238}U, $t_{\frac{1}{2}}=4.5\times10^{10}$ y; also called the $4n+2$ series;

3) the actinium series with parent ^{235}U, $t_{\frac{1}{2}}=7.07\times10^8$ y; also called the $4n+3$ series.

The missing $4n+1$ series was found after artificial transuranic nuclides could be made in nuclear reactors in the 1940's. It is the neptunium series based on ^{237}Np with $t_{\frac{1}{2}}=2.2\times10^6$ y. All the primeval neptunium disappeared from the earth a long time ago.

Tables 2.1–2.4 show the four series with the nuclide, its radiation, its half-life and its former designation. These names were used in the early period of radioactivity studies (1900–1935). A few of the collateral or secondary series of short-lived nuclides are also shown.

We note that the four series are terminated when they reach the stable nuclides of ^{208}Pb, ^{209}Bi, ^{206}Pb and ^{207}Pb. All the β-decays indicated are for β^--emission. This is because α-emission leaves the final nucleus slightly neutron-rich and this is then corrected by β^--emission which corresponds to turning a neutron into a proton, the resulting nucleus being more stable again. In a number of cases we see both α-emission and β-emission occurring as alternative modes of decay. In general the secondary series shown are initiated by formation of the parent in a nuclear reaction. The first such series in the $4n+3$ series is a famous one. ^{239}U is formed by neutron capture in ^{238}U, the abundant uranium isotope. After two β-decays the isotope ^{239}Pu is formed with a fairly long life. This artificial nuclide can be used as a nuclear fuel or nuclear explosive as an alternative to ^{235}U, the less abundant uranium isotope.

We can make use of our knowledge of half-lives to estimate the time since mineral formation. In a given sample of uranium ore we can assume that all the ^{206}Pb results from the uranium series. We also must assume that there has been no alteration of the decay rates or loss of any of the products of the chain. Then the ratio of the amount of ^{206}Pb to the amount of uranium gives the age.

$$\lambda_U N_U t = N_{Pb}$$

where N is the number present now. $\tau_U = \lambda_U^{-1}=7.5\times10^9$ y. Therefore the

$$\text{age} = \frac{\text{atoms } ^{206}\text{Pb}}{\text{atoms } ^{238}\text{U}} \times 7.5\times10^9 \text{ y.}$$

$$= 640\text{–}1380\times10^6 \text{ y.}$$

We can also estimate the time since the elements were formed by assuming that the isotopes ^{235}U and ^{238}U were originally formed in the same quantity and that the only reason we see much less now of ^{235}U is because of its shorter life. As $t_{\frac{1}{2}}(^{235}\text{U})=7.07\times10^8$ y, and $t_{\frac{1}{2}}(^{238}\text{U})=4.5\times10^{10}$ y, whereas the observed ratio of abundance now is ^{235}U$/^{238}$U$=\frac{1}{140}$, this gives a time since element formation $\approx5\times10^9$ y.

We see such nuclear instability when the change to residual nucleus + emitted particle from the parent nucleus results in a net release of energy. That is, when

TABLE 2.1   *4n* series (thorium series)

Primary series					Secondary series		
^{232}Th	α	1.39×10^{10} y	(Th)		^{240}U	β	14.1 h
^{228}Ra	β	6.7 y	(MsThI)		^{240}Np	β	60 m
^{228}Ac	β	6.13 h	(MsThII)		^{240}Pu	α	6.58×10^3 y
^{228}Th	α	1.90 y	(RaTh)		^{236}U	α	2.39×10^7 y
^{224}Ra	α	3.64 d	(ThX)		^{232}Th		
^{220}Rn	α	51.5 s	(ThEm)		^{228}Pa	α	22 h
^{216}Po	α, β	0.16 s	(ThA)		^{224}Ac	α	2.9 h
^{212}Pb	β	10.6 h	(ThB)		^{220}Fr	α	28 s
^{216}At	α	3×10^{-4} s			^{216}At		
^{212}Bi	α, β	60.6 m	(ThC)				
^{212}Po	α	3×10^{-7} s	(ThC′)				
^{208}Tl	β	3.1 m	(ThC′′)				
^{208}Pb		stable					

TABLE 2.2   *4n*+1 series (neptunium series)

Primary series			Secondary series						
^{241}Pu	β	10 y							
^{241}Am	α	500 y	^{237}Pu	e.c	40 d		^{237}U	β	6.75 d
^{237}Np	α	2.2×10^6 y	^{237}Np				^{237}Np		
^{233}Pa	β	27.4 d							
^{233}U	α	1.62×10^5 y	^{229}Pa	e.c, α	1.4 d				
^{229}Th	α	7.0×10^3 y	^{229}Th						
^{225}Ra	β	14.8 d							
^{225}Ac	α	10 d	^{225}Ac				^{229}U	α	58 m
^{221}Fr	α	4.8 m					^{225}Th	α	7.8 m
^{217}At	α	1.8×10^{-2} s					^{221}Ra	α	31 s
^{213}Bi	α, β	47 m					^{217}Rn	α	10^{-3} s
^{213}Po	α	4.2×10^{-6} s					^{213}Po		
^{209}Tl	β	2.2 m							
^{209}Pb	β	3.3 h							
^{209}Bi		stable							

4n+2 series (uranium series)   TABLE 2.3

Primary series				Secondary series		
^{238}U	α	4.5×10^{10} y	(UI)			
^{234}Th	β	24.1 d	(UX$_1$)			
^{234}Pa	β, i.c	1.14 m	(UX$_2$)			
^{234}Pa	β	6.7 h	(UZ)			
^{234}U	α	2.35×10^5 y	(UII)	^{230}Pa	β	17 d
^{230}Th	α	8×10^4 y	(Ionium)	^{230}U	α	20.8 d
^{226}Ra	α	1.62×10^3 y	(Ra)	^{226}Th	α	30.9 m
^{222}Rn	α	3.82 d	(RaEm)	^{222}Ra	α	38 s
^{218}Po	α, β	3.05 m	(RaA)	^{218}Rn	α	8.02 s
^{214}Pb	β	26.8 m	(RaB)	^{214}Po		
^{218}At	α	2 s				
^{214}Bi	α, β	19.7 m	(RaC)	^{226}Pa	α	1.7 m
^{214}Po	α	1.5×10^{-4} s	(RaC')	^{222}Ac	α	10 s
^{210}Tl	β	1.32 m	(RaC'')	^{218}Fr	α	10^{-2} s
^{210}Pb	β	22 y	(RaD)	^{214}At	α	10^{-6} s
^{210}Bi	α, β	5.0 d	(RaE)	^{210}Bi		
^{210}Po	α	140 d	(RaF)			
^{206}Tl	β	4.23 m				
^{206}Pb		stable	(RaG)			

4n+3 series (actinium series)   TABLE 2.4

Primary series				Secondary series		
^{235}U	α	7.13×10^8 y	(AcU)	^{239}U	β	23.5 m
^{231}Th	β	25.6 h	(UY)	^{239}Np	β	2.35 d
^{231}Pa	α	3.43×10^4 y	(Pa)	^{239}Pu	α	2.4×10^4 y
^{227}Ac	α, β	21.6 y	(Ac)	^{235}U		
^{227}Th	α	18.2 d	(RaAc)			
^{223}Fr	β	22 m	(AcK)			
^{223}Ra	α	11.7 d	(AcX)			
^{219}Rn	α	3.92 s	(AcEm)	^{227}Pa	α	38 m
^{215}Po	α, β	1.83×10^{-3} s	(AcA)	^{223}Ac	α	2.2 m
^{211}Pb	β	36.1 m	(AcB)	^{219}Fr	α	0.02 s
^{215}At	α	10^{-4} s		^{215}At		
^{211}Bi	α, β	2.16 m	(AcC)			
^{211}Po	α	5×10^{-3} s	(AcC')			
^{207}Tl	β	4.78 m	(AcC'')			
^{207}Pb		stable				

in the reaction

$$_{z}A^{A} \rightarrow \; _{z-2}B^{A-4} + _{2}\alpha^{4} + Q$$

the energy release Q is positive. We can also say that the difference of mass of A and B is in excess of the mass of an α-particle:

$$A - B \rightarrow \alpha + Q.$$

Thus when the binding energy curve becomes steep enough at high values of A we can get this condition and we observe natural radioactivity for $A > 210$.

Similarly

$$_{z}R^{A} \rightarrow \; _{z+1}S^{A} + \beta + Q.$$

We neglect the mass of the β-particle.

Finally for γ-emission

$$\underset{\text{excited state}}{_{z}X^{A}} \longrightarrow \underset{\text{lower state}}{_{z}X^{A}} + \gamma + Q.$$

We find a few other naturally radioactive nuclides through the periodic table outside the region of heavy nuclei. These are due to chance instabilities of various sorts.

Nd and Sm are weakly α-radioactive. We shall see that these are unusually deformed nuclei.

Re, Lu, La, In, Rb, V and K show weak β-radioactivity. The isotope of potassium which is radioactive, ^{40}K, only constitutes 0.01 % of ordinary potassium and has a half-life of 1.5×10^{9} y. However the rather large abundance of K in granite and such rocks means that an appreciable fraction of the heat of the earth results from the K radioactivity. It is also an ever-present background in some radiation measurements in the laboratory.

2   Laws of radioactive decay

Now let us return to the radioactive series. We want to consider the activities for successive transformations. Suppose we have a series

$$A \overset{\lambda_{A}}{\rightarrow} B \overset{\lambda_{B}}{\rightarrow} C \overset{\lambda_{C}}{\rightarrow} D \overset{\lambda_{D}}{\rightarrow} \ldots$$

with different decay constants.

Then $dN_{A}/dt = -\lambda_{A} N_{A}$ describes the decay of A. For growth and decay of B we have

$$dN_{B}/dt = \lambda_{A} N_{A} - \lambda_{B} N_{B}.$$

The first term describes the change of A to B and the second the change of B to C. Thus we can solve for N_A and N_B if we know the quantities present at time zero. However it is the *activity* or disintegration rate we are more interested in. The activities of A and B are, resp.,

$$\lambda_A N_A = \lambda_A N_A(0)e^{-\lambda_A t}$$

and

$$\lambda_B N_B = \lambda_B N_B(0)e^{-\lambda_B t} + \frac{\lambda_A \lambda_B}{\lambda_B - \lambda_A} N_A(0)(e^{-\lambda_A t} - e^{-\lambda_B t}) \tag{2.5}$$

where $N_A(0)$ means the number of A atoms present at time zero. We derive the second expression as follows. The activity of B

$$\begin{aligned} dN_B/dt &= \lambda_A N_A - \lambda_B N_B \\ &= \lambda_A N_A(0)e^{-\lambda_A t} - \lambda_B N_B(0)e^{-\lambda_B t}. \end{aligned} \tag{2.6}$$

Multiply through by $e^{\lambda_B t}$ and rearrange

$$(dN_B/dt)e^{\lambda_B t} + \lambda_B N_B e^{\lambda_B t} = \lambda_A N_A(0)e^{(\lambda_B - \lambda_A)t}$$

or

$$\frac{d}{dt}(N_B e^{\lambda_B t}) = \lambda_A N_A(0)e^{(\lambda_B - \lambda_A)t}. \tag{2.7}$$

Integrating,

$$N_B e^{\lambda_B t} = \frac{\lambda_A N_A(0)}{\lambda_B - \lambda_A} e^{(\lambda_B - \lambda_A)t} + C(\text{const. of integration}). \tag{2.8}$$

Multiply through by $e^{-\lambda_B t}$

$$N_B(t) = \frac{\lambda_A}{\lambda_B - \lambda_A} N_A(0)e^{-\lambda_A t} + Ce^{-\lambda_B t}. \tag{2.9}$$

For $t = 0$, $N_B = N_B(0)$, a constant; therefore

$$C = N_B(0) - \frac{\lambda_A}{\lambda_B - \lambda_A} N_A(0). \tag{2.10}$$

Substituting in eq. (2.9)

$$N_B(t) = \frac{\lambda_A}{\lambda_B - \lambda_A} N_A(0)(e^{-\lambda_A t} - e^{-\lambda_B t}) + N_B(0)e^{-\lambda_B t}, \tag{2.11}$$

so the activity is

$$\lambda_B N_B(t) = \lambda_B N_B(0)e^{-\lambda_B t} + \frac{\lambda_A \lambda_B}{\lambda_B - \lambda_A} N_A(0)(e^{-\lambda_A t} - e^{-\lambda_B t}). \tag{2.12}$$

If initially B is not present, i.e. $N_B(0)=0$, then

$$\frac{\text{activity of B}}{\text{activity of A}} = \frac{\lambda_B}{\lambda_B - \lambda_A}(1 - e^{-(\lambda_B - \lambda_A)t}). \tag{2.13}$$

Now let us consider three limiting cases.

1) $\lambda_B > \lambda_A$ but still comparable, i.e. the daughter is shorter lived than the parent. The activity ratio will tend to a constant value $\lambda_B/(\lambda_B - \lambda_A)$. Eventually both activities will decay with the parent lifetime.

When the ratio $\lambda_B/(\lambda_B - \lambda_A)$ is just reached, we call this a state of *transient equilibrium*. The daughter activity reaches a maximum at a time given by $dN_B/dt = 0 = \lambda_A N_A - \lambda_B N_B$. Thus at this instant $\lambda_A N_A = \lambda_B N_B$, that is, the activities are equal. Consider the example illustrated in fig. 2.3. ^{228}Th (RaTh) has a half-life of 1.90 y.

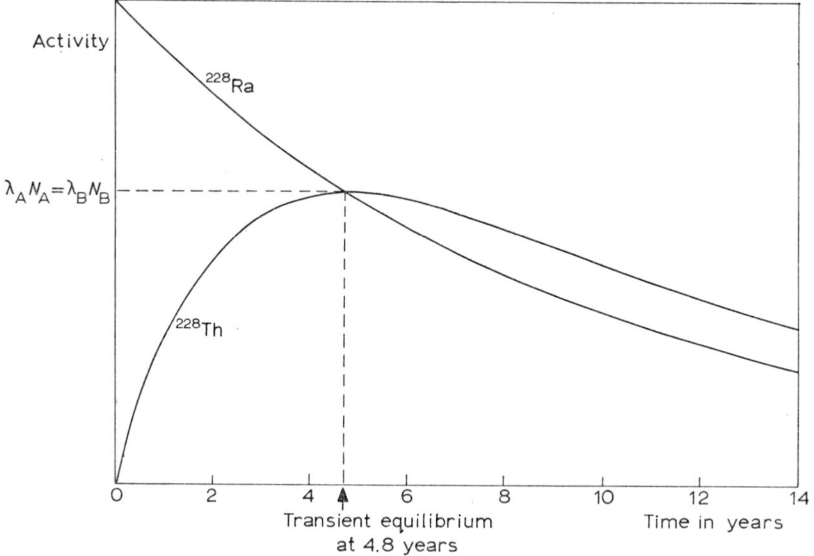

Fig. 2.3.   Transient equilibrium

It is the grand-daughter of ^{228}Ra (MsThI) with a half-life of 6.7 y. ^{228}Ac which links the two has a half-life of 6.1 h, so short that it can be considered prompt compared to the other activities. Thus $\lambda_A = 3.35 \times 10^{-9}$ sec^{-1} and $\lambda_B = 1.2 \times 10^{-8}$ sec^{-1}. Transient equilibrium occurs at 4.8 y.

2) $\lambda_B \gg \lambda_A$, that is, the parent is very long-lived compared to the daughter. In this case the ratio of the activities (since λ_A is very small) approaches 1 and thereafter B decays with the lifetime of the parent A. The activity of B grows until it equals the activity of the parent. This is called *secular equilibrium*. An example is shown

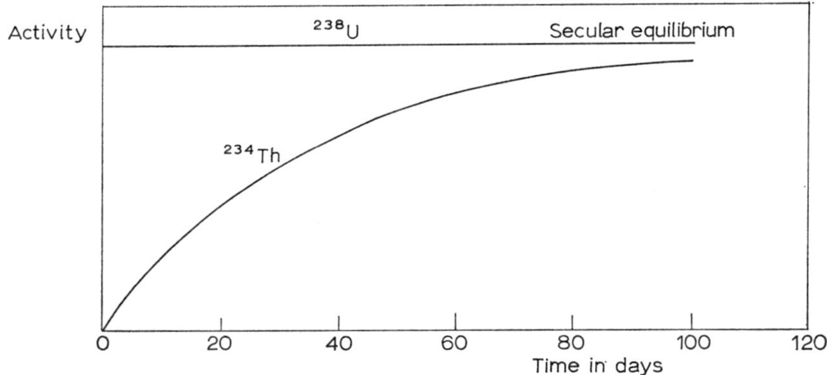

Fig. 2.4.   Secular equilibrium

in fig. 2.4. ^{238}U with $t_{\frac{1}{2}}=4.5\times10^{10}$ y decays to ^{234}Th with $t_{\frac{1}{2}}=24$ d. In fact for a series like the uranium series the primary parent has by far the longest life, hence all products of the series will be present in quantities given by

$$\lambda_A N_A = \lambda_B N_B = \lambda_C N_C = \lambda_D N_D = \text{constant}, \qquad (2.14)$$

that is, in quantities proportional to their lifetime and inversely proportional to their decay constant.

We have another important case of *secular equilibrium* when an activity is produced by an accelerator or reactor at a constant rate say p/sec. T is infinite (until we turn off our machine). Then $p=\lambda_B N_B$ if the irradiation is long compared to $t_{\frac{1}{2}}$ for B. If our irradiation time is not so long the number present at the end is $N_t = p\lambda^{-1}(1-e^{-\lambda t})$.

3)   Finally if $\lambda_B<\lambda_A$, i.e. the daughter is longer lived we get no equilibrium condition and the activity increases steadily.

3   *Measurement of decay constants*

We now consider the various methods for measuring decay constants or lifetimes:
1)   The most common method is simply to observe the activity at successive definite time intervals. If we then plot the activity vs. time on semi-log paper we observe a straight line from which the half-life can be read off directly (fig. 2.5). This method is applied to half-lives of minutes, hours or days.
2)   For shorter half-lives e.g. fractions of a minute we can use
a)   *Rotating disc.* ^{216}Po(ThA), e.g., is allowed to recoil onto the disc at one point. Detectors are placed at intervals around the periphery (fig. 2.6). The half-life in this case is 0.158 sec.

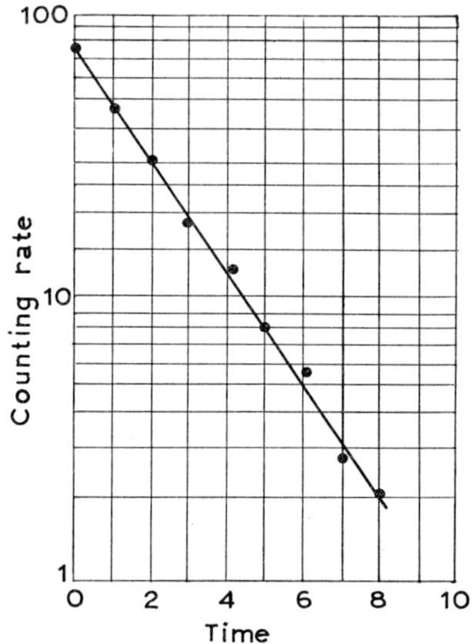

Fig. 2.5.   Radioactive decay

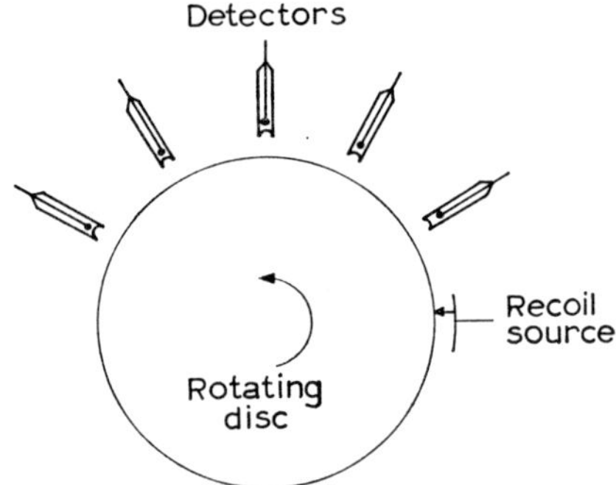

Fig. 2.6.   Lifetime determination by the rotating disc method

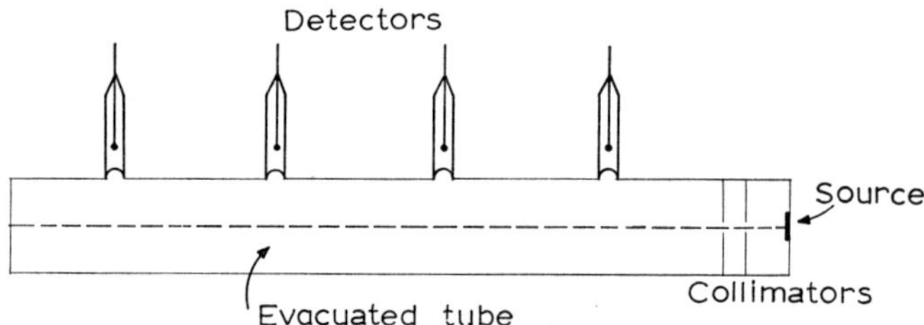

Fig. 2.7.   Lifetime determination by the flight distance method

b) *Recoil distance.* The recoil nucleus is allowed to pass down an evacuated tube along which detectors are placed (fig. 2.7), e.g. ^{214}Po (RaC$'$) ions recoil from the parent ^{214}Bi (RaC). The velocity of recoil can be determined and, with this special method, the half-life is found to be 1.6×10^{-4} sec.

c) *Delayed coincidence.* The parent may emit a β-particle while the short-lived daughter emits an α-particle. We measure the coincidence rate for β–α coincidences as a function of the delay introduced between the detectors. For example, ^{212}Bi$\rightarrow^{212}$Po, which latter has a half-life $t_{\frac{1}{2}} = 3 \times 10^{-7}$ sec.

3) For very long lifetimes it is more usual to detect the absolute number of disintegrations in a known mass, by equating it to the number of α-particles produced per second: $\lambda m N_0 / A$; m is the mass in grams, N_0 is Avogadro's number and A is the atomic weight of the parent. Thus we determine λ and hence T or $t_{\frac{1}{2}}$.

In a radioactive series if we know one λ we can determine others by assay of the relative amounts present. We can sometimes observe the *growth rate* of daughter products and thus determine the lifetime.

4) For shorter half-lives we can artificially pulse an accelerator beam and electronically observe the decay between beam pulses.

4   *Units of activity*

The basic unit is the *curie* which is defined to be the quantity of any radioactive nuclide in which the number of disintegrations/second is 3.700×10^{10}. The reason for this particular number is that it is close to the number of disintegrations/sec in 1 gram of ^{226}Ra. The curie is a rather large unit and in the laboratory we more frequently are dealing with millicuries or microcuries.

We now discuss briefly how we define the biological effects of radiation in order to establish health hazards. The biological effect is expressed in terms of the power of producing ionization in air or tissue.

The *röntgen* is defined as the quantity of X- or γ-radiation for which the associated corpuscular radiation in 0.001293 g of air (i.e. 1 cc at NTP) produces ions carrying 1 e.s.u. of charge of either sign. This is equivalent to an energy transfer of 83.8 erg/g in air and 93 erg/g in biological tissue.

A beam of any radiation which delivers this energy to 1 g has an intensity of 1 röntgen equivalent physical or *1 rep*. If the energy delivered is 100 erg/g the intensity is *1 rad*.

For radiation other than X- or γ-rays we must correct for the fact that a greater or less *ionization density* may result which could cause greater or less cell damage. This correction factor is called the *relative biological effectiveness* or RBE (cf. table 2.5).

TABLE 2.5

Radiation	RBE
β-, X- or γ-rays	1
α-particles	10
Fast neutrons (MeV range)	10
Heavy recoil nuclei	20

Finally the product rep × RBE is called röntgen equivalent man (rem). Permissible *doses* are thus quoted in rem. The usual permissible *weekly* dose at depths greater than 5 cm in the body is to be less than 100 millirem.

A rough rule is that at a distance of 1 foot from C curies of γ-radiation of energy E MeV, the intensity is $6CE$ rads per hour.

Neutrons of different energies have quite different RBE (cf. table 2.6).

TABLE 2.6

Neutron energy	RBE	Flux density in n · cm^{-2} · sec^{-1} equivalent to 2.5 millirem/h
Thermal	3	670
100 keV	8	80
1 MeV	10.5	18
10 MeV	6.5	17

Thus one would get his weekly allowed dose from exposure to a 10 MeV neutron flux of only 17 neutrons·cm^{-2}·sec^{-1} which is *very* few. This tends to make one treat fast neutrons with great respect.

5 *Alpha-decay*

5.1 Lifetime. We again consider the semi-empirical mass formula (1.6). We can compute the nuclei $M(A, Z)$ for which the separation energies of various particles are just zero. We can display this sort of result by plotting N/Z vs. A (fig. 2.8).

From this all nuclei with $A > 150$ might be expected to be α-unstable. While ^{144}Nd$(t_{\frac{1}{2}} = 10^{15}$ y$)$ and ^{147}Sm$(t_{\frac{1}{2}} = 10^{11}$ y$)$ are weakly α-active, the lightest of the normal radioactive series is $A = 210$. The reason is the electrostatic coulomb barrier which the α-particle must penetrate to escape.

Let us calculate the probability of an α-particle penetrating the barrier, i.e. we

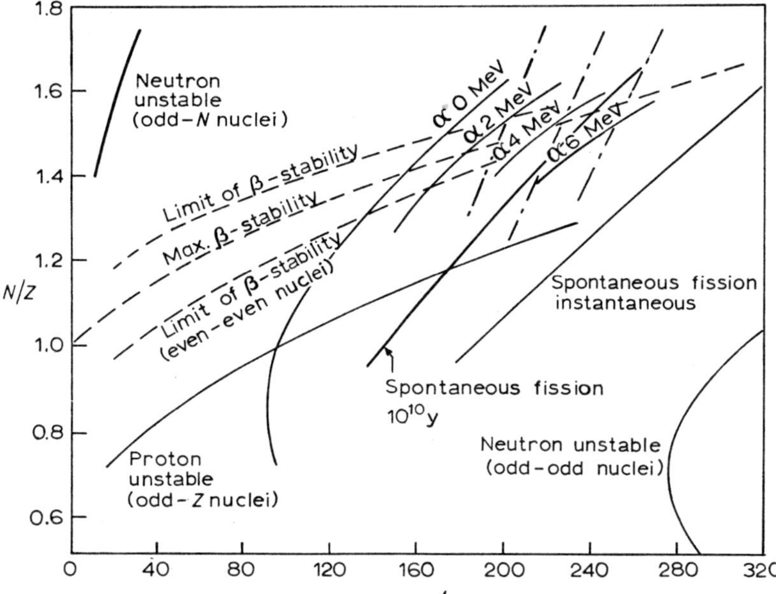

E. Segre (ed.), Experimental Nuclear Physics **3** (Wiley, New York 1959) p. 57

Fig. 2.8.   Nuclear decay stability limits

calculate the decay constant. The time between successive impacts on the barrier is

$$\tau_\alpha = 2R/v_\alpha \tag{2.15}$$

where v_α is the velocity and R is the nuclear radius. Thus the probability/unit time of α-emission is

$$\lambda = T_0/\tau_\alpha = v_\alpha T_0/2R \tag{2.16}$$

where $T_0(\leqq 1)$ is called the barrier transmission coefficient. The internal velocity v_α is related to the velocity v with which the α-particle emerges:

$$\tfrac{1}{2}m_\alpha v_\alpha^2 = \tfrac{1}{2}m_\alpha v^2 + V = E_\alpha + V \tag{2.17}$$

where V is the well depth and E_α is the observed α-energy outside.

We now wish to evaluate T_0. We can express T_0 as the product of the potential discontinuity factor (corresponding to reflection of the wave at the discontinuity) and a factor P_0 depending specifically on the barrier, called the barrier penetration coefficient, or penetrability.

The first factor can be shown to be $4k/K$ where k is the wave number of the α-particle just outside the well at R and K is the wave number just inside. Then

$$T_0 = P_0 k/K. \tag{2.18}$$

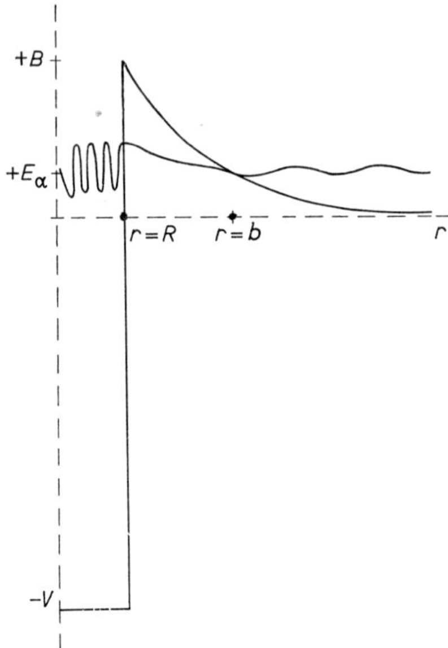

Fig. 2.9.   Alpha-particle emission through the coulomb barrier

The wave numbers are related to $E^{\frac{1}{2}}$, hence

$$T_0 = 4 \left(\frac{B-E_\alpha}{V+E_\alpha}\right)^{\frac{1}{2}} P_0 \qquad (2.19)$$

where $B=zZe^2/R$ is the barrier height. Now P_0 represents the decay of the intensity of the α-particle wave over the region of negative kinetic energy R to b. We have to find a solution of the wave equation

$$\frac{\mathrm{d}^2\psi}{\mathrm{d}r^2} + \frac{2m_\alpha}{\hbar^2}\left(E_\alpha - \frac{zZe^2}{r}\right)\psi = 0 \qquad (2.20)$$

or since $E_\alpha = \frac{1}{2}m_\alpha v^2 = zZe^2/b$ we can write

$$\frac{\mathrm{d}^2\psi}{\mathrm{d}r^2} + \frac{2m_\alpha zZe^2}{\hbar^2}\left(\frac{1}{b} - \frac{1}{r}\right)\psi = 0. \qquad (2.21)$$

For r lying between R and b the coefficient of ψ is negative. So we assume a solution like $\psi \propto e^{-\gamma(r)}$ where $\gamma(r)$ is a slowly varying function of r so that $\mathrm{d}^2\gamma/\mathrm{d}r^2$ can be neglected. Then we get

$$\frac{\mathrm{d}\gamma}{\mathrm{d}r} = \frac{1}{\hbar}(2m_\alpha zZe^2)^{\frac{1}{2}}\left(\frac{1}{r} - \frac{1}{b}\right)^{\frac{1}{2}} \qquad (2.22)$$

and we integrate between R and b using $\cos^2 x = r/b$. Then

$$\gamma = \frac{2zZe^2}{\hbar v}\left[\cos^{-1}\left(\frac{R}{b}\right)^{\frac{1}{2}} - \left\{\frac{R}{b}\left(1-\frac{R}{b}\right)\right\}^{\frac{1}{2}}\right] \qquad (2.23)$$

or in energy terms

$$\gamma = \frac{2zZe^2}{\hbar v}\left[\cos^{-1}\left(\frac{E_\alpha}{B}\right)^{\frac{1}{2}} - \left\{\frac{E_\alpha}{B}\left(1-\frac{E_\alpha}{B}\right)\right\}^{\frac{1}{2}}\right]. \qquad (2.24)$$

Then in this approximation

$$P_0 = e^{-2\gamma}. \qquad (2.25)$$

Thus the decay constant for an α-decay becomes

$$\lambda = \frac{v_\alpha}{2R}\,T_0 = \frac{2v_\alpha}{R}\left(\frac{B-E_\alpha}{V+E_\alpha}\right)^{\frac{1}{2}} e^{-2\gamma}. \qquad (2.26)$$

The main dependence is thus on the factor $e^{-2\gamma}$ and we can approximate by writing

$$\log \lambda = \text{const.} - 2\gamma. \qquad (2.27)$$

In the early period of research on α-decay a relation had been noted between half-life and the α-particle energy. When we plot the log of the half-life against the log of the observed energy, we obtain a series of straight lines. This was the empirical Geiger–Nuttall law,

$$\log \lambda = a + b \log v. \qquad (2.28)$$

Later it was found to apply only to a limited class of nuclides. However the approximate expression we derived above resembles it closely.

The observed decay constant is thus very sensitive to E_α (a 1 MeV change in E_α increases λ by a factor of about 10^5) and to the radius R (a 10 % increase in R, equivalent to a decrease in B, multiplies λ by a factor of 150).

There is a tendency also for α-particle energies to increase sharply for neutron number 126. There is a large energy gain for decay to a nuclide with this neutron number; this indicates special stability of such nuclei. $^{208}_{82}\text{Pb}_{126}$ is an example.

5.2   Alpha-particle spectra. When we look at α-particle spectra with a high resolution spectrometer we find energy distributions as shown in fig. 2.10.

The weak groups of lower energy are termed the *α-spectra fine structure* and are groups leading to different levels of the residual nucleus. They are weaker because of the dependence of the barrier penetration on energy.

The *excited* states as formed will typically decay to the *ground* state of the re-

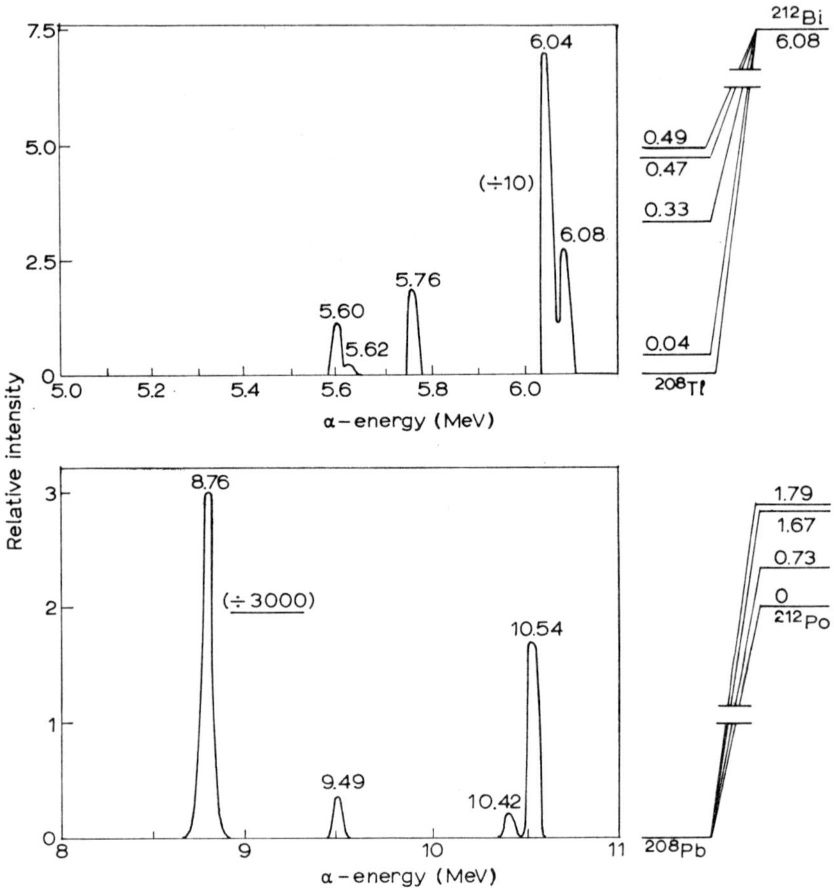

Fig. 2.10. Alpha-particle spectra of ^{212}Bi showing fine structure and of ^{212}Po showing long range α-particle groups

sidual nucleus by emission of γ-rays. This occurs very fast in times of the order of 10^{-13}–10^{-15} sec.

Another type of α-spectra structure was observed first by Rutherford and called by him the *long range α-particles*. These have ranges i.e. energies, *greater* than the main group. The explanation is that a β-emitting nucleus may populate excited states of its daughter nucleus as well as the ground state and sometimes the barrier is low enough so that α-decay of the *excited state* can compete with γ-ray emission which if the excited state energy is very low may be reduced in probability. The long range α-groups are always *very* weak, $\approx 10^{-4}$ of the main group, because of this competition.

6 *Beta-decay*

6.1   Energy relations. We have seen that emission of positive or negative electrons or capture of atomic electrons constitutes the main method for restoring nuclides to the stability line throughout the periodic table. This decay changes the atomic number by ± 1. The electrons or positrons are emitted with a *continuous* spectrum of energies. The electrons involved in β-decay have been shown to be identical in all respects to those in the atom.

We have seen that many β-active nuclides occur amongst the natural radioactive series. In addition they are commonly produced in nuclear reactions. For example a reaction like

$$\,^{7}_{3}\text{Li}_4 + \text{p} \rightarrow \,^{7}_{4}\text{Be}_3 + \text{n} - Q$$

requires energy to push us off the stability line. That is we are creating a nucleus with a proton excess. We return to stability via radioactivity of the ^{7}Be

$$\,^{7}_{4}\text{Be}_3 \overset{\beta^-}{\rightarrow} \,^{7}_{3}\text{Li}_4$$

thus restoring the situation by changing a proton into a neutron. In nuclear reactors a common reaction would be capture of a neutron

$$\,^{23}_{11}\text{Na}_{12} + \text{n} \rightarrow \,^{24}_{11}\text{Na}_{13} + \gamma.$$

We now have a nucleus with a *neutron* excess, and the product ^{24}Na is β^--active, turning the neutron back to a proton,

$$\,^{24}_{11}\text{Na}_{13} \overset{\beta^-}{\rightarrow} \,^{24}_{12}\text{Mg}_{12}.$$

The energy relations in β-decay can be somewhat confusing so we repeat them in more detail. Let us anticipate and assume that the continuous spectrum of β-particles implies emission of a further, massless particle called a neutrino ($\bar{\nu}$). Remember our atomic masses are masses of the *neutral* atom i.e. the nucleus plus its full complement of electrons. If e is an orbital electron and c.e a captured electron then we have for
β^--*decay*:

$$(\,^{11}_{4}\text{Be} + 4e) \rightarrow (\,^{11}_{5}\text{B} + 4e) + \beta^- + \bar{\nu} + Q_-;$$

neutral atom   positive ion   {leptons

in atomic masses:

$$M_\text{a}(^{11}\text{Be}) \rightarrow [M_\text{a}(^{11}\text{B}) - m_\text{e}] + m_\text{e} + Q_-$$

or

$$M_\text{a}(^{11}\text{Be}) - M_\text{a}(^{11}\text{B}) = Q_- = E_\beta\text{-(max)}. \qquad (2.29)$$

β^+-decay:

$$({}^{11}_{6}\text{C}+6\text{e}) \rightarrow ({}^{11}_{5}\text{B}+6\text{e})+\beta^+ +\nu+Q_+\,;$$
$$\underset{\text{neutral atom}}{} \qquad \underset{\text{negative ion}}{} \quad \underset{\text{leptons}}{}$$

in atomic masses:

$$M_{\text{a}}({}^{11}\text{C})-M_{\text{a}}({}^{11}\text{B}) = 2m_{\text{e}}+Q_+ = E_{\beta^+}(\text{max})+1.02 \text{ MeV}. \tag{2.30}$$

electron capture:

$$({}^{11}_{6}\text{C}+6\text{e})+\text{c.e} \rightarrow ({}^{11}_{5}\text{B}+6\text{e})+\nu+Q_{\text{EC}}\,;$$
$$\underset{\text{neutral atom}}{} \qquad \underset{\text{negative ion}}{}$$

in atomic masses:

$$M_{\text{a}}({}^{11}\text{C})-M_{\text{a}}({}^{11}\text{B}) = Q_{\text{EC}} = E_{\bar{\nu}}+E_{\text{K}} \tag{2.31}$$

where E_{K} is the binding energy of the captured K-electron.

If for a moment we consider *nuclear* masses we write eq. (2.29) as

$$\begin{aligned} M_{\text{N}}({}^{11}\text{Be})-M_{\text{N}}({}^{11}\text{B}) &= [M_{\text{a}}({}^{11}\text{Be})-4m_{\text{e}}]-[M_{\text{a}}({}^{11}\text{B})-5m_{\text{e}}] \\ &= Q_-+m_{\text{e}}. \end{aligned} \tag{2.32}$$

This is normally written as $W_0^{(-)}m_{\text{e}}c^2$, i.e. W_0 is the energy difference in units of $m_{\text{e}}c^2$ and is equal to

$$E_{\beta^-}(\text{max})+m_{\text{e}}c^2 \text{ where } m_{\text{e}}c^2 = 0.511 \text{ MeV}.$$

For β^+-decay we write eq. (2.30) as

$$M_{\text{N}}({}^{11}\text{C})-M_{\text{N}}({}^{11}\text{B}) = Q_+ +m_{\text{e}} = W_0^{(+)}m_{\text{e}}c^2 = E_{\beta^+}(\text{max})+3m_{\text{e}}c^2. \tag{2.33}$$

For electron capture we write eq. (2.31) as

$$M_{\text{N}}({}^{11}\text{C})-M_{\text{N}}({}^{11}\text{B}) = Q_{\text{EC}} = Q_+ +2m_{\text{e}}c^2 = (W_0^{(+)}+1)m_{\text{e}}c^2. \tag{2.34}$$

The true nuclear mass difference expressed as W_0 defines the *energy* available for the decay process.

6.2 Lifetime. As in the case of α-decay an empirical logarithmic relation was early recognised between the maximum kinetic energy of the β-particles and the decay constant. This is called the Sargent law. The points cluster around two more or less straight lines. Those corresponding to the shorter lifetimes are now called *allowed* transitions while the longer lifetime line refers to forbidden transitions.

We shall see later from the theory that the decay constant can be written

$$\lambda = \frac{0.693}{t_{\frac{1}{2}}} \propto f(\text{Z}, W_0),$$

W_0 being the total energy in units of mc^2.

Then a plot of $\log \lambda$ vs $\log (W_0 - 1)mc^2$ can be a straight line over a limited range of values of Z and W_0 for which we can assume $f(Z, W_0)$ is almost independent of Z and proportional to some power of $(W_0 - 1)mc^2$. This is the basis of the Sargent law.

It is customary nowadays, in order to classify β-transitions, to calculate the function $f(Z, W_0)$ for each nuclide according to the best theory we have and quote the *comparative half-life* $f(Z, W_0) \times t_{\frac{1}{2}}$. Normally the $\log_{10}$ of ft is quoted. From fig. 2.11 we note that these fall into a number of groupings.

$\log ft = 3\text{–}4$: a small group of especially probable transitions. Presumably very little change in nuclear wavefunction is involved. Termed *favoured* or *superallowed* transitions.

$\log ft = 4\text{–}5$: the largest group, called *allowed* transitions.

$\log ft = 6\text{–}10$: scattered over wide range of values called *forbidden* transitions.

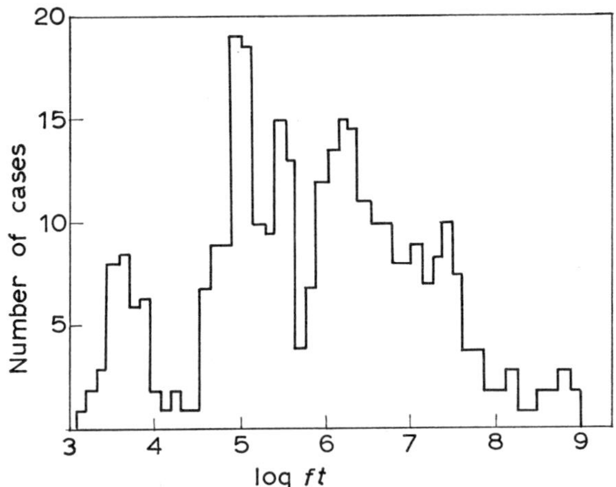

Rev. Mod. Phys. **22** (1950) 402

Fig. 2.11.   Distribution of values of $\log ft$ amongst β-emitters

We shall see that the divisions are not sharp because of the effect of the nuclear states between which the transition takes place.

Our theory will tell us that the *shape* of the β-spectrum should be of the form (for *allowed* spectra)

$$P(p) \propto p^2 F(Z, p)(W_0 - W)^2 dp \qquad (2.35)$$

where $P(p)$ is the probability of emission of an electron with momentum between p and $p + dp$, W_0 is the total kinetic energy as before in units of mc^2, W is the kinetic energy corresponding to momentum p in units of mc^2 and $F(Z, p)$ is the function as before in terms of momentum [compare $f(Z, W_0)$ in energy units].

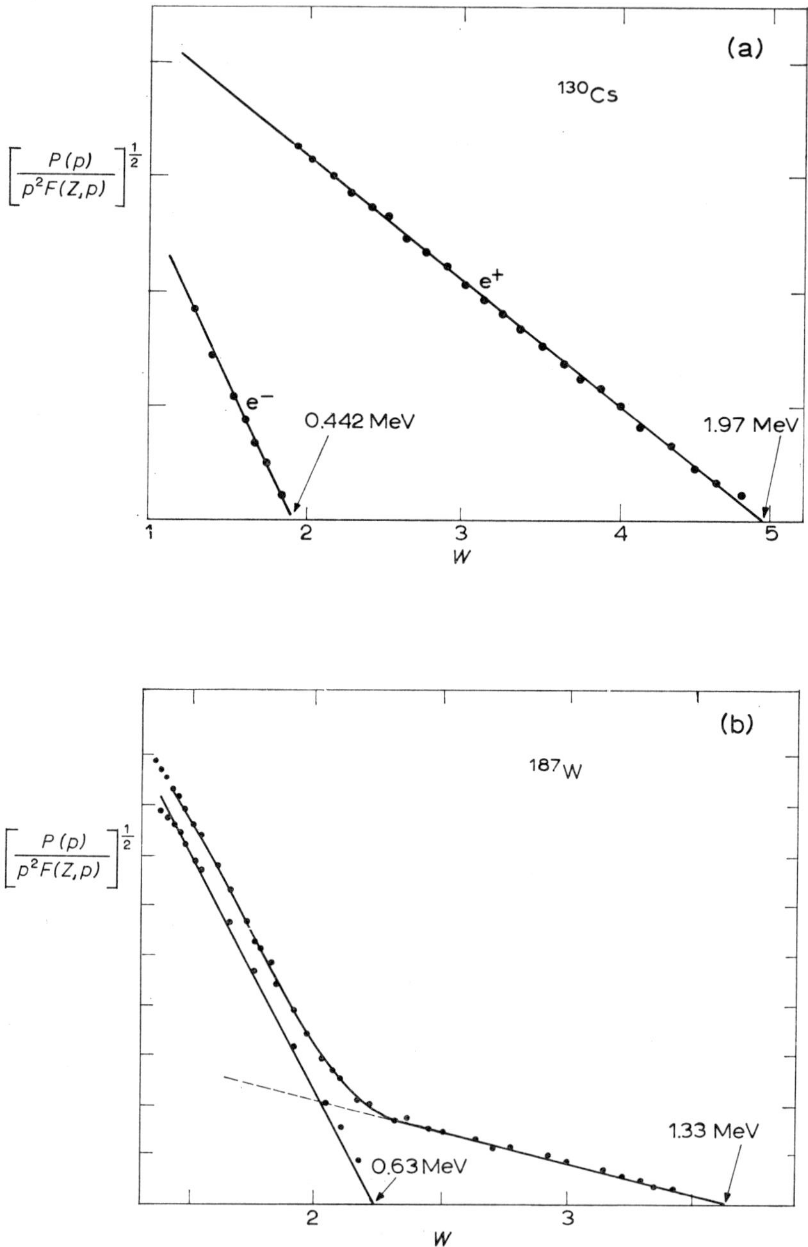

K. Siegbahn (ed.), Alpha-, Beta- and Gamma-ray Spectroscopy **1** (North-Holland, Amsterdam 1965) p. 489

Fig. 2.12.   Fermi–Kurie plot for (a) a simple allowed transition, (b) a complex spectrum with two transitions

If we set up a β-spectrometer we record through the slit set for a particular momentum a counting rate proportional to $pP(q)$. If we then vary the momentum setting we obtain $P(p)$ as a function of p. Then $\{P(p)/p^2 F(Z, p)\}^{\frac{1}{2}}$ plotted against energy W should be *linear*. Such a graph is called a Fermi or Fermi–Kurie plot. Two examples of Fermi plots are given in fig. 12a and 12b. We can extrapolate the line to cut the energy axis and so determine the maximum or end point of the spectrum W_0. This is the standard method for studying shapes and measuring end points. Deviations from a straight line may indicate

1)   A forbidden transition for which our theory was not correct. This results in a concave-up curve. We can apply a specific correction factor to restore linearity and determine the degree of forbiddenness.

2)   A complex spectrum involving two separate β-transitions.

6.3   The neutrinos. The continuum nature of the β-spectrum led to the realization by Fermi that a third particle must be involved to share the energy with the electron. It had to be neutral to conserve charge and massless to conserve momentum. So our β-decays are

$$^{24}\text{Na} \rightarrow {}^{24}\text{Mg} + \beta^- + \bar{\nu}$$

equivalent to $\qquad\qquad$ $\text{n} \rightarrow \quad \text{p} + \beta^- + \bar{\nu};$

$$^{22}\text{Na} \rightarrow {}^{22}\text{Ne} + \beta^+ + \nu$$

equivalent to $\qquad\qquad$ $\text{p} \rightarrow \quad \text{n} + \beta^+ + \nu.$

Note that the last decay written above is not energetically possible for a *free* proton. Note also that the number of heavy particles (protons or neutrons) remains constant. This is conservation of baryon number. We indicate by a different symbol our knowledge that the neutrino accompanying β^--decay is different from that accompanying β^+-decay. If we count

$$\beta^- \text{ as a particle with } L = +1, \quad \beta^+ \text{ as an antiparticle with } L = -1$$

and

$$\nu \text{ as a particle with } L = +1, \quad \bar{\nu} \text{ as an antiparticle with } L = -1$$

then we have conservation of lepton number L (light particles). We now summarize our knowledge of the neutrino properties.

1)   Zero charge (to conserve charge).

2)   Zero or nearly zero rest mass (to conserve momentum, experimentally observed; the measured mass is $<200 \text{ eV}$ from the shape of the ^3H decay spectrum).

3)   Half-integral angular momentum (to conserve angular momentum).

4)   Extremely small interaction with matter. Hence it must have essentially zero magnetic moment (experimentally no ionization is observed).

5)   They have also a new property, a definite *helicity*. This states whether the intrinsic spin vector (σ) is parallel ($+1$) or antiparallel (-1) to the direction of motion (p). It is defined as

$$\frac{\sigma \cdot p}{|\sigma||p|}.$$

Thus a right-handed screw has helicity $+1$ while a left-handed screw has helicity -1 (fig. 2.13).

Fig. 2.13.   Helicity definitions

6)   Electrons are accompanied by antineutrinos $\bar{v}$; positrons (antielectrons) are accompanied by neutrinos v.

Now let us review some of the experiments which demonstrate the nature of neutrinos. a) *Neutrino capture in* ^{37}Cl. A nuclear reactor producing fission products (which are all β^--emitters) emits large fluxes of antineutrinos which escape completely from the reactor. The reaction searched for was

$$^{37}_{17}\text{Cl} + \bar{v} \rightarrow {}^{37}_{18}\text{A} + \beta^-.$$

^{37}A decays by electron capture with $t_{\frac{1}{2}} = 34$ days,

$$^{37}\text{A} + e^- \rightarrow {}^{37}\text{Cl} + v.$$

But the first reaction is identical with

$$^{37}\text{Cl} \rightarrow {}^{37}\text{A} + \beta^- + v$$

that is two *light particles* are created which is not possible according to the lepton conservation rule. The target in this case was a tank car full of CCl_4. The reaction was *not* observed to a high order of accuracy. This proved that neutrinos and antineutrinos are really different.*

b)   *Double β-decay.* This in principle could happen between even mass isobars, for example

$$^{124}_{50}\text{Sn} \rightarrow {}^{124}_{52}\text{Te} + 2\beta^- + 2\bar{v}.$$

The lifetime is expected to be $\approx 10^{21}$ years. If however $\bar{v} \equiv v$ then we could have also

$$^{124}\text{Sn} \rightarrow {}^{124}\text{Sb} + \beta^- + \bar{v}$$
$$^{124}\text{Sb} + \bar{v} \rightarrow {}^{124}\text{Te} + \beta^-.$$

* R. Davis, Phys. Rev. **97** (1955) 766.

The antineutrino reabsorption is equivalent to

$$^{124}\text{Sb} \rightarrow {}^{124}\text{Te} + \beta^- + v.$$

If this reaction is possible we expect a much shorter lifetime of the order of 10^{17} years. All the evidence points toward the longer lifetime showing that the reabsorption is not allowed, so again β^- must be associated with $\bar{v}$ not with v.

c) *Inverse β-decay.* Can we make a β-decay go backwards? Reines and Cowan* again used antineutrinos from a reactor falling on a liquid scintillator of 1.4×10^3 litres volume, containing protons (triethyl benzene). Then

$$\bar{v} + \beta^- + \text{p} \rightarrow \text{n} - 780 \text{ keV}$$

is just the inverse of neutron decay. This is equivalent to

$$\bar{v} + \text{p} \rightarrow \beta^+ + \text{n} - 1800 \text{ keV}.$$

So in the liquid scintillator, coincidences between positrons and neutrons were sought (fig. 2.14). To aid in neutron detection some $CdCl_2$ was dissolved in the scintillator.

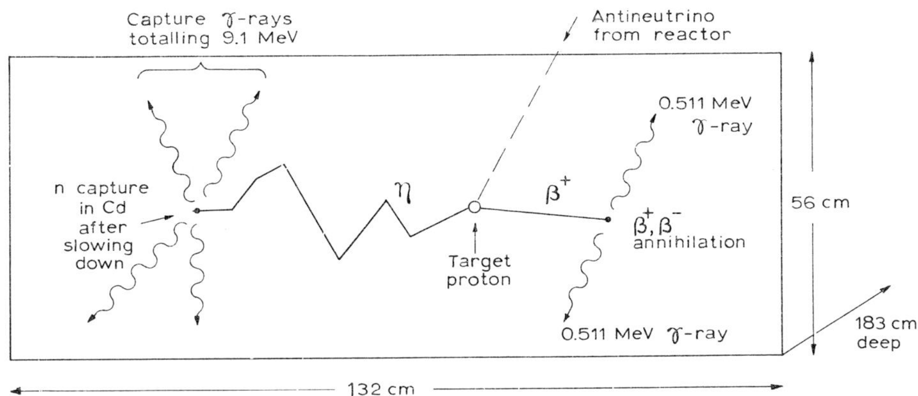

Ann. Rev. Nucl. Sci. **10** (1960) 9

Fig. 2.14.   The inverse β-decay experiment

We look for 1) a *prompt* signal from the ionization due to the positron passing through the scintillator along with the signal from the two 0.511 MeV γ-rays resulting from annihilation of the positron, followed by 2) a delayed signal from the cadmium capture γ-rays. The delay time is due to the slowing down time of the neutron, and is of the order of 30 μsec. The neutron capture reaction is

$$^{114}\text{Cd} + \text{n} \rightarrow {}^{115}\text{Cd} + \gamma + 9 \text{ MeV}.$$

* F. Reines and C. L. Cowan, Phys. Rev. **113** (1959) 273.

Thus 9 MeV of γ-energy is released. Reines and Cowan initially measured 36 ± 4 events associated with the reactor being on.

From this a cross section could be calculated for the inverse β-decay reaction or for neutrino capture of $(11\pm2.6)\times10^{-44}$ cm^2. This is consistent with the theoretical value. This was the first directly observed neutrino reaction.

Basically what we observe in a β^--decay is the change of a neutron into a proton. Thus we expect the free neutron to be itself radioactive, to be β^--unstable and to decay into a proton

$$n \rightarrow p+\beta^- +\bar{\nu}+780 \text{ keV}.$$

The energy available is just the neutron–proton mass difference. In the nuclear β^+-decay we observe the change of a proton into a neutron. Of course the *free* proton cannot do this as energy would be required,

$$p \rightarrow n+\beta^+ +\nu-1800 \text{ keV}.$$

The energy is made up of two electron masses, 1020 keV, plus the neutron–proton mass difference, 780 keV. However in the nucleus this extra energy can be taken up from the mass energy of the nucleus and β^+-emission is possible.

There has been much work on the precise measurement of the decay of the free neutron. Its half-life is found to be 11.7 ± 0.3 min. This decay was first measured in an elegant experiment by Robson at Chalk River in Canada.* At that time the highest neutron flux in the world was available in the Chalk River reactor.

The beam of slow neutrons from the reactor was carefully collimated and had an intensity of 1.5×10^{10} neutrons·cm^{-2}·sec^{-1}. The protons have initially a distribution of energies up to about 1 keV. Since as we shall see it is very difficult to detect such low energy particles they were accelerated by an electrode to about 13 keV and then focussed onto the sensitive surface of a detector called an electron multiplier. The decay electrons were also put through a β-spectrometer and detected in coincidence with the proton pulses with a delay to allow for the time of flight of the slow moving proton (fig. 2.15a). Even with the high flux of neutrons the chance of a neutron decaying in the small volume viewed was so small that only a few coincidences an hour were detected. The original half-life value was 12.8 ± 2.5 min. Later work at Oak Ridge in the U.S.A. and in Russia improved the accuracy and it is the Russian value quoted above.

The momentum spectrum of the β-particles could be recorded in the spectrometer and displayed as a Kurie plot (fig. 2.15b). This agreed with the expected end point energy.

* J. M. Robson, Phys. Rev. **83** (1951) 349.

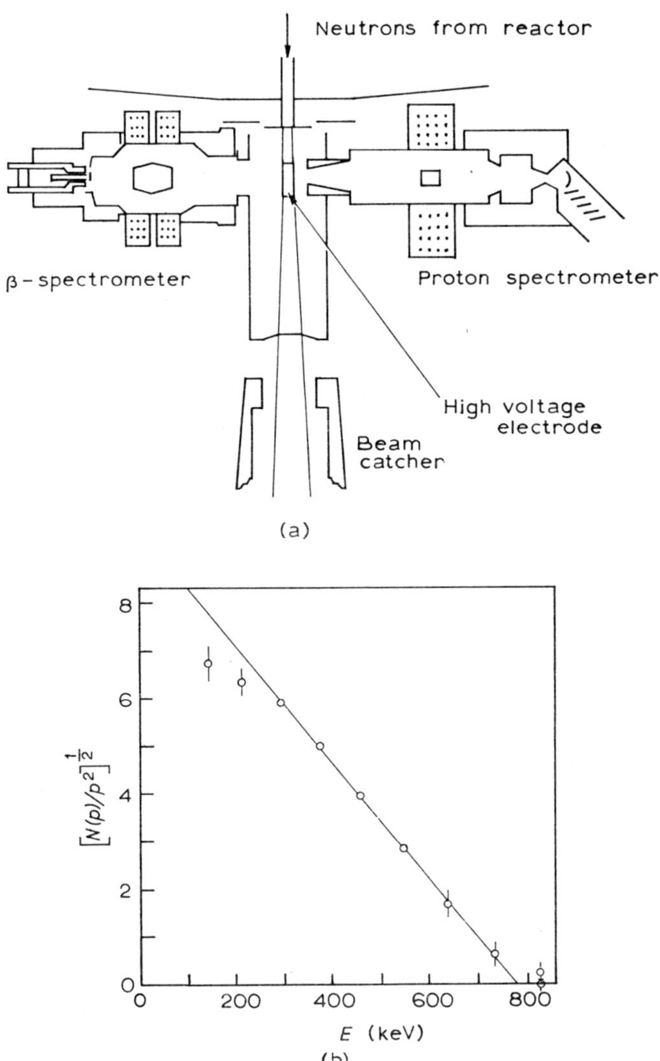

Phys. Rev. **83** (1951) 349

Fig. 2.15.   (a) The neutron lifetime experiment, (b) Fermi–Kurie plot of the β-spectrum from neutron decay

6.4 Beta-decay theory. We now consider the formal theory of β-decay due to Fermi. Consider the transformation of a neutron into a proton with creation of two light particles β^- and $\bar{\nu}$. The nucleon wavefunction ψ_i inside a volume Ω changes to the final wavefunction ψ_f and electron and neutrino waves emerge from the volume (fig. 2.16).

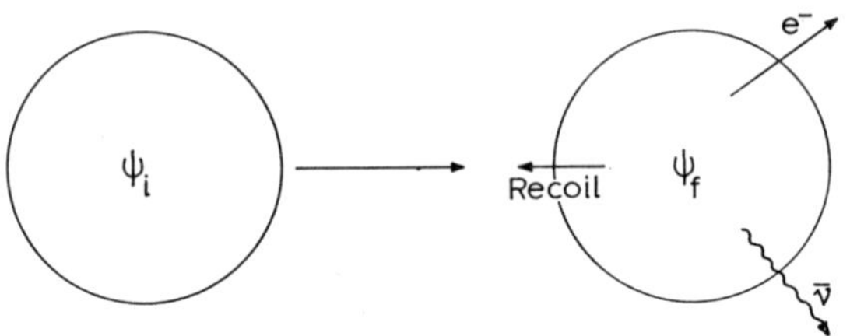

Fig. 2.16.   Initial and final states in β-decay

The probability per unit time for the transition can then be written

$$\frac{2\pi}{\hbar}|H_{if}|^2\rho(E) \tag{2.36}$$

where H_{if} is the potential or interaction relating the two nucleon states and $\rho(E)$ is the number of momentum states in the final system, per unit energy range. This is given by

$$\rho(E) = \frac{4\pi p_e^2 \mathrm{d}p_e}{(2\pi\hbar)^3} \frac{4\pi p_\nu^2 \mathrm{d}p_\nu}{(2\pi\hbar)^3} \frac{\Omega^2}{\mathrm{d}E_e} \tag{2.37}$$

where the momenta are related to total electron and neutrino energy via the relativistic relations

$$\begin{aligned}
E_e^2 &= p_e^2 c^2 + m^2 c^4 \\
E_\nu &= p_\nu c \qquad \text{(zero rest mass)} \\
E_0 &= E_\nu + E_e = E_\beta(\max) + mc^2
\end{aligned}$$

thus

$$\begin{aligned}
p_\nu &= E_\nu/c = (E_0 - E_e)/c \\
\mathrm{d}p_\nu &= -\frac{\mathrm{d}E_e}{c}.
\end{aligned}$$

Then the probability of emission of an electron of momentum p_e per unit time is given by

$$\begin{aligned}
P(p_e)\mathrm{d}p_e &= \frac{2\pi}{\hbar}|H_{if}|^2 \frac{p_e^2 \mathrm{d}p_e\, p_\nu^2 \mathrm{d}p_\nu}{4\pi^4 \hbar^6} \frac{\Omega^2}{\mathrm{d}E_e} \\
&= \frac{|\Omega H_{if}|^2}{2\pi^3 \hbar^7} \frac{p_e^2(E_0 - E_e)^2}{c^3} \mathrm{d}p_e. \tag{2.38}
\end{aligned}$$

This breaks into two parts.

1)   A statistical factor $p_e^2(E_0 - E_e)^2 \mathrm{d}p_e$ which determines the *shape* of the β-spectrum. Note that it goes to zero for both high and low electron momenta. If $m_e = m_\nu$

we should get a completely symmetric curve. But since $m_\nu = 0$ we get rather more low energy electrons (fig. 2.17).

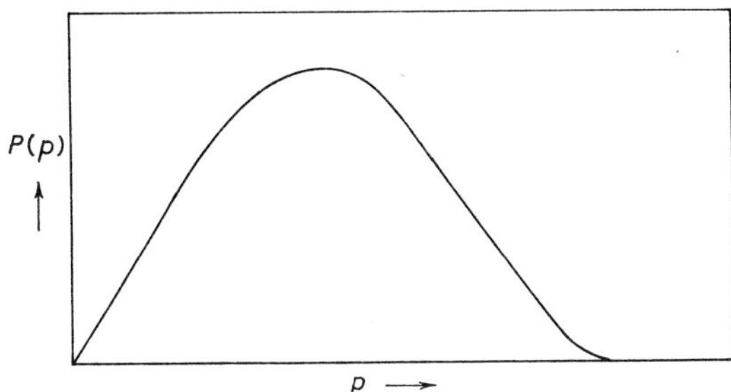

Fig. 2.17.   The β-spectrum

2) $|\Omega H_{if}|^2$ relates the electron and neutrino emission probability to the associated nuclear transformation. This is still uncertain in *form* but contains certainly the wavefunctions of nucleons, electron and neutrino normalized to a volume Ω so we can write

$$|\Omega H_{if}|^2 = g^2 |M|^2 \tag{2.39}$$

where g is an arbitrary constant and M contains the details of the interaction which we do not know yet. Our general formula is then

$$P(p_e)\,dp_e = \frac{g^2 |M|^2}{2\pi^3 \hbar^7 c^3}\, p_e^2 (E_0 - E)^2 \,dp_e. \tag{2.40}$$

We change to units p and W such that $p_e = p \cdot mc$ and $E_e = W \cdot mc^2$. Then

$$P(p_e)\,dp_e = \frac{m^5 g^2}{2\pi^3 \hbar^7}\, |M|^2 c^4 p^2 (W_0 - W)^2 \,dp$$

or collecting constants and changing to energy units

$$P(p_e)\,dp_e = \frac{G^2}{2\pi^3}\, |M|^2 p W (W_0 - W)^2 \,dW \tag{2.41}$$

where $G^2 = m^5 c^4 g^2 / \hbar^7$ is a constant. The assumption that this is a *universal* constant of nature like the gravitational constant is the basic assumption of a Universal Fermi Interaction which we discuss later.

We have not yet considered the electromagnetic effect of the electric charge involved. That is, in a β^--decay the lower energy negative electrons are held back by the coulomb field of the nucleus while in a β^+-decay the lower energy positive

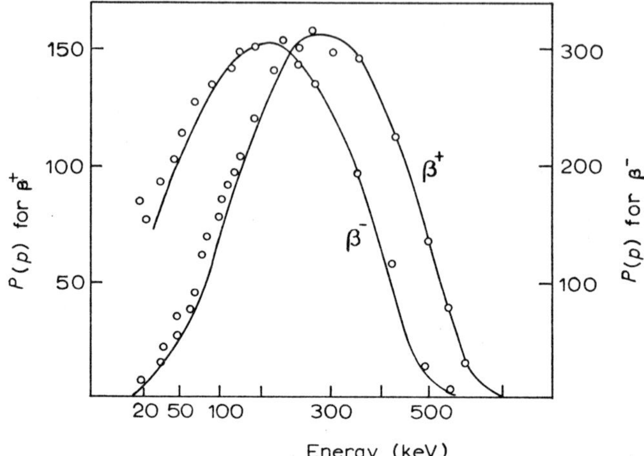

Phys. Rev. **75** (1949) 315

Fig. 2.18. β-spectrum shapes for positron and electron decay of ^{64}Cu

electrons are pushed out by the coulomb field of the nucleus (cf. fig. 2.18). We rewrite our spectral formula:

$$P(p_e)\mathrm{d}p_e = \frac{G^2}{2\pi^3}|M|^2 F_c(Z, p)p^2(W_0-W)^2\,\mathrm{d}p \tag{2.42}$$

where $F_c(Z, p)$ corrects the shape for nuclear charge effects. It is similar to a barrier penetration factor. Approximately for electrons

$$f(Z, E) = 2\pi\eta[1-\exp(-2\pi\eta)]^{-1} \tag{2.43}$$

where $\eta = Ze^2/\hbar v_e$ where Z is the atomic number of the residual nucleus and v_e is the electron velocity. We can integrate eq. (2.42) to give the total probability for decay of the nucleus per unit time i.e. the decay constant λ

$$\lambda = \frac{\log_e 2}{t_{\frac{1}{2}}} = \frac{1}{\tau} = \int_0^{po} P(p)\mathrm{d}p$$

$$= \frac{G^2}{2\pi^3}|M|^2 F(Z, p_0)$$

$$= \frac{G}{2\pi^3}|M|^2 f(Z, W_0). \tag{2.44}$$

The Fermi function (which includes energy factors and coulomb effects) is often referred to as f and we have for the comparative half-life or ft-value:

$$ft_{\frac{1}{2}} = \frac{2\pi^3}{G^2}\frac{\log_e 2}{|M|^2}\text{ sec.}$$

For *large* values of W_0, $f(Z, W_0) \propto W_0^5$.

The quantity $|M|^2$, we recall, depended on the nuclear wavefunctions. We now have two possibilities for emission of the light particles.

1)   If electron and neutrino are emitted with their spins opposed ($\uparrow\downarrow$) and no net spin angular momentum is carried away, this is termed a *Fermi transition*.

2)   If electron and neutrino are emitted with their spins aligned ($\uparrow\uparrow$) and a net spin angular momentum of $1\hbar$ is carried away, this is termed a *Gamow–Teller transition*. Allowed transitions are those in which the light particles do not carry away any orbital angular momentum i.e. s-wave. This means the parity of the decaying nucleus does not change. We can see that Fermi allowed means $\Delta I = 0$, no parity change, GT allowed means $\Delta I = 0, \pm 1$, no parity change but $I_i = 0 \rightarrow I_i = 0$ is not permitted since spin momentum *must* be carried away. Hence we find a pure Fermi transition only for $0 \rightarrow 0$, *no*; we find a pure GT transition only for $1 \rightarrow 1$, *no*; and we find mixed Fermi–GT for all other allowed transitions.

We write thus the matrix element $|M|^2$

$$|M|^2 = |C_F|^2|M_F| + |C_{GT}|^2|M_{GT}|^2 \tag{2.45}$$

where $|C_F|^2 + |C_{GT}|^2 = 1$ and M_F and M_{GT} are integrals over nuclear wavefunctions for the two types of interactions. We can write

$$ft = \frac{B}{(1-x)|M_F|^2 + x|M_{GT}|^2} \tag{2.46}$$

where $x = |C_{GT}|^2$ and then evaluate the constants B, x from pure Fermi and pure GT decays. We find $B = 2787 \pm 70$ sec, $x = 0.56$. From the value of B we can derive the coupling constant $g = 1.4 \times 10^{-49}$ erg·cm^3.

If in the decay the parity of the *nuclear* state changes, then the light particles cannot be emitted as an s-wave. They 'originate' effectively at a distance λ from the centre of the nucleus where λ is the wavelength of the electron or neutrino. Then the probability of emission is reduced by a factor $(R/\lambda)^2$. We call such a transition *first forbidden*, and on the average it will be $(R/\lambda)^2$ of the probability of an allowed transition. $(R/\lambda)^2$ is of the order of 10^{-2} so for first forbidden transitions the log ft value will be ≈ 2 units greater. So we have the selection rules:

	Fermi	GT
Allowed	$\Delta I = 0$, no	$\Delta I = 0, \pm 1$, no (not $0 \rightarrow 0$)
First forbidden	$\Delta I = 0, \pm 1$, yes (not $0 \rightarrow 0$)	$\Delta I = 0, \pm 1, \pm 2$, yes

We can display this more graphically as follows. Recall that allowed means

$$L = 0, \qquad \text{Fermi } S = 0, \qquad \text{GT } S = 1,$$

while first forbidden means

$$L = 1, \qquad \text{Fermi } S = 0, \qquad \text{GT } S = 1.$$

Then

$$J_i + S + L = J_f.$$

So

$$J_i + \uparrow\downarrow(S = 0) + L = 0 = J_i \qquad \qquad \Delta J = 0, \text{ no, Fermi allowed.}$$

$$
\begin{aligned}
J_i + \uparrow\uparrow(S = 1) + L = 0 &= J_i - 1 \text{ or} \\
&= J_i \text{ or} \qquad \quad \Delta J = 0, \pm 1, \text{ no, GT allowed.} \\
&= J_i + 1
\end{aligned}
$$

$$
\begin{aligned}
J_i + \uparrow\downarrow(S = 0) + L = 1 &= J_i - 1 \text{ or} \\
&= J_i \text{ or} \qquad \quad \Delta J = 0, \pm 1, \text{ yes, Fermi first forbidden.} \\
&= J_i + 1
\end{aligned}
$$

$$
\begin{aligned}
J_i + \uparrow\uparrow(S = 1) + L = 1 &= J_i - 2 \\
&= J_i - 1 \qquad \Delta J = 1, \pm 1, \pm 2, \text{ yes, GT first forbidden.} \\
&= J_i \\
&= J_i + 1 \\
&= J_i + 2
\end{aligned}
$$

Some example of β-decay are presented in fig. 2.19.

Now let us see how electron capture is treated in this framework. No positron rest mass has to be created and the rest mass of the captured electron is added to the energy release. Thus electron capture is favoured over positron decay by an energy $2mc^2 = 1.02$ MeV. The energy is entirely removed by the neutrino e.g.

$$^{7}_{4}\text{Be} + e^{-}_{\text{K}} \rightarrow {}^{7}_{3}\text{Li} + \nu + 0.86 \text{ MeV}.$$

The neutrino energy is slightly less than 0.86 MeV because the absorption of the K-electron from ^{7}Be leaves the ^{7}Li with a K-shell vacancy. Thus energy is radiated equivalent to the K-shell binding by the valence electron falling down into the K-shell vacancy. This energy is radiated as X-rays or Auger electrons.

We note that electron capture is a two-body process unlike electron decay, hence the neutrino has a unique energy. Our formulae will change since only a neutrino phase space factor is required. However the electron factor is now replaced by a term representing the density of the bound electron wavefunction at the nucleus. So

$$\lambda_{\text{K}}(\text{allowed}) = \frac{G^2}{2\pi^3} |M|^2 f_{\text{K}} \qquad\qquad (2.47)$$

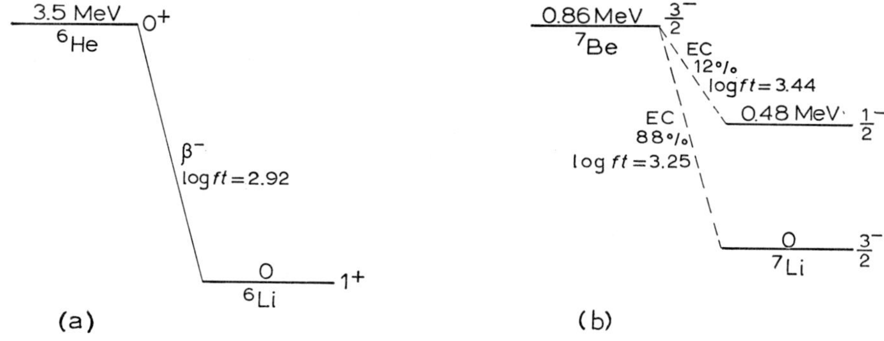

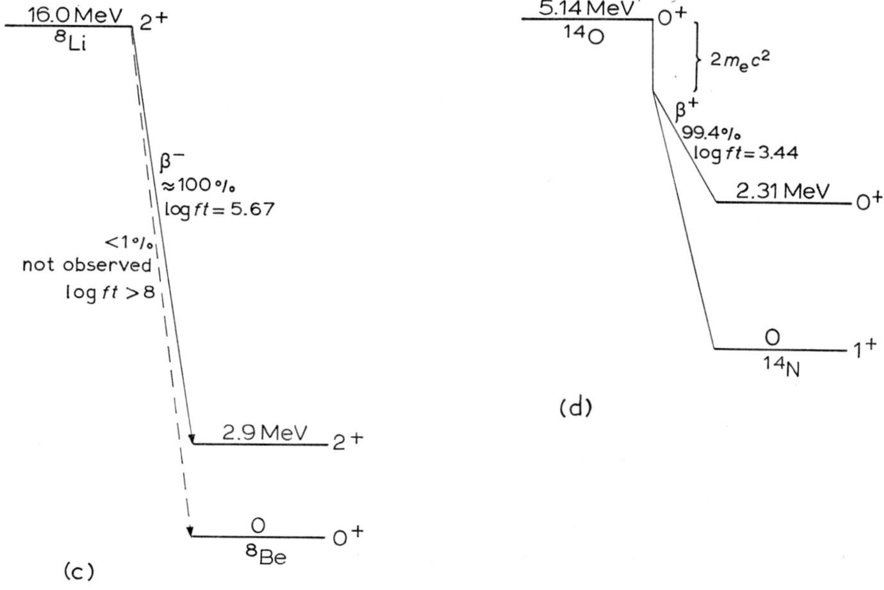

Fig. 2.19. Examples of β-decay.   (a) $\Delta I=1$, no, pure GT allowed. The ft-value indicates a superallowed transition, i.e. M is a maximum. (b) Ground state transition $\Delta I=0$, no, mixed F and GT transition, superallowed; excited state transition $\Delta I=1$, no, pure GT allowed. (c) Ground state transition $\Delta I=2$, no, forbidden; excited state transition $\Delta I=0$, no, mixed F and GT allowed. (d) Ground state transition $\Delta I=1$, no, pure GT allowed. The value of M is unusually small; excited state transition $\Delta I=0$, no, pure F superallowed

where

$$f_K \approx 2\pi(\alpha Z')^3 (W_0 + 1)^2$$

where $\alpha = \frac{1}{137}$ is the fine structure constant, W_0 is the total energy available for positron decay and Z' is the atomic number of the decaying nucleus. We omit in this approximation the K-shell binding.

Hence for heavy neutron-deficient nuclei we can see that K-capture will be strongly favoured over β^+-decay because

1)   there is a more favourable energy release (1.02 MeV $\equiv 2mc^2$ is not required),
2)   the large potential barrier for escape of the positron is not present,
3)   for high Z there will be a high electron density at the nucleus.

In cases where the total energy available is only a little more than the K-shell binding, L-electron capture may then be observed because of the lower binding energy.

The ratio of K-capture probability to β^+-decay probability can often be measured in the same nucleus

$$\frac{\lambda_+}{\lambda_{EC}} \approx \frac{f(Z, W_0)}{2\pi(\alpha Z')^3 (W_0 + 1)^2} \tag{2.48}$$

which is independent of the unknown nuclear factor $|M|^2$. So this is a good test of our theoretical expressions for the remaining parts as it excludes specifically nuclear effects.

This is one of the few cases where the chemical state of the atom may influence a nuclear happening. This is because the electron density at the nucleus may vary from compound to compound, e.g. λ for ^{7}Be metal and for ^{7}BeO differ by about one part in 10^{-4}.

Electron capture is most frequently indicated by observation of the K X-rays or Auger electrons resulting from the optical transition in the residual nucleus when the K-shell vacancy is filled. Of course this merely indicates that K-capture has occurred and does not show anything about the energy change involved. We can find this by measuring the recoil energy of the product nucleus. This is typically only a few eV. We can measure it either by 1) applying a retarding potential just sufficient to turn back the recoil, 2) measuring the time of flight of the recoil timed from the emission of the X-rays, which is essentially prompt.

Finally there is another effect which can be used to find the energy involved. This is by observing the so-called inner bremsstrahlung. This is continuous radiation of energy extending from zero up to the maximum electron energy radiated by the captured electron as it changes energy in the coulomb field of the nucleus.

6.5   Parity in β-decay. We have said that parity is conserved in strong interactions and hence is a good quantum number; that is, one which gives some definite characteristic of a state.

We have also mentioned that it is not conserved in weak interactions. The idea that it might not be arose about ten years ago when the new fundamental particles were being identified in cosmic rays. Two types of new particle of similar mass ($\approx 1000\, m_e$) were seen, called the τ and θ mesons. They were observed to decay as follows:

$$\tau^+ \to \pi^+ + \pi^0$$
$$\theta^+ \to \pi^+ + \pi^0 + \pi^0$$
$$\text{or } \pi^+ + \pi^+ + \pi^-.$$

Now also it was known that in each of these *weak* decays there was zero orbital angular momentum involved. Since the pions have spin 0^- it was concluded that the θ^+ must have spin 0^- and the τ^+ spin 0^+. However as experiments improved it was found that the masses of these two were more and more nearly identical.

So Lee and Yang* suggested that it was really one particle decaying two ways and that possibly in weak decays parity was not conserved. There was plenty of evidence that parity *was* conserved in strong and electromagnetic interactions but they could find no experiment which proved that it was conserved in weak interactions such as β-decay.

An experiment to test the possibility of parity violation was quickly devised and performed by Wu et al.** They chose ^{60}Co as the β-emitter; its decay scheme is shown in fig. 2.20. A special technique was used to align the spins of the ^{60}Co nuclei in a known direction. This is done by cooling down the cobalt nuclei by a process

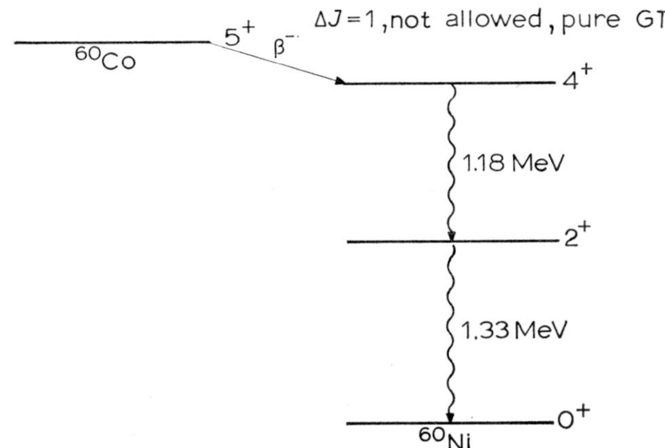

Fig. 2.20. ^{60}Co decay scheme

* T. D. Lee and C. N. Yang, Phys. Rev. **108** (1957) 1611.
** C. S. Wu, E. Ambler, R. W. Hayward, D. D. Hoppes and R. P. Hudson, Phys. Rev. **105** (1957) 1413.

known as adiabatic demagnetization to about $0.1°K$. The nuclei can then be polarised by a weak magnetic field. The alignment can be confirmed by the angular distribution of the γ-rays emitted. The experiment consisted of measuring the emission probability of the β-particles as a function of the cobalt spin-direction. It was found that more β-particles were emitted in the direction *opposite* to the aligned spin direction than were emitted in the same direction. That is, an asymmetric angular distribution was observed of a form $a+b\cos\theta$. What does this tell us about the parity?

$$^{60}\text{Co} \rightarrow {}^{60}\text{Ni} + \beta^- + \bar{v}$$

spin: 5^+ 4^+ $\frac{1}{2}$ $\frac{1}{2}$ $+0$ orbital (allowed).

The lepton part of the system would have positive parity if parity is conserved. Positive parity means the system can be reflected about a symmetry axis without change. The spins are both in the same sense as they must add to one unit for a GT transition. Since there is an observed preference for the direction of electron emission, parity must be violated.

Another way of interpreting the experiment is that the observation of an asymmetric angular distribution must involve interference between two amplitudes of opposite symmetry. Observation of the β-asymmetry for the ^{60}Co decay of the form $a+b\cos\theta$ implied that the lepton pair were emitted in both odd and even parity states and that the amplitude of the two states interfered.

We can account for what happens by noting that the neutrino being massless must possess a definite helicity. Helicity is simply related to the polarisation of the spinning particle, and is defined as

$$\mathcal{H} = \frac{\boldsymbol{\sigma} \cdot \boldsymbol{p}}{|\boldsymbol{\sigma}||\boldsymbol{p}|} \tag{2.49}$$

where $\boldsymbol{\sigma}$ and $\boldsymbol{p}$ are the spin vector and linear momentum vector of the particle. If the particle has mass and thus $v<c$, special relativity does not allow a definite helicity. Suppose we observe a particle passing our space ship with spin vector in the direction of its motion (right-handed screw) and with velocity v. However if $v<c$ we can always speed up the space ship until its velocity exceeds v and observe the particle moving to the rear with opposite polarisation (left-handed screw). Its helicity is thus only defined as v/c. For the massless neutrino which moves always with velocity c the helicity is definite and equal to ± 1. Thus in the ^{60}Co experiment the neutrinos refuse to emerge with the wrong helicity! (fig. 2.21).

The helicity of the neutrino has been directly measured in an experiment by Goldhaber et al.* They used a source of ^{152}Eu which decays by K-capture to an

---

* M. Goldhaber, L. Grodzins and A. W. Sunyar, Phys. Rev. **109** (1958) 1015.

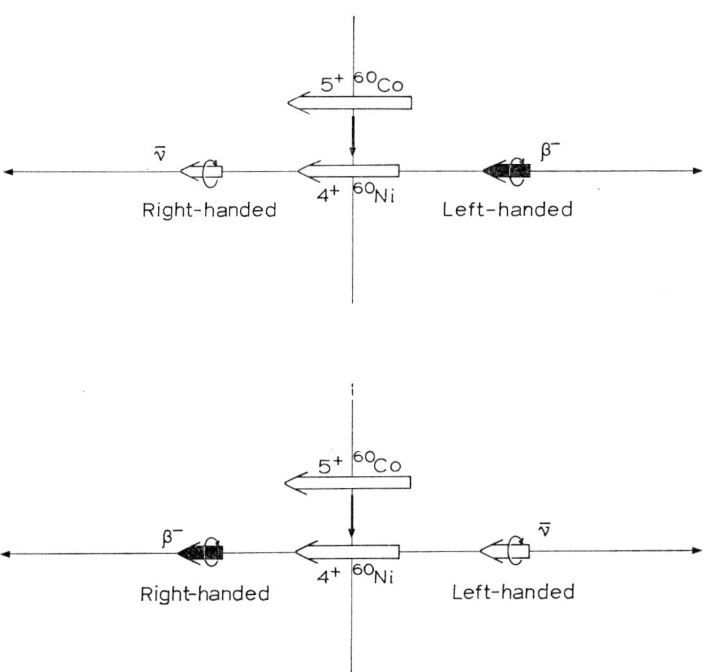

Fig. 2.21.   The parity violation experiment

excited state of ^{152}Sm which then decays by γ-emission to the ground state by a 961 keV E1 transition (fig. 2.22)

$$^{152}\text{Eu} + e^- \rightarrow {}^{152}\text{Sm} + \nu.$$

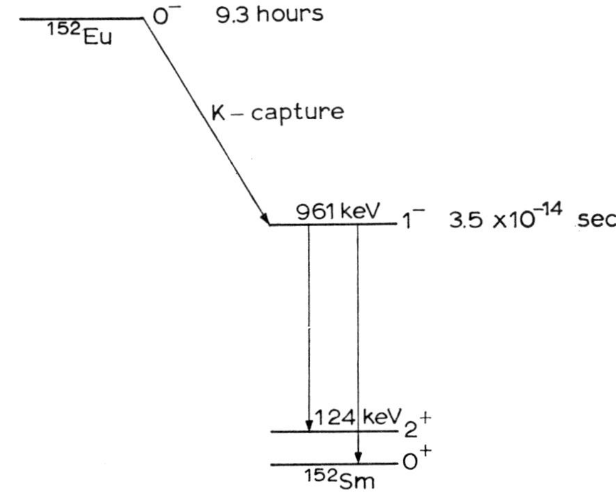

Fig. 2.22. ^{152}Eu decay scheme

The lifetime of the 1^- state is so short (the state is so broad), that resonance scattering of the decay γ-ray is observable if the neutrino is emitted opposite to the γ-ray thus compensating the recoil energy.

If resonant scattering takes place the neutrino must have gone *up* so as to compensate the recoil. The magnetised iron polarimeter (fig. 2.23a) would only transmit γ-rays with circular polarisation such that their spin was up and hence the spin of the 1^- state of ^{152}Sm was also up. So the decay of the 0^{-152}Eu to the 1^- state has added one unit of angular momentum. But note that it is an allowed GT transition, that is, no orbital angular momentum is involved and the captured electron was in an s-state. Thus the electron spin must have been up and the neutrino spin must have been down.

$$^{152}\text{Eu} + e^- \rightarrow\ ^{152}\text{Sm}^* + \nu$$

spin: $0 \quad \tfrac{1}{2}\uparrow \quad 1\uparrow \quad \tfrac{1}{2}\downarrow$

The neutrino thus had spin down but travelled upward, so its helicity is negative like a left-handed screw. Fig. 2.23b demonstrates schematically the sequence of events.

We also get helicity information by studying the electron–neutrino angular correlations for Fermi allowed transitions, $\Delta J = 0$, and GT allowed transitions, $\Delta J = 1$. We infer these correlations from measurements of the electron–recoil correlations. In the case of the Fermi transition the electron and neutrino tend to come out in the same direction, while in the GT case the electron and neutrino tend to come out in opposite directions.

TABLE 2.7

Helicities

Neutrinos	-1	(100 % polarised as left-handed screw)
Particles with mass as electrons	$-v/c$	(polarised v/c as left-handed screw)
Antineutrinos	$+1$	(100 % polarised as right-handed screw)
Anti-particles with mass as positrons	$+v/c$	(polarised v/c as right-handed screw)

Our world of particles thus tends to have this left-handed symmetry. An analogous situation is found in the giant molecules of biological importance. Only those proteins twisted into a left-handed helical form take part in life processes. Right-handed twisted molecules are quite inert.

It should be noted that the initial and final nuclear wavefunctions involved in the β-decay have perfectly definite parity being strongly interacting systems. Thus

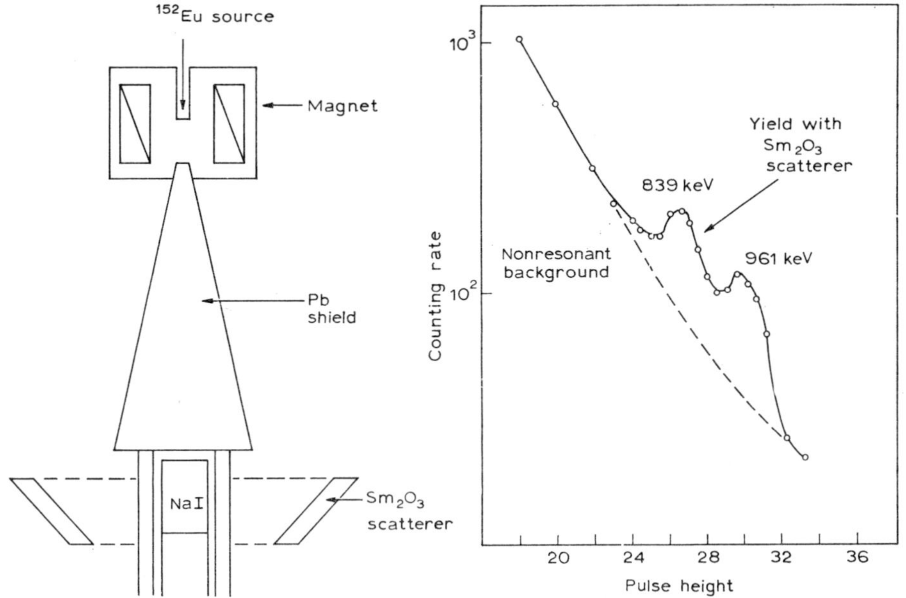

Phys. Rev. **109** (1958) 1015

Fig. 2.23a.   Neutrino helicity: experimental set-up and results

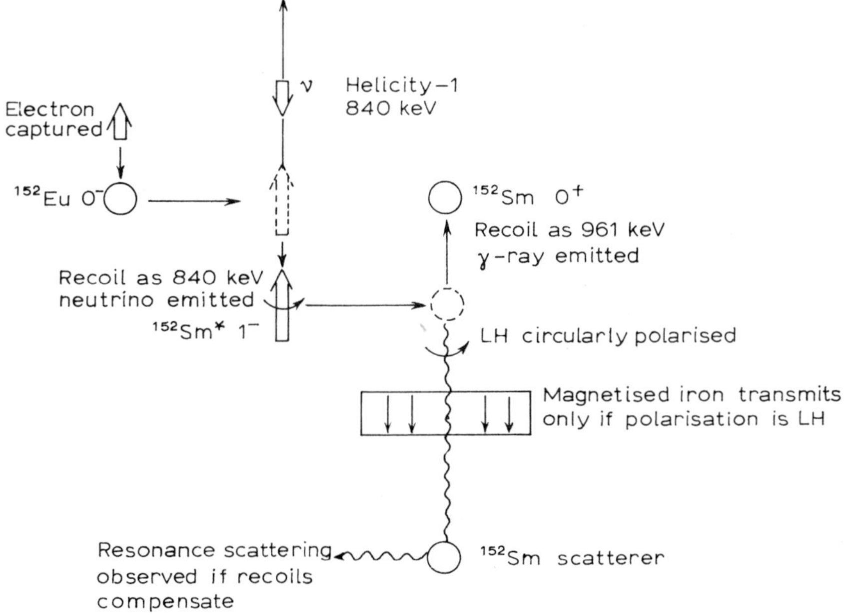

Phys. Rev. **109** (1958) 1015

Fig. 2.23b.   Neutrino helicity: schematic

the selection rules which related the *change* in angular momentum and parity to decay probability are still completely valid. The parity violation is in the system of weakly interacting particles and their angular momentum.

We have mentioned the symmetry operation of charge conjugation which changes all particles into antiparticles. It is easy to see that if we reflect the ^{60}Co decay system and simultaneously perform charge conjugation then no change is observed. In other words the weak decay is still invariant under *CP* but is not invariant under *P* alone and presumably under *C* alone. We have already remarked that conservation of *CP* is equivalent to invariance under time reversal. We return in ch. 13 to these points as in some weak decays *CP* is not *exactly* conserved.

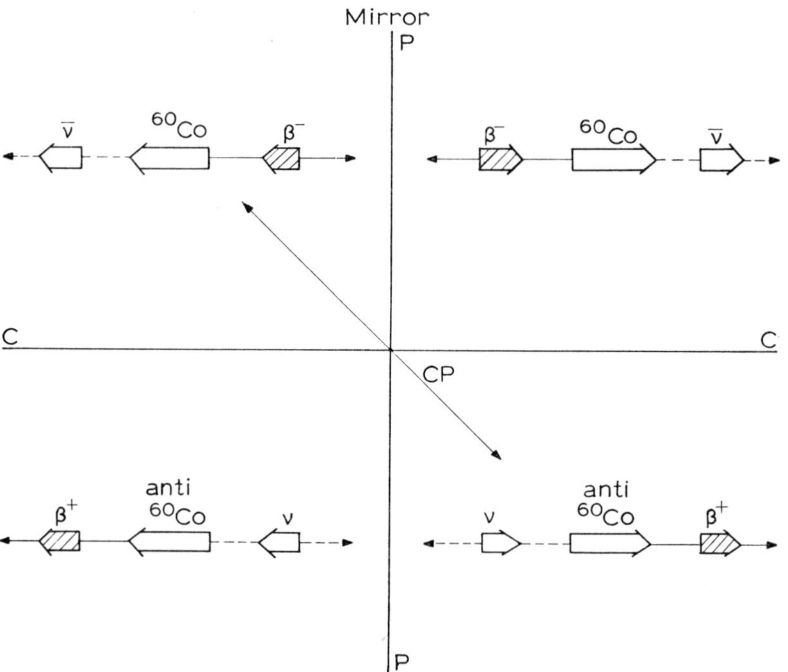

Fig. 2.24.   Parity and charge conjugation violation; *CP*-invariance.

The arrows shown in fig. 2.24 indicate the *preferred* direction of emission. The mirror images in both cases reverse the preferred direction (violation of parity). The charge conjugated system also reverses the preferred direction (violation of *C*). The *CP* operation leaves the preferred direction the same (*CP*-invariance).

7   *Radiative decay*

7.1   Gamma-emission. We have noted that when an α-decay or a β-decay or a nuclear reaction leaves a nucleus in an excited state it normally releases the excitation energy by emission of electromagnetic radiation, the γ-rays. Since emission and absorption of radiation is an electromagnetic phenomenon it is relatively well understood in terms of electrodynamics.

Consider two states as shown in fig. 2.25. We define an energy uncertainty for the excited state which arises from the Heisenberg uncertainty principle. This states

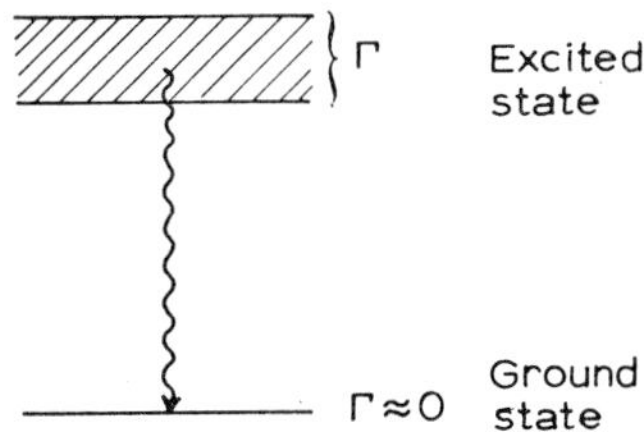

Fig. 2.25.   Radiative decay

that there is a fundamental limit to the accuracy of definition of related quantities such as

$$\Delta p \cdot \Delta x \geqq \hbar \qquad \text{(momentum–position)}$$
$$\Delta E \cdot \Delta t \geqq \hbar \qquad \text{(energy–time)}.$$

where $\hbar$ is Planck's constant divided by 2π. Thus

$$h = 6.625 \times 10^{-27} \text{ erg} \cdot \text{sec}, \qquad \hbar = 1.054 \times 10^{-27} \text{ erg} \cdot \text{sec}.$$

Thus we see that for any quantum mechanical state of motion there will be a limiting relation between the precision of its energy and its mean lifetime. We call this energy width Γ measured in eV. Then its mean lifetime τ is given by

$$\tau \Gamma = \hbar, \qquad \tau = \hbar/\Gamma.$$

The transition probability per sec analogous to λ is

$$T = \tau^{-1} = \Gamma/\hbar \text{ sec}^{-1}.$$

We can divide up the transition probability according to the *mode* of decay. Suppose a state can decay by α-emission or β-emission or γ-emission. Then

$$T_\alpha + T_\beta + T_\gamma = \frac{\Gamma_\alpha + \Gamma_\beta + \Gamma_\gamma}{\hbar} = \frac{\Gamma}{\hbar} = \frac{1}{\tau} \tag{2.50}$$

where $T_\alpha = \lambda_\alpha$ and $T_\beta = \lambda_\beta$, τ is then the observed *total* mean life and Γ is the total width. The Γ_α are partial widths.

In energy units $\hbar = 6.56 \times 10^{-16}$ eV · sec. We consider now only the radiative transition probability T_γ. The radiative width of a bound state is of the order of 0.1 eV or less. Hence

$$\tau_\gamma \approx \frac{6.56 \times 10^{-16}}{0.1} = 6.56 \times 10^{-15} \text{ sec (or more)}.$$

For unbound states i.e. those above a particle separation energy, a particle will generally be emitted as soon as it comes to the nuclear surface. The lifetime is thus comparable to the time for a particle to cross the nucleus $\tau \approx R/v$. Assuming a typical particle velocity inside the nucleus is $\approx 0.1c$ we have

$$\tau = 10^{-12}/10^9 = 10^{-21} \text{ sec, hence } \Gamma = \frac{6.56 \times 10^{-16}}{10^{-21}} = 6.56 \times 10^5 \text{ eV} = 0.65 \text{ MeV}.$$

Then when we speak of *widths* of nuclear levels we refer simultaneously to decay probability and mean lifetime. A wide state implies a short lifetime, a narrow state implies a long lifetime, while an infinitesimal width implies an infinite lifetime i.e. stable such as the ground states of the stable nuclides.

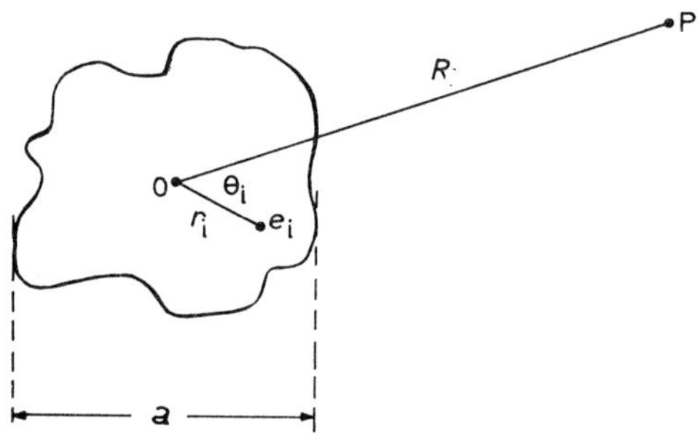

Fig. 2.26.   Potential due to charge distribution

Let us consider the potential at a point P due to a charge distribution (fig. 2.26):

$$\Phi_P = \sum_i \frac{e_i}{|\mathbf{R} - \mathbf{r}_i|} = \sum_i \frac{e_i}{(R^2 + r_i^2 - 2Rr_i \cos \theta_i)^{\frac{1}{2}}}. \qquad (2.51)$$

R and r_i are the position vectors with angle θ_i for point P and charge e_i respectively. If the distance to the point P is large compared to the dimension of the charge distribution, a, that is $R \gg a$ or $R \gg r$, we can expand the denominator in powers of r/R:

$$\Phi_\mathrm{P} = \frac{\sum e_i}{R} + \frac{\sum e_i r_i \cos\theta}{R^2} + \frac{\sum e_i r_i^2 \tfrac{1}{2}(3\cos^2\theta - 1)}{R^3}$$

$$\ldots + \frac{\sum e_i r_i^L P_L(\cos\theta)}{R^{L+1}} + \ldots$$

$$= \Phi_0 + \Phi_1 + \Phi_2 + \ldots + \Phi_L + \ldots. \qquad (2.52)$$

Hence Φ_L is proportional to $1/R^{L+1}$. Then Φ_0 is just the potential due to the charge $\sum e_i$; Φ_1 is $\sum e_i r_i$ grad $1/R$, the potential due to the dipole moment $\boldsymbol{D} = \sum e_i r_i$, the static dipole moment; Φ_2 is the potential due to the quadrupole moment; Φ_L is the potential due to the 2^Lth moment. When we consider now the time varying moment we find the second term corresponds to the *field* of an electric dipole (E1) while the next term corresponds to the field of an electric quadrupole (E2) and also to a magnetic dipole (M1). In general the term Φ_L corresponds to EL and M$(L-1)$ radiations. Because of the successively more symmetric oscillations represented we might expect the terms to be successively smaller. The L is called the multipolarity of the radiation.

Radiation of a given multipolarity may be even or odd parity unlike the particle case where the angular momentum determines the parity. This is because the electric field is a vector of even intrinsic parity, while the magnetic field is an axial vector of odd intrinsic parity (cf. fig. 2.27). The multipolarity corresponds to the orbital angular momentum.

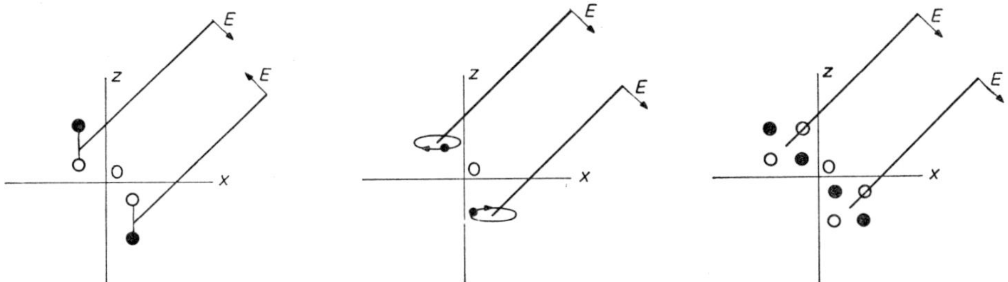

W. E. Burcham, Nuclear Physics (Longmans, London 1963) p. 98

Fig. 2.27.   Parity of radiations E1, M1, and E2

For example the electric dipole reverses the sign of the electric vector upon reflection (odd parity) while the magnetic dipole does not reverse upon reflection (even parity). The electric quadrupole also does not reverse the electric field sign upon reflection (even parity).

Thus we can make a table of parity change due to an electromagnetic transition of given multipolarity; cf. table 2.8.

TABLE 2.8

Multipole	Parity change
E1	yes
M1, E2	no
M2, E3	yes
M3, E4	no
M($L-1$), EL	yes for L odd
	no for L even

The angular momentum change must be given by $J_i - J_f = L$ so that $J_i + J_f \geqq L \geqq |J_i - J_f|$. Note that the projection of L on the space axis is just $|m_i - m_f| = M$, as shown in fig. 2.28.

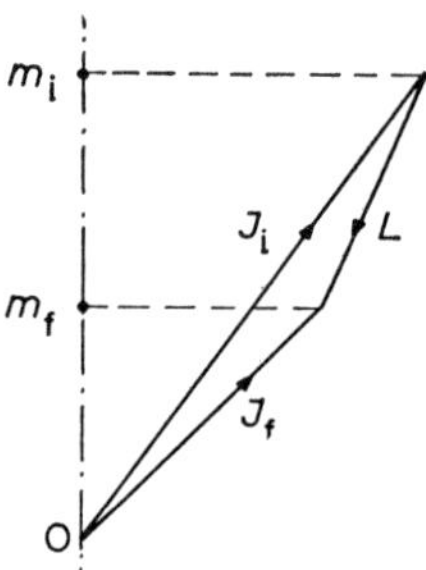

Fig. 2.28.   Angular momentum change with radiation

When we use the two rules together we can predict the type of radiation allowed for a given change of angular momentum (cf. table 2.9). We get no radiation corresponding to $L=0$ since there is no monopole radiation. The monopole operator corresponds simply to a translation of the whole charge distribution, and would result in a longitudinal wave. However the electromagnetic wave can be only transverse. Thus there is no $M=0$ (M is a substate of L) component, only $M= \pm 1$. We return later to decay processes for $0 \rightarrow 0$ transitions.

We can conclude several things about the relative intensity of the different multipoles. Each term in our potential changed by a factor of R^{-1} where $R \gg a$. Thus for the *intensity* of the multipole fields, each is smaller than the preceding term by a factor $(a/\lambdabar)^2$ which is of the order 10^{-3} for $E_\gamma = 0.5$ MeV, $a \approx 10^{-12}$ cm. Also the intensity of a magnetic multipole is smaller than the electric multipole of the same order by a factor again of order 10^{-3} for $E_\gamma = 0.5$ MeV. This can be seen from the fact that the magnetic moment of a single particle is

$$\pi r^2 i = \pi r^2 \frac{e\omega}{2\pi} = \tfrac{1}{2} erv/c \propto vD/c,$$

TABLE 2.9

Multipole radiation

$J_i \rightarrow J_f$	Parity change	Radiation
$0 \leftrightarrow 0$	yes	none
	no	none
$1 \leftrightarrow 0$	yes	E1
	no	M1
$2 \leftrightarrow 0$	yes	M2
	no	E2
$1 \leftrightarrow 2$	yes	E1 (M2, E3)
	no	(M1, E2) M3
$1 \leftrightarrow 3$	yes	(M2, E3) M4
	no	E2 (M3, E4)
$2 \leftrightarrow 3$	yes	E1 (M2, E3) (M4, E5)
	no	(M1, E2) (M3, E4) M5

where D is the electric dipole moment. From these arguments the intensity will be *roughly* as shown in table 2.10.

TABLE 2.10

Multipole	Relative intensity
E1	1
M1 or E2	10^{-3}
M2 or E3	10^{-6}
M3 or E4	10^{-9}
etc.	

The expression we derive from a detailed calculation for the transition probability for multipolarity L and energy E_γ is

$$T(L) = \frac{8\pi(L+1)}{L\{(2L+1)!!\}^2} \frac{1}{\hbar} \left(\frac{E_\gamma}{\hbar c}\right)^{2L+1} B_{if}(L) \; \text{sec}^{-1} \tag{2.53}$$

where $(2L+1)!! = 1 \cdot 3 \cdot 5 \cdot 7 \cdots (2L+1)$ and $B_{if}(L)$ is called the reduced transition probability and contains all the specifically nuclear factors depending on the character of initial and final states. We have separated out the energy dependent factors and those depending on combinations of angular momenta.

We shall see later how models of nuclear structure can suggest values for the static moments and hence for $B_{if}(L)$. We use it for the present as an energy independent parameter.

We show in fig. 2.29 some illustrative examples of radiations expected in decay schemes. Only the strongest multipoles are shown.

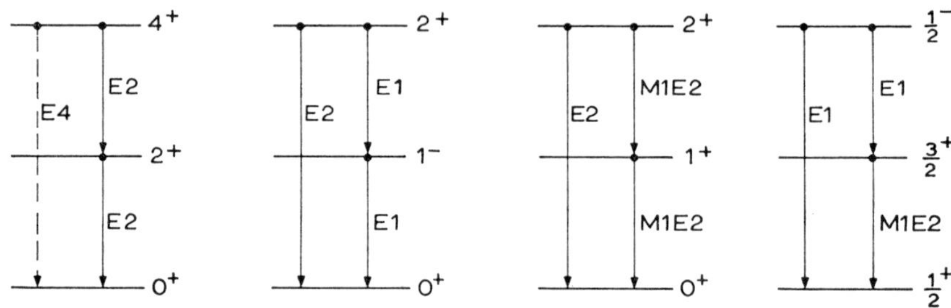

Fig. 2.29.   Examples of radiative decay

When we consider certain nuclear models we shall find further more restrictive selection rules for L when we evaluate $B_{if}(L)$. However the general rules quoted above must always be obeyed (conservation of angular momentum and parity).

7.2 Internal conversion. The formula we wrote down is just the probability of decay for a bare nucleus with no atomic electrons playing a role. But when such electrons are near the nucleus, energy can also be removed from the nucleus by giving it to the atomic electron which is then ejected from the atom. This process is called *internal conversion* (a rather misleading term). The total probability of decay of a state is thus $\Gamma = \Gamma_\gamma + \Gamma_e$. They are competing alternatives. Both are electromagnetic effects. We define the *internal conversion coefficient* as

$$\alpha = N_e/N_\gamma = \Gamma_e/\Gamma_\gamma. \tag{2.54}$$

N is the number observed per excited nucleus and α may take any value from 0 to ∞. If we fix the total width from the lifetime of the state $\tau = \hbar/\Gamma$ then $\Gamma_\gamma = \Gamma/(1+\alpha)$. These electrons are emitted in discrete energy groups corresponding to the nuclear excitation energy. These are often seen together with β-spectra. An example is shown in fig. 2.30. We find the kinetic energy of the electron is just $E_e = E_\gamma - E_K$, where E_K is the binding energy of an electron in the K-shell.

Measurement of internal conversion coefficients is a useful spectroscopic tool since they depend on the amplitude of the multipole fields near the nucleus. The value of α increases with the value of L and decreases with increased transition energy (fig. 2.31). It also depends on Z since this determines the electron density at the nucleus. Thus it is most important for heavy nuclides. Discrimination in L-value is best at low energy since the wavelength of the radiation is longer and extends further

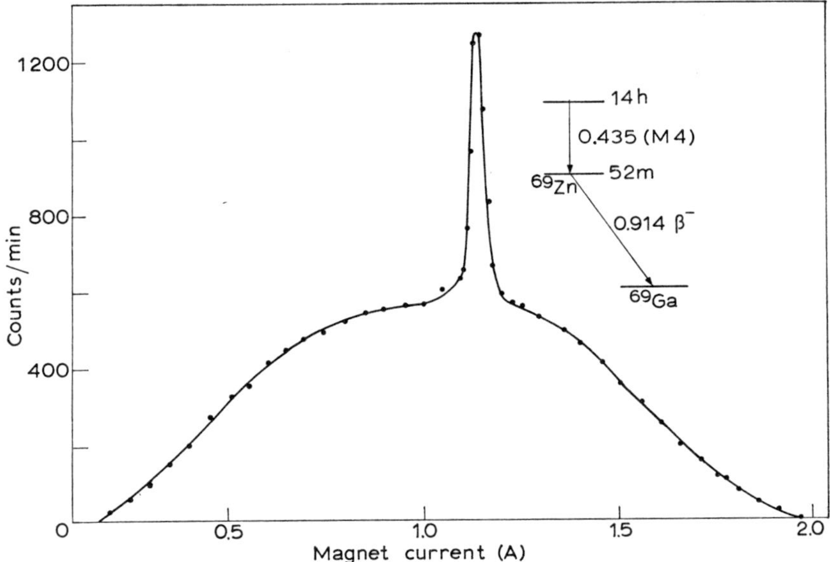

K. Siegbahn (ed.), Alpha-, Beta- and Gamma-ray Spectroscopy **1** (North-Holland, Amsterdam 1965) p. 487

Fig. 2.30.   A conversion electron line superimposed on a β-spectrum in the electron spectrum from ^{69}Zn decay

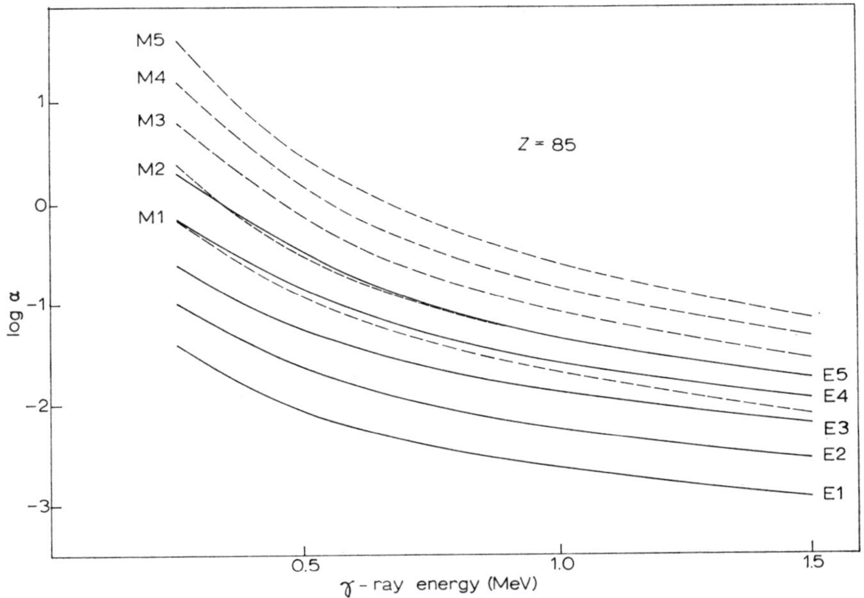

E. Segre (ed.), Experimental Nuclear Physics **3** (Wiley, New York 1959) p. 342

Fig. 2.31.   Internal conversion coefficients as a function of energy and multipolarity

from the nucleus. The value of α is largest for low energy and high multipolarity. As the energy increases the wavelength decreases, and becomes short compared to the region with appreciable electron density.

7.3 Pair internal conversion. If the transition energy exceeds $2m_e c^2 = 1.02$ MeV, it is possible for an electron–positron pair to be produced as an alternative to γ-emission or internal conversion. This is a field effect, the pair being created in the electric field of the nucleus. It is therefore not dependent on electron density, that is, not on Z. Its probability increases with increasing transition energy and is also greatest for small multipolarities (cf. fig. 2.32). Thus it will occur in just the cases where internal conversion will not.

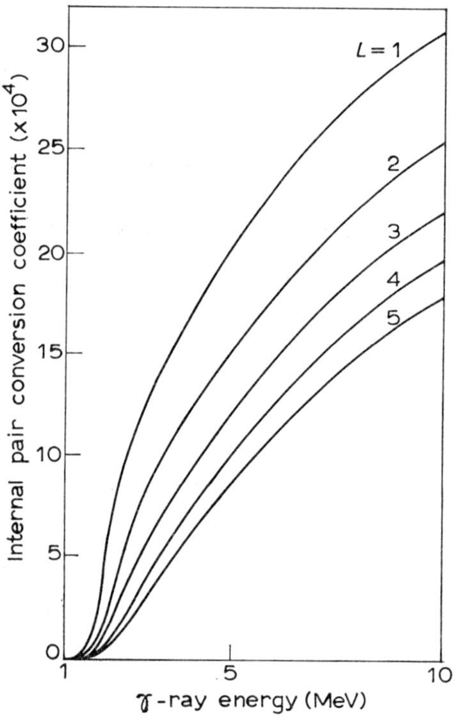

E. Segre (ed.), Experimental Nuclear Physics 3 (Wiley, New York 1959) p. 368

Fig. 2.32.   Pair internal conversion coefficients as a function of energy and multipolarity

It is necessary to distinguish between the pairs due to pair internal conversion and those due to pair conversion of γ-rays in other nuclei (external conversion). Pair internal conversion is easiest to observe in light nuclei.

7.4 Zero-zero transitions. As was mentioned single quantum radiation is forbidden between states of spin zero. However if the electron wavefunction penetrates the nucleus we can have total internal conversion ($\alpha = \infty$) and a single electron line of energy $E - E_K$ will be observed. The lifetime of the level is generally quite long. Such transitions are seen in heavy nuclei for low transition energies. An example is the 0.72 MeV 0^+ state in ^{72}Ge decaying to the 0^+ ground state.

In light nuclei pair internal conversion is more probable and the first excited state of ^{16}O at 6.06 MeV (0^+) decays to the ground state (0^+) in this way.

No clear case of a $0^- \rightarrow 0^+$ transition is known. It could not decay by either of the processes described above but would have to decay by an intermediate state.

7.5   Nuclear isomerism. We have noted that electromagnetic transitions with high multipolarity, i.e., high spin difference and low energy tend to be slow, i.e., $T(L)$ is small, τ_γ is long, Γ_γ is small. This gives rise to a type of γ-ray radioactivity in which the lifetime is that of the excited level itself. Since the nucleus is unchanged in Z or A after the transition such emitters are called *isomers*. Since the conditions for isomerism are exactly those for internal conversion we expect most isomeric transitions to be highly internally converted. The observed lifetime will be due to γ-radiation and internal conversion in competition. If we observe a lifetime τ then the partial γ-lifetime will be $\tau(1 + \alpha)$. We show some examples of isomerism in fig. 2.33. There is no essential difference between isomeric states and all other excited states. It is merely that they have unusually long lifetimes. A lifetime longer than 10^{-5} sec might be termed an isomer.

7.6   Measurement of radiative widths. Since typical radiative lifetimes range from 10^{-10}–10^{-15} sec special methods for measuring these short lifetimes have been developed.

a)   *Comparison with other transitions.* For example in ^{212}Po (ThC′) the long range α-group competes with the γ-transition (fig. 2.34). The ^{212}Po ground state decay has a half-life of 3×10^{-7} sec. Comparing the intensities of the two α-groups gives us the partial decay constant for the long range group, together with the ratio of parent β-transitions. Comparing the long range α-particle intensity with the γ-ray intensity gives us the partial decay constant for the γ-transition. The lifetime of the state is thus $\tau = (\lambda_\gamma + \lambda_{\alpha'})^{-1}$ and is about 10^{-13} sec for the 1.76 MeV state.

b)   *Delayed coincidence method.* As mentioned before we can study α–γ or β–γ coincidence rates as a function of delay between the detectors by a set-up as in fig. 2.35a. Comparison with the delay curve (fig. 2.35b) for prompt coincidences can yield a value for the lifetime of the γ-emitting state. The method can be used down to about 10^{-11} sec.

Alternatively we can use a pulsed accelerator to produce the excited state. For example bursts of protons a few nanoseconds long were used* to produce the 0.73 MeV state in ^{10}B by the ^{10}B(p, p′)^{10}B* reaction. The decay of the state between

---

* E. K. Warburton, D. E. Alburger and D. H. Wilkinson, Phys. Rev. **129** (1963) 2180.

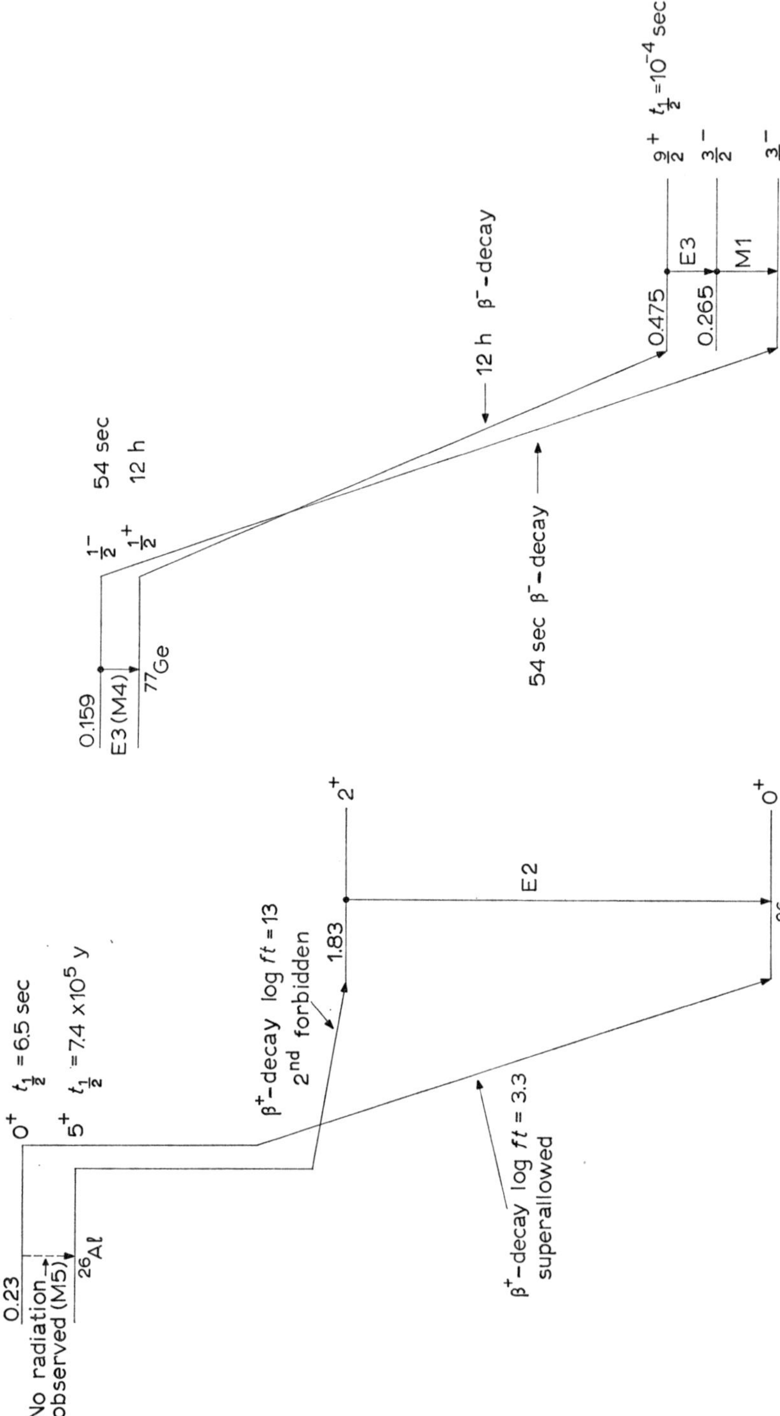

Fig. 2.33b. An isomeric state in which both radiation and β-decay is observed

Fig. 2.33a. An isomeric state in which no radiative decay is observed

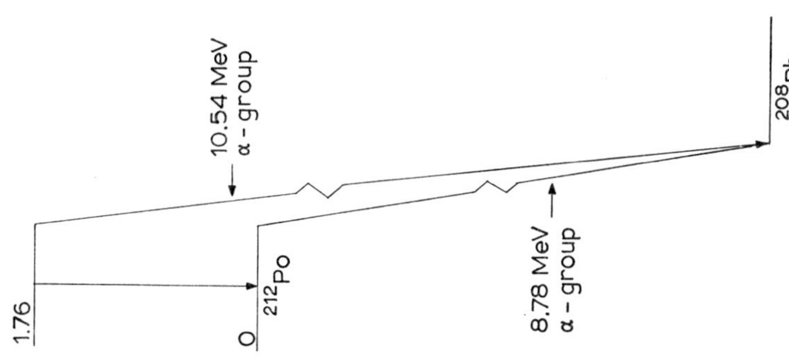

Fig. 2.34.   Decay of ^{212}Po(ThC′)

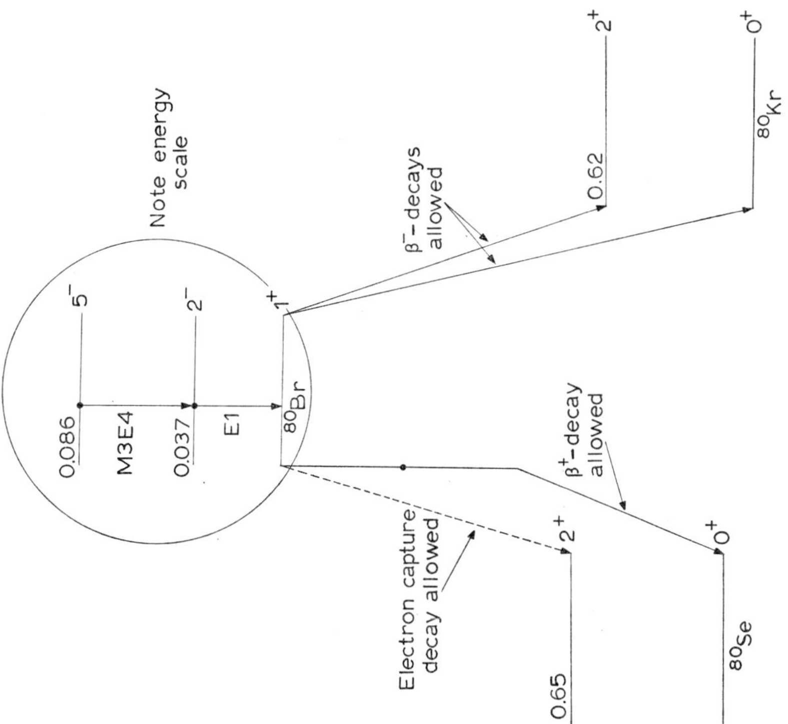

Fig. 2.33c.   An isomeric state in which only radiation is observed

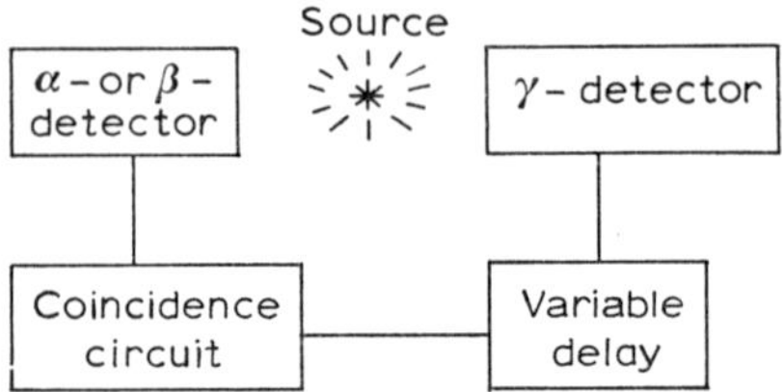

K. Siegbahn (ed.), Alpha-, Beta- and Gamma-ray Spectroscopy **2** (North-Holland, Amsterdam 1965) p. 913

Fig. 2.35a.   Delayed coincidence: experimental set-up

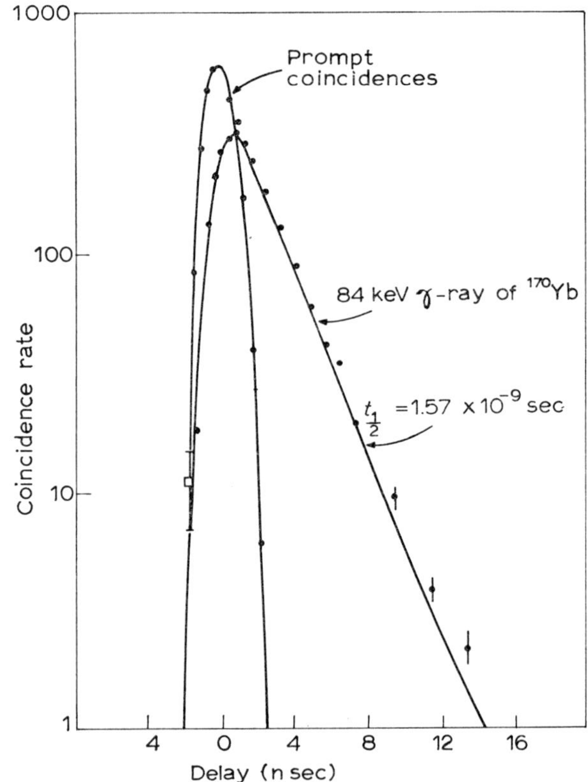

K. Siegbahn (ed.), Alpha-, Beta- and Gamma-ray Spectroscopy **2** (North-Holland, Amsterdam 1965) p. 913

Fig. 2.35b.   Delayed coincidence: results

beam pulses could be studied by plotting the number of γ-counts as a function of time from the end of the beam burst. A lifetime of 1.05×10^{-9} sec was measured in the case mentioned. Beam pulses less than about 5×10^{-10} sec wide are difficult to achieve.

c)   *Recoil distance method.* This was already mentioned in connection with the α-

lifetimes. An example* is the measurement of excited state lifetimes in ^{16}O (fig. 2.36) produced by the reaction ^{19}F + p → ^{16}O* + α. A thin target is used and the recoiling ^{16}O nucleus leaves the target with a velocity of about 10^8 cm/sec. The flight distance before decay occurs is then determined as shown in fig. 2.37 by moving the target with respect to the corner of the shield.

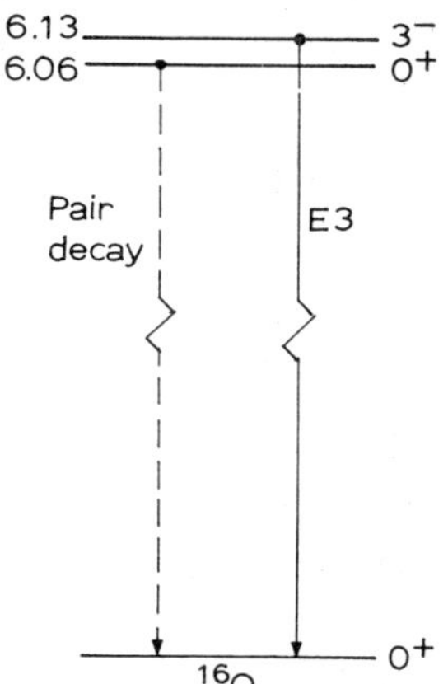

Fig. 2.36.   Excited states of ^{16}O

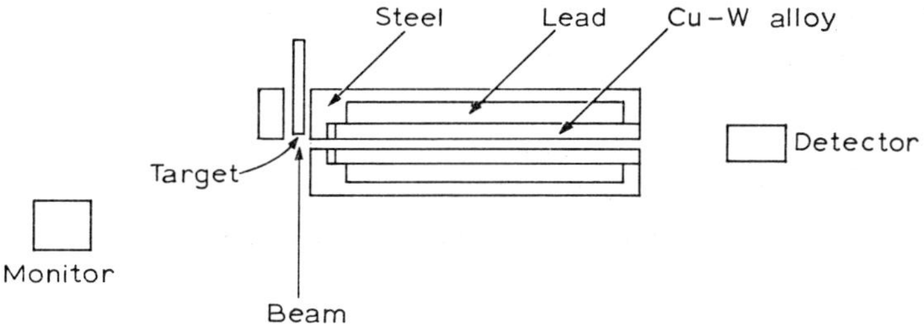

F. Ajzenberg-Selove (ed.), Nuclear Spectroscopy **A** (Academic Press, New York 1960) p. 529

Fig. 2.37.   Recoil distance lifetime measurements

Values of lifetimes of 7×10^{-11} sec for the 6.06 MeV state and of $\tau \leq 1.4 \times 10^{-11}$ sec for the 6.13 MeV state of ^{16}O were reported.

d)   *Recoil Doppler method.* If we try to slow down the recoiling nucleus in a time comparable to the lifetime of the state then a Doppler shift of the radiation frequency (energy shift) can be observed. For example the γ-ray energy when the recoil is travelling through vacuum can be compared with the γ-ray energy when the recoil travels through a slowing down material of known properties (fig. 2.38). This

* S. Devons, G. Goldring and G. Lindsey, Proc. Phys. Soc. London **A67** (1954) 134.

has been applied* to the ^{19}F(p, α)^{16}O reaction mentioned above. This gave a lifetime for the 6.13 MeV state of ^{16}O of $\tau \geq 2 \times 10^{-12}$ sec. Combined with the recoil
distance measurement this bracketed the lifetime value for this state.

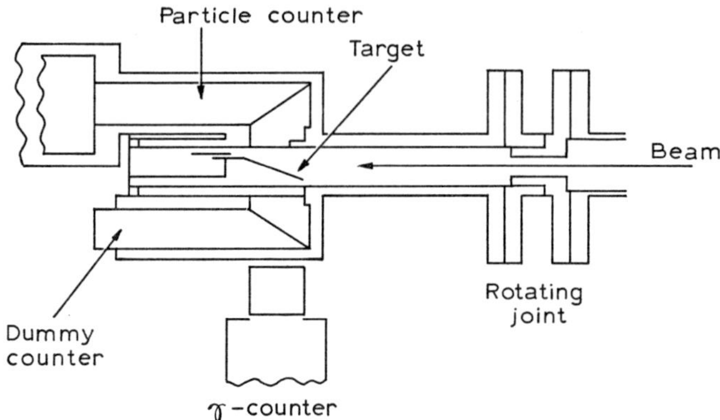

Fig. 2.38.　Recoil Doppler lifetime measurements

With the availability of accelerated heavy ions such as ^{12}C and ^{16}O and with high
resolution solid state γ-ray spectrometers this method is now widely used. The heavy
ion beam results in a higher velocity recoil and shorter lifetimes can be measured.
e)　*Resonance fluorescence.* In certain cases excited states can be resonantly excited by incident γ-radiation of the correct frequency. If the frequency can be varied
the width of the excited state and hence its lifetime can be determined.
f)　*Width of excited states.* As we have seen, a measurement of the width of an
excited state is equivalent to a measurement of its lifetime since $\tau = \hbar/\Gamma$. This is the
only method at present feasible for inferring lifetimes shorter than about 10^{-12} sec.

8　Other nuclear decay modes

We have discussed α-decay, β-decay and γ-decay processes. In addition some nuclei
can undergo *fission.* Also excited states above a nucleon or α-particle binding energy decay with a typical lifetime of the order of 10^{-21} sec. These decay modes are
more properly discussed under nuclear reactions.

There are a few special cases of neutron or proton radioactivity. In *all* cases
these result from β-decays leading to an unbound state of the daughter nucleus
(fig. 2.39). This state then emits the proton or neutron promptly (10^{-21} sec). The
observed lifetime is just the lifetime of the parent β-emitter. No true neutron or pro-

---

* S. Devons, G. Manning and D. St. P. Bunbury, Proc. Phys. Soc. London **A68** (1955) 18.

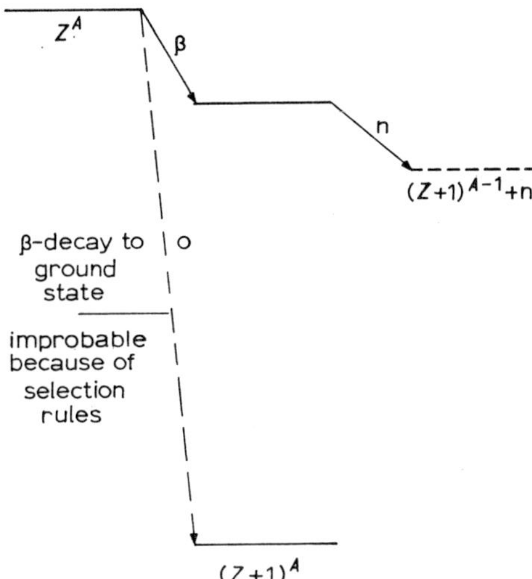

Fig. 2.39.   Neutron 'radioactivity'

ton radioactivity comparable to the α-decay process is ever observed as it is energetically impossible.

9   Summary

We have seen that a nucleus is composed of nucleons and is held together by short-range charge-independent strong forces. Nuclei are characterised by
1)   *their mass*, which is intimately related to the binding energy. We have seen that the binding energy curve indicates a saturation of the nuclear force, and that the analogy between nuclear matter and an incompressible liquid drop gives a rough guide to how nuclei are built up. Nuclei seek to maximise their binding energy by various decay processes.
2)   *their charge*, which is equal to the number of protons in the nucleus.
3)   *their size*. Radius measurements indicate a charge and mass distribution which falls off exponentially at the surface while they indicate a potential shape extending beyond the mass or charge boundary by an amount consistent with the expected short range of the nuclear force.
4)   *their angular momentum*, made up of the resultant of the intrinsic spins of the constituent nucleons together with their orbital angular momenta.
5)   *their parity*, a property specifying the spatial symmetry of the assembly.
6)   *their statistics*, a property specifying the exchange symmetry of the assembly and associated with integral or half-integral total angular momentum.

7) *their moments,* arising from non-uniform charge distributions giving electric moments and non-uniform current distributions giving magnetic moments.

We then discussed nuclear instabilities starting with the naturally radioactive nuclei. Natural radioactivity was seen to be nuclear stabilization which occurs with a probability small enough to give a lifetime comparable to the time since element formation. Daughter products of shorter life will also be found in equilibrium with the long lived parent.

Alpha-decay was shown to be quantum mechanical penetration of the coulomb barrier of otherwise α-unstable nuclei and occurs at a point on the binding energy curve when emission of an α-particle results in a more stable product. The decay probability is controlled by the energy available and the coulomb barrier height.

Beta-decay was shown to be release of energy associated with leptons characteristic of the weak interaction. It is an improbable process compared to nuclear transitions or electromagnetic transitions. It is the most common type of radioactivity however. Parity is violated for the lepton system.

Gamma-radiation was shown to be release of energy in the form of photons; electromagnetic radiation associated with the electromagnetic force. It occurs with higher probability than β-decay but with lesser probability than nuclear force transitions. Various multipoles are emitted depending on the spin and parity change involved. Alternatives to release of radiation is ejection of an atomic electron or production of an electron–positron pair in the field of the nucleus.

3

Interaction of

Radiation with Matter

1 Collisions

Collisions of nuclear particles are governed by the laws of energy and momentum conservation. We call collisions *elastic* when there is no change in total kinetic energy E_{kin}. We call collisions *non-elastic* when either or both particles absorb or emit energy by, for example, internal motions or radiation. We define the energy change $Q = E_{kin,f} - E_{kin,i}$ where f and i refer to the final and initial systems resp. Hence for elastic events $Q = 0$ while for non-elastic events Q is positive or negative. We consider first elastic collisions in the non-relativistic approximation. Conservation of linear momentum in the direction of v_0 then gives (cf. fig. 3.1a)

$$M_1 v_0 = M_2 v_2 \cos \psi_2 + M_1 v_1 \cos \psi_1 . \tag{3.1}$$

Conservation of linear momentum in the direction at right angles to v_0 gives

$$0 = M_1 v_1 \sin \psi_1 - M_2 v_2 \sin \psi_2 . \tag{3.2}$$

Conservation of energy gives

$$\begin{aligned} 0 &= \tfrac{1}{2}M_1 v_1^2 + \tfrac{1}{2}M_2 v_2^2 - \tfrac{1}{2}M_1 v_0^2 \\ &= E_{kin,f} - E_{kin,i} . \end{aligned} \tag{3.3}$$

Manipulation of these equations gives the relations

$$v_2 = 2v_0 \frac{M_1}{M_1 + M_2} \cos \psi_2 \quad \text{and} \quad \frac{M_2}{M_1} = \frac{\sin \psi_1}{\sin(2\psi_2 + \psi_1)} . \tag{3.4}$$

93

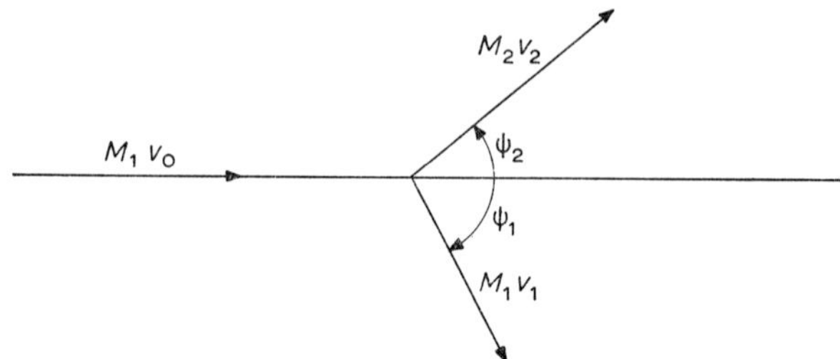

Fig. 3.1a.   Kinematics of nuclear collisions: laboratory system

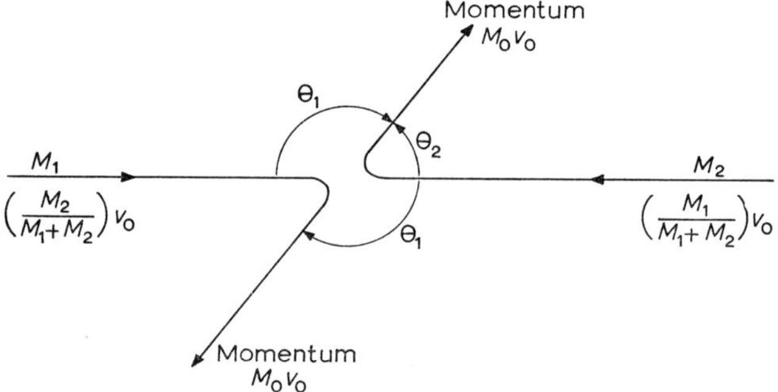

Fig. 3.1b.   Kinematics of nuclear collisions: centre of mass system

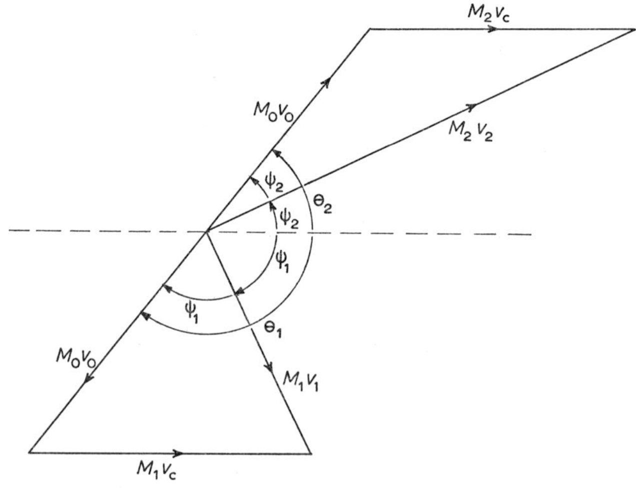

Fig. 3.1c.   Relations of angles in the two systems

It is more convenient to work in the centre of mass coordinate system and remove the asymmetry which appears in the equations expressed in the laboratory system of coordinates used above. We change to the coordinate system in which the vector sum of *all* momenta is zero (fig. 3.1b, c).

The velocity of the centre of mass is $v_c = M_1 v_0/(M_1 + M_2)$. The two particles approach each other with equal and opposite momenta equal to

$$\frac{M_1 M_2}{M_1 + M_2} v_0 = M_0 v_0$$

where $M_0 = M_1 M_2/(M_1 + M_2)$ is called the reduced mass.

In an elastic collision the particles must be turned through equal centre of mass angles θ_i. Then

$$\tan \psi_1 = \frac{\sin \theta_1}{\cos \theta_1 + M_1/M_2} \tag{3.5}$$

$$\psi_2 = \tfrac{1}{2}\theta_2 = \tfrac{1}{2}\pi - \tfrac{1}{2}\theta_1 .$$

Note that we use ψ for laboratory angles and θ for centre of mass angles.

In any collision process, elastic or inelastic, there is a kinetic energy

$$\tfrac{1}{2}(M_1 + M_2)v_c^2 = \frac{\tfrac{1}{2}M_1^2 v_0^2}{M_1 + M_2}$$

which is associated with the centre of mass motion which is not available for internal excitation effects in the system. The remaining kinetic energy is thus

$$\tfrac{1}{2}M_1 v_0^2 - \frac{\tfrac{1}{2}M_1^2 v_0^2}{M_1 + M_2} = \frac{\tfrac{1}{2}M_1 M_2 v_0^2}{M_1 + M_2} = \tfrac{1}{2}M_0 v_0^2 \tag{3.6}$$

where M_0 is the reduced mass.

For inelastic collisions equation (3.3) above becomes

$$Q = \tfrac{1}{2}M_1 v_1^2 + \tfrac{1}{2}M_2 v_2^2 - \tfrac{1}{2}M_1 v_0^2 . \tag{3.7}$$

The subsequent relations become more complicated.

2   *Cross section*

We now consider the probability of a certain collision process occurring. We express this as a *cross section*.

Consider a beam of n_0 particles per second incident on a slab of target material of thickness t containing N scattering or absorbing centres per unit volume. Then

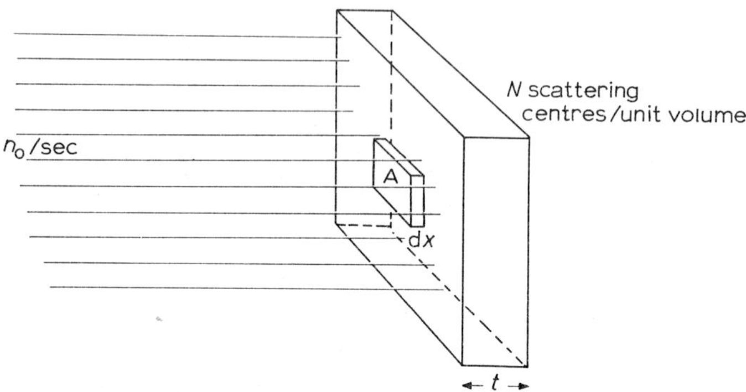

Fig. 3.2.   Cross section

the chance of colliding while passing through a lamina of thickness dx and area
A (fig. 3.2) is

$$\frac{NA\,dx\,\sigma}{A} = N\sigma\,dx = dx/\lambda \qquad (3.8)$$

where σ has the dimensions of area and is called the cross section and λ has the
dimension of length and is called the mean free path for collision. Obviously
$\lambda = 1/N\sigma$. The attenuation of the beam is then

$$dn = -nN\sigma\,dx$$

or

$$n_t = n_0 e^{-N\sigma t} = n_0 e^{-\mu t} \qquad (3.9)$$

where $\mu = N\sigma = 1/\lambda$ has dimensions cm^{-1} and is called the *linear attenuation coefficient* for collisions of this beam in this material. The quantity $\mu_m = \mu/\rho$ where
ρ is the material density is called the *mass attenuation coefficient* and is measured
in cm^2/g.

$$\mu_m = \mu/\rho = N\sigma/\rho = \sigma/m_A \qquad (3.10)$$

where m_A is the mass of a scattering centre. The ratio $n_t/n_0 = e^{-N\sigma t}$ is called the
transmission of the material. The number of collisions which take place is called the
yield

$$Y = n_0 - n_t = n_0(1 - e^{-N\sigma t})$$
$$\approx n_0 N t \sigma \text{ for small attenuation.} \qquad (3.11)$$

The cross section is thus

$$\sigma = \frac{Y}{n_0 Nt} . \qquad (3.12)$$

The cross section is equal to the probability of one event occurring for one particle incident on a material containing 1 scattering centre per cm^2 of projected area. The cross section is measured in cm^2 or for nuclei in units of 10^{-24} cm^2 called barns.

If we record the yield of resultant particles emerging into a small solid angle $d\Omega$ at a definite angle ψ we define a *differential* cross section, per atom or nucleus,

$$d\sigma = \sigma(\psi)d\Omega. \qquad (3.13)$$

If the yield is independent of azimuthal angle

$$d\Omega = 2\pi \sin \psi \, d\psi$$

and

$$d\sigma = 2\pi \sin \psi \, \sigma(\psi)d\psi. \qquad (3.14)$$

The total cross section is

$$\sigma = 2\pi \int_0^\pi \sigma(\psi) \sin \psi \, d\psi. \qquad (3.15)$$

$\sigma(\psi) = d\sigma/d\Omega$ gives the angular distribution of the events, and is expressed in cm^2 per scattering centre per steradian.

We relate the angular distribution $\sigma(\psi)$ measured in the laboratory system to the angular distribution in the centre of mass system $\sigma(\theta)$ as follows

$$\sigma(\theta) \sin \theta \, d\theta = \sigma(\psi) \sin \psi \, d\psi. \qquad (3.16)$$

That is, the same number are seen per unit solid angle at corresponding angles in each system. For elastic scattering

$$\sigma(\psi) = \frac{(1+2\gamma \cos \theta + \gamma^2)^{\frac{3}{2}}}{1+\gamma \cos \theta} \sigma(\theta) \qquad (3.17)$$

where $\gamma = M_1/M_2$. For non-elastic events of type $M_1 + M_2 \rightarrow M_3 + M_4$

$$\gamma = \left(\frac{M_1 M_3}{M_2 M_4} \frac{E}{E+Q}\right)^{\frac{1}{2}} \qquad (3.18)$$

where $E = \frac{1}{2}M_0 v_0^2$.

3   *Charged particle collisions*

Consider a particle of mass M_1, charge $Z_1 e$ and velocity v_1 interacting with a target particle of mass M_2, charge $Z_2 e$ and velocity zero (fig. 3.3). Again we assume non-relativistic velocities $v_1 \ll c$. The coulomb force is then $F = Z_1 Z_2 e^2/r^2$. Suppose first that $M_2 \gg M_1$ and that the force is repulsive (charges same sign).

Then the orbit of M_1 is a hyperbola with M_2 at the outer focus S and an angle of deflection $\psi = \pi - 2\phi$ where 2ϕ is the angle between the asymptotes to the hyperbola. The perpendicular drawn from S to the asymptote, p, is called the impact

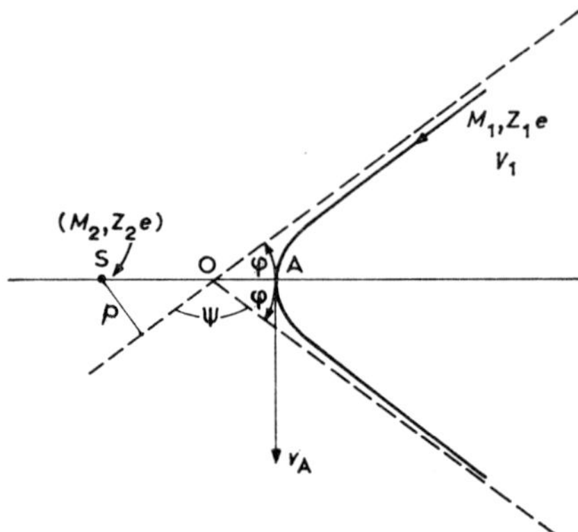

Fig. 3.3.   Rutherford or coulomb scattering

parameter. The velocity of the particle at the apex A, v_A, is then given by the conservation of angular momentum as $pv_1 = v_A \times SA$. The conservation of energy gives

$$\tfrac{1}{2}M_1 v_1^2 = \tfrac{1}{2}M_1 v_A^2 + \frac{Z_1 Z_2 e^2}{SA} \tag{3.19}$$

or

$$v_1^2 = v_A^2 + \frac{bv_1^2}{SA} \tag{3.20}$$

where $b = 2Z_1 Z_2 e^2 / M_1 v_1^2$ is called the collision diameter. It is the distance of closest approach for a head-on collision ($\phi = 0$). Solving we get $p^2 = SA(SA - b)$. Since the orbit is a hyperbola of eccentricity sec $\phi = SO/OA$

$$SA = SO + OA = SO(1 + \cos \phi) = p \operatorname{cosec} \phi(1 + \cos \phi) = p \cot \tfrac{1}{2}\phi. \tag{3.21}$$

Then

$$\begin{aligned} p^2 &= p \cot \tfrac{1}{2}\phi(p \cot \tfrac{1}{2}\phi - b) \\ &= p^2 \cot^2 \tfrac{1}{2}\phi - bp \cot \tfrac{1}{2}\phi \end{aligned} \tag{3.22}$$

or

$$b = 2p \cot \phi. \tag{3.23}$$

We can relate this to the angle of deflection $\psi = \pi - 2\phi$. Thus

$$\begin{aligned} b &= 2p \tan \tfrac{1}{2}\psi \\ p &= \tfrac{1}{2}b \cot \tfrac{1}{2}\psi. \end{aligned} \tag{3.24}$$

If M_2 is not infinite we use the reduced mass for M_1 in the formulae and work in the centre of mass system. Then

$$b = \frac{2Z_1 Z_2 e^2}{M_0 v_1^2} \quad \text{where} \quad M_0 = \frac{M_1 M_2}{M_1 + M_2}$$

$$p = \frac{Z_1 Z_2 e^2}{M_0 v_1^2} \cot \tfrac{1}{2}\theta. \tag{3.25}$$

We can now calculate the probability of scattering through an angle between θ and $\theta + \mathrm{d}\theta$. The differential cross section is $\mathrm{d}\sigma = 2\pi p \mathrm{d}p = \sigma(\theta)\mathrm{d}\Omega$. Thus

$$\sigma(\theta) = 2\pi \cdot \tfrac{1}{8}b^2 \cot \tfrac{1}{2}\theta \, \mathrm{cosec}^2 \tfrac{1}{2}\theta \, \mathrm{d}\theta/\mathrm{d}\Omega = \tfrac{1}{16}b^2 \, \mathrm{cosec}^4 \tfrac{1}{2}\theta$$

$$= \left(\frac{Z_1 Z_2 e^2}{2M_0 v_1^2}\right)^2 \mathrm{cosec}^4 \tfrac{1}{2}\theta. \tag{3.26}$$

This is the Rutherford or coulomb scattering cross section in centre of mass coordinates. The number of scatterings observed $\mathrm{d}Y$ per unit solid angle $\mathrm{d}\Omega$ is

$$\mathrm{d}Y = n_0 N t \left(\frac{Z_1 Z_2 e^2}{2M_0 v_1^2}\right)^2 \mathrm{cosec}^4 \tfrac{1}{2}\theta \, \mathrm{d}\Omega. \tag{3.27}$$

4   *Energy loss for charged particles*

4.1   Introduction. The main mechanism for energy loss is through interaction with atomic electrons by the coulomb force. When the energy needed to excite them to new levels or to remove them from the atom is small compared to the total energy, the mechanism is almost elastic. A more minor process is through nuclear interactions but this is generally neglibible except for very heavy ions such as fission fragments.

Electrons ejected directly from the atom (the *primary* ionization) may have energies up to $4m_e E/M$, i.e. about 0.002 of the initial energy of the ionizing particle. Such ejected electrons (called δ-rays) may in turn produce further ionization (the *secondary* ionization). If the δ-ray range is short only the *total ionization* is observed. When the energy of the incident particle has decreased below that necessary to ionize, this is said to be the end of its *range* and it reverts to a neutral atom.

Although nuclear encounters are unimportant as an energy loss mechanism, they may be significant in removing particles from the beam by scattering or reactions. When this is an important effect, a range is no longer observed; instead we define an attenuation coefficient μ by $n_x = n_0 e^{-\mu x}$.

There are also radiative energy loss mechanisms in addition to ionization. Sudden decelerations of the charged particle result in electromagnetic radiation being emitted. This is called the *bremsstrahlung* or braking radiation. Polarization of the medium when a charged particle passes through with a velocity greater than the velocity of light in the medium can also result in radiation called *Cerenkov radiation*.

If E is the kinetic energy of the charged particle then the *specific energy loss* is $-dE/dx$ and the absolute stopping power is $(-dE/dx)_{collision} + (-dE/dx)_{radiation}$. Energy loss by collision is approximately the same for all particles of the same charge and velocity and reaches a minimum at relativistic energies. At lower energies it varies as $1/v^2$. Above the minimum, bremsstrahlung loss becomes dominant for electrons (fig. 3.4). For heavy particles radiative loss is negligible and before the minimum is reached, loss of particles by reactions begins to dominate. The relative importance of the effects for electrons and protons in energy loss/cm is shown in table 3.1.

We can define a critical energy ξ when $(dE/dx)_{coll} = (dE/dx)_{rad}$. In the region where $E \gg \xi$ when radiation predominates the absorber thickness for the particle

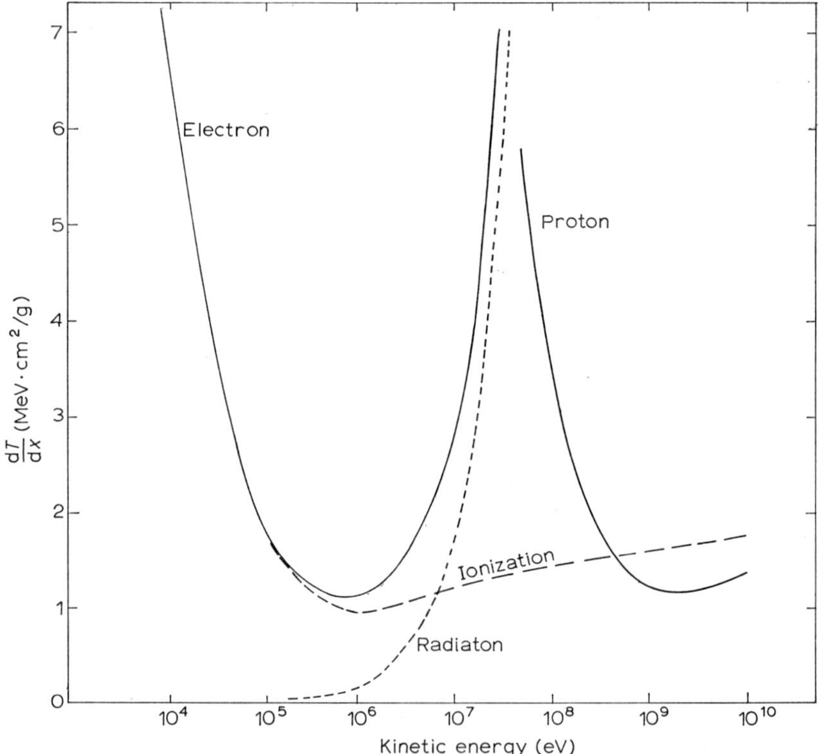

Fig. 3.4.   Energy loss mechanisms

TABLE 3.1

	Energy	$\beta = v/c$	Collision loss	Bremsstrahlung	Cerenkov
Electron	100 MeV	≈ 1	2 MeV	2 MeV	2.7 keV
Proton	1000 MeV	0.87	2 MeV	0.01 keV	1.65 keV

to emerge with e^{-1} of its energy is called the *radiation length* X_0. In the region where $E \gg \xi$

$$(dE/dx)_{rad} \propto EZ^2$$
$$(dE/dx)_{coll} \propto Z \text{ (independent of } E).$$

4.2   Collision energy loss. The energy transferred from a heavy particle of mass M to an electron mass m can be derived from our kinematic equations

$$v_e = 2v_0 \frac{M}{M+m} \sin \tfrac{1}{2}\theta \tag{3.28}$$

where v_e, v_0 are electron and heavy particle initial velocities respectively and θ is the centre of mass deflection angle of the heavy particle. Then the energy transfer for $M \gg m$ is

$$\Delta T = \tfrac{1}{2}mv_e^2 = 2mv_0^2 \sin \tfrac{1}{2}\theta. \tag{3.29}$$

If T is the kinetic energy of the heavy particle

$$\Delta T = \frac{4mT}{M} \sin \tfrac{1}{2}\theta. \tag{3.30}$$

Since $p = \tfrac{1}{2}b \cos \tfrac{1}{2}\theta$

$$\Delta T = \frac{4mT}{M} \frac{\tfrac{1}{4}b^2}{p^2 + \tfrac{1}{4}b^2} \tag{3.31}$$

where p and b are impact parameter and collision diameter as before.

The cross section for transfer of energy between ΔT and $\Delta T + d\Delta T$ is

$$d\sigma = 2\pi p dp = -2\pi \frac{z^2 e^4}{mv^2} \frac{d\Delta T}{\Delta T} \tag{3.32}$$

where z is the heavy particle charge (electron charge is unity). Thus as a particle of mass M passes through a thickness of absorber Δx which has N atoms of atomic number Z per cm^3 the number of collisions in which energy between ΔE and $\Delta E + d\Delta E$ is transferred is $NZ\Delta x d\sigma$ and the total loss by collision is

$$-dT = \int_{\Delta T_{\min}}^{\Delta T_{\max}} \Delta T\, NZ\Delta x\, d\sigma = -2\pi NZ\Delta x\, \frac{z^2 e^4}{mv^2} \int \frac{d\Delta T}{\Delta T}$$

$$= -4\pi NZ\Delta x\, \frac{z^2 e^4}{mv^2} \int_{p_{\max}}^{p_{\min}} \frac{p\, dp}{p^2 + \tfrac{1}{4}b^2}$$

$$= 2\pi NZ\Delta x\, \frac{z^2 e^4}{mv^2} \log \frac{p_{\max}^2 + \tfrac{1}{4}b^2}{p_{\min}^2 + \tfrac{1}{4}b^2}; \qquad (3.33)$$

$p_{\min}$ can be set equal to zero for a head-on collision and $p_{\max}$ is set equal to v/ω after the suggestion of N. Bohr, where ω is the characteristic frequency of the electron in its parent atom.

The *stopping power* is then written

$$-\frac{dT}{dx} = 2\pi NZ \frac{z^2 e^4}{mv^2} \log \frac{4 p_{\max}^2}{b^2}$$

$$= 4\pi \frac{z^2 e^4}{mv^2} NZ \log \frac{mv^3}{\omega z e^2} \text{ erg cm}^{-1}. \qquad (3.34)$$

This was Bohr's original formulation. Later Bethe and Bloch followed a quantum mechanical treatment to derive

$$-\frac{dT}{dx} = 4\pi \frac{z^2 e^4}{mv^2} NZ \log \frac{2mv^2}{I} \qquad (3.35)$$

where they interpreted the parameter I as the mean atomic excitation potential. It has a value $I \approx 11.5Z$ eV. Note that $-dT/dx \propto z^2/v^2$ and is independent of the particle mass M. We have here derived the energy loss due to *collision* of charged particles with electrons.

4.3 Radiation energy loss. A charged particle of mass M and charge ze approaching a nucleus of charge Ze in the absorber will suffer an acceleration proportional to zZ/M. According to classical electrodynamics the resulting radiation will have an intensity proportional to $(zZ/M)^2$. This radiation is the *bremsstrahlung*. Because of the mass factor its intensity is much smaller for heavy particles like protons or mesons than it is for light particles like electrons. With low energy electrons it is the process for producing the continuous X-ray spectrum as discovered by Röntgen in 1895. At high energies the radiation initiates showers as observed in cosmic rays. Because of the Z^2 factor it is predominantly a particle–nuclear charge effect rather than a particle–electron charge effect.

For a particle of kinetic energy T, any energy loss between 0 and T is possible. However small angle events corresponding to low energy transfers are more probable and low energy quanta are favoured. The energy loss per unit frequency of the radiation is roughly independent of energy that is $\mathrm{d}T \propto 1/h\nu$. The total loss is

$$(-\mathrm{d}T/\mathrm{d}x)_{\mathrm{brems}} = NEZ^2 f(Z, T) \tag{3.36}$$

where N is the number of atoms/cm^2 in the absorber, $E = T + m_e c^2$ is the total energy of the incident electron, Z is the absorber atomic number, or nuclear charge, and $f(Z, T)$ is a slowly varying function of Z, T; $f(Z, T)$ rises more slowly for heavy elements due to screening of the nuclear charge by the electrons as shown in fig. 3.5.

The radiation is emitted into a cone of semi-angle θ where $\theta \approx m_e c^2 / (m_e c^2 + T)$

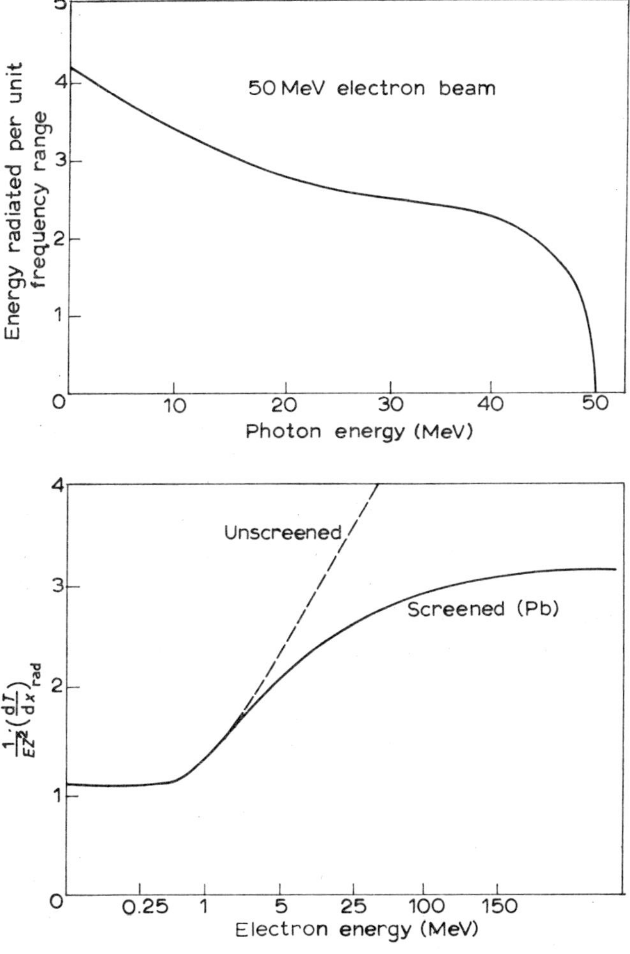

W. E. Burcham, Nuclear Physics (Longmans, London 1963) p. 157

Fig. 3.5.   Bremsstrahlung energy spectrum and the function $f(Z, T)$

Since T is generally much greater than $m_e c^2$, this is a very small angle. As T increases the radiation is emitted into a smaller forward cone.

Bremsstrahlung emission is thus mainly due to interaction with the *nuclear charge*.

The emission of *Cerenkov radiation* is a property of the gross structure of the absorber materia. As a charged particle passes through the bound electrons of a material it induces a polarisation of the medium along its path (fig. 3.6a). The

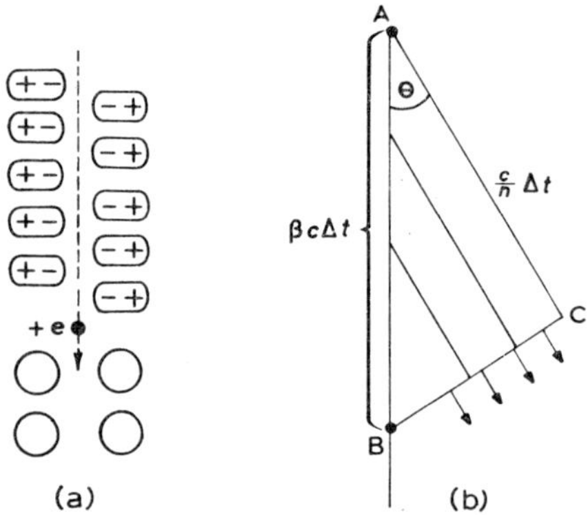

Fig. 3.6. Cerenkov radiation, (a) polarisation of the medium, (b) propagation of wave front

time variation of this polarisation can lead to radiation. If the charge is moving slowly the phase relations will be random for radiation from different points along the path and no coherent wave front will be formed. However if the particle is moving faster than the velocity of light in the medium, $v > c/n$ where n is the index of refraction, then a coherent wave front can be formed and radiation emitted (fig. 3.6b). Radiation from A will reach the wave front CB after travelling a distance $(c/n)\Delta t$ while radiation will start from B after the particle has travelled a distance AB $= v\Delta t = \beta c\Delta t$ where $\beta = v/c$. The radiation will be in phase and propagated at an angle CAB to the axis where

$$\cos(\angle\,\mathrm{CAB}) = \frac{c}{n}\,\frac{\Delta t}{\beta c\Delta t} = \frac{1}{\beta n} = \cos\theta. \tag{3.37}$$

So the radiation is emitted in a forward cone of semi-angle θ. As the velocity increases, $\cos\theta$ decreases and θ increases. Thus the cone opens out in contrast to the behaviour of bremsstrahlung emission. Also there is a threshold for emission of Cerenkov radiation given by $\beta \geq 1/n$. For velocities less than this we get no coherence and hence no emission. The energy loss is

$$\left(-\frac{\mathrm{d}T}{\mathrm{d}x}\right)_{\mathrm{Cerenkov}} = \frac{4\pi^2 z^2 e^2}{c^2}\int\left(1 - \frac{1}{\beta^2 n^2}\right)v\,\mathrm{d}v. \tag{3.38}$$

The integration extends over all frequencies such that $\beta n > 1$. The radiation is mainly in the blue end of the visible spectrum and in the ultraviolet. Ultraviolet absorption bands in the material place an upper limit on the frequency observable. Note that the intensity is proportional to $v \, dv$ while the intensity in the bremsstrahlung spectrum is proportional to dv.

The relative magnitude of the energy loss due to this effect is small compared to collision, or bremsstrahlung effects. Typically it is of the order of 0.1 % of the ionization loss. However it is of great importance in particle detection and momentum analysis technique.

4.4 Stopping power.   We define the *absolute stopping power* of a material for a heavy charged particle as $-dT/dx$ measured in ergs $\cdot$ cm^{-1} or in keV $\cdot$ cm$^{+2} \cdot$ mg^{-1}.

We can write eq. (3.35)

$$-dT/dx = 4\pi \frac{z^2 e^4}{mv^2} NB \tag{3.39}$$

where N is the number of stopping atoms/cm^3. Then B is called the *atomic stopping power*. From the Bethe–Bloch formula we can write

$$B = Z \log\left(2mv^2/I\right) \tag{3.40}$$

where I is the atomic excitation potential and Z is the atomic number of the stopping atom.

The *stopping cross section per atom* is defined as

$$\varepsilon = -\frac{1}{N}\frac{dT}{dx}\,\text{keV} \cdot \text{cm}^2 \text{ per atom.} \tag{3.41}$$

We also use the *relative stopping power S*, the inverse ratio of the thickness of a given material for a given energy loss to the thickness of a standard material for the same energy loss. Air was formerly used as the standard; aluminium is more frequently used nowadays.

$$S_1 = \frac{(dT/dx)_1}{(dT/dx)_{\text{Al}}} = \frac{N_1 B_1}{N_{\text{Al}} B_{\text{Al}}} = \frac{\rho_1 B_1 A_{\text{Al}}}{\rho_{\text{Al}} B_{\text{Al}} A_1} \tag{3.42}$$

where ρ is the density and A is the mass number.

The *relative stopping power per atom* is just the ratio of the atomic stopping power to that for the standard

$$S_{\text{At}} = B_1/B_{\text{Al}}. \tag{3.43}$$

The *relative stopping power per electron* is similarly

$$S_{\text{elect}} = \frac{B_1}{Z_1}\frac{Z_{\text{Al}}}{B_{\text{Al}}}. \tag{3.44}$$

A rough rule to estimate or interpolate stopping power is to assume that $B \propto A^{\frac{1}{2}}$ and that the B's are additive. Thus for a molecule containing n atoms of mass number A per molecule $B \propto \sum n A^{\frac{1}{2}}$, the sum extending over the different kinds of atoms in the molecule. As an example consider lithium oxide, Li_2O: $S_{Li}=0.5$, $S_O=1.07$, $S_{Li_2O}=0.69$ by the additive rule. Very approximately $S=0.25A^{\frac{1}{2}}$.

Since $B \propto \log I^{-1} \propto \log Z^{-1}$, B decreases somewhat with Z. This expresses the fact that electrons in light elements are somewhat more effective in stopping than for heavy elements since in heavy elements the inner electrons are more tightly bound.

4.5   Range. For heavy particles the *range* is simply obtained from

$$R = \int_0^T dT \left(\frac{dT}{dx}\right)^{-1} \tag{3.45}$$

that is, it is just equal to the path length for all the energy to be lost since scattering is relatively unimportant. Since at low energies our expressions for dT/dx become too inaccurate, we generally determine ranges empirically by measuring path lengths.

We might expect that since

$$-dT/dx = \frac{z^2}{v^2} f(v) \quad \text{or} \quad Mv \frac{dv}{dx} = \frac{z^2}{v^2} f(v)$$

and upon integration

$$R \propto M v^4 / z^2. \tag{3.46}$$

However it was found empirically by Geiger that for α-particles the range is more nearly proportional to v^3. We note that for different particles of the *same velocity* the ranges are proportional to M/z^2. Thus if we know the range–energy relation for protons we can derive the relation for deuterons or α-particles with fair accuracy.

Thus if for protons of energy E_p, charge $+1$ the range is R, then deuterons (charge $+1$) of energy $E_d = \sqrt{2}E_p$ have the same velocity and their range will be $M/z^2 = 2$ times as great i.e. $2R$. Alpha-particles of energy $E_\alpha = \sqrt{4}E_p$ also have the same velocity and charge $+2$ and their range will be $M/z^2 = 4/2^2 = 1$ times the proton range R. Singly charged He^3 ions with energy $\sqrt{3}E_p$ will have a range equal to $3R$ and so on.

In the case of electrons large angle coulomb scattering by other electrons is much more probable and the range is much less sharply defined (fig. 3.7). The extrapolated range for electrons is fairly reproducible and an empirical formula has been given by Feather

$$R(mg/cm^2 \ Al) = 543E(MeV) - 160.$$

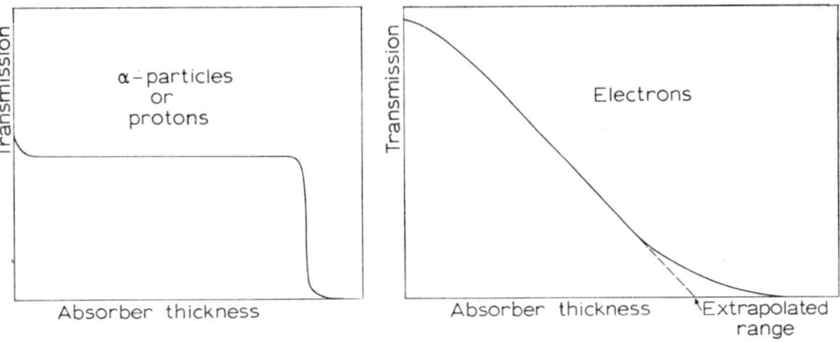

Fig. 3.7. Ranges of heavy particles and electrons

This is useful in determining the maximum energy of a β-spectrum by an absorption experiment.

4.6   Effects in the absorber. The ionization of a gas by a charged particle is beautifully illustrated in a cloud chamber photograph or in a photograph of an accelerator beam being stopped in air. What we observe is the *specific ionization,* the number of ion pairs per cm of path. The peak of the α-particle curve corresponds to about 6600 ion pairs/mm while the peak of the proton curve corresponds to about 2750 ion pairs/mm (fig. 3.8). The four to one ratio for specific ionization between α-particles and protons is just a result of the z^2 dependence of the energy loss. The peaking is the result of the $1/v^2$ dependence.

From measurements of specific ionization and of energy loss we can derive the

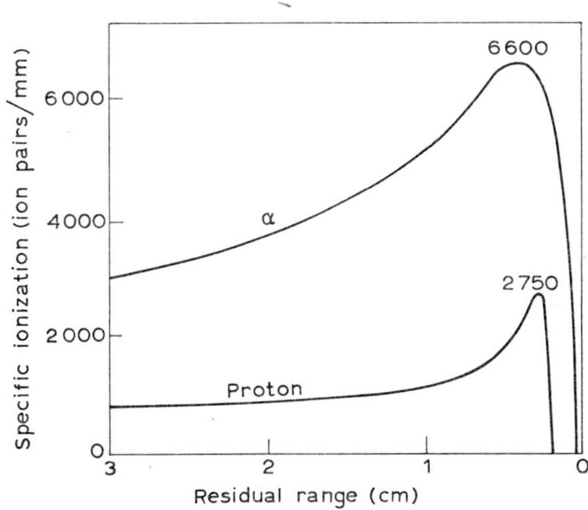

R. D. Evans, The Atomic Nucleus (McGraw-Hill, New York 1955) p. 655

Fig. 3.8.   Specific ionization near the end of the range

average energy expended to form one ion pair. This is found to vary little with type of particle or its energy or the gas being ionized, as is shown in table 3.2.

TABLE 3.2

Specific ionization

Gas	5 keV electron	5 MeV α-particle	340 MeV proton
H_2	38 eV	37 eV	35.3 eV
Air	35	35.2	33.3
Argon	27	25.9	25.5

The ionization in a gas can be measured by measuring the ion current directly. The specific ionization can be determined by counting droplets or bubbles on ions along the path of a particle in a cloud or bubble chamber or by measuring grain density along a path in a photographic emulsion.

The energy loss mechanism depends of course on individual collisions with many electrons. A 5 MeV α-particle coming to rest makes about 10^6 collisions. Thus there will be statistical fluctuations about the point corresponding to the average range. The standard deviation in range σ_R is given by

$$\sigma_R^2 = 4\pi z^2 e^4 N Z \int_0^T (\mathrm{d}T/\mathrm{d}x)^{-3} \, \mathrm{d}T$$

or approximately

$$\frac{\sigma_R^2}{R^2} = \frac{2m}{M} \frac{1}{\log(2mv^2/I)} . \tag{3.47}$$

Thus for 5 MeV α-particles $\sigma_R/R \approx 0.9\%$ while for protons of the same velocity ($E_p = 1.25$ MeV) σ_R/R is just twice as great because of the $M^{\frac{1}{2}}$ factor in the denominator.

If we assume a normal distribution for the straggling we get the number with range between x and $x + \mathrm{d}x$

$$\mathrm{d}n = \frac{n_0}{\sigma_R \sqrt{2\pi}} \exp - \frac{(x-R)^2}{2\sigma_R^2} \, \mathrm{d}x$$

$$= \frac{n_0}{\alpha \sqrt{\pi}} \exp - \left(\frac{x-R}{\alpha}\right)^2 \mathrm{d}x \tag{3.48}$$

where $\alpha = \sqrt{2}\sigma_R$ is called the range straggling parameter and R is the mean range.

If we integrate we get the number–distance curve. The inflection point gives the

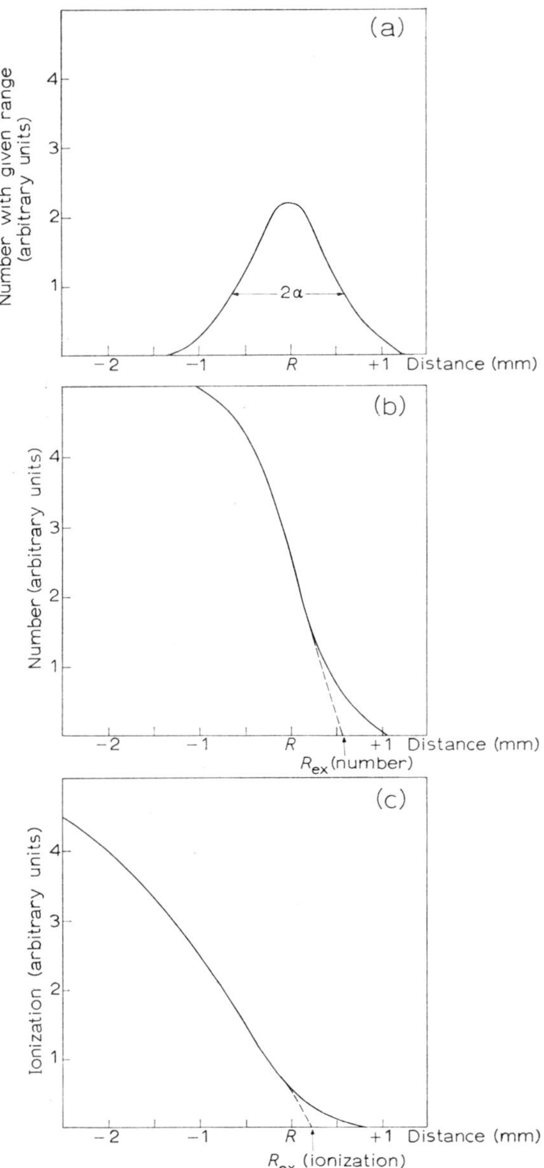

W. E. Burcham, Nuclear Physics (Longmans, London 1963) p. 178

Fig. 3.9.   (a) Straggling distribution. (b) Number–distance curve. (c) Average ionization curve

mean range R, that is the range reached by half the particles. The extrapolated number range $R_{ex}(n)$ is larger than the mean range by $\sqrt{\frac{1}{2}\pi}\,\sigma_R$.

We can also plot the specific ionization vs. distance. If we multiply this by the straggling distribution we get a curve of average ionization usually called a Bragg curve. The extrapolated ionization range is also shown in fig. 3.9.

Finally we consider the charge on ions near the end of their range. It is observed for example that α-particles possessing initially 2 positive charges, near the end of their range are a mixture of doubly charged and singly charged ions and neutral atoms. This is due to the pickup of atomic electrons and then reionization of the particle in a subsequent collision. If we pass α-particles through an absorber and then magnetically deflect them we can see the three charge states, as in fig. 3.10.

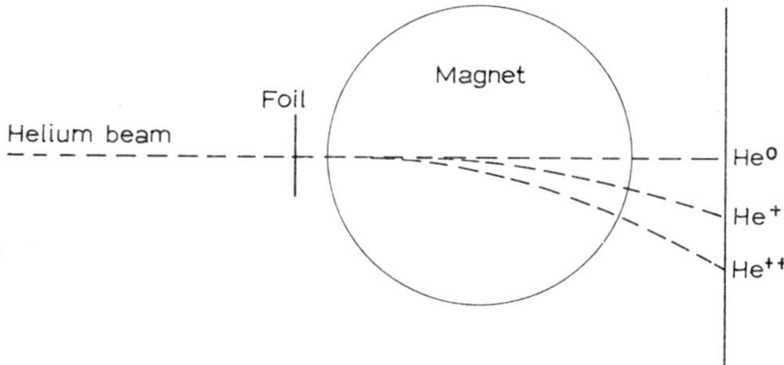

Fig. 3.10.   Charge states of helium

For particles of higher Z we see a whole series of charge states. We can define mean free paths for the capture and loss processes

$$\mathrm{He}^{++} + \mathrm{e}^- \rightarrow \mathrm{He}^+$$
$$\mathrm{He}^+ \rightarrow \mathrm{He}^{++} + \mathrm{e}^-$$

as $\lambda_{\mathrm{loss}}/\lambda_{\mathrm{cap}} = \sigma_{\mathrm{cap}}/\sigma_{\mathrm{loss}}$ the ratio of the cross sections/atom for the two processes. For α-particles the capture and loss mean free paths are shown in table 3.3. This

TABLE 3.3

E	λ_{cap} (mm)	λ_{loss} (mm)	Average charge Z_{av}
6.78	2.2	0.01	1.995
4.43	0.5	0.008	1.985
1.70	0.04	0.005	1.883
0.65	0.003	0.003	1.500

interchange of charge may take place 1000 times in the last centimetre of range. For an α-particle emerging from a source of even a few μg/cm^2 thickness some of the particles will be in the He$^+$ state.

5   Interaction of neutrons with matter

Since neutrons have no charge there will be no electromagnetic effect of importance as an energy loss mechanism. (We neglect the interaction of the neutron
magnetic moment.) Only nuclear encounters can affect the neutron energy. The
cross sections for nuclear scattering or reactions are of the order of a few barns
compared with ionization cross sections for charged particles which are of the
order of 10^8 barns. This is simply because an atom is about 10^4 times the size of
a nucleus. For the fastest neutrons (>10 MeV) the main energy loss mechanism is
through nuclear reactions resulting in charged particles which then lose energy in
the usual way. For intermediate energy neutrons (a few eV to 10 MeV) elastic and
inelastic scattering are the main mechanisms. Elastic scattering will be most effective if the scattering nuclei are the same mass as the neutron so that on average
the neutron loses half its energy per collision. Thus materials containing hydrogen
(protons) are effective attenuators. For slow neutrons in thermal equilibrium with
their surroundings there is little energy loss for long periods as the neutrons elastically scatter from the surrounding molecules. Eventually the slow neutron is captured by a nucleus with emission of the binding energy as γ-radiation.

6   Interaction of electromagnetic radiation with matter

6.1   Introduction. Since the reaction of radiation with electrons and nuclei
has the effect of removing photons from the beam we cannot define a range for
such radiation but rather speak of attenuation coefficients. The two types of processes are 1) absorption in which the photon energy is converted wholly or partially into charged particle energy and 2) scattering in which the photon is deflected
from the beam without appreciable loss in energy. We divide our attenuation coefficient $\mu = \mu_a + \mu_s$ where μ corresponds to the diminution of the particles in the beam
while μ_a corresponds only to those processes which deposit energy in the medium.
We further subdivide the processes as follows:

1)   *Elastic scattering*   a) *Rayleigh scattering*: scattering of photons by bound
     electrons.
     b) *Thomson scattering*: scattering of photons by free electrons or nuclei.
     c) *Nuclear resonant scattering*: a nuclear reaction initiated by a photon.
     d) *Delbrück scattering*: scattering of photons by the electromagnetic field of
     the nucleus.

2)   *Photoelectric effect*   an inelastic process in which a bound electron takes up
     the full energy of the photon.

3) *Compton effect*   an inelastic scattering process in which a photon scatters from a single electron.

4) *Pair production*   an inelastic process which can occur for photon energies in excess of $2m_e c^2 = 1.02$ MeV in which the photon energy is taken up by a positron–electron pair in the field of a nucleus.

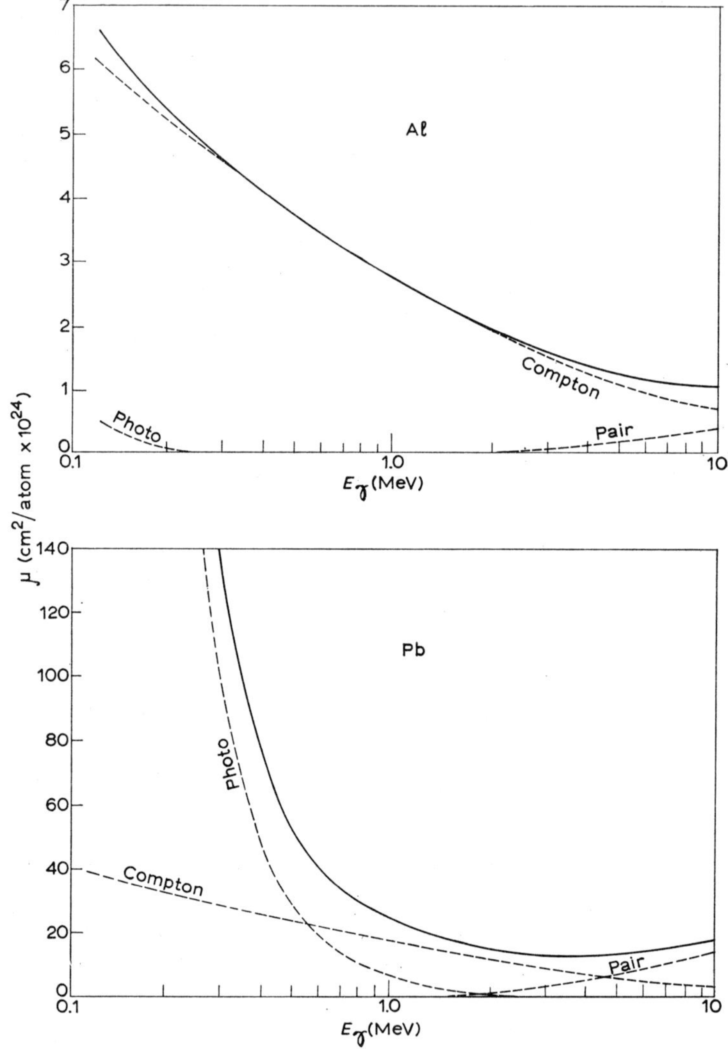

K. Siegbahn (ed.), Alpha-, Beta- and Gamma-ray Spectroscopy **1** (North-Holland, Amsterdam 1965) p. 75

Fig. 3.11.   Absorption coefficients for photons in Al and Pb

The attenuation coefficient can then be written

$$\mu = N(\sigma_S + \sigma_{PE} + \sigma_{PP}) + ZN\sigma_C \tag{3.49}$$

where N is the number of atoms/cm^2 and the factor Z multiplying the Compton cross section implies that this is an individual-electron effect, unlike the others which depend on numbers of atoms or nuclei.

In general then we observe some mixture of all these effects. In fig. 3.11 we plot the mass attenuation coefficient $\mu_m = \mu/\rho$ where ρ is the absorber density as a function of energy. We note the regions where the different inelastic effects dominate.

6.2   Elastic scattering of photons. If a plane polarised electromagnetic wave falls on a free electron the latter gets accelerated due to its charge

$$\ddot{z} = (eE_0/m) \cos \omega t \tag{3.50}$$

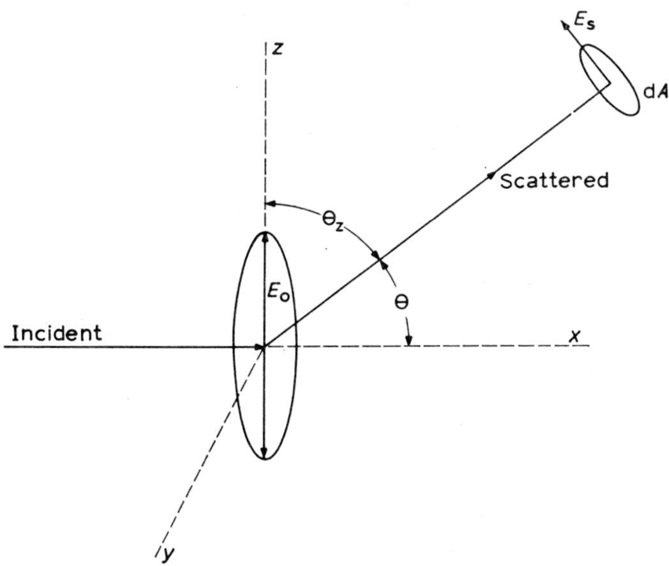

Fig. 3.12.   Elastic scattering of a photon

where the electric field $E_z = E_0 \cos \omega t$. The solution to eq. (3.50) is $z = -(eE_0/m\omega^2) \cos \omega t$, that is, the electron moves out of phase with the electric vector (fig. 3.12). The accelerated electron creates a time-varying dipole ez and a radiation field results

$$E_s = -\frac{e^2 E_0}{mc^2}\frac{\sin \theta_z}{R} = f(\theta_z)\frac{E_0}{R} \tag{3.51}$$

at a distance R from the dipole and with an angle θ_z between the polarisation vector and the direction of observation; $f(\theta_z)$ is called the scattering amplitude.

Over a small area dA perpendicular to R the flux of radiation energy is

$$dW = \frac{e^4 \sin^2\theta_z \, E_0^2}{8\pi R^2 m^2 c^3} \, dA$$

$$= \frac{e^4 \sin^2\theta_z \, E_0^2}{8\pi m^2 c^3} \, d\Omega \; \text{erg} \cdot \text{sec}^{-1}. \tag{3.52}$$

The incident intensity of radiation is

$$I = \frac{cE_0^2}{8\pi} \; \text{erg} \cdot \text{sec}^{-1} \cdot \text{cm}^{-2}.$$

We can thus define a cross section

$$dW = I \, d\sigma = I\sigma(\theta_z) d\Omega$$

$$\sigma(\theta_z) = \left(\frac{e^2}{mc^2}\right)^2 \sin^2\theta_z = |f(\theta_z)|^2. \tag{3.53}$$

The incident radiation in general is unpolarised so we must average over all possible orientations of the electric vector. We then derive the differential cross section for scattering of the photon through angle θ by a single electron.

$$d\sigma = \left(\frac{e^2}{mc^2}\right)^2 \frac{1+\cos^2\theta}{2} \, d\Omega. \tag{3.54}$$

To derive the total cross section we substitute $d\Omega = 2\pi \sin\theta d\theta$ and integrate whence

$$\sigma_\text{T} = \frac{8\pi}{3} \left(\frac{e^2}{mc^2}\right)^2 = \frac{8\pi}{3} r_\text{e}^2 \tag{3.55}$$

where v_e is the classical electron radius e^2/mc^2. This expression was first derived by J. J. Thomson and is generally called the Thomson cross section. Note that it is independent of the initial γ-ray frequency or energy.

This expression is valid for $h\nu \ll mc^2$ i.e. it is non-relativistic. The Rayleigh scattering by the bound electrons in which they are not ejected from the atom is obtained by averaging the Thomson cross section over the amplitudes due to all electrons. This leads to the inclusion of an atomic form factor, f_θ, by integration over the electron distribution,

$$d\sigma_\text{R} = \left(\frac{e^2}{mc^2}\right)^2 \frac{1+\cos^2\theta}{2} |f_\theta|^2 \, d\Omega. \tag{3.56}$$

The scattering of X-rays from crystals is one manifestation of this Rayleigh scattering. For scattering from the *nuclear charge* the Thomson cross section is seen to be valid since $hv \ll Mc^2$ where M is the nuclear mass. The f_θ factor causes the Rayleigh scattering to be strongly peaked in the forward direction. Because this does not really remove photons from a wide beam, it is not usually included in the mass absorption coefficient. The Thomson scattering from the nuclear charge is independent of angle but is less than the Rayleigh scattering by the factor $(m/M)^2$. Typically at $90°$ Rayleigh scattering is about 50 times the Thomson scattering and Compton scattering is about 50 times the Rayleigh scattering (cf. fig. 3.13).

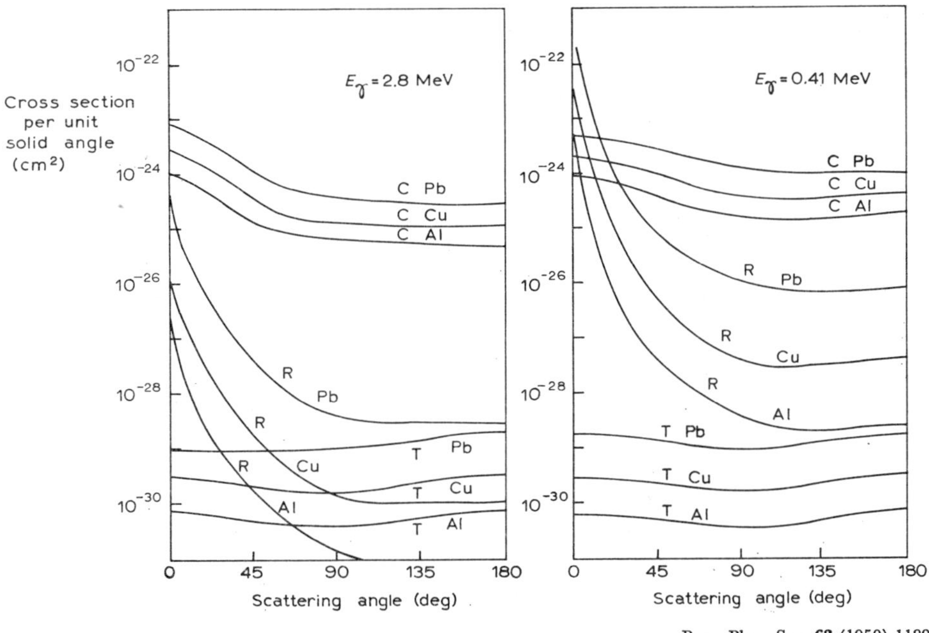

Proc. Phys. Soc. **63** (1950) 1189

Fig. 3.13.   Compton (C), Rayleigh (R) and Thomson (T) scattering cross sections for two photon energies

6.3   Photoelectric effect. A free electron cannot take up the complete photon energy since it is impossible to conserve both energy and momentum because of the zero rest mass of the photon. However it is possible for a bound electron to do so since the ion can recoil and take up the required momentum. Energy must be supplied to ionize the atom, so the energy of the ejected photoelectron is $T = hv - E_K$ where E_K is the ionization energy for the K-electron. The electron is ejected in the direction of the electric vector. The photoelectric cross section for low energies $\sigma_{PE} \propto Z^5 \lambda^{\frac{7}{2}}$ or $Z^5 E^{-\frac{7}{2}}$ where λ is the photon wavelength while for higher energies $\sigma_{PE} \propto Z^5 \lambda$ or $Z^5 E^{-1}$ and the photoelectrons are emitted at more forward angles.

There is a sharp increase in cross section as the ionization energy for another electron shell is exceeded.

The small energy (E_K, E_L etc.) is released as X-rays or as Auger electrons. Auger electrons are released as an alternative to X-rays when there is a vacancy to be filled.

For example if there is a K-shell excitation one can get an electron from the L-shell of energy $h\nu_K - E_L = E_K - 2E_L$ where $h\nu_K = E_K - E_L$ is the K X-ray energy. The ratio of the K X-ray quanta to K-shell vacancies is called the fluorescence yield. Its variation with Z is shown in fig. 3.14. Thus the Auger effect predominates for light elements. Auger electrons are also seen accompanying electron capture decay.

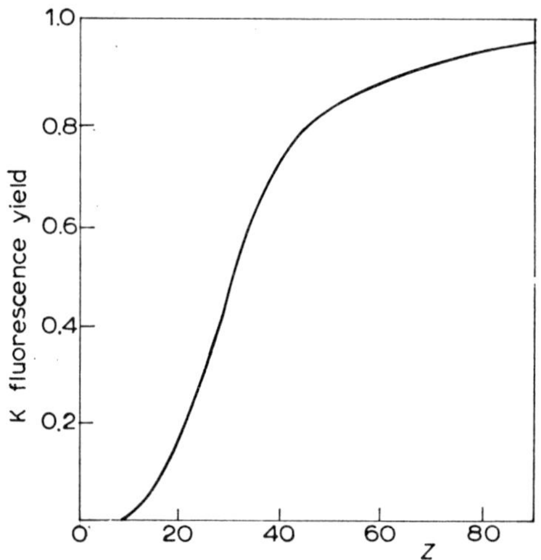

R. D. Evans, The Atomic Nucleus (McGraw-Hill, New York 1955) p. 565

Fig. 3.14.   Fluorescence yield as a function of Z

6.4   Compton effect. At higher energies the photon wavelength is shorter and hence interaction with individual electrons is more probable. The Thomson theory being non-relativistic is not appropriate and instead we consider essentially a particle scattering, i.e. photon–electron interaction. It is included as an inelastic process since although the energy of the whole system is not changed, the scattered photon has less energy (longer wavelength) than the incident photon. It is thus distinguished from Rayleigh scattering which is elastic in this sense. It is also very different from Rayleigh scattering in its dependence on angle and on Z. Note that the Compton energy loss is greater for greater scattering angle. The Rayleigh scattering increases as Z^2 while the Compton increases only as Z, the number of electrons.

Let an incident photon of frequency v_0 interact with an electron which is scattered through an angle ϕ while the photon after scattering through an angle θ has frequency v' (fig. 3.15). From momentum conservation

$$\frac{hv_0}{c} = \frac{hv'}{c} \cos\theta + p \cos\phi$$

$$0 = \frac{hv'}{c} \sin\theta - p \sin\phi \qquad (3.57)$$

where p is the electron momentum. From energy conservation

$$T = hv_0 - hv' \qquad (3.58)$$

where T is the electron kinetic energy. Thus $p^2 c^2 = T(T + 2mc^2)$ and the change in frequency or wavelength is given by

$$\frac{c}{v'} - \frac{c}{v_0} = \lambda' - \lambda_0 = \frac{h}{mc}(1 - \cos\theta). \qquad (3.59)$$

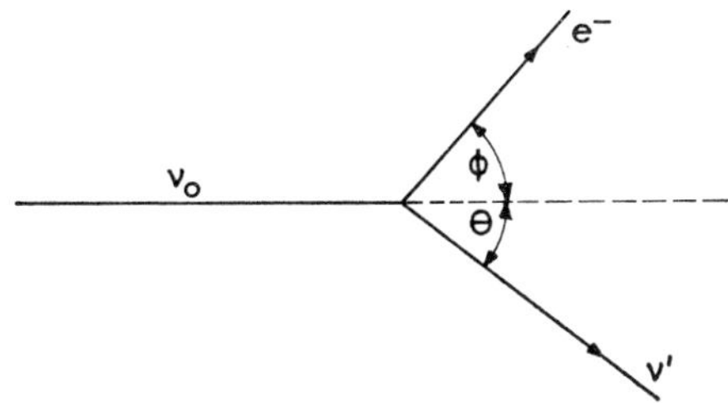

Fig. 3.15.   Compton scattering

The Compton shift in wavelength is thus independent of incident wavelength so that high energy, short wavelength photons can lose a large amount of energy. The shift is also independent of the scattering material. The fundamental constant h/mc appearing in the expression is just the de Broglie wavelength of an electron travelling at velocity c. This is called the Compton wavelength and is also the wavelength of a photon of energy equal to mc^2 the rest energy of the electron.

The energy of the scattered quantum is

$$hv' = \frac{hv_0}{1 + hv_0(1 - \cos\theta)/mc^2}. \qquad (3.60)$$

The kinetic energy of the recoil electron is

$$T = h(v_0 - v') = hv_0 \frac{hv_0(1 - \cos\theta)/mc^2}{1 + hv_0(1 - \cos\theta)/mc^2}. \qquad (3.61)$$

The electron energy is zero for $\theta = 0$ and maximum for $\theta = \pi$. Note also that for $h\nu_0 \gg mc^2$ the scattered photon at $90°$ always has energy $mc^2 = 0.511$ MeV. All these consequences of the simple theory have been carefully verified.

The probability for Compton scattering is more complicated to derive since it depends on details of the photon–electron interaction. The expression for the cross section is generally called the Klein–Nishina formula. It applies to polarised incident radiation

$$d\sigma_C = \frac{r_0^2}{4} d\Omega \left(\frac{\nu'}{\nu_0}\right)^2 \left(\frac{\nu_0}{\nu'} + \frac{\nu'}{\nu_0} - 2 + 4\cos^2 \Phi_p\right) \tag{3.62}$$

where $r_0 = e^2/mc^2$ is the classical electron radius and Φ_p is the angle between the polarisation vectors of incident and scattered radiation. The scattered photon and the recoil electron lie in a plane perpendicular to the electric vector of the incident polarised radiation. We see later how this is used to detect γ-ray polarisation.

For unpolarised radiation we average over polarisation angles and derive

$$d\sigma_C = \frac{r_0^2}{2} d\Omega \left(\frac{\nu'}{\nu_0}\right)^2 \left(\frac{\nu_0}{\nu'} + \frac{\nu'}{\nu_0} - \sin^2 \theta\right) \tag{3.63}$$

where θ is the angle through which the photon is scattered.

The total Compton cross section integrated over angle decreases with incident energy and increases with Z the number of electrons.

6.5 Pair production. Dirac's electron theory states that the electron occupies states of negative energy as well as states of positive energy. For a particle of rest energy mc^2 the total energy is

$$E^2 = (T + mc^2)^2 = p^2 c^2 + m^2 c^4$$

or

$$E = \pm(p^2 c^2 + m^2 c^4)^{\frac{1}{2}}. \tag{3.64}$$

Classically the negative sign has no meaning but quantum mechanically transitions between the two states could take place. In order to preclude such transitions occurring all the time Dirac proposed that *all* the negative energy states were filled and hence no transitions can take place spontaneously. However if we apply a field electrons can be dug out of the sea of filled negative energy states and moved up to a positive energy state (fig. 3.16). This creates a hole which behaves as if it had energy $-(-E)$ and momentum $-(-p)$ and charge $-(-e)$ that is as if it were a positively charged electron of energy E and momentum p. These holes were called positrons, and have been subsequently observed.

We see that the electron dug out must occupy some state of energy greater than mc^2 and the external field must supply an energy of at least $2mc^2$. The electron

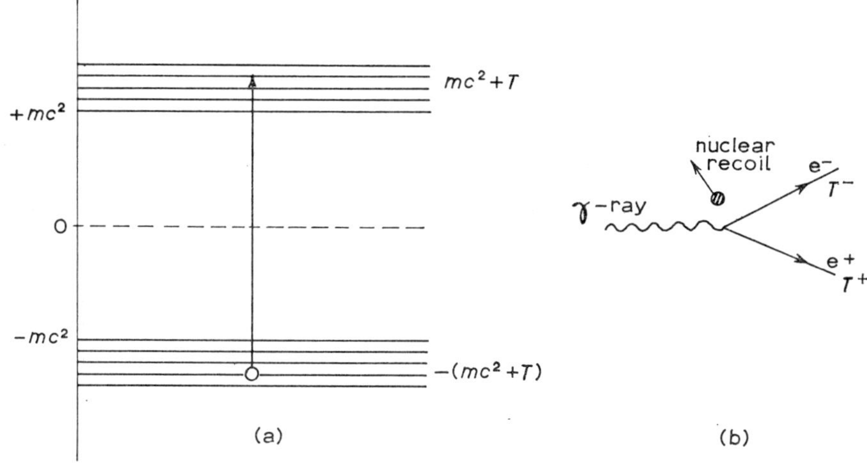

Fig. 3.16.   Pair production

and its hole form an electron–positron pair and the process will have an energy threshold $2mc^2 = 1.02$ MeV.

When formed by a photon the pair must be produced in the field of a nucleus or electron which can take up the linear momentum involved. The energy absorbed from the photon will be $2mc^2 + T^- + T^+$, if T^- and T^+ are the kinetic energies of electron and positron.

We can also observe the inverse process of annihilation in which a positron–electron pair disappears and results in radiation. The electron makes a transition to the negative energy state and electron and hole disappear with release of energy. For free positrons at rest annihilating with atomic electrons (almost at rest) we get $2mc^2$ appearing as photon energy. This normally appears as two γ-ray quanta each of energy $mc^2 = 0.511$ MeV directed at 180° from each other. (Two quanta must appear for charge conjugation invariance. The photon is odd under C while the electron–positron pair is even in the $^1S_0$ state). This process does not have to occur in the field of a nucleus. It is exactly analogous to emission of bremsstrahlung except that in the latter a transition occurs between two positive energy states.

The differential cross section for pair production is

$$d\sigma_{pp} = \frac{\sigma_0 Z^2 P(h\nu, Z)}{h\nu - 2mc^2}\, dT^+ \tag{3.65}$$

where $\sigma_0 = \frac{1}{137}(e^2/mc^2)^2 = \alpha r_0^2 = 5.8 \times 10^{-28}$ cm^2 and $P(h\nu, Z)$ is a slowly varying function. Pair production increases at first as the energy exceeds $2mc^2$ since production can take place further and further from the nucleus. Eventually the increase becomes less fast due to screening of the nuclear charge by the electrons. In the absence of screening the cross section varies as Z^2.

To summarize the various effects,

$$\sigma_{\text{PE}} \propto Z^5 \lambda, \qquad \sigma_{\text{C}} \propto Z, \qquad \sigma_{\text{PP}} \propto Z^2.$$

We note that the photoelectric effect is dominant at low energy, while pair production dominates at high energy (fig. 3.17).

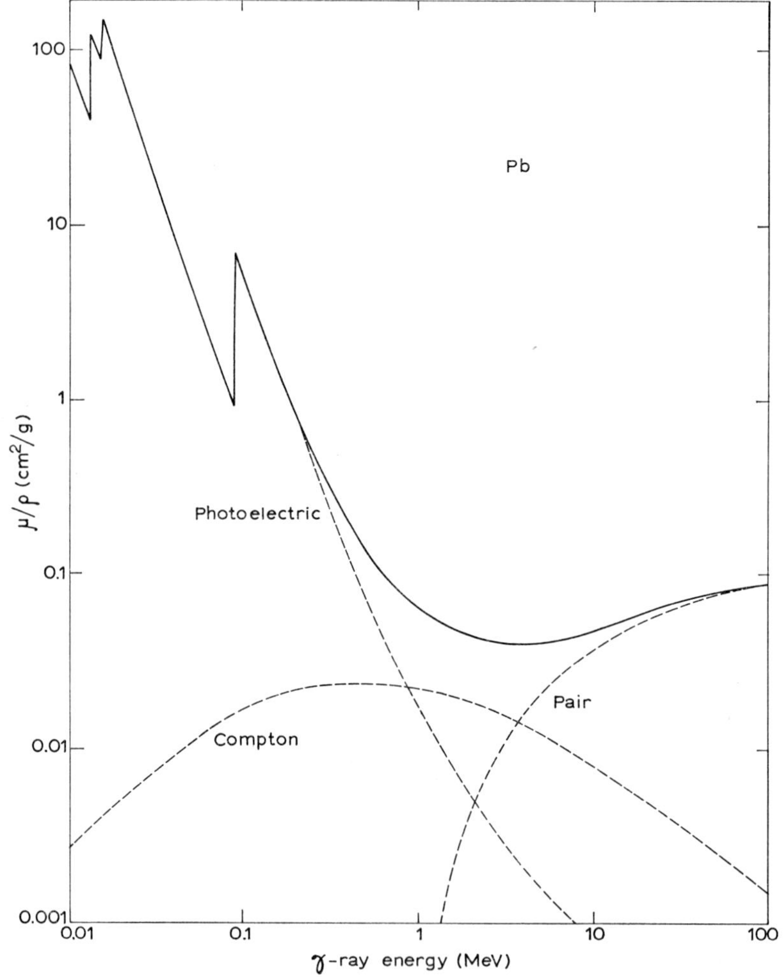

R. D. Evans, The Atomic Nucleus (McGraw-Hill, New York 1955) p. 716

Fig. 3.17.   Mass absorption coefficient for Pb as a function of photon energy

We must distinguish this *external* pair production from the *internal pair production* type of electromagnetic decay (ch. 2 § 7.3). The latter was very dependent on the angular shape of the field of the emitting nucleus, i.e. the multipolarity. External pair production occurs typically in a nuclear field far from the nucleus emitting the γ-ray, so there can be no dependence on the multipolarity.

4

Detection and

Measurement of Radiation

1   Introduction

The loss of energy by ionization or radiation of a charged particle in matter is the basic detection process. The ionization may be converted in a variety of ways to an electrical pulse which in some arrangements may have a size proportional to the energy deposited by the charged particle in the detecting medium. This is then an electrical detector. Alternatively the ionization along the particle path may provide nuclei upon which droplets, bubbles, spark discharges or photographic developable grains form to create a visible track. If the detecting medium is in a magnetic field, the track curvature can enable us to determine particle momentum, the track range gives us the energy and the ionization density per unit track length gives us the energy loss, which is related to particle mass.

Momentum or energy measurements of charged particles may also be carried out using magnetic or electrostatic deflection elements. The orbit of a particular particle is recorded using one of the detectors mentioned above. The resolution of these spectrometers is generally superior to the energy resolution of the detectors themselves.

In the case of uncharged radiation, such as neutrons or photons a variety of arrangements are used to convert the energy of the original particle into energy of a charged particle which can then be detected by one of the methods above. For energy determination the neutron momentum may be directly measured by timing its flight over a known distance or by relating the charged particle energy to the neutron energy by knowledge of the reaction involved. In the case of the photon one of the various mechanisms in which photon energy is converted to electron energy may be used in conjunction with an electron spectrometer.

2   Charged particle detectors: electrical

2.1   Ion chambers. In this device a chamber is filled with some gas and an electric field is applied. When ionization occurs in the gas because of the passage of charged particles, the resulting current can be measured. This current exhibits saturation above a certain field strength E (fig. 4.1). The saturation current is

$$i_s = qne \qquad (4.1)$$

where q ionizing events produce n ion pairs per second and e is the electronic charge. The current i_s corresponds to *all* of the primary ions being collected before re-combination; therefore it is proportional to the total energy deposited. The current

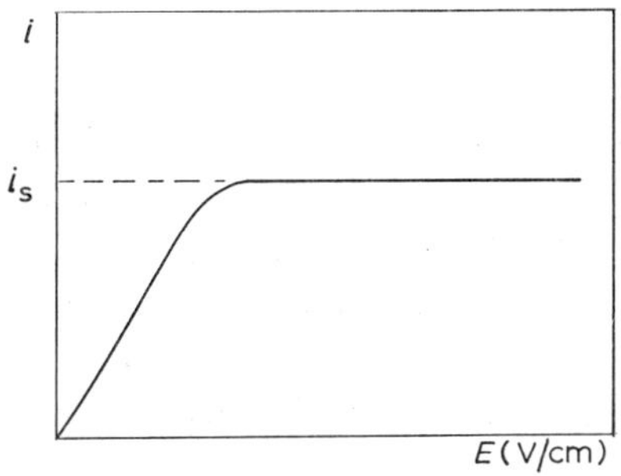

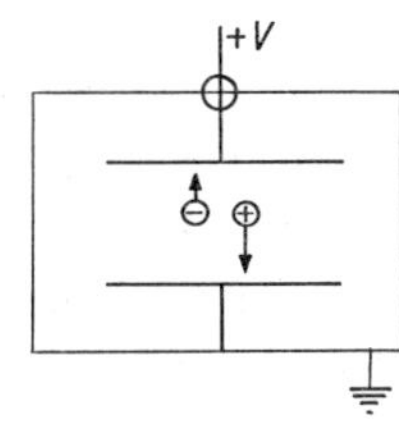

Fig. 4.1.   The ionization chamber

is of the order of 10^{-14}–10^{-16} A which can be detected using a high impedance output circuit (10^{10}–10^{13} ohms). Different gases are used depending on the application. For example high pressure argon is used for γ-rays, hydrogen for fast neutrons (proton recoils detected) or BF_3 for slow neutrons [the ^{10}B (n, α) reaction].

Alternatively, individual pulses can be recorded due to the ionization of a single particle. For example a 6 MeV α-particle produces 2×10^5 ion pairs in coming to rest (about 30 eV per pair). This is equivalent to 9.6×10^{-5} e.s.u. of charge. Then if the chamber capacity is 10 pF (10^{-11} farad), we can get a 3.2 millivolt signal which is easy to amplify and detect. One can detect a pulse due to as few as 1000 ion pairs.

The electrons produced by ionization in a field of 100 V/cm in a 1 cm deep chamber will be collected in about 10^{-6} sec while the slower moving positive ions will be collected in about 10^{-3} sec. This would make a very slow pulse which would make coincidence time resolution very poor and also for high counting rates would lead to overlapping of successive pulses with a resulting incorrect pulse height. To avoid these problems a gridded ion chamber is often used (fig. 4.2). In this case the collector plate is shielded from the effects of the positive ions and only the fast electron pulse is collected.

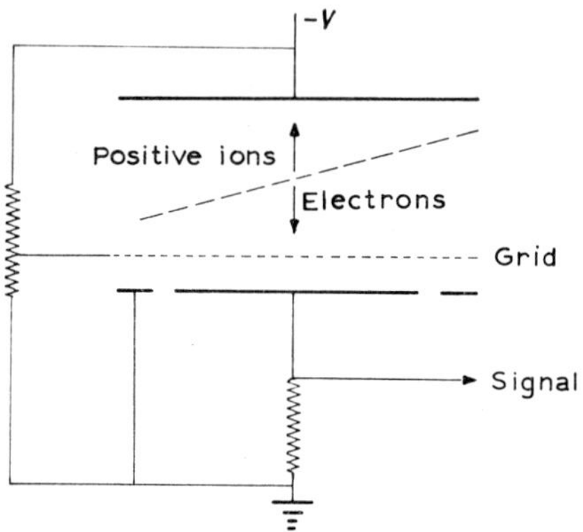

Fig. 4.2.   Gridded ionization chamber

The gas must be chosen so that negative ions are not formed. The drift velocity of the ions is $eE\lambda/Mv$ where E is the electric field, v is the rms thermal velocity, λ is the mean free path for collision and M is the ion mass. Thus for a fast electron pulse we must keep the electron temperature low, that is the random velocity v for the electrons small, so that the collision loss will be small. To achieve this we add methane or CO_2 to the filling gas (typically argon). These gases have low-lying electronic states to which the electrons easily lose energy. This keeps v low and the drift velocity high.

2.2 Proportional counters. If we increase the field strength in the ion chamber somewhat further we reach a point where secondary ionization by collision takes place (fig. 4.3). This critical voltage gradient in argon is about 10 kV/cm.

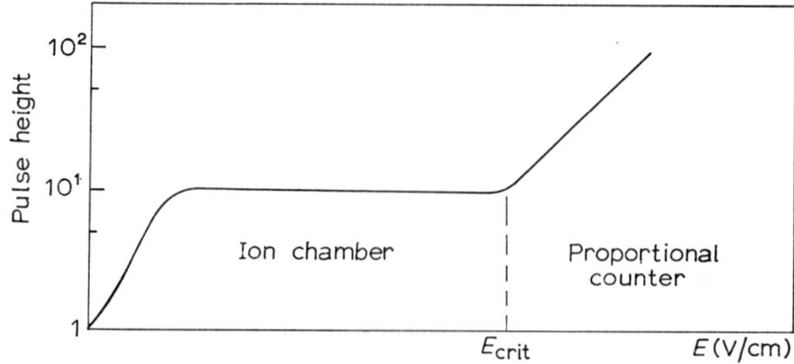

Fig. 4.3.   Ion chamber and proportional operating region

To achieve this high field gradient it is usual to use cylindrical geometry (fig. 4.4) so that a large field gradient is set up near the axial wire electrode. The field is

$$E = \frac{V}{r \log_e b/a} \tag{4.2}$$

where V is the voltage applied, a is the wire radius and b is the shell radius. Thus for $V=1000$ V, $a=0.01$ cm, $b=1$ cm the critical field is exceeded inside a distance of 0.02 cm from the wire. Thus in this small cylindrical region we get *multiplication* or an avalanche of secondary ionization. Multiplication factors of 10–1000 are used. The pulse will be independent of the position of the initiating track in the chamber, since the multiplication region is so small compared to the shell radius.

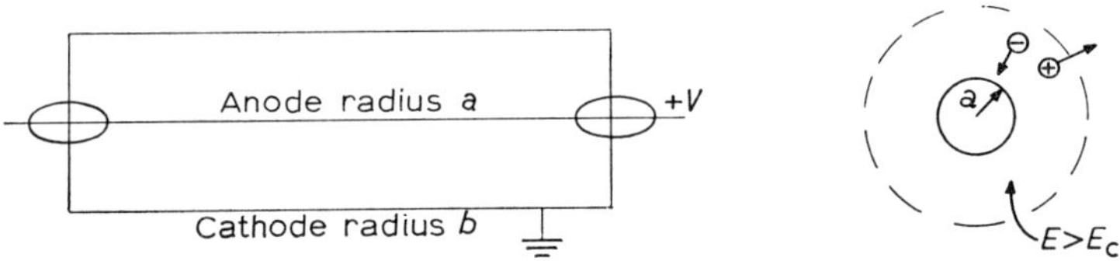

Fig. 4.4.   The proportional counter

The electrons will take about 1 μsec to cross the 0.02 cm region and reach the wire. Thereafter the pulse shape is controlled by the movement of the positive ions away from the avalanche region which removes the neutralizing effect of the positive ions. The final pulse rises to half height in about 2.5 μsec. In such a counter the avalanche only occurs in the limited region along the wire where the electrons from the original ionizing track fall. Hence the pulse resulting will again be proportional to the energy deposited by the particle, but the pulse will be larger than that from an ion chamber by the multiplication factor.

Proportional counters are especially useful for lightly ionizing particles such as electrons, mesons or fast protons, whose small specific ionizations make the use of an ion chamber difficult. However proportional counters are also used for all types of charged particles.

2.3   Geiger counters. If in a proportional counter the voltage is increased so that the multiplication exceeds about 10^3, proportionality will no longer obtain and in the limit the pulse height will be independent of the primary ionization and only dependent on details of the multiplication process, geometry of the chamber etc. This is called the Geiger region (fig. 4.5).

The operation of a Geiger counter can be summarized as follows (fig. 4.6):

a)   Ion pairs are produced in the gas by a charged particle.

b)   The electrons move toward the central wire in the field.

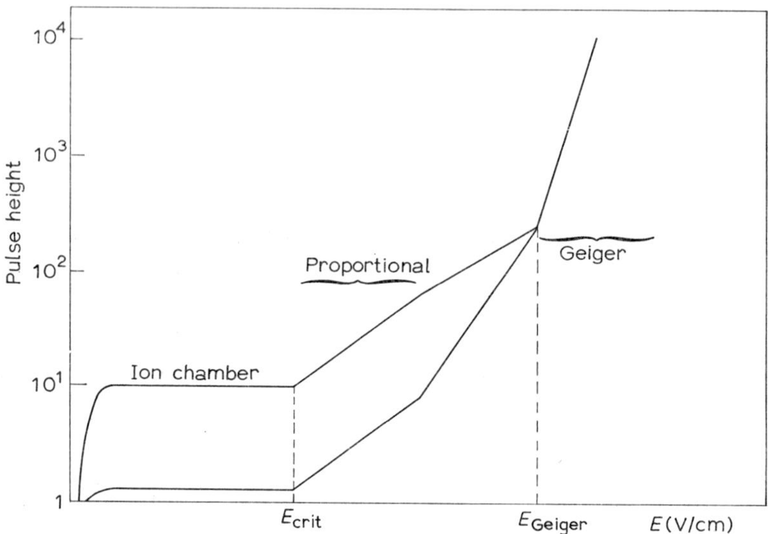

Fig. 4.5. The Geiger operating region. The upper curve gives the pulse height for α-particles; the lower curve, for β-particles.

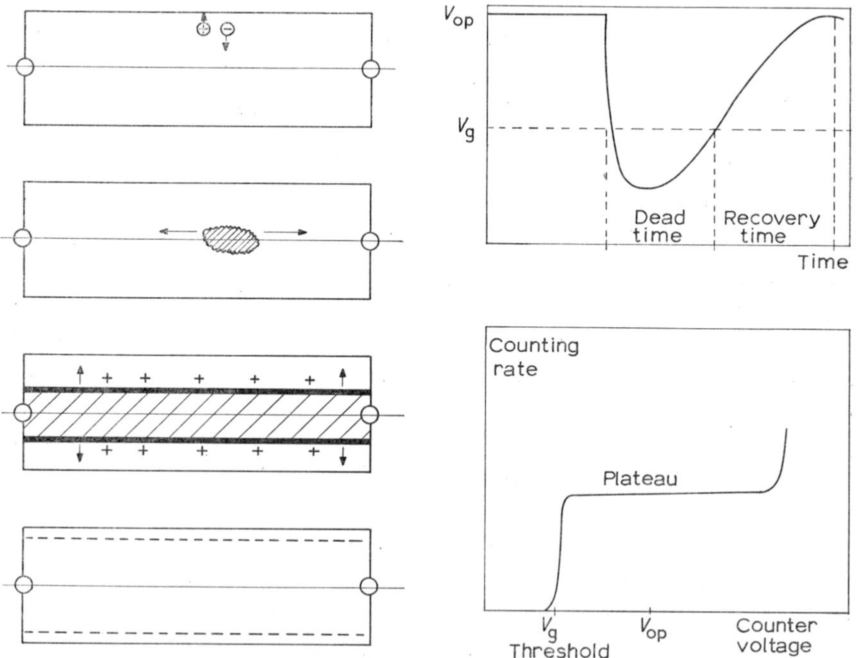

Fig. 4.6. The Geiger counter

c)   Positive ions of argon (for example) move toward the cylindrical cathode wall.

d)   A violent ionization avalanche starts near the central wire as the electrons approach.

e)   This avalanche is propagated along the wire toward each end by ultraviolet emission creating fresh electrons. The avalanche propagation velocity is about 10^6–10^7 cm/sec.

f)   When the avalanche reaches the ends of the wire a charge is collected of size independent of the initial ionization.

g)   Meanwhile the sheath of positive ions move out from the wire toward the wall and screen the wire from the cathode. The field near the wire is decreased below E_c and the avalanche stops.

h)   However when the positive ion sheath reaches the wall and strikes the cathode surface it can eject fresh electrons which see the full field, move to the wire and start a fresh avalanche. A second pulse would be created and so on.

i)   This effect is avoided by introducing a *quenching* gas or vapour which is usually alcohol. This has an ionization potential of 11.3 eV which is less than that for argon, 15.7 eV. The argon positive ions are therefore neutralized by collisions with alcohol molecules before they reach the wall and no secondary pulse is produced.

j)   Alcohol ions however will now strike the cathode but these prefer to absorb energy by dissociation of the alcohol molecule rather than by producing secondary electrons.

k)   The alcohol vapour also serves to absorb ultraviolet radiation produced at the cathode and prevents further ionization due to this.

Thus in a Geiger counter we have 1) a pulse size independent of the primary ionization, 2) sensitivity to a single ion pair, 3) an insensitive time or dead time following the pulse while the positive ion sheath is shielding the wire.

An external circuit is generally arranged such that the insensitive time is a constant value in excess of the dead time plus recovery time. This is called the *paralysis time*. If we plot the counting rate from a constant source against voltage, the Geiger counter exhibits a plateau in which the pulse rate is independent of voltage. This is also shown in fig. 4.6.

2.4   Solid state ion chambers. It is not essential to use a gas in an ion chamber. Indeed liquid xenon and argon ion chambers have been built. In 1945 it was shown that diamond, ZnS and $AgCl_2$ crystals could also respond to ionization.

The mechanism in a solid semiconductor is as follows. These materials exhibit an energy gap between the filled electronic levels and the conduction electron levels (fig. 4.7). The crystal must have very low conductivity, that is there must be very few electrons in the conduction bands. Thus if we apply a field to the crystal there will be a very small conduction current. Now if an ionizing event occurs in the semiconductor, electrons are released which enter the conduction band and move

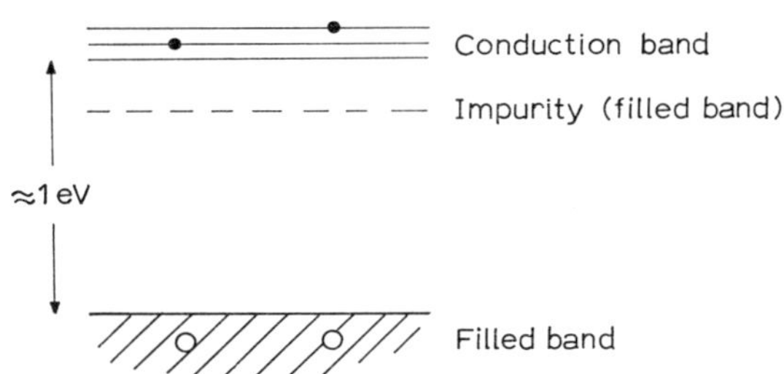

Fig. 4.7.   The solid state ion chamber

under the applied field. Then the charge collected will be proportional to the primary ionization.

However in practice this ideal situation is rarely achieved. This is because there are in fact many more charged carriers present, either electrons or positive holes, than we assumed above. These arise because a) there is thermal excitation across the gap, b) there are impurities present whose filled level position is much closer to the conduction band, c) there are also impurities with many holes, which serve to trap electrons, impede conduction and collection and polarise the crystal.

In recent years some of these difficulties have been overcome by a) use of ultra-pure silicon or germanium, b) cooling the crystal to low temperatures to reduce thermal excitation, c) adding selected impurities which cancel out the effects of the natural impurities. This is called doping or compensating. Under these circumstances *bulk conduction counters* can be made.

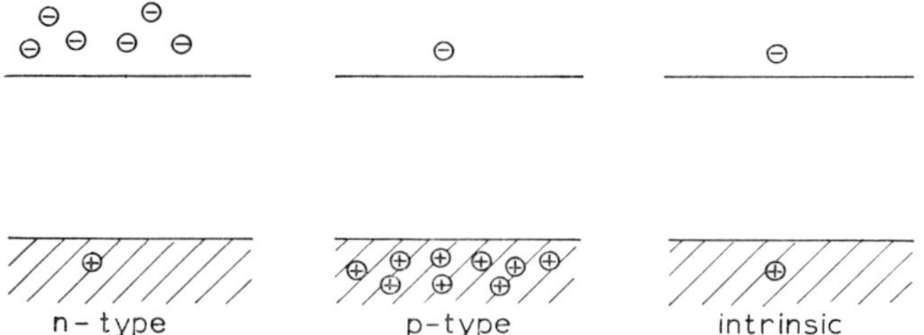

Fig. 4.8.   Doped and intrinsic semiconductor material

Another approach is to make limited volume counters or *junction counters*. There are two types of semiconductor material (fig. 4.8)

a)   n-type where conduction is due to movement of *negative* electrons in the conduction band.

b)   p-type where conduction is due to movement of *positive* holes, caused by electron vacancies in the filled band. Pure semiconductor material containing minimum natural impurities is called *intrinsic*.

The two types can be prepared from intrinsic semiconductor material by doping (adding) with electron donating (n-type) or electron accepting (p-type) impurities.

Now at a boundary or junction between two slabs of different types n and p there will be a resultant electric field which will clear *all* carriers from an intermediate layer at the interface. This is called the *depletion layer*. The depth of this layer can be further increased by an applied external field (fig. 4.9). Thus one has created a region of high quality intrinsic semiconductor free of carriers and this thin layer can then act as a bulk conduction chamber. Ionization by a particle produces electron–hole pairs which are swept out and the pulse can be detected.

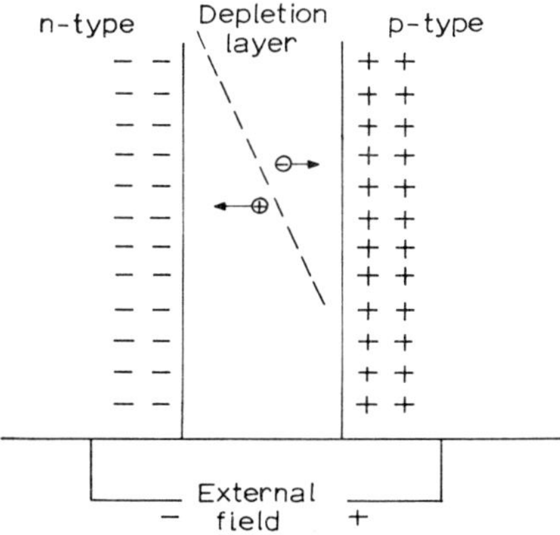

Fig. 4.9.   The junction counter

Two types of junction counters are used. In the *diffused junction* counter one starts with p-type material for example boron-doped silicon since boron with electron structure $(1s)^2(2s)^2(2p)^1$ is electron accepting. Then phosphorus is diffused in from one face of the slab (fig. 4.10). Phosphorus is an electron donating impurity [structure $(1s)^2(2s)^2(2p)^6(3s)^2(3p)^3$]. The region containing phosphorus is then n-type material. So at the plane where phosphorus diffusion stops we have a junction and a depletion layer is formed. The disadvantage of this type of counter is that the particle to be counted must penetrate the rather thick window of n-type material to reach the depletion layer.

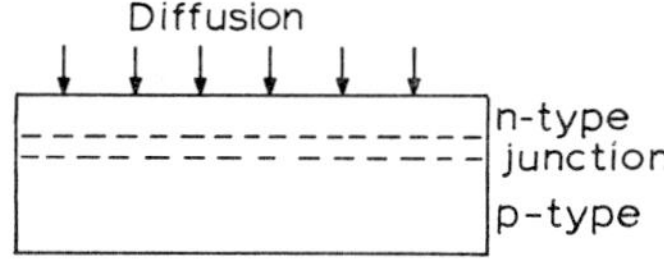

Fig. 4.10.   The diffused junction counter

In the second type called a *surface barrier counter* (fig. 4.11) one starts with n-type silicon generally of rather low conductivity. The oxide layer which forms on the surface of a fresh crystal exposed to air is p-type. So between the oxide layer and the rest of the crystal we get a depletion layer. The layer of oxide is then a very thin window.

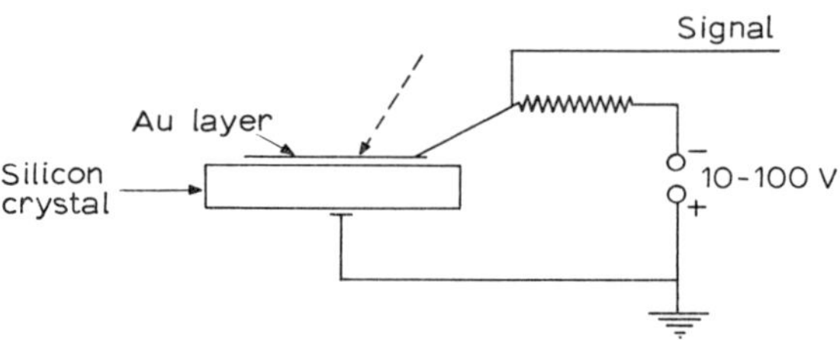

Fig. 4.11.   The surface barrier counter

Electrical connection is made to both types of junction counter by evaporating a thin layer of gold onto the surface and connecting to it and to the rear surface of the crystal.

Germanium can be used instead of silicon but as it has more thermally excited carriers than silicon it must be operated at liquid N_2 temperature.

Greater thicknesses of p-type silicon or germanium may be compensated by drifting lithium (electron donating) through the material. In this way thicknesses of depleted material up to several centimetres can be produced and bulk conduction counters of large volume can be constructed from the compensated material. Germanium–lithium drifted counters make excellent spectrometers for γ-rays or electrons.

Since there is no multiplication as in the gas ion chamber the electrical pulse is very small and must be amplified by a large factor. Noise in the electronic circuits frequently sets a limit on the resolution of such counters.

In a gas each ion pair represented an energy loss of about 30 eV. In silicon each ion pair represents a loss of only about 3 eV. Hence we get ten times as many ion pairs in silicon as in a gas and since the energy resolution is proportional to $n^{\frac{1}{2}}$, the intrinsic resolution is about 3 times better. In a gas counter under ideal conditions we can approach 1 % energy resolution. In a solid state counter we can achieve 0.3 % energy resolution. Collection times are also very much faster in solid state devices being of the order of 10^{-10} sec.

2.5 Electron multipliers. This is a device which can multiply the electrons resulting from an ionizing event but which does not depend on properties of the gas discharge. Thus it can be operated inside a high vacuum region and does not require windows for the particles to enter.

It consists of a series of electron multiplying surfaces or dynodes with voltage between them (fig. 4.12). Of the order of 2–5 electrons are released for each electron incident per stage. Up to 14 multiplying stages are used, and overall multiplying factors up to about 10^9 are possible. The pulse is formed in about 10^{-9} second. The electron multiplier will respond to low energy particles as only the work

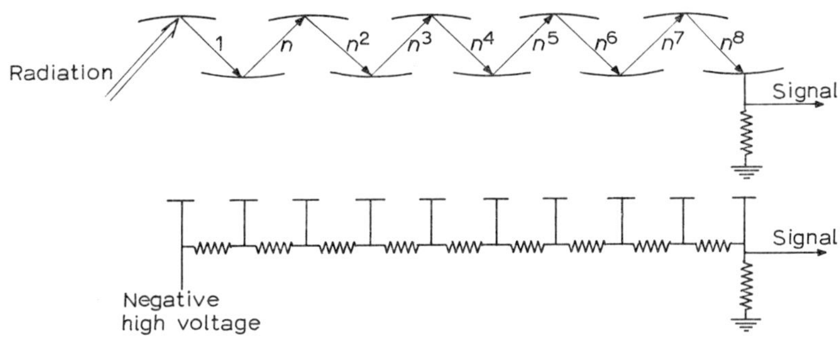

Fig. 4.12.   The electron multiplier

function of the first surface has to be exceeded. The multiplication is eventually limited by the space charge resulting from large interstage currents which have a defocussing effect on the electrons.

2.6 Scintillation counters. In 1903 Crookes introduced the spinthariscope, in which the light flashes resulting from a charged particle striking a ZnS screen could be counted through a low power microscope. This technique was of great importance in the early period of radioactivity research. However it was exceedingly laborious and gas counters soon supplanted it when they became available.

Since 1945 however the technique has been revived because of the development of photomultipliers to replace the eye and of new scintillation materials which are transparent to their own radiation. ZnS is not and so only thin layers could be used. A modern scintillation counter is shown in fig. 4.13. The mechanism is as follows. A charged particle produces an ionization trail in about 10^{-9} or 10^{-10} sec. In inorganic crystals to which has been added an activator impurity there are atomic or molecular groupings in which excitation energy is liberated by radiation rather than by mechanical interaction with the lattice. These are called luminescent centres. So the primary ionization removes electrons from the filled bands in the crystal and these holes are filled by electrons from a nearby luminescent centre. The centre then radiates a photon. The bulk of the crystal is however transparent to the luminescent centre radiation. The efficiency for production of photons depends on the density of luminescent centres and is of the order of 5–20 %. The time for the light to appear depends on the time to transfer energy to the luminescent centre and the decay time of the centre.

Pure organic crystals also scintillate although the mechanism is somewhat dif-

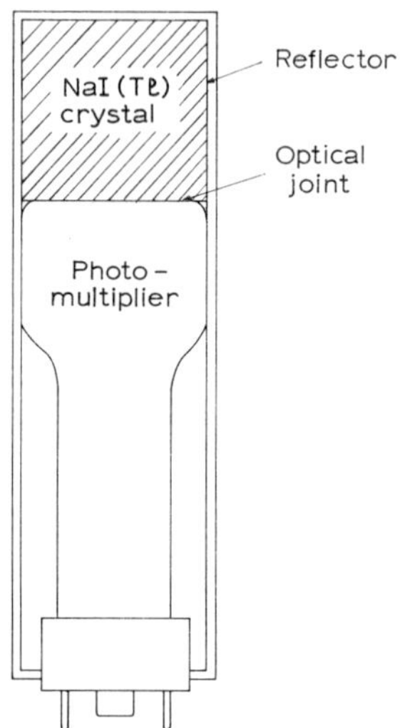

Fig. 4.13.   The scintillation counter

ferent. The light production is considerably faster giving a narrower pulse. Plastic or liquid scintillators are solutions of organic materials in an organic solvent. The excitation energy is transferred from solvent to solute which then emits light for which the solvent is transparent. Plastic and liquid scintillators can be made almost any size or shape. All organic scintillators however have poorer photon efficiency than the inorganic ones.

The noble gases will also scintillate and give very fast pulses. However the emitted light is in the ultraviolet and a wavelength shifting material is used to coat the walls and window of the container so the light finally emitted is in a wavelength range to which the photomultiplier is sensitive.

In table 4.1 are listed some properties of scintillator materials. The element shown in brackets after the inorganic scintillators is the activator, usually in about 0.1 % molar concentration.

In some scintillators there are effects due to the ionization density. In organic scintillators this is a large effect. Electrons give much larger pulses than heavy particles of the same energy. In inorganic scintillators there is a small effect, in gas scintillators no effect. The ionization density effect in organic scintillators also affects the *decay time* of these scintillators. This is made use of to discriminate between heavily ionizing particles such as protons and α-particles which lead to long decay times and lightly ionizing particles such as electrons which show a short decay time. Since this affects the pulse shape electronic circuits can be devised to distinguish the two types of pulse. Pure organic crystals like stilbene show the effect

TABLE 4.1

Scintillator	Density (g/cm³)	Photon efficiency (%)	Decay time (μsec)
NaI(Tl)	3.7	20	0.25
CsI(Tl)	4.5	—	1.1
ZnS(Ag)	4.1	20	10
Anthracene	1.25	10	0.03
Stilbene	1.16	6	0.008
Terphenyl in toluene (liquid)	0.86	3.5	0.002
Terphenyl in polystyrene (plastic)		3.9	0.004
Xenon gas		10	0.01–0.1

most strongly. It is more difficult but still possible to use the effects in liquid scintillators (if free of dissolved oxygen) but almost impossible to use in plastics.

We have to conduct the light to the photosensitive surface with as high efficiency
as possible. This requires good optical matching between crystal and photo-surface, and good reflectivity at the other surfaces of the crystal. If the scintillator must
be at some distance from the photomultiplier a light pipe is used to connect them
in which the light is totally reflected successively down the pipe. A light pipe cannot
be 100 % efficient however. The photomultiplier is simply an electron multiplier
in which the cathode electrode is a transparent photosensitive surface. The quantum efficiency of the photo-surface (usually Sb-Cs) is about 20%, that is, we get one
electron per 5 photons. The photomultiplier must have a low 'dark current', that
is the current due to spontaneous thermal emission of electrons from the photo-
surface. For this reason the photo-surface is often cooled to a low temperature.

The number of electrons is proportional to the number of photons which in
turn is proportional to the energy loss in the crystal so the uncertainty in proportionality or the energy resolution is proportional to $E^{-\frac{1}{2}}$.

The photomultiplier tube generally has a system of focussing electrodes to transfer the photoelectrons to the first electron multiplying surface. The photo-tube
must also be shielded from stray magnetic fields which deflect electrons off the
multiplying plates and so reduce the gain.

2.7   Cerenkov counters. We have seen that when a fast moving charged particle
traverses a medium of refractive index n, radiation is emitted on the surface of a
forward cone of semi-angle θ where $\cos \theta = 1/\beta n$, $\beta = v/c$, provided the particle velocity $v > c/n$, the velocity of light in the medium. The effect is independent of

particle mass; the angle of emission increases with particle velocity; the electric vector is perpendicular to the cone surface, that is, the light has a characteristic polarisation; there is a critical velocity c/n below which no light is seen; it is a fast effect, $\approx 10^{-10}$ sec; the spectral distribution is $\nu d\nu$ so it is most intense in the blue and ultraviolet parts of the spectrum. It is for this latter reason that a swimming pool reactor core glows a deep blue when seen through the water. We are seeing the Cerenkov radiation from fast β-particles in the water.

When used as a detector of particles, a relativistic particle of unit charge passing through 10 cm of glass produces about 2500 photons in the visible region of the spectrum. We can view these with a photomultiplier and get a few hundred photoelectrons. So our electrical pulse will have about 10 % spread. We can devise either a threshold detector so that we detect all particles whose velocity exceeds c/n or a Cerenkov spectrometer in which only light emitted in a given angular range is detected when $\beta = 1/n \cos \theta$.

Cerenkov counters are also used to measure the total energy in electromagnetic cascades resulting from high energy photons, or particles. If the cascade is completely absorbed the light output is proportional to the total energy in the cascade and hence to the energy of the initiating particle.

3   Charged particle detectors: visual

3.1   Wilson cloud chambers. The Wilson expansion chamber or cloud chamber was devised in 1912 by C. T. R. Wilson. If we start with a container of gas containing a saturated vapour at temperature T_1 and then perform an adiabatic expansion from volume V_1 to V_2, the temperature will fall to T_2 where

$$T_2 = T_1(V_1/V_2)^{\gamma-1} \tag{4.3}$$

where γ is the ratio of the specific heats of the gas–vapour mixture. The initial vapour pressure was p_1 at temperature T_1. After expansion the vapour pressure is

$$p_f = p_1(V_1/V_2)^{\gamma} \tag{4.4}$$

which is greater than p_2, the saturated vapour pressure at temperature T_2.

We now consider the relation between the vapour pressure p_0 in equilibrium with a plane liquid surface and the vapour pressure p near a convex surface of radius r, that is, of a droplet.

The dotted curve in fig. 4.14 represents the equilibrium condition. Above this curve drops grow in size, below it the drops evaporate.

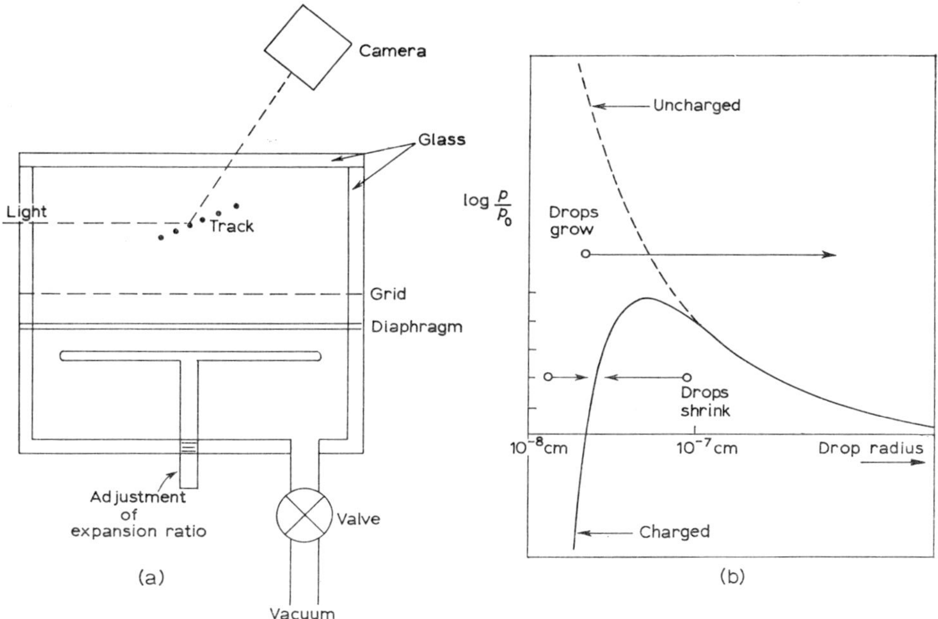

W. E. Burcham, Nuclear Physics (Longmans, London 1963) p. 241

Fig. 4.14.   The expansion cloud chamber

If the droplet has an electric charge q, then

$$\frac{RT\rho}{M} \log_e \frac{p}{p_0} = \frac{2\delta}{r} - \frac{q^2}{8\pi r^4} \tag{4.5}$$

where R is the gas constant, T is the absolute temperature, δ is the surface tension, M the molecular weight, ρ the liquid density and r the drop radius. This gives the solid curve shown in fig. 4.14. Thus if the supersaturation $S = p_f/p_2$ achieved in the expansion, exceeds the maximum of the curve of p/p_0, then the droplets will grow from 10^{-8} cm to 10^{-3} cm in about half a second.

For water vapour at 0 °C, the critical supersaturation $S = 4.2$ and the drop radius at the maximum of the curve is 6×10^{-8} cm. The droplets produced by the ionization are about 3×10^{-8} cm. During the half-second growth period the droplets may be photographed when suitably illuminated.

The expansion ratio V_2/V_1 must be somewhat greater for positive ions than for negative ions. The upper limit on the ratio is set by the point at which condensation occurs on uncharged centres. The *sensitive time* is set by the time taken for the temperature to rise from T_2 to a value for which the supersaturation is too low. Counter control is often used in which counters are placed above and below the chamber. When a coincidence count is recorded indicating that a charged particle has passed through, the chamber is expanded and the track photographed. If there is too much delay between the particle passage and the expansion the ions will diffuse and contribute to *track broadening*.

3.2 Diffusion cloud chambers. In these chambers, devised by Langsdorf in 1936, a temperature gradient is set up in a gas–vapour mixture. A mixture of CO_2 and alcohol is typical. The sensitive region as shown by the solid curve in fig. 4.15 for supersaturation vs. temperature is 5–10 cm deep. However it is *continuously* sensitive in contrast to the expansion chamber. A disadvantage is that a burst of intense ionization can easily deplete the sensitive region of vapour and it may take 10–20 sec for it to refill. A clearing field of a few hundred volts is continuously applied to sweep ions out of the sensitive layer.

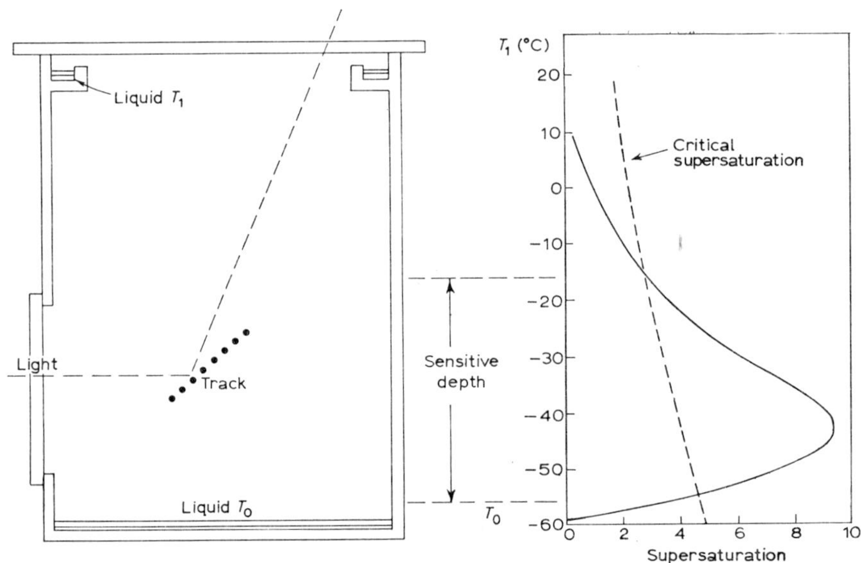

W. E. Burcham, Nuclear Physics (Longmans, London 1963) p. 245

Fig. 4.15.   The diffusion cloud chamber

Diffusion cloud chambers are convenient to use with pulsed accelerators. They can also be operated at higher pressures than the expansion chamber and thus more stopping material is available in the sensitive region.

3.3 Bubble chambers. These were developed by Glaser in 1952. Cloud chambers are limited to gas fillings and consequently have a rather low stopping power in the sensitive region. The bubble chamber (fig. 4.16) uses a superheated liquid in which bubbles grow from the ions rather than as in the cloud chamber a supersaturated vapour–gas mixture in which droplets grow on the ions. Since the bubbles are formed along the ionization path they can be photographed before uncontrolled boiling occurs. The operating cycle is as follows:

1) The liquid is heated to a temperature above the boiling point. 2) It is kept in the liquid phase by a pressure which exceeds the saturation vapour pressure.

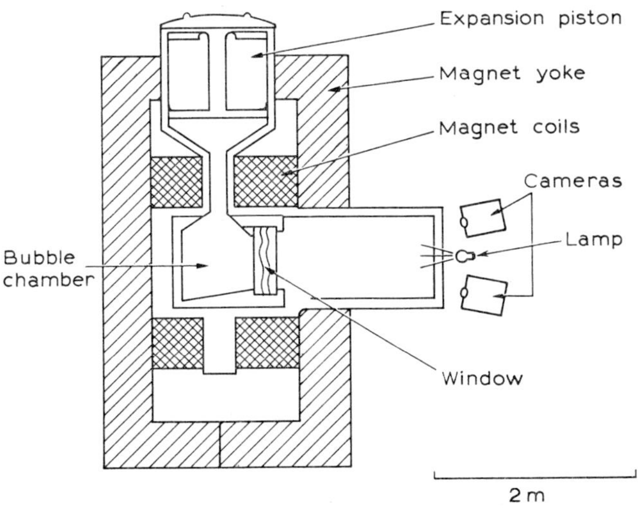

Ann. Rev. Nucl. Sci. **10** (1960) 132

Fig. 4.16.   The bubble chamber

3) The pressure is suddenly reduced and the liquid is then superheated. 4) The chamber is now sensitive to ionizing particles for several seconds before general boiling takes place. That is, bubbles will form preferentially on ionized centres and grow. 5) In practice the sensitive time is limited to a few milliseconds after which recompression takes place and general boiling is prevented. 6) The cycle can then be repeated.

The bubbles are initiated by the local heating due to short δ-rays (secondary electrons) resulting from the primary ionization. Bubble growth takes place in about 10 msec. The bubble forming centres last for about the same time. The cycle cannot be initiated fast enough to make counter control practical. However the chamber may be recycled rather rapidly. In use with high energy pulsed accelerators it is synchronised with the beam pulse.

Liquids used are hydrogen, deuterium or helium which have the advantage of being simple nuclei but the difficulty that the boiling point is at a very low temperature. Heavy liquids such as pentane, propane or xenon have the advantage of higher stopping power, though the targets are complex nuclei. Bubble chambers especially for hydrogen have been made in very large sizes up to 2 metres diameter and 50 cm deep. Some data for bubble chamber liquids are listed in table 4.2.

TABLE 4.2

Bubble chamber liquids

	Operating temperature (°C)	Pressure (atm)	Density (g/cm³)	Mean free path of 100 MeV γ-rays (cm)
H_2	−246	5	0.06	2700
D_2	−241	7	0.13	2000
He	−369	1	0.13	1800
Propane	58	21	0.43	220
Pentane	157	23	0.5	
Xenon	−20	26	2.3	6.6

3.4   Nuclear emulsions. The blackening of photographic film by nuclear radiation was of course first observed by Becquerel. However its use as a practical measurement technique dates from about 1940 when Powell persuaded film manufacturers to produce very thick emulsion layers with very fine grain. Such emulsions are referred to as nuclear emulsions.

The ionization of a particle passing through a grain of silver bromide renders it developable. The grain size must be small to give good track resolution and there must be a low density of background developable grains. Much early work on fast neutron spectra was done by analysing the proton recoil tracks in emulsions. The pion and the first strange particles were observed in emulsions flown at high altitudes. It is a simple technique since it stores up data and integrates over a long period. Thus in spite of its relatively small sensitive volume it has been of value in discovering rare events.

The target material is fairly invariant as it is difficult to introduce other materials into the emulsion. The composition is as follows (percentages): H 37, C 19, Br 15, Ag 15, N 4, O 10. Deuterium can be substituted for the hydrogen quite readily. The emulsions are generally 20–600 μm thick. The range of a 10 MeV proton in the emulsion is about 600 μm.

To increase the sensitive volume, a whole set of emulsion layers without backing can be stacked together to make an emulsion stack. After exposure the emulsions are developed separately. Events occurring in a stack several inches thick can then be reconstructed. Particle mass, charge and energy can be determined from a) range, b) grain density, c) δ-ray production and d) multiple scattering.

Magnetic fields of sufficient strength and stability cannot be used with emulsions which is a serious limitation on momentum determination.

3.5 Track scanning. Methods for automatic scanning and measurement of the photographs of visual chamber events are in an advanced state of development. The photographs which are generally stereophotographs are reprojected. The coordinates of tracks present are digitized and stored in a computer. Track patterns which correspond to nuclear events are recognised and measurements of curvature, range etc. made. The computer then does the necessary kinematic calculations to identify the particles involved. Almost all of these steps can now be taken automatically, which is essential since a bubble chamber experiment may involve several million photographs.

4 Combinations of visual and electrical techniques

4.1 Counter hodoscopes. This is simply an array of counters whose outputs are displayed as neon lights in the same spatial relation as the counters (fig. 4.17). By indicating which counters fire together the path of the particle is displayed. It can indicate little more than the direction of a particle.

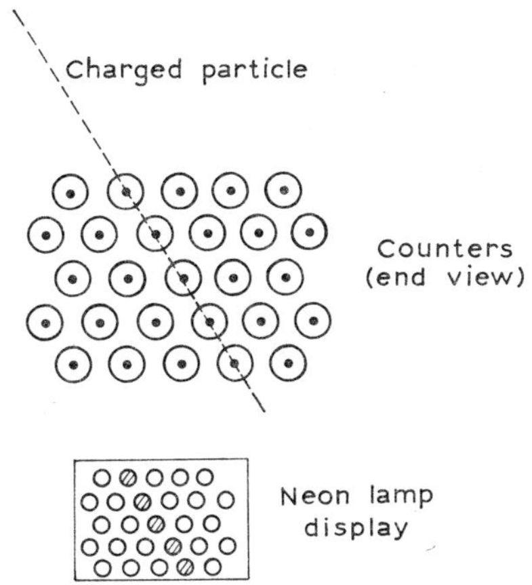

Fig. 4.17.   The counter hodoscope

4.2 Spark chambers. In this device a series of parallel metal plates are arranged with the gaps between filled with a noble gas e.g. neon (fig. 4.18). When triggered by counters which show that an ionizing particle has passed through, about 15–20 kV is applied between the plates in about 300 nsec. About 100 nsec later a spark appears at the point where ions have been deposited by the charged particle passing through the gas. When viewed from the edge the spark positions indicate the path of the particle. A clearing field is then applied to sweep the gas free of ions and the chamber is ready for firing again in a few msec. The sensitive time is about 0.5 sec and the dead time is 1–10 msec. The spark position may be a) photographed and the photographs analysed as for a bubble chamber, b) the chamber may be viewed with an image tube so the track coordinates are directly digitized,

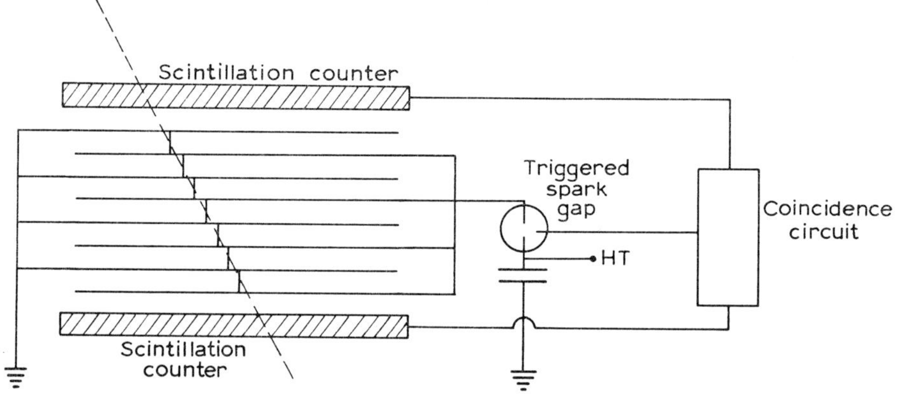

Fig. 4.18.   The spark chamber

c) the sound of the spark may be picked up by transducers at the edge of the chamber and the coordinates determined by sound ranging, d) the plates may be made up of two sets of parallel wires at right angles. The currents flowing in the wires near the spark indicate the position and the coordinates then fed directly into a core store.

Spark chambers are relatively cheap and simple to construct, and have been made up to 1 metre square and weighing many tons. Spark positions can be determined to about 0.25 mm.

5   Electrical counting circuits

The electrical pulse produced by a detector contains several kinds of information. 1) The presence of the pulse indicates a charged particle has deposited energy in the detector. We can *count* the number of such particles. 2) The size of the pulse indicates, if the detector is proportional, the value of the energy loss in the detector. We can find the *energy* of the particle counted from the pulse height. 3) The *time* the pulse rises to a certain value is related to the time the particle passed through the detector. The accuracy will depend upon how fast the pulse rises.

If we wish only to count the number of pulses we need only amplify the pulse, not necessarily linearly, so that it can be recorded in a scaler and register. If we wish only energy information we must use amplifiers which are accurately linear but which have only a relatively narrow bandwidth of the order of 1–3 MHz. If we wish timing information for coincidence counting or time-of-flight measurement we require amplifiers with very wide bandwidth of the order of 100–300 MHz in order to have a rise time of a few nanoseconds.

The combined requirement of good linearity and fast rise time is frequently difficult to achieve. A double amplifying system – fast–slow system – is often used. Many circuit arrangements are used. A typical one is shown in fig. 4.19.

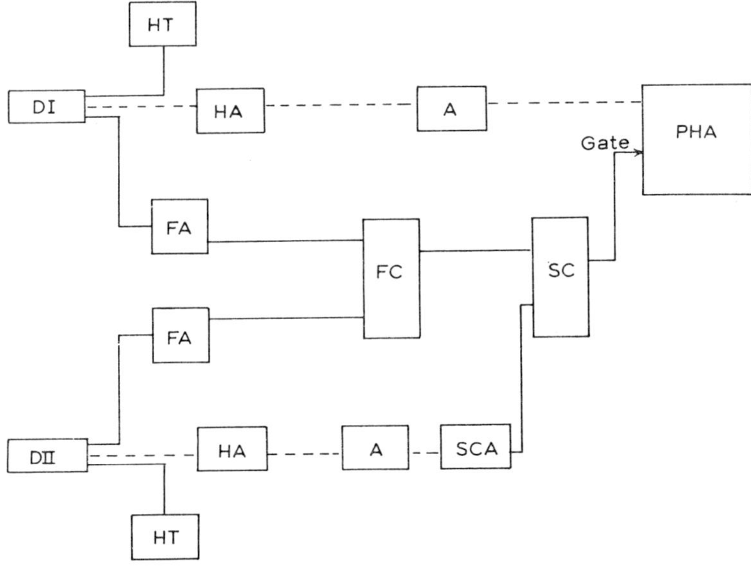

Fig. 4.19.   A typical electronic arrangement. DI and DII are detectors, HT are high voltage supplies, HA are head amplifiers, A are main amplifiers, FA are fast amplifiers, FC is a fast coincidence circuit, SC is a slow coincidence circuit, SCA is a single channel analyser and PHA is a multichannel pulse height analyser. Linear signals are shown dashed

DI and DII are two detectors (for example scintillation counters). The linear parts of the circuit are shown as hatched lines. The detectors must be connected to stable HT supplies. The detector signal (generally at high impedance) of a millivolt or so is amplified by a head amplifier HA or a cathode or emitter follower may be used to give no amplification but to give an impedance match for the low impedance signal output. This signal may then be fed down a long line to the main amplifier A which produces an output signal of a few volts. These signals may be sorted according to pulse height by a pulse height analyser, PHA. A detector signal may also be derived from each detector, amplified through a fast amplifier FA and a fast coincidence detected in FC. A linear signal from detector II is put through a single-channel analyser which gives an output pulse only when the input pulse height lies between selected limits. A discriminator could be used instead which gives an output pulse of fixed size only when the input pulse height exceeds a selected level. A single-channel analyser consists of two discriminators set at different levels the outputs being placed in anticoincidence. The output of this SCA is put in coincidence with the fast coincidence output via a slow coincidence circuit SC. The output of this circuit is used to gate the PHA. A pulse from the linear side of detector I will only be recorded when there is a gate pulse in coincidence.

All of this then ensures that the pulse height spectrum recorded from DI corresponds to those particles which were in coincidence to a high precision with particles in DII which had a specified energy.

The multichannel pulse height analyser PHA performs the same function as the SCA except that particles having many different pulse heights may be recorded simultaneously. The number of channels may be many thousands. Many schemes have been devised to do this. The earliest were simply multidiscriminators but about 30–50 channels is the maximum feasible for this method. Modern analysers depend

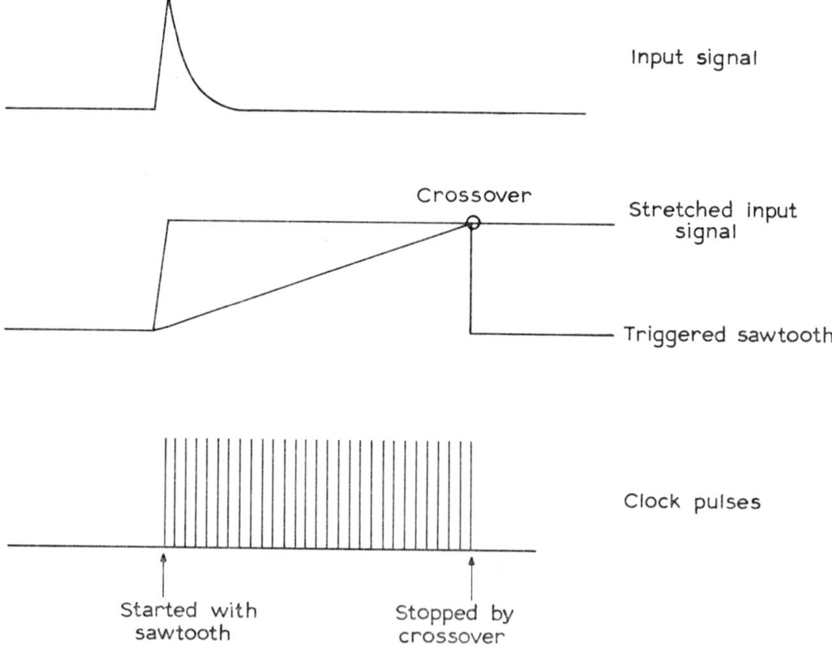

Fig. 4.20. Analog to digital converter

on converting the pulse height into a time interval. When the incoming pulse starts to rise a ramp pulse is started which rises at a constant rate (fig. 4.20). When the ramp pulse attains the height of the input signal, the ramp pulse is stopped. Its duration is then proportional to the pulse height. When the ramp pulse starts, a clock running at a few MHz is also started and stopped when the ramp pulse stops. A binary counter counts the number of clock pulses and this number then corresponds to the pulse height. One is then added to the appropriate position in a matrix core store and the pulse has been analysed into its appropriate channel or bin. A circuit for converting the pulse height into a binary number is called an analog to digital converter.

6   Statistical considerations in counting

In general we are dealing with random events and the distribution follows the Poisson distribution $W(n) = N^n \mathrm{e}^{-N}/n!$ where $W(n)$ is the probability of observing n particles in a given time interval, N being the average number expected in that interval. Thus the number of events recorded in a given interval will be distributed about the mean number N with a variance given by

$$\sigma^2 = \sum_0^\infty (n-N)^2 \frac{N^n \mathrm{e}^{-N}}{n!} = N. \tag{4.6}$$

Thus the variance or standard deviation σ in a count of N particles (when N is large) is simply $N^{\frac{1}{2}}$ and our observation with error will be $N \pm N^{\frac{1}{2}}$.

If our counter is insensitive after each count we may lose events which occur in this interval. If we count n/sec then the apparatus is insensitive for a time $n\tau_\mathrm{D}$ each second and the number of counts lost is $Nn\tau_\mathrm{D}$ where τ_D is the dead time. Thus

$$n = N - Nn\tau_\mathrm{D}$$

or

$$N = \frac{n}{1 - n\tau_\mathrm{D}} \approx n(1 + n\tau_\mathrm{D}). \tag{4.7}$$

We must make sure of course that τ_D is a constant. If it is not we sometimes introduce electronically a fixed paralysis time longer than the greatest value of τ_D.

7   Coincidence counting

If we observe coincidences between two truly coincident radiations and plot the coincidences as a function of delay we ideally get a curve as shown in fig. 4.21b.

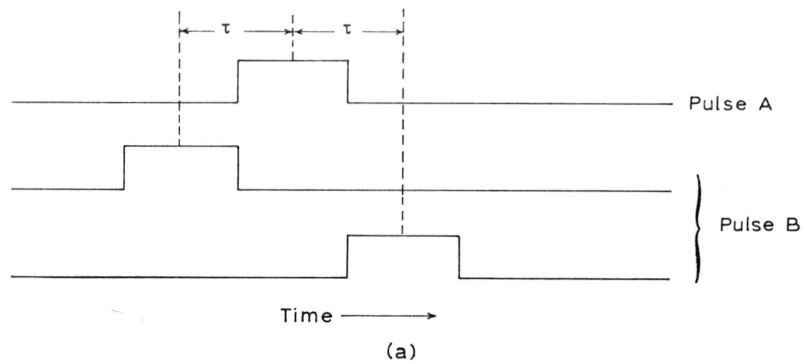

Fig. 4.21a. Coincidence resolving time

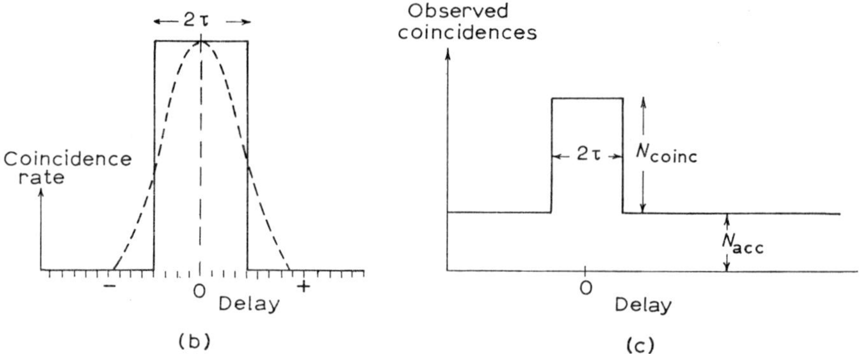

Fig. 4.21b, c. Coincidence rate as a function of delay and delay curve showing accidental coincidences

If the *rates* in the two counters are N_A and N_B per second and the counts are *not* coincident we will still see an *accidental* coincidence rate $N_{acc} = 2\tau N_A N_B$. This is just the number of B which fall within $\pm\tau$ of an A count and the true coincidence rate must be corrected for this accidental rate. We must measure N_A, N_B and τ and compute N_{acc} or more usually we put in a large delay so that the counts cannot be true coincidences.

8   *Energy resolution*

Even if we achieve high statistical accuracy with regard to random nuclear events we may still get uncertainty in pulse height for a given energy due to the statistical nature of the basic detection process e.g. ionization, photon production or electron multiplication. If a given event produces x electrons and the multiplication factor is f then the variance in the final number of electrons is

$$\sigma^2 = f^2\sigma_x^2 + x^2\sigma_f^2 \tag{4.8}$$

where σ_f, σ_x are variances in f and in x.

The final signal is proportional to fx and its spread to σ. So the energy resolution is proportional to

$$\frac{\sigma}{fx} = \left(\frac{\sigma_f^2}{f^2 x} + \frac{\sigma_x^2}{x^2}\right)^{\frac{1}{2}}. \tag{4.9}$$

For *ion chambers* or *semiconductor counters* there is no multiplication so $f=1$, $\sigma_f=0$. For a Poisson distribution $\sigma_x^2 = x$ and then $\Delta E/E \propto x^{-\frac{1}{2}}$. But $x = E/($energy per ion pair$)$. Thus for the semiconductor the denominator is 3 eV/ion pair while for the gas ion chamber it is 30 eV/pair. Thus the energy spread of the semiconductor is $10^{-\frac{1}{2}} \approx \frac{1}{3}$ that for the gas chamber. For proportional counters and photomultipliers f is large and for both these

$$\sigma_f^2/f^2 \approx 1 \quad \text{hence} \quad \Delta E/E \approx (2/x)^{\frac{1}{2}}.$$

Thus the resolution of a proportional counter is at least $2^{\frac{1}{2}}$ times worse than an ion chamber.

In a scintillation counter the energy required to give one photon is $\approx 50\text{–}100$ eV. This together with light collection efficiency and photocathode efficiency gives an overall scintillation counter resolution about 10 times worse than for ion chambers.

Note also that in all detectors $\Delta E/E \propto x^{-\frac{1}{2}} \propto E^{-\frac{1}{2}}$ since $x = E/($energy per ion pair or per photon$)$.

9   Measurement of charged particle energies

9.1   Detector spectrometers. We have considered a number of electrical detectors in which an electrical pulse is derived which is proportional to the energy deposited in the detector. If the particle is stopped within the detector the pulse is then proportional to the particle energy and a differential pulse spectrum is directly interpretable as an energy spectrum. In visual detectors in which the particle stops, the range may be measured directly and the energy derived from empirical range–energy curves. The precision is limited in this case by the range straggling but may approach 1 %.

The mass of a particle cannot be directly inferred from the total energy or the range or the momentum as measured in a visual chamber with magnetic field. Additional information is required; in particular a measurement of the energy loss

dE/dx in addition to energy E fixes the mass of the particle (see ch. 3). In some visual chambers the ionization density per unit path length can be derived by counting individual droplets, bubbles or developed grains. In a counter experiment a thin counter in which the energy loss is small compared to the range is used in front of another thick counter. The pulse from the thin counter is proportional to dE/dx while the sum of the two counter pulses gives the total energy.

Together these serve to identify the mass of the particle. An alternative scheme is to measure the time of flight over a known distance and the total energy of the charged particle. The product Et^2 is then directly proportional to the mass.

In a Cerenkov counter since the particle velocity fixes the minimum angle of emission of the Cerenkov light, the particle velocity may be determined by a differential Cerenkov detector.

9.2 Magnetic spectrometers. For more accurate determinations, magnetic or electrostatic spectrometers are used. Let us define first some parameters. ΔE is the full width at half maximum of a line recorded in a spectrometer corresponding to a fixed particle energy E; Δp is similarly the momentum width, δE is the uncertainty in the absolute energy scale of the spectrometer. $\Delta E/E$ or $\Delta p/p$ is called the resolution. T is the transmission: $N_{\text{detected}}/N_{\text{incident}}$.

The simplest type of spectrometer is a $180°$ magnetic spectrometer. This type is used for both electrons and heavy particles. The source and detector are inside a homogeneous magnetic field.

The radius of curvature in the magnetic field will be proportional to the particle momentum, $R = pc/eH$ where p is the momentum and H the magnetic field. Thus in a detector such as a photographic plate we record a momentum spectrum (fig. 4.22a). Note however that for particles of the same momentum emerging from a point source and limited by a slit to angles less than α with the perpendicular these are brought to a focus at the detector plane (fig. 4.22b). If the source size is small the line width is

$$s = 2R(1-\cos\alpha) = R\left(\alpha^2 - \frac{\alpha^4}{12} + \frac{\alpha^6}{360}\cdots\right) \tag{4.10}$$

$$\approx R\alpha^2 \text{ (since } \alpha \text{ is small).}$$

The momentum resolution is thus

$$\frac{\Delta p}{p} = \frac{\Delta R}{R} = \frac{s}{2R} = \frac{\alpha^2}{2} \tag{4.11}$$

while the transmission is

$$T = \frac{\Omega}{4\pi} = \frac{1}{4\pi}2\pi(1-\cos\alpha) \approx \frac{\alpha^2}{4}. \tag{4.12}$$

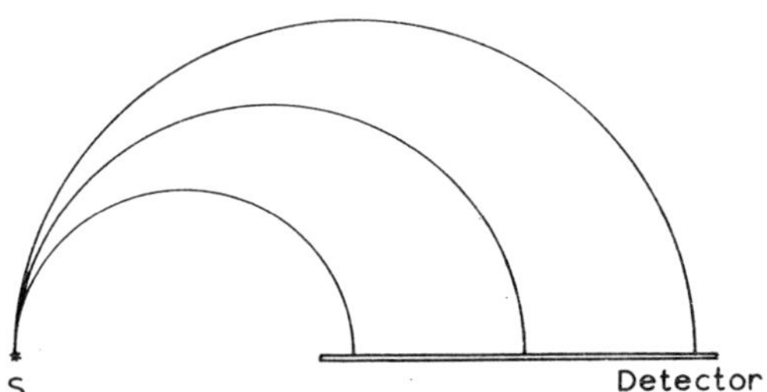

Fig. 4.22a.   The 180° magnetic spectrometer

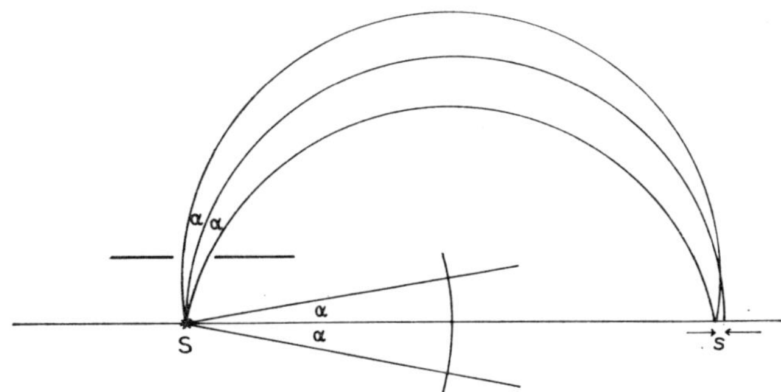

Fig. 4.22b.   Focussing in the 180° spectrometer

Typical values are $\Delta R/R = 1\%$, $T = 0.1\%$; δE will depend on the precision with which the magnetic filed can be measured and on its uniformity.

As mentioned a photographic plate could be used as a detector. But since this requires long exposures and hence extreme stability of the magnetic field the technique is usually only employed where permanent magnets are used to supply the magnetic field. It is more usual to use a single fixed counter detector and to vary the magnetic field. By plotting N/p vs H we derive a momentum spectrum (since Δp, the detector aperture, is proportional to p).

In magnetic lens spectrometers, much used for β-ray spectroscopy, the magnetic field is produced by a solenoid with source and detector on the axis (fig. 4.23). The path of a particle emerging from source S at angle α to the axis is a helix which again cuts the axis at the focus F. The angular velocity of the particle about the line of force is

$$\frac{v \sin \alpha}{R} = \frac{eH}{mc}.$$

(4.13)

Hence during the time τ to sweep out an angle of 2π the particle moves parallel to

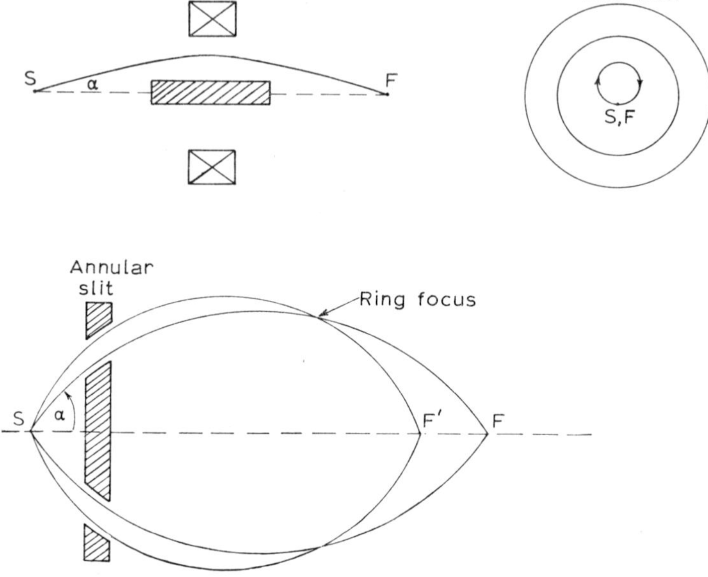

Fig. 4.23.   The lens spectrometer; side view and end view of orbit; focussing

the field a distance

$$\text{SF} = (v \cos \alpha)\tau = \frac{2\pi mvc}{eH} \cos \alpha = \frac{2\pi pc}{eH}\left(1 - \frac{\alpha^2}{2} + \frac{\alpha^4}{24}\cdots\right). \qquad (4.14)$$

Thus as α increases the distance travelled will be shorter for a given particle energy, and the focal region will be spread along the axis.

Note that there is a ring focus before the trajectories reach the axis. A further annular slit is often placed at this point.

Electron spectrometers may be of *long lens* type in which the solenoid reaches from S to F or of the *thin lens* type in which the length of the solenoid is short compared to SF. The *intermediate image* spectrometer of Siegbahn and Slätis uses two thin solenoids. The first produces a ring focus and the second focusses this ring focus image onto a point on the axis. This avoids the aberration of the single-lens type and permits better resolution.

In the semicircular magnetic spectrometer described above, particles emerging with different angles α in the plane at right angles to the field direction were focussed but for those travelling out of the plane at right angles to the field (at angle ϕ say) there was no focussing. This is called *single focussing* (single-plane focussing). In the thin lens spectrometer with an annular slit at the ring focus, particles of the same momentum are brought to this focus independent of α and also of ϕ because of the cylindrical symmetry. This is called *double focussing* (double plane). In all of these cases the proportionality factor between R and the particle momentum contained a function of α. Focussing was only obtained on the assumption that α was small enough so that higher terms than the first could be ignored in the series expansion. This is termed *first order* focussing.

Another method for achieving double focussing is to use magnetic spectrometers with inhomogeneous fields. For example if a flat gap spectrometer is constructed such that there is a field gradient $H_R = H_0 (R_0/R)^{\frac{1}{2}}$ then focussing in both planes will take place and particles emerging from a point in the field will refocus to a point after deflection through $\pi\sqrt{2} = 254.6°$. This has the effect of increasing the transmission of the spectrometer without worsening the resolution. When such instruments are used for studying heavy charged particle spectra it is usual to use

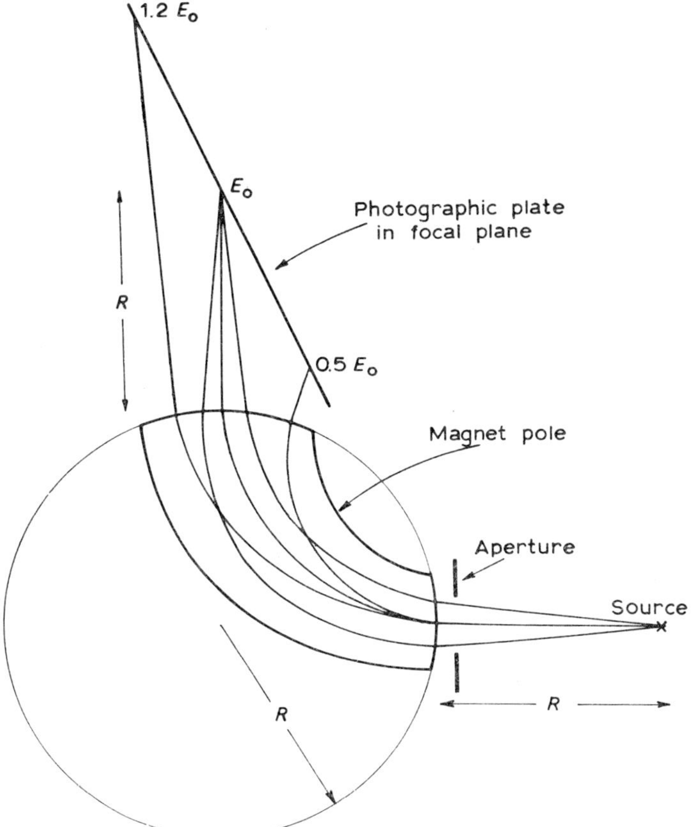

F. Ajzenberg-Selove (ed.), Nuclear Spectroscopy A (Academic Press, New York 1960) p. 64

Fig. 4.24.   The wide range magnetic spectrometer

a smaller angle of deflection such as 180°. The source and image points then lie well outside the magnetic field region which is more convenient.

In most of these spectrometers a momentum spectrum is obtained by measuring the counting rate for successive values of the magnetic field. Another type of homogeneous field spectrometer has been developed particularly for studying nuclear reactions in which a *wide range* of momenta can be recorded simultaneously. Such a device (single focussing, cf. fig. 4.24) can record over about a factor of 3 in energy. Photographic plates are normally used for recording and individual tracks in the emulsion are scanned and counted. The resolution can be made about 0.06 %. In all such devices in which the source and detector are not inside the field the fringing field effects make absolute calibration difficult. Calibration with particles of known energy is used.

Broad range spectrometers have also been made in which a number of magnet gaps are arranged side by side around a circle with the source at the centre. The energy spectra as a function of angle can then be recorded simultaneously. Such multigap or 'orange' spectrometers are used for both electrons and heavy particles.

9.3   Electrostatic spectrometers. Both cylindrical and spherical geometry electrostatic spectrometers have been constructed. Although it is mechanically difficult to mount insulated curved electrodes with the required precision, once this is achieved it is relatively easy to achieve high absolute accuracy in energy measurements.

9.4   Calibration of spectrometers. Certain α-particle energies from radioactive sources have been measured to high precision (1 part in 5000) in simple absolute spectrometers such as homogeneous magnetic field or electrostatic devices. These are ^{210}Po, $E_\alpha = 5.30$ MeV, ^{212}Po (ThC$'$), 8.78 MeV and ^{212}Bi (ThC), 6.04 and 6.08 MeV. These in turn have been related to the threshold proton energy for the reaction:

$$^7\text{Li}(\text{p, n})^7\text{Be}, \qquad Q = -1.6449 \text{ MeV}, \qquad E_{\text{th}} = 1880.7 \pm 0.4 \text{ keV}.$$

These are used as primary standards to calibrate spectrometers of all kinds and also accelerator energy scales.

10   Measurement of absolute disintegration rate

Several methods have been devised to measure the absolute rate in particular from radioactive sources.

10.1   Absolute counting. Absolute counting of α- or β-particles may be carried out in which the solid angle of the detector is carefully defined. Alternatively the source may be placed in the centre of a double counter arrangement, the so-called 4π-counter since almost all particles emerging into the sphere should be counted. However careful corrections for absorption and scattering must be made for even the thinnest 'weightless' source. Source and backing must be 10–20 $\mu g/cm^2$ or less.

10.2 Coincidence methods. When two radiations in coincidence occur as in β–γ or γ–γ decays, coincidence methods may be used. Two counters 1 and 2 with efficiencies ε_1 and ε_2 for the radiations record the number of coincidences, $C = N\varepsilon_1 \varepsilon_2$ where N is the number of source disintegrations. The counts recorded in counter 1 in the same period are $C_1 = N\varepsilon_1$ and in counter 2 are $C_2 = N\varepsilon_2$. Then $C/C_1 = \varepsilon_2$ and $C_2/\varepsilon_2 = N$. Corrections must be made as described earlier for accidental coincidences. The angular correlation of the two radiations must also be determined and corrected for.

10.3 Annihilation radiation. In order to measure absolute rates for positron emitters, annihilation radiation may be used. The source is surrounded by a foil of thickness such that all positrons are stopped and annihilate. Two γ-rays each of energy 0.511 MeV are emitted in opposite directions. Two identical counters are arranged to be sensitive to this energy only. The coincidence rate is then $N d\Omega/4\pi$ where $d\Omega$ is the solid angle subtended by each counter.

10.4 Flow counters. When the radioactive isotope is gaseous it can be used as the counter filling gas. Tritium (^{3}H) is frequently counted in this way. Its maximum β-energy is only 18 keV and all particles which result from decay in the counter are recorded with 100 % efficiency. With the flow rate and sensitive volume of the counter the specific activity can be derived from the counting rate.

11 Detection and energy measurement of neutrons

11.1 Introduction. Since neutrons interact with matter through the short range nuclear force and only negligibly via the electromagnetic force we have to depend on secondary charged particles or photons to signal a nuclear encounter involving a neutron.

The detection or measurement technique will be very dependent on the neutron energy. Neutron energies are classified as follows.

1) *Slow* (0–100 eV).

a) *Cold* neutrons ($<$0.025 eV) in equilibrium with matter at less than normal temperature.

b) *Thermal* neutrons ($\approx$0.025 eV) in equilibrium with matter at about 20 °C.

c) *Epithermal* neutrons (0.025–1 eV).

Neutrons of a), b) and c) exhibit a cross section for nuclear interaction which is smooth and proportional to $E^{-\frac{1}{2}} \propto v^{-1}$.

d)   *Resonance* neutrons (1 eV–100 eV).

For these neutrons the cross section frequently shows numerous resonances.
2)   *Intermediate* (100 eV–500 keV). In this region most of the resonances are becoming smoothed out as they become broader and more numerous.
3)   *Fast* (> 500 keV). The cross section is again quite smoothly varying but different experimental techniques from those used for the intermediate energy neutrons are now appropriate.

The general classes of techniques used for neutron measurement can be grouped as follows:
1)   *Activation* (§ 11.2). The nuclear reaction results in a radioactive product and the β- or γ-radiation is detected. Only crude neutron energy measurement is possible.
2)   *Nuclear reaction product* (§ 11.3). The product radiation of (n, γ), (n, α), (n, p) or (n, fission) reactions is measured directly. In general it is not useful for neutron spectroscopy except in one or two cases.
3)   *Hydrogen recoil* (§ 11.4). The recoil proton resulting from elastic scattering by the neutron is detected. When its energy and the reaction angle are defined the neutron energy is determined from the kinematics.
4)   *Direct velocity measurement* (§ 11.5). The time-of-flight of the neutron over a known distance is measured electronically and a momentum spectrum can be derived $(v = L/t)$.
5)   *Wavelength measurement* (§ 11.6). The angle for Bragg diffraction (in a crystal) of the de Broglie waves associated with the neutron can be related to the de Broglie wavelength $\lambda = h/mv$ so a momentum spectrum can again be determined.

11.2   Activation methods. A number of elements have very large cross sections for capture of thermal neutrons which result in radioactive products. For example

$$^{115}\text{In}(n, \gamma)^{116}\text{In}$$

$$^{116}\text{In} \xrightarrow{\beta} {}^{116}\text{Sn}, \ t_{\frac{1}{2}} = 54 \text{ min.}$$

Thus a foil of indium containing N atoms of ^{115}In in a neutron flux of $f/\text{cm}^2 \cdot \text{sec}$ will show a rate of increase of ^{116}In atoms

$$dn/dt = \sigma N f - \lambda n \tag{4.15}$$

where σ is the cross section for the capture reaction and λ is the decay constant of ^{116}In. Thus

$$n(t) = \frac{\sigma N f}{\lambda}(1 - e^{-\lambda t}). \tag{4.16}$$

The *activity* of the ^{116}In is λn and the β-counter with which we count the activity of the foil will give a rate varying with time proportional to $n(t)$. If we irradiate for a time $t_0 \gg 1/\lambda$ (equilibrium condition) then $n(t) \to \sigma N f/\lambda$. So if we stop the irradiation at time t_0 the ^{116}In will start to decay

$$N_{t+t_0} = N_{t_0} \mathrm{e}^{-\lambda t}. \tag{4.17}$$

Thus by measuring the decay rate we can derive N_{t_0} from which we can deduce the flux f if we know σ, N, λ and the efficiency of the β-detector.

Such activation measurements are widely used to determine flux distributions through a reactor core and shield. A variety of detectors are used with various half-lives and cross sections. One must avoid reactions with low lying resonances as these will falsely enhance neutrons of the resonant energy. To ensure that only thermal neutrons are detected the foil may be surrounded by Cd which has a strong capture reaction cross section (3000 b) for neutrons of energy less than about 0.4 eV. Any activity observed in this case must be due to neutrons above 0.4 eV, that is, epithermal neutrons, and correction can be made.

Normally activation tells us nothing about the neutron energy. However, various detectors can be chosen which have a *threshold* energy for production of a certain activity, generally a (n, p), (n, α) or (n, 2n) reaction. If we chose a series of these whose cross section as a function of neutron energy is known, we can use their relative activations to derive a shape for the neutron spectrum. This so-called *spectral index* method is used in the study of fast reactor spectra.

11.3 Nuclear reaction product method. In this case a nuclear reaction is again induced by the neutron but instead of detecting the subsequent decay of the product, we detect promptly the charged particle or γ-ray emitted in the reaction. In this case neutrons may be said to be 'counted' although it is the reaction product which produces the electrical pulse.

Several neutron reactions with light elements result in a charged particle and have large cross sections. The commonest are

$$^{10}\mathrm{B} + \mathrm{n} \to {}^{7}\mathrm{Li} + \alpha + 2.8 \text{ MeV},$$
$$^{6}\mathrm{Li} + \mathrm{n} \to {}^{3}\mathrm{H} + \alpha + 4.6 \text{ MeV}.$$

Natural boron contains 20 % of ^{10}B and 80 % of ^{11}B. It is the ^{10}B which has the large cross section for absorption of thermal neutrons by boron. For natural boron the cross section is 755 b; the isotopic ^{10}B cross section is thus $\frac{100}{19} \times 755 = 4000$ b. The boron can be introduced in gaseous form as $\mathrm{BF_3}$ into ion chambers or proportional counters or mixed with a scintillation material. The resulting α-particle is detected with about 100 % efficiency through its ionization. The cross section for slow neutrons varies as $1/v$ where v is the neutron velocity. Separated ^{10}B is often used to increase the efficiency.

In the case of the lithium reaction again only one isotope ^{6}Li is responsible for the large cross section. Its isotopic cross section is 945 b (abundance 7.4 %). Again ^{6}Li may be used to coat the inside of an ion chamber, as a scintillator constituent or mixed into a nuclear emulsion.

Although counters based on these reactions are most efficient for slow neutrons because of the $1/v$ dependence of the cross section, they can also be used to detect faster neutrons by surrounding them with materials which contain hydrogen (e.g. paraffin wax or polythene) so that elastic collisions of the neutrons with hydrogen nuclei reduce the neutron energy to a point where the counters are efficient.

Another reaction is the fission process. Here we detect the heavily ionizing fission fragments as a signal that a neutron has induced a fission. ^{235}U coated on the plates of an ion chamber will make the chamber sensitive to slow neutrons ($\sigma_f \approx 600$ b). ^{238}U similarly used will be sensitive only to neutrons of energy greater than about 1 MeV. Fission chambers can be made very small (diameter 1 mm) since the range of the fission fragments is so short and the dE/dx is so large.

None of the reactions mentioned is very convenient for neutron energy measurement, since the charged particle energy even for zero energy neutrons is already several MeV. Hence the difference in the α-energy due to the kinetic energy of the neutron is a small fraction of the total and we also have to know the angle at which the α-particle emerged with respect to the neutron direction. However the ^{6}Li reaction has been so used in photographic plates where energies and angles for the two products can be determined and in solid state ion chambers where both the ^{3}H and ^{4}He particles are absorbed and the energy resolution is good.

A more suitable reaction for energy determination is ^{3}He$+$n$\rightarrow$^{3}H$+$p; $Q=$ $+0.77$ MeV. Here the Q-value is smaller and the ^{3}He cross section is large. The ^{3}He (a rare isotope of helium which can be artificially produced) gas is contained in an ion chamber or proportional counter or between two solid state ion chambers in coincidence. Neutron energy resolution of 5–10 % is possible for fast neutrons in the energy range 0.5–5 MeV.

11.4   Hydrogen recoil method. When a neutron of energy E_n collides with a proton, the proton will recoil at angle ψ in the laboratory system, with energy $E_p = E_n \cos^2\psi$. For energies less than about 10 MeV the n–p scattering cross section is isotropic in the centre of mass system. Thus the differential cross section for scattering through angle θ is

$$d\sigma = \frac{\sigma(\theta)}{4\pi}\,d\Omega = \tfrac{1}{2}\sigma(\theta)\sin\theta\,d\theta. \tag{4.18}$$

But we recall that $\theta = \pi - 2\psi$. Then in the laboratory system

$$d\sigma = -2\sigma(\theta)\sin\psi\cos\psi\,d\psi = \frac{\sigma(\theta)}{E_n}\,dE_p. \tag{4.19}$$

Remember that $\sigma(\theta)$ is a constant as the angular distribution is isotropic.

Thus the *proton* energy distribution from a thin target containing hydrogen bombarded by monoenergetic neutrons is uniform up to the neutron energy E_n (fig. 4.25). Hence we can determine the neutron energy.

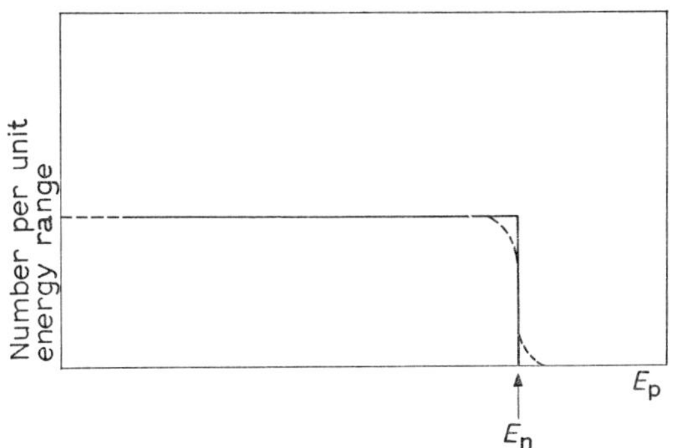

Fig. 4.25.   Hydrogen recoil energy distribution for monoenergetic neutrons

By observing only protons projected into a small solid angle near 0° we essentially differentiate this curve and can observe a neutron energy spectrum.

Many schemes based on hydrogen recoil have been used to study fast and intermediate energy neutrons. Hydrogen filled ion chambers, hydrogen-containing scintillators or organic crystals or plastics, hydrogen in nuclear emulsions or cloud chambers have all been used. Thin hydrogenated foils viewed by a counter telescope have been used in which the telescope defines a small solid angle. The main difficulty in hydrogen-filled ion chambers is correction for the wall effect in which protons which strike the wall give a smaller pulse size. High pressure ion chambers reduce this effect.

For flux measurement the *current* in an ion chamber (rather than individual pulses) may be measured in which the walls of the chamber have the same composition as the filling gas and have a thickness greater than the proton range. For example the chamber may be filled with ethylene and lined with polyethylene. In this case the ion current is

$$i = f \frac{\varepsilon \Omega}{\omega} N \sigma \bar{E} \tag{4.20}$$

where f is the neutron flux density, $\bar{E}$ the mean recoil energy, Ω the chamber volume, N the number of hydrogen nuclei/cm^3, σ the n–p scattering cross section for the given neutron energy, and ω is the energy for forming one ion pair.

11.5   Direct velocity measurement. This is usually referred to as the time-of-flight technique. It is used throughout the neutron energy range. Basically a pulsed source of neutrons, a flight path and a detector of neutrons is used. The time between the neutron burst and the detector pulse then gives the velocity since the source–detector distance L is known, $v = L/t$. In the case of slow neutrons the pulsing may be accomplished mechanically by chopping the beam using a rotating collimator (fig. 4.26). The neutrons can only pass through when the slit is in line with the

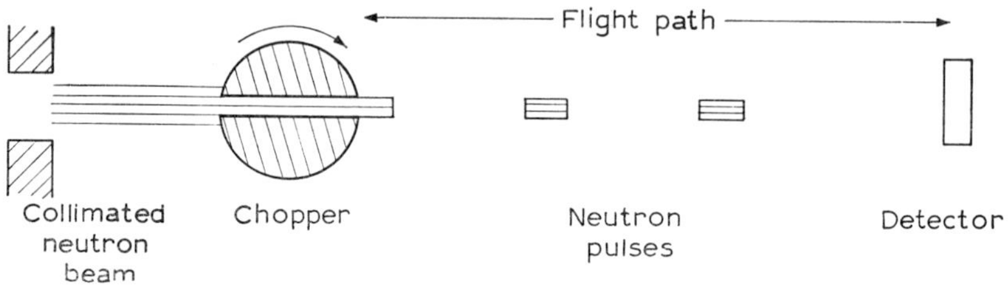

Fig. 4.26a.   The neutron chopper: experimental layout

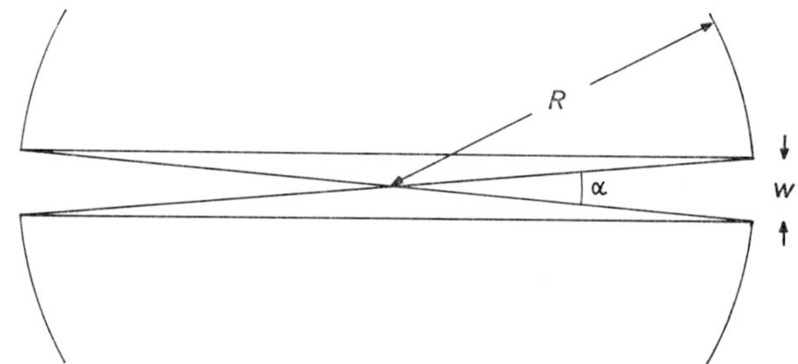

Fig. 4.26b.   The neutron chopper: the rotating collimator

beam direction. For example with a rotor diameter of 20 cm and slit width of 0.1 mm the angle α is 10^{-3} radian. At an angular velocity of 30000 rpm

$$\Delta t = \frac{10^{-3}}{2\pi} \times \frac{60}{30000} = 0.3 \times 10^{-6} \text{ sec.}$$

The flight paths are generally 10–100 m in length. The detector is typically an array of BF_3 counters. Since thermal neutrons have a velocity of 2200 m/sec, the time delays to be measured are in the range 5–50 milliseconds. The energy resolution is

$$\frac{\Delta E}{E} = \frac{2\Delta p}{p} = \frac{2\Delta t}{t} = \frac{2(\Delta t_{\text{pulse}}^2 + \Delta t_{\text{detector}}^2)^{\frac{1}{2}}}{T} \tag{4.21}$$

where Δt_{pulse} is the duration of the neutron burst and $\Delta t_{\text{detector}}$ is the uncertainty in the timing measurement. If many energies are present in the beam we can sort out the neutrons according to their arrival time. $\Delta t_{\text{detector}}$ is thus the time channel width. If $\Delta t_{\text{pulse}} = \Delta t_{\text{detector}} = \frac{1}{3}$ μsec the energy resolution can be a few parts in 10^4 or 10^5.

Rather than producing the neutrons continuously as in a reactor and chopping them, we can produce the neutrons by a reaction with an accelerator beam that is pulsed. Electrons from a linear accelerator or protons from a synchrocyclotron are produced in short pulses. The neutrons produced will extend up to high energies but they can be slowed down in hydrogenous material to give a spectrum of lower mean energy. The intensities of slow neutrons from such arrangements are comparable to or larger than those available from the reactor–chopper arrangement.

Fast neutrons can also be studied by the time-of-flight method, but now the velocities involved are much greater and the times involved are much shorter, since as we shall see the sources of fast neutrons are much less intense so flight distances cannot be made longer. Indeed they are generally much shorter (< 10 m). Table 4.3 shows the time of flight in time/metre equivalent to the inverse velocity.

TABLE 4.3

E_{n}	Time of flight
thermal	455 μs/m
1 eV	72 μs/m
100 eV	7.2 μs/m
10 keV	720 ns/m
1 MeV	72 ns/m
10 MeV	23 ns/m

This means that for neutrons of a few MeV energy, the pulse width must be a few nanoseconds so electrostatic deflection chopping or time bunching must be applied.

For the timing of slow neutrons in the μsec range, electronic scalers may be turned on after the delay time appropriate to the neutron energy a given scaler is to record, or more usually the burst mechanism starts a clock and when an event is recorded the clock is stopped and the binary number of clock pulses is recorded in a computer or on magnetic tape. For the faster timing in the nsec range, various systems have been devised which convert the time delay between start and stop pulses into a pulse height. This can be displayed on a pulse height analyser so that a time spectrum is recorded. For a fixed flight path the time scale is proportional to velocity (or momentum).

An alternative to the pulsed beam system, is to signal the start of the neutron flight by a count from perhaps a charged particle resulting from the neutron production. For example $D+T \rightarrow n + {}^4He$ is a frequently used source of neutrons. We can count the ${}^4He$ particles emerging from the target and this constitutes a start signal for the neutron flight. This is called the *associated particle technique*.

11.6 Wavelength measurement. The wavelength of a neutron in equilibrium with its surroundings of temperature $T\,°K$ is

$$\lambda = \frac{h}{Mv} = \frac{h}{(3MkT)^{\frac{1}{2}}} . \tag{4.22}$$

For thermal neutrons the wavelength is 1.8 Å. (Angstrom unit $= 10^{-8}$ cm). These neutron waves can be diffracted by a crystal just as X-rays except that in this case the neutrons are elastically scattered by the nuclei in the crystal while X-rays are Rayleigh scattered by the atoms of the crystal. Neutrons of wavelength λ obey the Bragg law $\lambda = 2d \sin \theta$ where θ is the diffraction angle and d is the crystal lattice spacing.

Thus by studying the variation of neutron intensity with angle we can determine a neutron spectrum or by working at a fixed angle we can pick out neutrons of a definite energy, using the crystal as a monochromator. However by noting the angles involved as a function of neutron energy we see that the technique can be used only for the slowest neutrons. In table 4.4 we assume a NaCl crystal is used with $d=2.8$ A.U. The method is only practical for energies up to a few tens of eV.

TABLE 4.4

E_n	λ	θ (first order)
0.01 eV	3×10^{-8} cm	31°
1.0 eV	3×10^{-9} cm	3°
100 eV	3×10^{-10} cm	0.3°
10 keV	3×10^{-11} cm	0.03°
1 MeV	3×10^{-12} cm	0.003°

11.7 Flux measurement. We have noted that for most of the methods we have mentioned, the detection efficiency will be dependent on the energy. Frequently we wish to measure the number of neutrons $\cdot$ cm$^{-2} \cdot$ sec^{-1} independent of energy. One technique is to position the neutron source to be measured in the centre of a tank containing an aqueous solution of $MnSO_4$. Essentially all neutrons are slowed down by collisions and, by the reaction ${}^{55}Mn(n, \gamma){}^{56}Mn$, captured in the manganese nuclei. ${}^{56}Mn$ is radioactive. In equilibrium the rate of production equals the rate of absorption. This can be determined by stirring the solution to make the ${}^{56}Mn$ concentration uniform and counting a sample of known volume. By this method source strengths can be compared with an accuracy of 1 % or better.

Another common instrument for energy independent flux measurement is the *long counter* (fig. 4.27). This is simply a BF_3 counter (sensitive to thermal neutrons) placed along the axis of a cylinder of paraffin. Neutrons entering the plane end will travel a mean distance dependent on their energy but will be counted by the counter on the axis with roughly equal efficiency. The response is approximately flat (to about 5 %) up to 5 MeV neutron energy and thereafter falls by about 15 % up to 20 MeV. Such a device can be calibrated quite accurately.

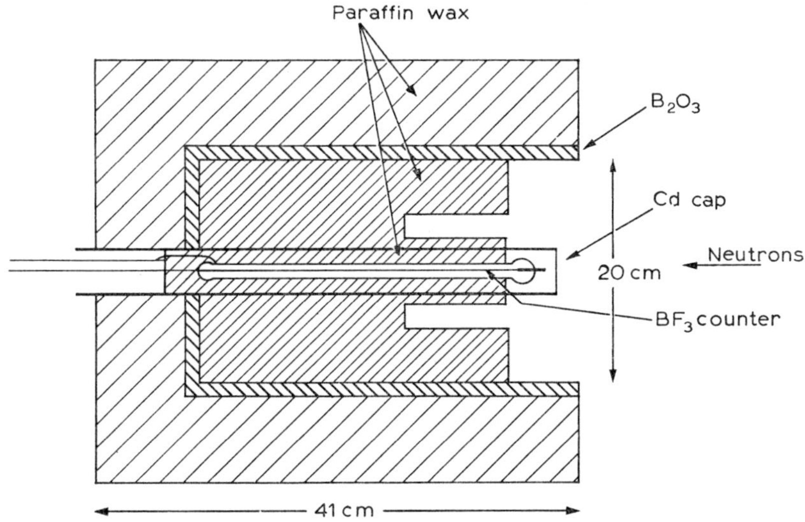

Fig. 4.27.   The long counter

In many of the devices mentioned the detection efficiency for thermal neutrons is very large indeed. Thus if the source under study emits neutrons of higher energy it is important to exclude background thermal neutrons. These may have been slowed down in nearby material in the laboratory such as floor, walls or roof, since wood or concrete contain hydrogen and other light elements. Such thermal neutrons are usually excluded by shielding with cadmium. As mentioned earlier the large capture cross section implies that a thin layer will be an effective shield. The cadmium reaction also leads to a non-radioactive product which is an additional advantage.

12 *Energy and intensity measurement of γ-rays*

12.1 Wavelength methods. From the complexity of the absorption phenomena for γ-rays it is easy to see that absorption measurements are not very useful for estimating energy.

Since they are electromagnetic waves they should be diffracted by crystals and this was indeed used by Rutherford's group at an early stage. However because the wavelengths are very short, e.g. a 1 MeV γ-ray has $\lambda = 0.01$ Å compared to the lattice spacings of a few Å units, it is only useful for the lower energies.

The main application has been by DuMond and his collaborators*. In fig. 4.28,

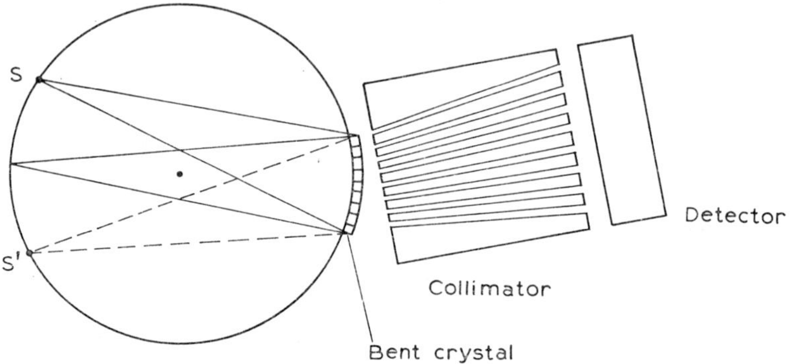

Phys. Rev. **88** (1952) 775

Fig. 4.28.   The bent crystal γ-ray spectrometer

the source S is moved with high precision around the focal circle and the curved crystal (of radius twice that of the focal circle) is also moved in order that the massive collimator–detector arrangement can be left fixed. Only certain crystals can be bent in this way, for example in quartz with $d = 1.178$ Å and for $E_\gamma = 511$ keV, the angle $\theta = 34$ minutes. However extreme accuracy is possible. $R \approx 1\%, \delta E/E \approx 0.04\%$. The efficiency is $\approx 5 \times 10^{-8}$ so the method is only applicable if *very strong* sources are available.

12.2   Magnetic spectrometers. All of these devices are intended to analyze magnetically the electrons resulting from photon interactions: photoelectric, Compton or pair production.

a)   Early work detected the electrons in cloud chambers with a magnetic field.

b)   A lens type β-spectrometer can be used with a radiator at the source position. A foil of a heavy element such as Pb will emphasize the photoelectrons, a foil of a light element will emphasize the Compton electrons.

c)   The intermediate image type of β-spectrometer can be used to measure pairs (fig. 4.29). Again a radiator is in place of the source. The positrons spiral one way, the electrons the other. They can be arranged to focus at slightly different points so we can detect them in coincidence.

d)   180° magnetic pair spectrometer. This is used for energies above ≈ 3 MeV. The momenta of the two particles add up to $E_\gamma - 2mc^2$. The resolution $\delta E/E \approx 0.2\ \%$; the efficiency $\approx 10^{-7}$.

* J. DuMond, Ann. Rev. Nucl. Sci. **8** (1958) 163.

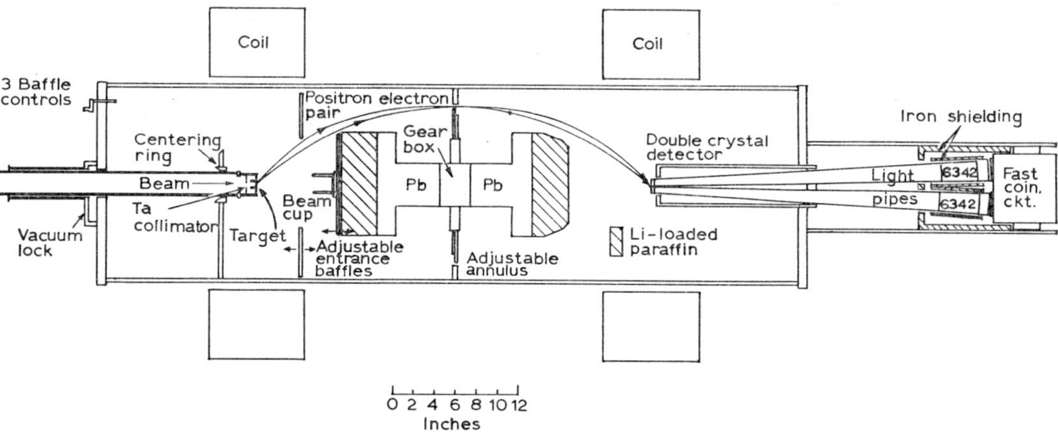

K. Siegbahn (ed.), Alpha-, Beta- and Gamma-ray Spectroscopy **1** (North-Holland, Amsterdam 1965) p. 758

Fig. 4.29.   The intermediate image electron spectrometer used as a γ-spectrometer

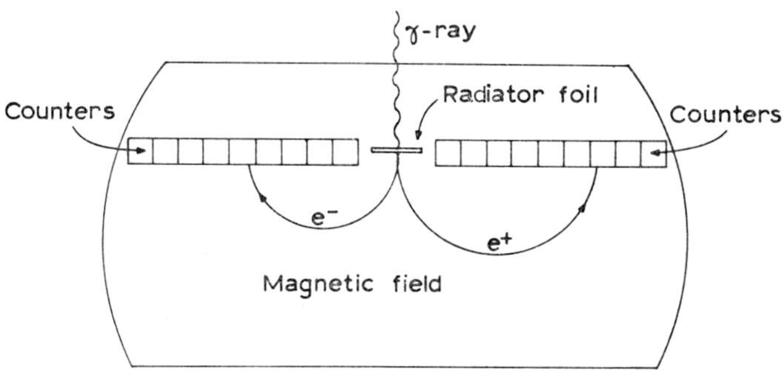

Fig. 4.30.   The pair spectrometer

12.3   Total absorption spectrometers. These are the most commonly used type of spectrometer because of their high efficiency. In general however the energy resolution is worse than that of the instruments already mentioned. Single crystals of Tl-activated NaI coupled to a photomultiplier give energy resolutions of 5–8 % and efficiencies of $\approx 50\%$, depending on the size of crystal and the γ-ray energy. For energies of 1 MeV or less in crystals of 5–10 cm dimension we see (cf. fig. 4.31a) a full energy peak made up of contributions from the photoelectric process and from Compton processes in which the scattered photon is subsequently absorbed. In addition we see a Compton distribution of lower energy pulses corresponding to cases where the scattered photon escapes from the crystal. The scattered photon may carry away up to the full energy of the photon and so the distribution extends to zero pulse height. For energies of several MeV in a similar crystal (cf. fig. 4.31b) we see a full energy peak containing contributions from the photoelectric effect (less probable at high energy), total absorption of Compton processes and total absorption of pair production processes. In the latter case the pulse is due to the ionization of initial electron and positron and to absorption of the two 0.511 MeV γ-rays resulting from annihilation of the stopped positron. We also see peaks at lower energy, one at $E_\gamma - 0.511$ MeV corresponding to the escape of one of the two annihilation photons (single-escape peak) and another at $E_\gamma - 1.02$ MeV corresponding to escape of both of the annihilation quanta (double-escape peak).

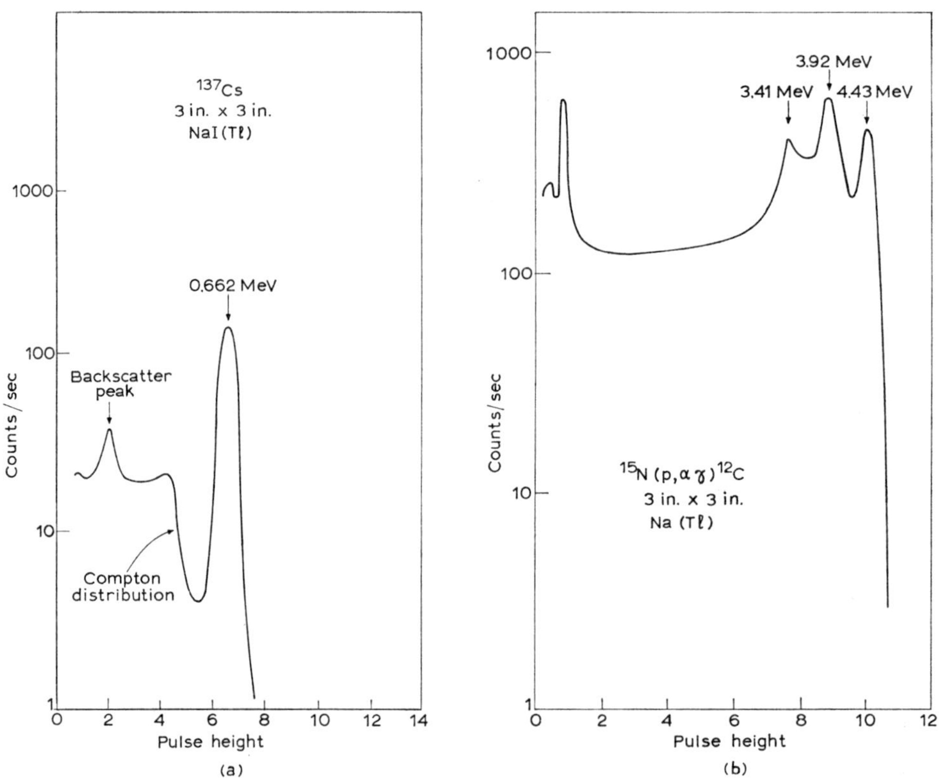

K. Siegbahn (ed.), Alpha-, Beta- and Gamma-ray Spectroscopy **1** (North-Holland, Amsterdam 1965) p. 261

Fig. 4.31.   NaI scintillation counter spectrum for a γ-ray of energy (a) less than $2m_0c^2$, (b) greater than $2m_0c^2$

These escape peaks are superimposed on the Compton escape distribution as before. Obviously as the size of the crystal is increased, complete absorption becomes more and more probable and more of the events result in a full energy pulse. Crystals up to 25 cm dimension have been used. In the case of complex γ-ray spectra in which lines of many different energies are present, the resulting spectra in smaller crystals can be exceedingly complicated and difficult to analyse.

Many schemes have been used involving multiple scintillation detectors to detect the escaping radiations and electronically simplify the pulse spectrum. For

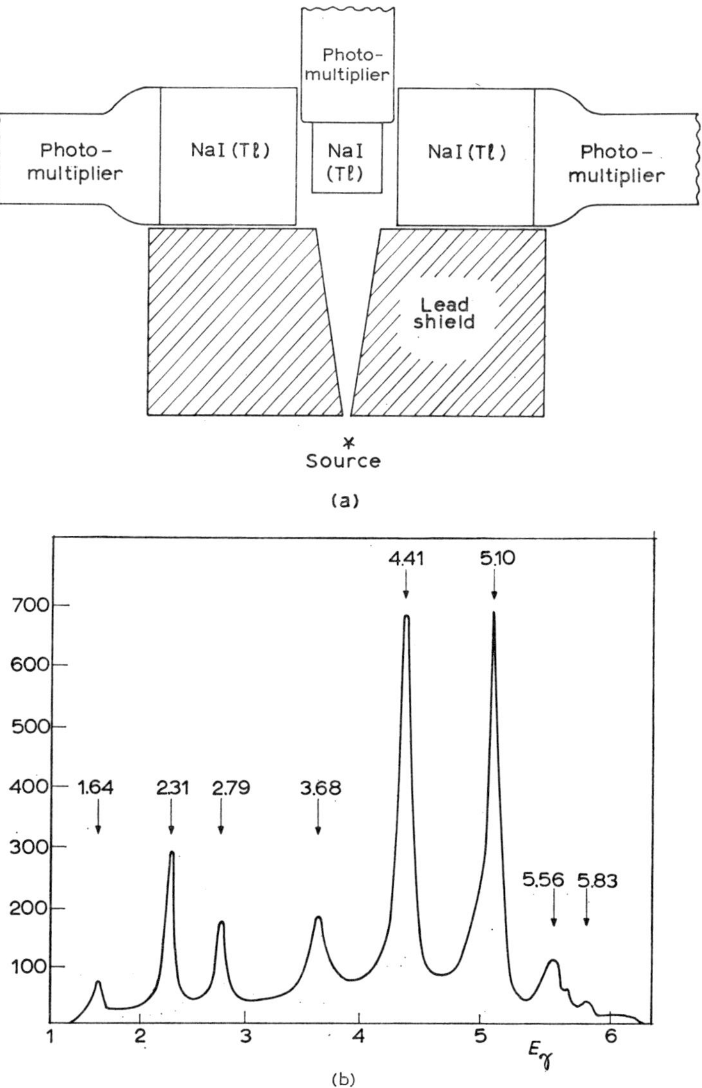

K. Siegbahn (ed.), Alpha-, Beta- and Gamma-ray Spectroscopy 1 (North-Holland, Amsterdam 1965) p. 760, 761

Fig. 4.32. (a) Three-crystal γ-ray spectrometer, (b) Typical spectrum obtained with the three-crystal spectrometer

example three crystals may be used, a small central one and two large crystals, on either side (fig. 4.32a). The side crystals may be arranged to detect only 0.511 MeV radiation resulting from pair production escape. If a triple coincidence is demanded, the gated spectrum from the central crystal will show only the double-escape peak. If a side crystal anticoincidence is demanded the spectrum will show only the full-energy peak. The efficiency is of course reduced to about 10^{-4} and energy resolution may be $\approx 5\,\%$. Multi-crystal Compton spectrometers can be arranged in a similar way.

More recently it has been found possible to make bulk solid state ionization detectors consisting of lithium-doped germanium which are sufficiently large in volume to make γ-ray detection possible with reasonable efficiency (see ch. 4 § 2.4). Detectors of up to 100 cm^3 volume have been produced although volumes of 10–30 cm^3 are more usual. The energy resolution is excellent, $\approx 1\,\%$. A 100 cm^3 Ge(Li) detector would be roughly equivalent in efficiency to a $3''$ diam. $\times 3''$ long NaI(Tl) crystal.

Only one *reaction* has been used extensively to study γ-rays. This is the photo-disintegration of deuterium.

$$^2\mathrm{H} + \gamma \rightarrow \mathrm{n} + \mathrm{p} - 2.23 \text{ MeV.}$$

The cross section is accurately known and thus a high pressure deuterium-filled ionization chamber is useful as a γ-ray flux measuring instrument. If we also arrange to measure the proton energy in the chamber we can deduce something about the γ-ray energy.

5

Accelerators

1 Introduction

We consider now various systems for the artificial acceleration of charged particles to high energies. Many different schemes are in use, each with certain advantages and disadvantages. We require a variety of accelerators to meet the various experimental requirements of nuclear physics.

In 1930 the first serious attempt to observe artificial nuclear disintegration was made by Cockcroft and Walton. Work had been in progress for the previous five years on methods of achieving high energy and high voltage. The aim had been to achieve at least 5 MeV as it was expected that at least this energy would be required to overcome the coulomb repulsion of the nuclear charge. But in 1928 Gamow, Gurney and Condon had proposed the quantum mechanical penetration of potential barriers and Rutherford realised that much more modest energies could suffice. He therefore encouraged Cockcroft and Walton to build a device to produce protons of only a few hundred kilovolts energy. They were able to observe the α-particles from the reaction ^{7}Li(p, α)^{4}He by bombarding lithium with 300 keV protons.

2 Direct current accelerators

2.1 Cockcroft–Walton accelerators. To produce the required high voltage Cockcroft and Walton used a multiplying circuit now named after them (fig.

165

5.1a). Its operation may be seen as follows. The ac voltage of peak voltage V produced by the transformer charges up C_1 to voltage V through rectifier R_1. The voltage across R_1 varies from 0–$2V$ during the cycle and this charges C_2 up to $2V$ via rectifier R_2. Again the voltage across R_2 varies from 0–$2V$ and the process continues. Points P_1, P_2 and P_3 have fixed potentials with respect to earth. Hence we have $2V$ at P_1, $4V$ at P_2 and $6V$ at P_3. We get a final dc potential of $2nV$ where n is the number of stages ($n=3$ in our example).

If current i is drawn from P_3 we get a voltage drop across the last condenser of $(\tfrac{2}{3}n^3 + \tfrac{1}{2}n^2 + \tfrac{1}{3}n)i/fC$ where f is the frequency and C is the capacitance value. There will be a voltage ripple in the output of magnitude $\tfrac{1}{2}n(n+1)i/fC$. These features limit the number of stages n, although by increasing the frequency this limit can be extended.

This type of generator is very suitable for delivering large currents and many are in use up to about 1 MeV. For higher potentials the physical size becomes unreasonable. Sometimes the whole device is enclosed in high pressure gas to reduce the size and increase the possible voltage gradient. However 2–3 MV seems to be the highest practical potential for such accelerators.

The Cockcroft–Walton generator was connected to a three stage acceleration tube with an ion source at the top and three acceleration gaps (fig. 5.1b). The tube was evacuated and connected to the doubling circuit via a filtering network to reduce the voltage ripple.

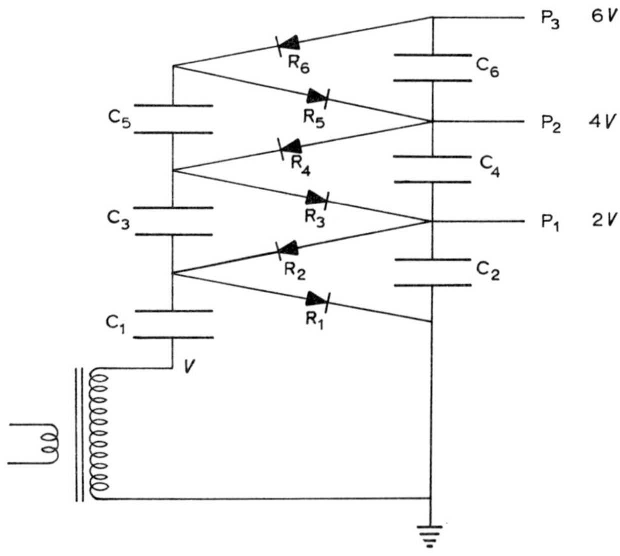

Fig. 5.1a.   The Cockcroft–Walton accelerator: circuit

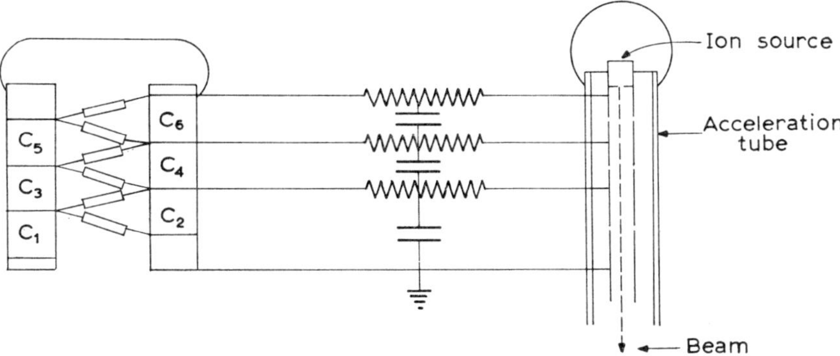

Fig. 5.1b.   The Cockcroft–Walton accelerator: lay-out

The Cockcroft–Walton is one of the class of accelerators in which a potential V is produced by some means and the ions fall through the potential thus acquiring energy zeV where ze is the ionic charge. The beam so produced is obviously dc. We call such devices dc or potential drop accelerators. Another class of accelerators achieves the high ion energy by applying successive small accelerations to the ion until it reaches energy $NzeV$ where V can now be quite small if the number of accelerations N is large. The beam current may still be continuous but in general will be 100 % modulated by the frequency of alternating voltage applied to the accelerator gaps. We call this a C.W. beam. It may also be that the conditions for repetitive acceleration can only be achieved for a particular group of ions at a time. The beam is then pulsed with a repetition frequency depending on how often a new group of ions can be started out. Such machines have a duty cycle which is $\frac{\text{pulse time width}}{\text{repetition time}}$ and is less than 100 %. Dc machines have a 100 % duty cycle.

2.2   Van de Graaff accelerators. These were developed at about the same time, 1930, by R. J. Van de Graaff. The high voltage is produced by a modification of the Wimshurst electrostatic disk machine. Lord Rayleigh had originally suggested using a fast moving belt of insulating material as a means of transferring charge, and it was this method that Van de Graaff adopted. Other methods of charge transfer have been tried, rotating disks or cylinders, falling steel balls or dust blown through insulated channels. The rotating cylinder method has been developed by a French company for accelerators up to about 2.5 MeV and these are quite successful. The dust scheme has been used once or twice for accelerators. However the belt system has been by far the most successful and practical.

The principles of operation are as follows (fig. 5.2). The moving belt passes over pulleys, one at ground potential, the other at the top of an insulated stack. Sharp points placed near the belt at the earth end are maintained at a positive potential of about 50 kV. Ionization of the gas occurs at the high field region near the point and the belt acquires a positive charge on the outer surface of the belt. This charge is then transferred on the moving belt to the inside of the high voltage terminal. Once inside the conducting terminal another set of points collect the charge from the belt and the terminal becomes positively charged with respect to earth. If the capacity of the terminal with respect to earth is C and charge q is deposited, the potential is $V=q/C$.

The charge from the terminal will leak off in various ways a) through supporting insulators   b) through the surrounding gas   c) down the belt itself   d) via the accelerated ion beam if connected to an accelerator. When the current being supplied by the charging supply is equal to the various leakage currents we get equilibrium.

The whole device is normally enclosed in a pressure vessel containing up to 20 atm of very dry nitrogen and CO_2 to reduce breakdown to the surroundings (fig. 5.2b). The insulated stack is normally subdivided into many short insulators separated by conductors. The conductors are connected by high resistances so that the potential gradient is kept constant down the stack. This is necessary to avoid insulator breakdown.

The accelerator is arranged as follows. The voltage is controlled by providing a variable load current from corona discharge of needle points facing the terminal. This is controlled by the orbit of the accelerated beam in a magnetic field. If the

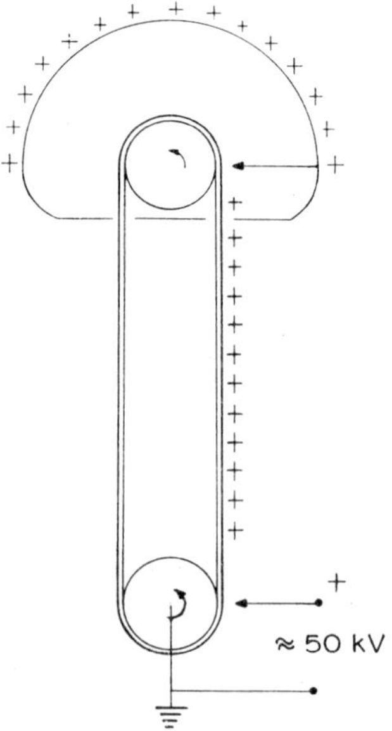

Fig. 5.2a.   The Van de Graaff accelerator: schematic

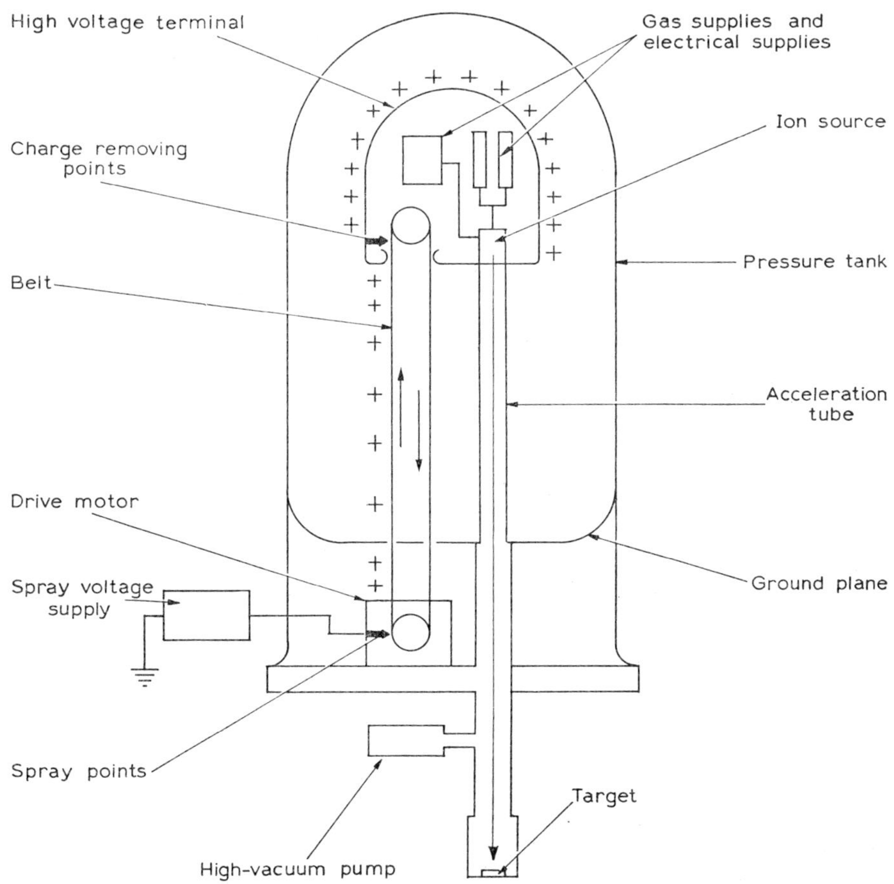

I. Kaplan, Nuclear Physics (Addison-Wesley, 1963) p. 682

Fig. 5.2b.   The Van de Graaff accelerator: layout

energy rises the beam will strike the lower slit plate and the load current is increased. If the energy falls the beam will strike the upper slit plate and the corona current is decreased to compensate. In this way the beam energy can be controlled to 1 part in 5000 or better. Continuously variable energy settings are thus possible by varying the magnetic field and hence the chosen ion orbit.

The Van de Graaff accelerator is the precision machine in low energy nuclear physics and more than 200 are in use. The limit on the energy available is the voltage which can be applied to the accelerator tube without getting multiplication of secondary electrons. This is in the neighbourhood of 7 MeV although recent designs have reached 10 MeV. Beam currents of some tens of microamperes are normal. The spacing between the equipotential planes in the stack and acceleration tube is about 2 cm. This is shorter than the mean free path for ionization by a secondary electron in the acceleration tube vacuum. The Van de Graaff stack may be vertical or horizontal.

2.3   Tandem Van de Graaff accelerators. These were introduced in about 1954, to enable higher energy particles to be produced. It did not seem feasible to increase the high voltage on the terminal for the reasons already mentioned. So a trick that had been suggested earlier was adopted. Let us start with negative ions, for example hydrogen nuclei with *two* attached electrons. These can be accelerated up to the positive high voltage terminal of a Van de Graaff (fig. 5.3). If at this point they are passed through some matter so that all the electrons are stripped off, the ion left is a positive proton, which can be accelerated back down to earth in the same potential. Thus we have two Van de Graaff stacks with a common high voltage terminal.

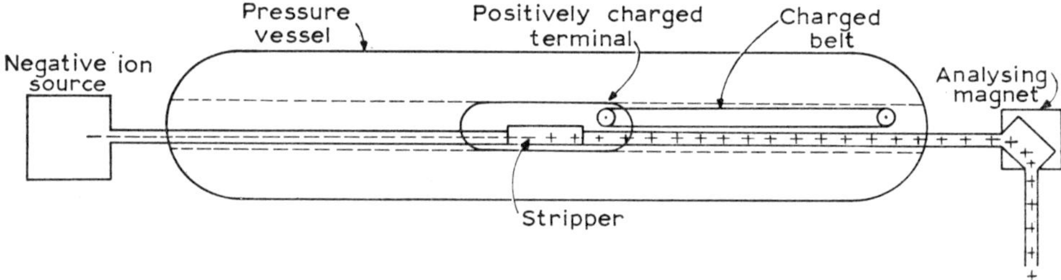

Nucl. Instr. Methods **8** (1960) 195

Fig. 5.3.   The Tandem Van de Graaff accelerator

If the terminal is at potential V, the negative ion gets energy eV and the positive ion thus attains energy $2eV$. Hence if $V = 5$ MV we get 10 MeV protons. For heavier ions like oxygen we might start with O^- but strip to O^{5+}. We would then have $eV + 5eV = 6eV$ or 30 MeV oxygen ions. Energy control is still as precise as in the single stage Van de Graaff. Both vertical and horizontal tandems are in operation. The latest versions can develop 10 MV on the terminal. Some 30 tandem accelerators are in use.

3   Ion sources

3.1   Introduction. We should now consider how positive (or negative) ions can be produced for acceleration. In general some form of gas discharge is set up and the required ions are then extracted from it by an electrostatic field, focussed and injected into the acceleration tube. We want the ion production process to be fairly efficient as all unused gas has to be pumped away often down the acceleration tube itself. Many schemes have been devised but the most successful are the following.

3.2   Cold arc sources. [or PIG (Philips Ion Gauge) sources]. A chamber (fig. 5.4a) is maintained at about 20 torr pressure by leaking in gas. An axial magnetic field of about 500 gauss is supplied. Electrons are produced by cold emission from the cathode and spiral along the magnetic lines of force down the axis of the anode cylinder. When they reach the other end they decelerate and reverse direction. So we get a region of intense ionization along the axis. Ions drift out of the small aperture and are drawn via the extractor electrode into the acceleration region. This source is simple and rugged, can deliver large currents and is easily pulsed.

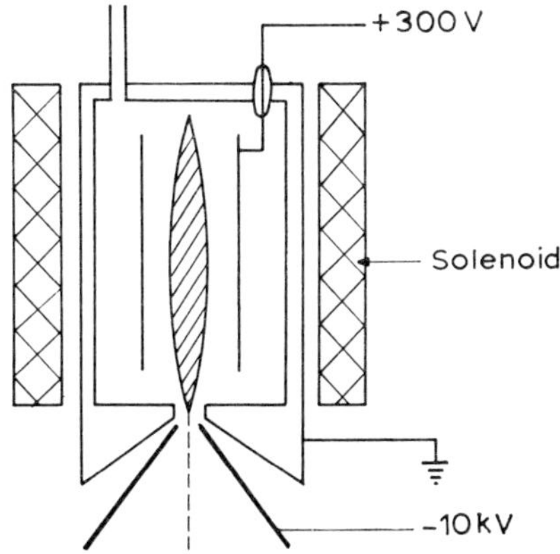

Fig. 5.4a.   Ion sources: the arc source (the intensely ionised region is shown hatched)

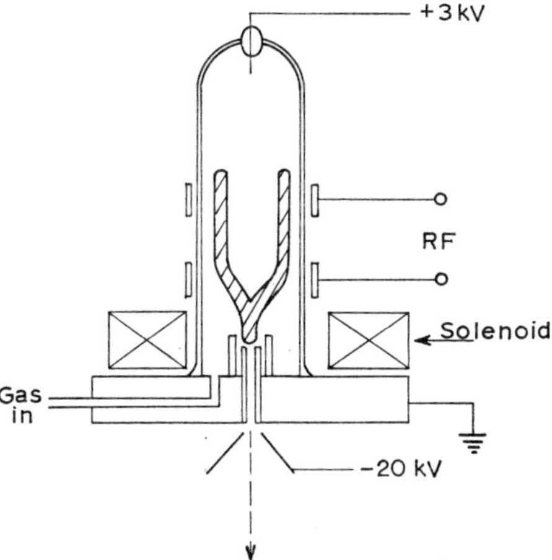

Fig. 5.4b.   Ion sources: the rf source (the intensely ionised region is shown hatched)

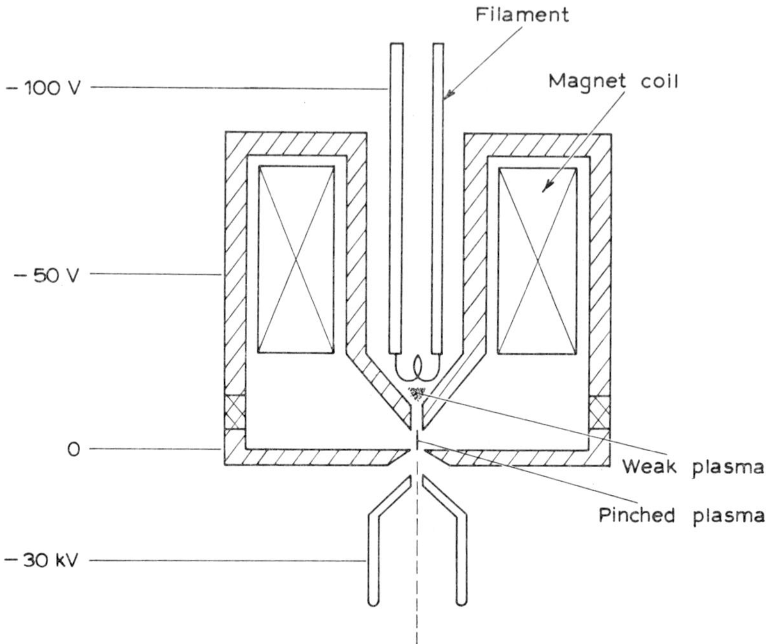

Fig. 5.4c.   Ion sources: the duoplasmatron source

However it produces a rather large percentage of molecular rather than atomic ions.

3.3   Radiofrequency sources. See fig. 5.4b. A glass bottle is kept at a pressure of 10–20 torr. An electrodeless discharge is set up in the gas by exterior electrodes or a coil connected to an rf oscillator running at some tens of MHz. The discharge near the walls is concentrated near the axis at the canal mouth by an electric field between the electrode at the top and the canal tip and by an axial magnetic field. Because of reduced recombination at the glass walls this source gives a high percentage of atomic ions.

3.4   Duoplasmatron sources. See fig. 5.4c. This is a most efficient version of the arc source devised by M. van Ardenne. Electron emission by the hot filament creates a weak discharge in the inner chamber. A strong inhomogeneous magnetic field with axial symmetry is created in the gap between the conical electrode and the flat plate carrying a small aperture. This field pinches the plasma discharge to small dimensions so a filament of 100 % ionized plasma fills the aperture. Extremely large ion currents can then be extracted. Sources for 500 mA have been constructed. The efficiency is in the region of 50 %.

3.5   Negative ion sources. In the usual arrangement a high intensity positive ion beam, often from a duoplasmatron source, is passed through a gas cell at about 50 keV energy. Charge exchange takes place and at this energy a few per cent of negative ions are present, the remainder being positive or neutral. The negative ions are accelerated further and injected into the tandem.

It has recently been found possible to extract negative ions directly from an ionized plasma. In general many electrons are also mixed with the negative ion beam and these must be removed by magnetic analysis before injection.

4   Repetitive accelerators

We now turn to the class of accelerators where many repeated accelerations are produced of the same particle to give a resultant high energy. The acceleration field must be in synchronism with the arrival time of the particle.

4.1   The cyclotron. See fig. 5.5. The cyclotron was developed by Lawrence and Livingston in 1930–31 at the same time as the Cockcroft–Walton and Van de Graaff machines. The particles move in circular orbits in a uniform magnetic field and pass repeatedly through an accelerating region of electric field. As the

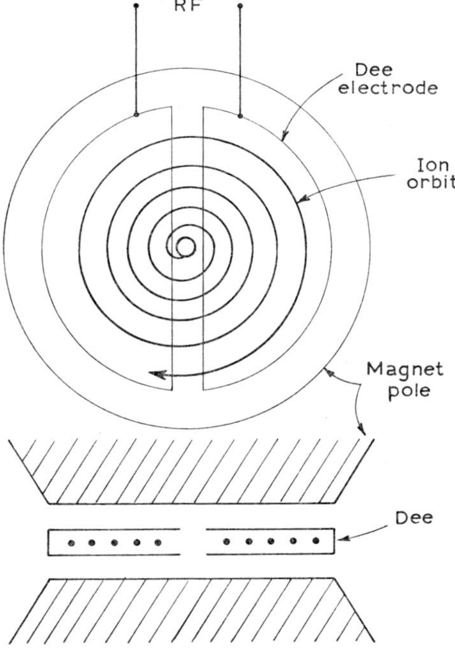

Fig. 5.5.   The cyclotron: schematic

energy is increased the particles spiral outwards in larger and larger orbits. In a magnetic field H a particle of charge ze, mass M and velocity v moves in a circle of radius r such that

$$\frac{Hzev}{c} = \frac{Mv^2}{r};$$

(5.1)

its angular velocity

$$\omega = v/r = zeH/Mc \text{ radians/sec}$$

(5.2)

its momentum

$$p = Mv = zeHr/c.$$

(5.3)

These relations are true for a relativistic particle provided we put $M = M_0/(1-v^2/c^2)^{\frac{1}{2}}$ where M_0 is the particle rest mass. The *total* energy is thus

$$E = Mc^2 = (p^2c^2 + M_0^2 c^4)^{\frac{1}{2}} = [(zeHr)^2 + M_0^2 c^4]^{\frac{1}{2}}$$

(5.4)

and the kinetic energy T is given by

$$T + M_0 c^2 = E$$
$$T(T + 2M_0 c^2) = p^2 c^2 = (zeHr)^2.$$

(5.5)

When $T \ll M_0 c^2$ (non-relativistic)

$$T \approx p^2/2M_0;$$

(5.6)

when $T \gg M_0 c^2$ (relativistic)

$$T \approx pc = zeHr.$$

(5.7)

Now we note that in the non-relativistic case the angular velocity in the field is independent of radius, $\omega = zeH/M_0 c$. The time for one revolution, $2\pi M_0 c/zeH$, is the same no matter what the orbit radius is. This is the important *isochronous* condition. The number of revolutions/sec in the field is $f_0 = \omega/2\pi = zeH/2\pi M_0 c$. This has the value 1.525 MHz per kilogauss for protons. If we reverse the electric field on the dee electrodes with frequency f_0 any particle, no matter what its velocity and orbit radius, will cross the dee gap just as the field is accelerating. We have synchronism. Acceleration continues until it reaches some maximum radius R at the edge of the magnetic field. The kinetic energy it has attained is then

$$T_R = \frac{p^2}{2M_0} = \frac{z^2 e^2 H^2 R^2}{2M_0 c^2} = 2\pi^2 f_0^2 M_0 R^2.$$

(5.8)

All this analysis is however for the non-relativistic case ($v \ll c$) but as we increase the particle velocity its mass is increasing also: $M = M_0(1-v^2/c^2)^{-\frac{1}{2}}$. Thus we can only preserve the resonant condition

$$f_0 = \frac{zeH}{2\pi Mc} \tag{5.9}$$

by *decreasing* the frequency for constant H or by *increasing* the field H with radius at constant frequency.

For a *fixed frequency cyclotron* we might arrange that H increases with radius from its value H_0 at the centre

$$H = \frac{2\pi Mcf_0}{ze} = \frac{H_0}{(1-v^2/c^2)^{\frac{1}{2}}} = H_0 \frac{E}{M_0c^2}. \tag{5.10}$$

M_0c^2 for protons is about 1000 MeV so for a 10 MeV proton, H would have to increase by about 1 %. The field shape might be as shown in fig. 5.6a. However we have to consider what this field gradient will do to the directions of the ions. In fact they will experience forces pushing them away from the median plane, that is, there will be a defocussing. We can only have focussing action if the magnetic field *decreases* with radius (fig. 5.6b).

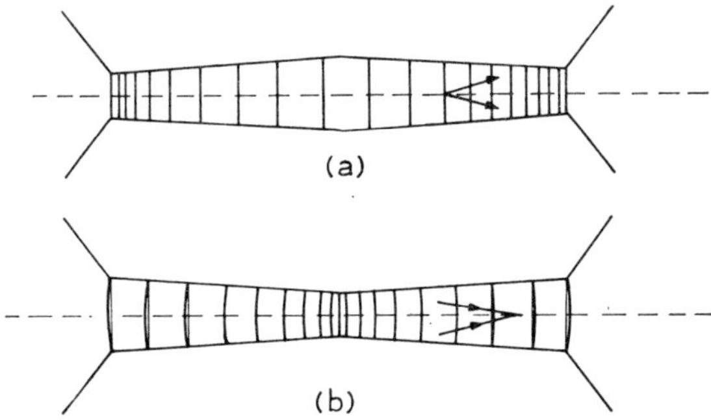

Fig. 5.6.   The cyclotron pole gap, (a) vertical defocussing, (b) vertical focussing

Hence for the fixed frequency cyclotron we have to use a magnetic field gradient such that focussing is achieved and be content with an energy limited by the ions getting out of synchronism as the mass increases.

When the ions start out from the ion source (usually an arc source) at the centre, the magnetic field is a little too large for resonance, hence the revolution time is too short and the ions arrive at the dee gap too early (fig. 5.7). However eventually the field drops off and the ions move back in phase to resonance and past it. Now the time taken is too long and the particles arrive later and later as the mass increases. Eventually when they have slipped 90° in phase, there is no further acceleration. We arrange to extract the beam just before this point is reached.

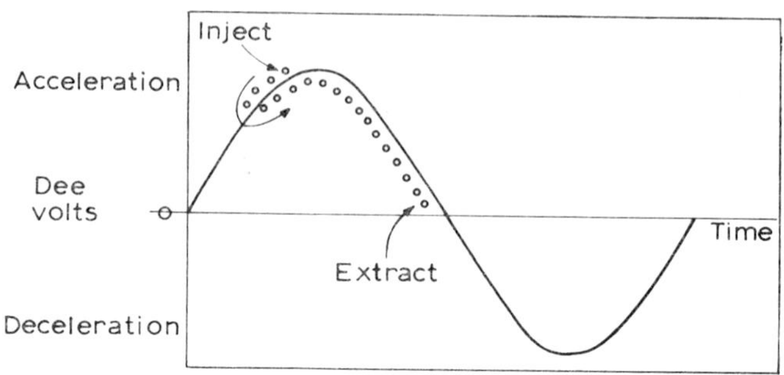

Fig. 5.7. The cyclotron acceleration cycle

This limits the ordinary fixed frequency cyclotron to maximum velocities of about $0.2c$ which corresponds to 18 MeV protons, 36 MeV deuterons or 72 MeV α-particles.

4.2 Oscillation resonances. In a device where charged particles are moving in a magnetic field we can get oscillations about the stable orbit. When the field gradient is $H(r) = H_0(r_0/r)^n$ the frequency of such oscillations in the vertical plane is

$$f_v = f_0 n^{\frac{1}{2}} \tag{5.11}$$

where f_0 is the orbit frequency and in the horizontal plane (radial oscillations)

$$f_r = f_0(1-n)^{\frac{1}{2}} \tag{5.12}$$

We see that for $n>1$ the radial oscillation orbit is not closed while for $n<0$ the vertical oscillation runs away; that is, we cannot have a field increasing with radius. Hence n must lie between 0 and 1. Furthermore when the frequencies of the two oscillations are equal or are small multiples of each other or of f_0 we can get resonances when energy from one oscillation is fed into the other again leading to instability. For example

$$
\begin{aligned}
n &= \tfrac{1}{2}, & f_v &= f_r \\
n &= \tfrac{1}{4}, & f_v &= \tfrac{1}{2}f_0 \\
n &= \tfrac{3}{4}, & f_r &= \tfrac{1}{2}f_0 \\
n &= \tfrac{1}{5}, & f_r &= 2f_v
\end{aligned}
$$

will all give rise to instability and must be avoided.

Notice that the n-value chosen for the double focussing magnetic spectrometer $n=\tfrac{1}{2}$ is just such that the oscillations in the two planes just reach a node at $\pi\sqrt{2}$ radians or 254.6°.

4.3 The frequency modulated or synchrocyclotron. The alternative method for extending the energy range was to decrease the frequency to match the change in mass. This technique is possible because of the principle of *phase stability*.

Consider ions crossing the dee gap of a fixed frequency cyclotron. They will have a range of phase angles with respect to the rf voltage. Ions crossing the gap at zero field (point A in fig. 5.8) will, if the field index is correct so that there are no spatial oscillations, continue in this orbit indefinitely and be phase stable. Ions

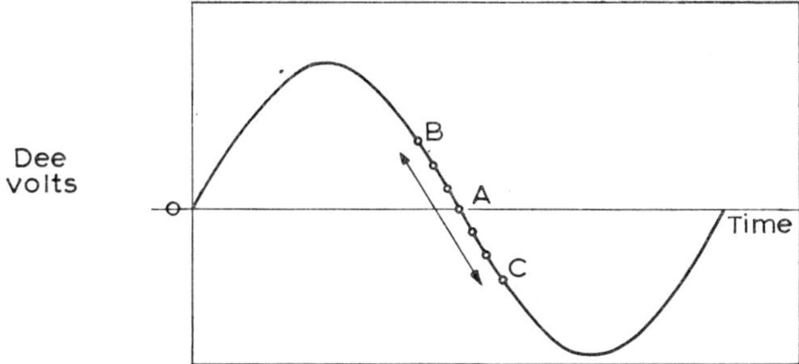

Fig. 5.8.   Phase stability

crossing earlier (point B) will get an acceleration, their energy will be greater than that for resonance and their orbit frequency will be less. Thus the time to the next crossing will be longer and they will arrive at a phase position closer to A. Similarly ions arriving late (C) will arrive earlier next time, and also move toward A in phase. Thus a stable *phase oscillation* will be set up about point A.

If now we decrease the frequency applied to the dees, the phase stable bunch will again see a net acceleration (fig. 5.9) and will start to move toward the new phase stable position. So if the change in frequency is slow compared to the phase bunching effect we can get an indefinite increase in energy limited only by beam blow-up when we reach a resonant *n*-value. The existence of phase stability means many more turns can be made in a synchrocyclotron, so that the dee voltages can be smaller. The beam can be extracted by introducing small regions with different

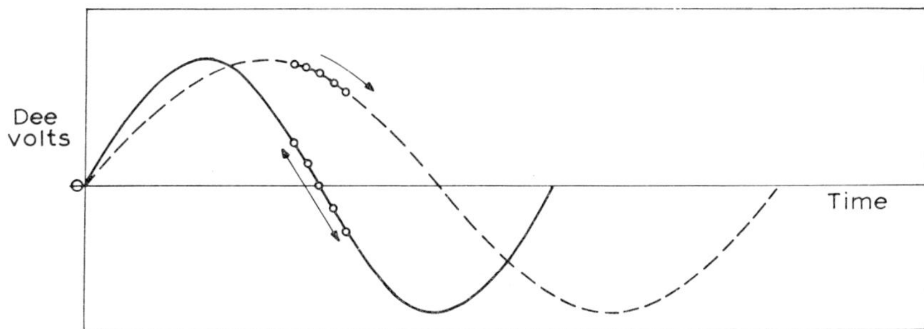

Fig. 5.9.   The synchrocyclotron acceleration cycle

field index so the radial oscillations are enhanced and the beam spirals out into a deflector. The orbit spacings are very small. The limit on synchrocyclotrons is mainly size and cost of the magnet. Machines for proton energies up to 700 MeV have been built. The ions are now delivered in bursts as only one phase stable bunch can be accelerated at a time. After extraction the frequency has to be returned to its original value and we can start another acceleration cycle. Pulse repetition rates of 100/sec are typical with a duty cycle of about 1 % and mean currents of a few μA.

4.4 The fixed frequency azimuthal gradient cyclotron. Still another approach to the problem of extending the cyclotron range has been to let the magnetic field *increase* with radius as we originally suggested so that the synchronous condition is maintained with increase in mass. To keep the beam focussed we vary the field gradient with azimuth rather than with radius by the use of ridges on the pole faces. The ridges may be radial or spiral (fig. 5.10). These field variations can be arranged to compensate for the defocussing effect of the radial variation. The device is thus operated at fixed frequency like a normal cyclotron. The ions now no longer lose phase as in the radially decreasing field of the ordinary cyclotron, and the upper limit on energy is much higer, perhaps 750 MeV. Also it is rather easy to change the energy of the ridged cyclotron so variable energy cyclotrons, even of modest size, are now constructed in this way.

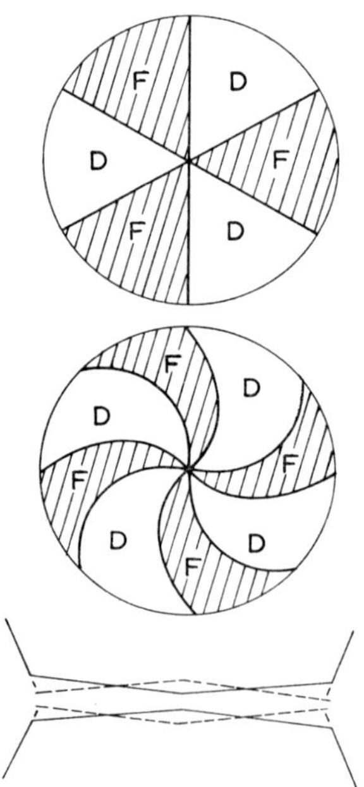

Fig. 5.10.   Fixed frequency azimuthal gradient cyclotron magnet with radial or spiral ridges

4.5   The betatron. This accelerator is characterised by making use of the electric field induced by a changing magnetic field (fig. 5.11). As it is only suitable for electron acceleration it is termed a betatron. The electrons are injected into a toroidal ceramic vacuum chamber with a conducting coating placed in the gap of a specially shaped ac magnet operated at about 100 Hz. The magnetic flux between the pole tips alternates with this frequency.

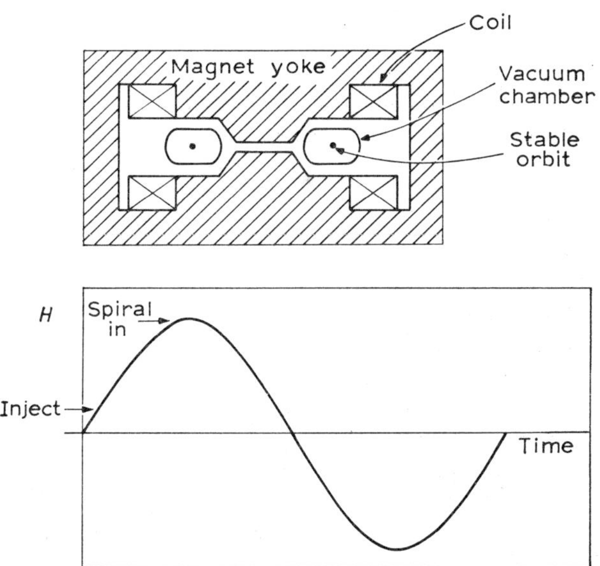

Fig. 5.11.   The betatron and its acceleration cycle

If the electrons move in an orbit of radius R, the emf E acting tangentially around the orbit is related to the change in magnetic flux ϕ through the orbit by $2\pi Re = \phi/c$. This is just the emf found in a loop of wire surrounding the magnet (as in a transformer). If H is the magnetic field at the orbit and p is the electron momentum then $p = eHR/c$ and $\dot{p} = Ee$ or $\phi = 2\pi R^2 \dot{H}$. That is, the electrons will remain in a stable orbit of radius R provided the magnetic flux linking the orbit is changing at just *twice* the rate corresponding to a uniform field across the gap. Thus inside the toroid the gap is smaller so the field is larger than the guide field at the toroid. Saturation of the central section will occur first so the guide field can only rise to half the saturation value. For this reason the maximum betatron energy is limited to about 30 MeV. About 1000 V per turn is acquired. There is no relativistic limit as the particle mass is not included in the resonance condition.

The beam spirals inwards at the end of acceleration. Pulses of beam are delivered at magnet frequency. Internal targets for production of bremsstrahlung are normally used since beam extraction is almost impossible as the orbits are all at the same radius.

4.6   The electron synchrotron. This is closely related to the betatron. Suppose that in a betatron we place an accelerating gap as well (fig. 5.12). Thus we make something like an electron cyclotron. The relativistic cyclotron equation is

$$ f = \frac{ecH}{2\pi E} = \frac{ecH}{2\pi[m^2c^4+(eHR)^2]^{\frac{1}{2}}}. $$

At very high energies where the kinetic energy eHR is very much greater than the rest mass mc^2 (0.511 MeV for the electron) the resonant frequency $f \to c/2\pi R$. That is, the electrons move around in a stable orbit of radius R, with constant frequency.

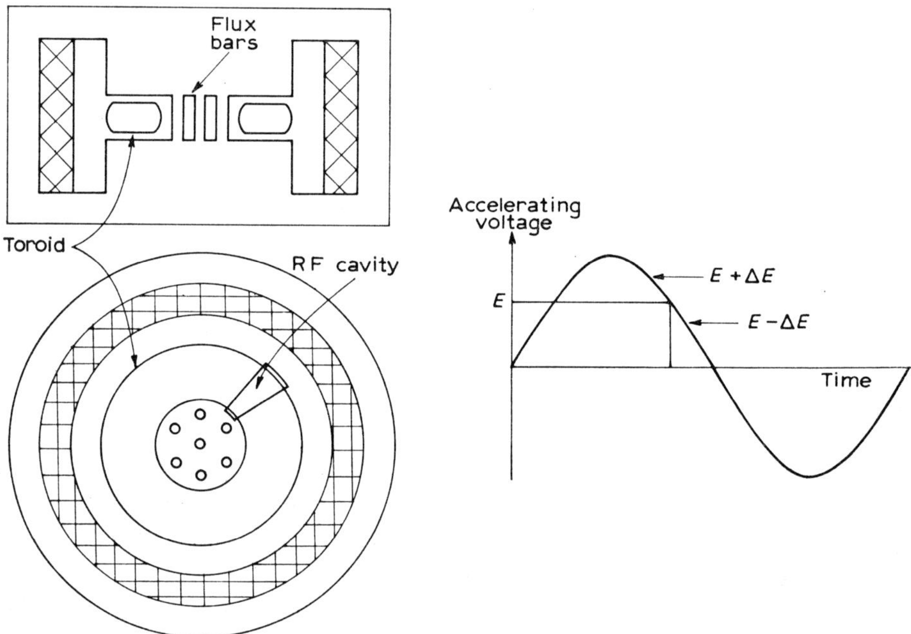

Fig. 5.12.   The electron synchrotron and its acceleration cycle

Now if we increase the energy E by an accelerating element (generally a rf cavity with the electric vector along the axis) the orbit will tend to expand but if we increase H in proportion the orbit radius will remain fixed.

Electrons are injected in a low magnetic field as in a betatron and are trapped into stable betatron orbits. Initial acceleration up to a few MeV takes place by betatron action. Then the flux bars at the centre saturate and the flux linkage rises too slowly thereafter to satisfy the betatron condition. However the electrons are now moving with a velocity of $\approx 0.98c$ so the fixed frequency cyclotron condition applies and the accelerating cavity with frequency $f = c/2\pi R$ is switched on.

It is now phase stable as an electron crossing the cavity too early gets too much energy, will describe an orbit of larger radius and next time arrives later as it is travelling at fixed velocity $\approx c$. Electrons arriving too late will next time arrive earlier. Thus we get phase bunching.

The final energy is $\approx eH_{max}R$ where H_{max} is the peak field. The beam is difficult to extract as in the betatron so again the final orbit is allowed to spiral inwards and strike a bremsstrahlung target on the inner wall of the toroid. The main limit on energy is the energy loss due to radiation by the radially accelerated electrons. This loss increases as $(E/mc^2)^4/R$ and eventually becomes too great per turn for the accelerating cavity to supply.

Many electron synchrotrons in the 100 MeV range are in use. The largest are 6 GeV machines at Cambridge, USA and Hamburg, Germany and a 5 GeV machine at Daresbury, England.

4.7   The proton synchrotron. The synchrotron action can be applied to acceleration of heavy ions if we maintain resonance not only by varying the magnetic field but also by varying the frequency. The protons are not relativistic and so the change in velocity with acceleration is appreciable. We now no longer need the saturation effect so we can use a ring magnet which is very economical when a large machine is built. It may be a continuous ring or an arrangement of sectors with straight sections in between (fig. 5.13). In one class of proton synchrotron the field index is 0.7 with no azimuthal variation to give focussing (so-called weak focussing)

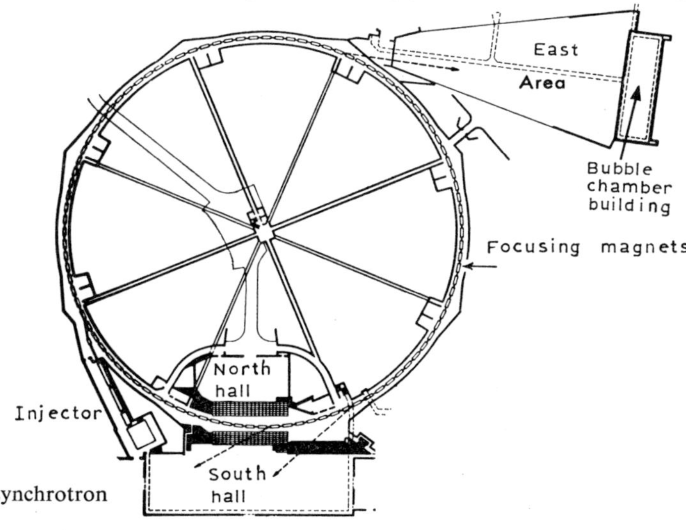

Fig. 5.13.   The proton synchrotron at CERN, Geneva

of the beam. The magnet current is supplied by an ac alternator supplying recti-
fiers. When the magnet is connected the rate of rise of current is given by $V = L\,di/dt$. Usually about 1 second is required to reach maximum current and during
this period acceleration takes place. As the current decays it can be arranged to
feed back into the alternator and speed it up again. Thus only resistive losses have
to be supplied and we can repeat the cycle every few seconds. A number of such
weak focussing proton synchrotrons in the 3–7 GeV range have been built. The
largest is the synchrophasotron in Dubna, USSR of 10 GeV.

The ions are injected at 5–20 MeV by a Van de Graaff or linear accelerator. The
beam may be extracted as in the synchrocyclotron, by introducing small magnets
of suitable n-value.

Still larger proton synchrotrons are constructed using the principle of strong
focussing, i.e. azimuthal field variations. It was discovered in 1952 that a combination of alternate magnet sectors in which n was alternately $+n$ and $-n$ resulted in
a net focussing effect. Moreover n could now be made very large.

An application familiar now in accelerator laboratories is the strong focussing
quadrupole lens. If magnets are arranged as shown in fig. 5.14 then positive ions
will be focussed in the plane BB′ and defocussed in the plane AA′ in the left hand
drawing. If another element as shown in the right hand drawing is placed also in
the beam line then we get focussing in AA′ and defocussing in BB′. The net result
is focussing in both planes as in a converging–diverging achromatic lens combination in optics. The fixed frequency azimuthal gradient cyclotron also involves the
same principle of strong focussing.

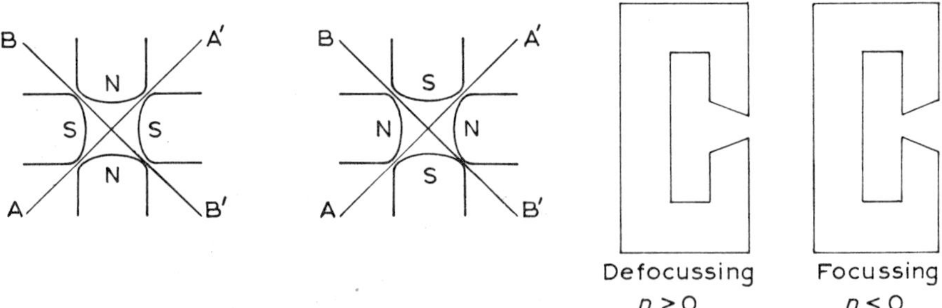

Fig. 5.14.   Strong focussing

In the strong focussed synchrotron the magnet is built of alternate sectors. The
main advantage is that since n can be greater (≈ 300) the amplitudes of vertical and
radial oscillations are much less and the aperture required in the pole gap can be
much smaller. The consequent saving in size and cost of the magnet is enormous.
The largest accelerators are of this type. These include the 28 GeV machine at

CERN, Switzerland, the 33 GeV machine at Brookhaven, USA and the 70 GeV machine at Serpukov, USSR. Some characteristics of the CERN proton synchrotron are listed below:

Magnet ring 200 m diameter

Injection at 50 MeV

Ions acquire 50 keV per turn

H varies from 150–1200 gauss

Ions describe 500 000 turns taking 1 second

About 10^{12} particles/pulse

Repetition rate 12 pulses/minute

Magnet aperture 8×16 cm

Dimensional accuracy of magnet 1 in 10^5, i.e. $\frac{1}{2}$ mm in radius of 100 metres.

4.8   Linear accelerators. We turn now to another class of accelerator in which a particle is given successive accelerations except that separate gaps are in a straight line instead of using one gap through which the ions pass in a magnetic field. The earliest linear accelerator (linac) was devised in 1928 and operated as follows (cf. fig. 5.15).

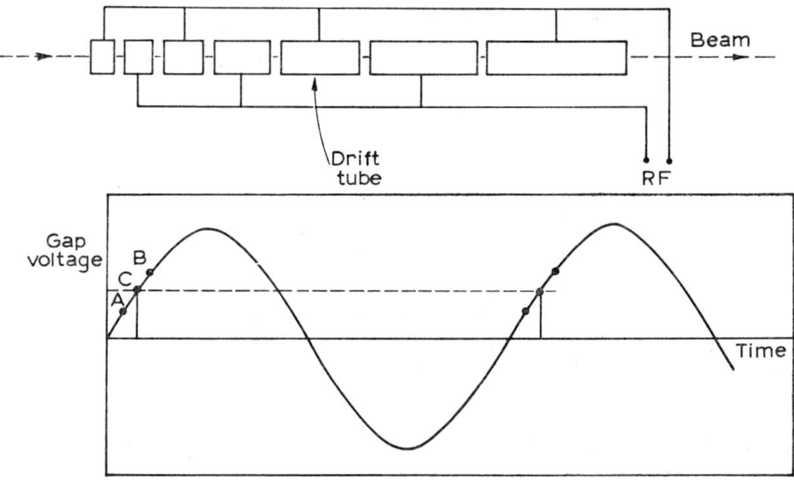

Fig. 5.15.   The linear accelerator and its phase stability

It consists of a series of hollow cylinders called drift tubes with acceleration gaps between. We want the rf voltage to reverse just in time for the particle to reach the next gap. If the particle sees a voltage V at each gap its energy as it reaches the nth drift tube is just $nzeV$ and its velocity $v_n = (2nzeV/M)^{\frac{1}{2}}$. If the rf oscillator frequency is c/λ so the flight time is just one half cycle, the drift tube length $L_n = \frac{1}{2}v_n\lambda/c = \frac{1}{2}\beta_n\lambda$. Thus $L_n \propto n^{\frac{1}{2}}$ and the length of the machine is controlled by λ as well as V. In the early days the frequencies available were not very high and hence only relatively slow ions could be accelerated for practical values of L. But after the development in World War II of high power centimetre wavelength sources for radar, the linear accelerator again became practical. It was also realised at this time that phase stability obtains also in linear accelerators and so the machine could accept particles over a range of phase angles. If the particle arrives early, A, it will get a smaller acceleration, its velocity in the next drift tube will be less and it will arrive later at the next gap. If too late, it will get increased velocity and will arrive early next time. So we get phase bunching about the stable phase position. Note however that the phase stable position is on the rising side of the voltage curve, unlike the synchrotron. Unfortunately the condition for phase stability is not compatible with radial stability of the beam – it results in defocussing.

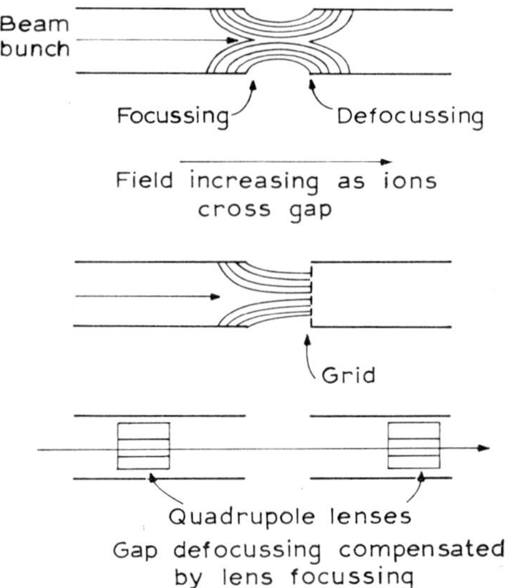

Fig. 5.16.   Focussing in the linear accelerator

The defocussing effect upon leaving the gap is greater than the focussing effect upon entering the gap since the field has increased.

Two methods have been used to get round this difficulty (fig. 5.16):

1)   Grid focussing. In this case a wire grid is placed on the entrance face of the drift tube thus weakening the defocussing effect. It has the disadvantage that beam is lost to the grids.

2)   Quadrupole focussing. In this case quadrupole magnet elements are installed inside each drift tube which compensate for the defocussing at the gaps. This has the disadvantage that the high power magnets inside the vacuum have to be elaborately cooled.

Standing wave linacs. The proton linear accelerator developed by Alvarez in 1948 is arranged as shown in fig. 5.17. The tank containing the drift tubes is excited to give standing waves with the *E* vector parallel to the axis of the tank.

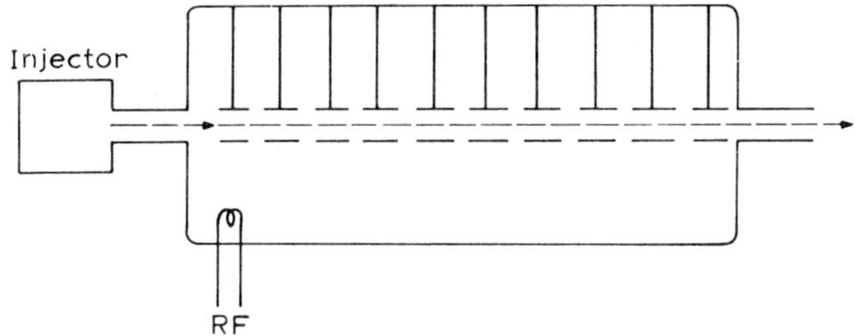

Fig. 5.17.   The standing wave linear accelerator

Such accelerators have been built for protons up to several hundred MeV and for heavier ions up to 10 MeV/nucleon. Frequencies of the order of 100–200 MHz are used. These are pulsed machines with 10–50 pulses per sec. Injectors are Cockcroft–Walton or Van de Graaff accelerators. Starting at fairly high energies means that the variation in drift tube length is not excessive.

Travelling wave linacs. The standing waves in the Alvarez tank can be considered as two waves travelling in opposite directions, one of which accelerates the particles when they appear in the gaps. This suggests setting up a progressive wave in a metal wave guide and accelerating bunches of particles 'on' them. If the particles are relativistic (constant velocity) the wavelength of the accelerating field can then be constant. Hence this system can be applied to electrons.

In the usual waveguide the *phase velocity* exceeds the velocity of light, but this can be reduced by loading the waveguide with, for example, apertures suitably spaced. The electron bunch is then injected just before the crest of the wave at several MeV. They gain energy (mass, not velocity) continuously from the wave. The beam can be easily focussed from outside the waveguide by solenoids.

The frequencies involved are several thousand MHz with wavelength of ≈ 10 cm. Pulse repetition rates of a few hundred a second are usual. Very large peak electron currents of several amperes can be achieved. Many electron linacs are in use in the 30–100 MeV range, to produce pulsed neutrons by (γ, n) and $(\gamma, \text{fission})$ reactions for neutron time-of-flight spectrometry. Larger machines up to 2 GeV are also in use and much of the work on nuclear charge radius measurement has been done with these accelerators. An electron linac 2 miles long is now operating at Stanford, USA, for 50 GeV electrons.

5   *Accelerated beams*

We must now consider how such accelerated beams are used in practice. A most important point is the provision of targets in which the energy loss is equal to or less than the energy resolution desired. We note first the range of protons in air and copper (cf. table 5.1). For 1 % energy resolution our targets must be less than 0.01 of this range.

TABLE 5.1

Ranges of protons

Energy	R_{air}	R_{Cu}
1 MeV	2.2 cm	0.005 mm
10 MeV	110 cm	0.25 mm
100 MeV	67 m	1.4 cm
1 GeV	2.8 km	54 cm
30 GeV	110 km	21 m

To avoid energy losses in transmitting beams from accelerator to target we have to conduct all beams less than about 50 MeV in evacuated pipes. Of course all accelerators must start the beam out in a vacuum and in multiple passage accelerators, like the proton synchrotron (where it travels some 200 000 miles during acceleration) the vacuum must be very good. For a 'thin' target, that is, with energy loss comparable to energy resolution, we need a few micrograms/cm^2 of Cu for a few MeV while for 1 GeV a few millimeters thickness would be thin.

In low energy nuclear physics we frequently wish to study resonances, that is, peaks in the curve of counting rate vs. bombarding energy. We need to establish the positions as well as the width of such resonances, as the width is a measure of decay probability. The *observed* width will depend upon 1) the natural width of the state proportional to decay lifetime, 2) the energy spread or inhomogeneity of the accelerator beam, 3) the spread introduced by finite energy loss in the target. Thus the success of an experiment often depends critically on the ability to prepare a target of acceptable thickness, chemical purity and stability under bombardment by a beam of particles.

6

Nuclear Reactions

1 Introduction

We have referred many times to nuclear reactions. Let us recall the general features. We are primarily referring to changes involving the strong nuclear force. We do not normally include purely electromagnetic interactions such as coulomb scattering or the changes involving the weak force such as β-decay. However changes in nuclear state resulting from the electromagnetic effect of γ-rays are included such as 'γ-ray in – nuclear particle out' or the decay of one nuclear state to another with emission of a γ-ray.

The general form is $A + B \rightarrow C + D + Q$.

We list some general rules.

1) Mass and/or energy is conserved, i.e.

$$(E_A + M_A c^2) + (E_B + M_B c^2) = (E_C + M_C c^2) + (E_D + M_D c^2). \tag{6.1}$$

The energy balance Q is given by

$$Q = (M_A + M_B - M_C - M_D)c^2 = E_C + E_D - E_A - E_B. \tag{6.2}$$

If Q is positive the reaction is said to be *exoergic*, that is kinetic energy is gained, $E_C + E_D > E_A + E_B$. If Q is negative the reaction is said to be *endoergic* that is kinetic energy must be added to make the reaction go, $E_C + E_D < E_A + E_B$. Such a reaction will exhibit a threshold energy below which it will not occur

$$E_{A,\,\text{thresh}} = |Q|\,\frac{M_A + M_B}{M_B}. \tag{6.3}$$

187

2) Charge is conserved.

3) The number of heavy particles (baryons), e.g. protons and neutrons remains constant.

4) Linear momentum is conserved.

5) Angular momentum is conserved. If a spin angular momentum in units of $\frac{1}{2}\hbar$ is associated with each constituent and orbital angular momentum in units of $\hbar$ associated with the relative motion, then the total angular momentum $l+s$ is conserved.

6) Parity is conserved.

The reaction products may be stable or radioactive.

In studies of nuclear reactions we make the following types of measurement.

1) The energy balance Q by a study of the kinetic energies involved.

2) The yield of a given reaction as a function of bombarding energy, often called an 'excitation function' (fig. 6.1a). Features of this tell us about excitations of the intermediate combination system $(M_A + M_B)$ or $(M_C + M_D)$.

3) The energy spectrum of a product particle (fig. 6.1b). Features of this tell us about excitations of the other product or residual nucleus.

4) The angular relation of one product relative to the incident beam direction, termed an *angular distribution* (fig. 6.1c).

5) The angular relation of one product relative to another product termed an *angular correlation* (fig. 6.1d).

We generally classify nuclear interactions into types as follows.

1) $x+X \rightarrow X+x+0$. This is *elastic scattering*. There is no energy change as $Q=0$ and there is no change in particles. An example is $^{12}C(p, p)^{12}C$. This is *nuclear* scattering to be distinguished from electric field scattering called coulomb or Rutherford scattering.

2) $x+X \rightarrow X^*+x+Q$. This is *inelastic scattering*. There is no change in particles but there is a change in energy: $Q \neq 0$, and is negative. The final nucleus is in an excited state (indicated by an asterisk). An example is $^{12}C(p, p')^{12}C^*$ (in the 4.4 MeV first excited state for example).

3) $x+X \rightarrow Z+\gamma+Q$. This is *radiative capture*. The nucleus Z is formed of a combination of X and x. $Q \neq 0$ and is equal to the separation energy of x from Z or to the binding energy of x in Z. An example is $^{12}C(p, \gamma)^{13}N$.

4) $x+X \rightarrow Y+y+Q$. This is the *reaction* $X(x, y)Y$. Both Y and y differ from X, x. Q may have any value. Y is often left in an excited state. Examples:
$^{12}C(p, n)^{12}N$, $^{12}C(p, \alpha)^9B$, $^{12}C(p, d)^{11}C$, $^{12}C(p, t)^{10}C$, $^{12}C(p, {}^3He)^{10}B$,   etc.

5) $x+X \rightarrow Y+y_1+y_2+Q$. This is a *three-body reaction*. Q may have any value, Y is often in an excited state. Examples: $^{12}C(p; p'\gamma)^{12}C$, $^{12}C(p, 2p)^{11}B$, $^{12}C(p; p, n)$ ^{11}C, $^{12}C(p, 2n)^{11}N$ etc.

6) $\gamma+X \rightarrow Y+y+Q$. This is a *photo-reaction*. $Q \neq 0$ and is equal to the separation

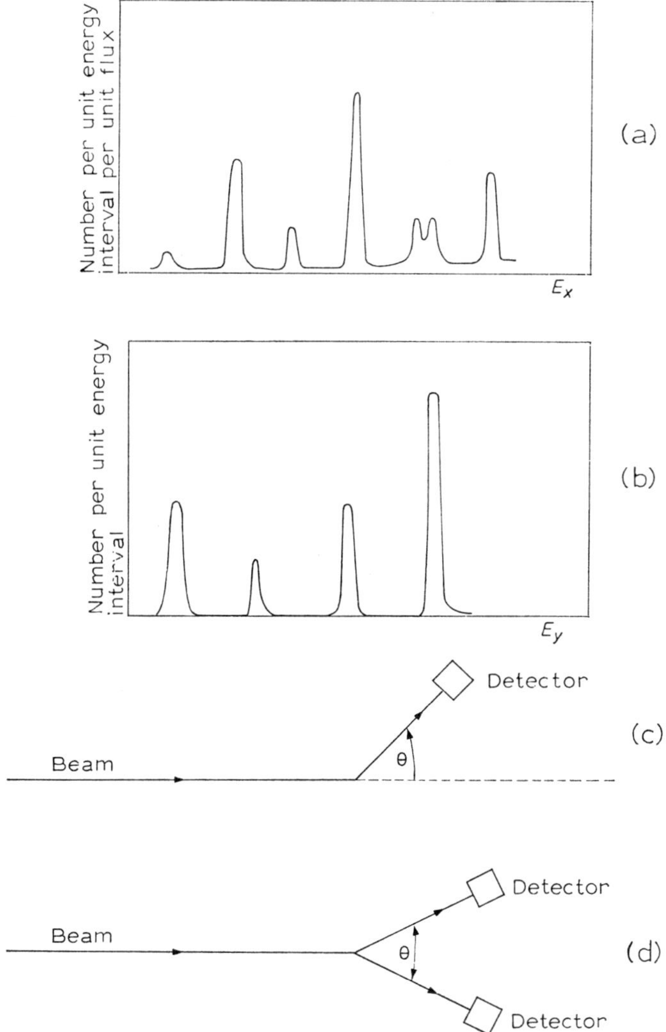

Fig. 6.1.   Observation of nuclear reactions: a) excitation function, b) product particle spectrum, c) angular distribution, d) angular correlation

energy of y from X or to the binding energy of y in X.

When X is a heavy nucleus and y and Y are comparable in mass the reaction is referred to as *fission*. In this case the products will not be unique but will exhibit a distribution in mass and charge. An example is $^{235}U(n, f)$.

When the energy is very high we may get a very large range in the number of reaction products, from 2–30 for example. This is referred to as *spallation* and will only occur when the energy available is an appreciable fraction of the total binding energy, i.e. $\approx 8A$ MeV.

Any of these reactions will compete (go on at the same time) provided they are energetically possible (threshold, if any, exceeded). Their relative probability will depend on details of the interaction mechanism. For a particular reaction we write the cross section as σ_{xy}. The total cross section for interaction of x with X can then be written

$$\sigma_T = \sigma_x = \sigma_{xx} + \sigma_{xx'} + \sigma_{x\gamma} + \sigma_{xy} + \ldots$$
$$= \sigma_{el} + \sigma_{inel}$$

The cross sections for individual types of reactions are called partial cross sections. σ_{inel} is often called $\sigma_{non\text{-}elastic}$ to distinguish it from $\sigma_{xx'}$. It is also referred to as the reaction cross section or the absorption cross section to distinguish it from σ_{el}.

We often refer to cross sections in terms of the interaction channels – the various possibilities of specified energy and angular momentum and radius. Thus $x + X$ for specific $E_x (E_x = 0)$ is the entrance channel. For elastic scattering obviously the exit channel is identical with the entrance channel. Other channels are specified as $X^* + x'$ for specific $E_{x'}$ or $Y + y$ for specific E_y etc.

2   General theory

2.1   Geometrical theory. Let us now consider what we can say about reaction cross sections using simple geometrical arguments.

We consider the incident beam to be composed of *particles* striking a completely 'black' (absorbing) sphere (fig. 6.2). For the present we neglect spin effects, that

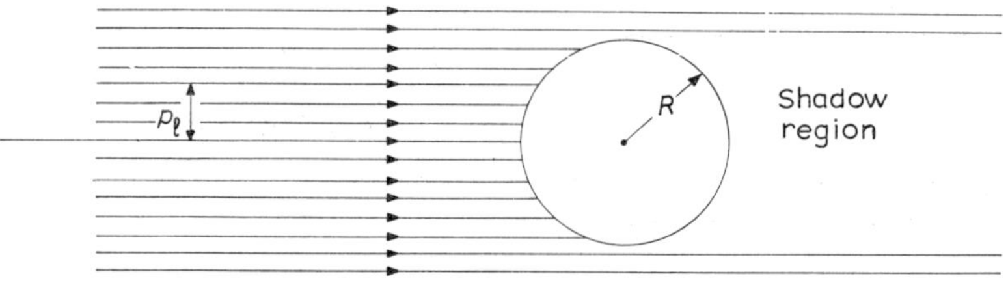

Fig. 6.2.   Cross section of an absorbing sphere

is, we assume that both bombarding particles and target have zero spin angular momentum. The incident particle, however, has angular momentum Mvp_l about the centre. But this angular momentum has to be quantized i.e. it can only occur in integral multiples of $\hbar$; $l\hbar = Mvp_l$ or $p_l = l\hbar/Mv = l\lambda$ since $\hbar/Mv$ is the de Broglie wavelength of the particle. We thus introduce the wave nature of the particle.

Thus we say that all particles with impact parameters between p_l and p_l+1 have angular momentum $l\hbar$. The cross section for these particles to interact with the nucleus cannot *exceed*

$$\sigma_l \leqq \pi(p_{l+1}^2 - p_l^2) = (2l+1)\pi\lambdabar^2. \tag{6.4}$$

If we have very slow particles such as thermal neutrons we can get very large cross sections since $\lambdabar$ is large. For thermal neutrons of average energy 0.025 eV or velocity 2200 m/sec

$$\lambdabar = \frac{\hbar}{Mv} = \frac{1.05 \times 10^{-27}}{1.67 \times 10^{-24} \times 2.2 \times 10^5} = 2.85 \times 10^{-9} \text{ cm}$$

and

$$\pi\lambdabar^2 = 2.56 \times 10^7 \text{ barns.}$$

For example, σ_{abs} for ^{135}Xe is 3.4×10^6 barns for thermal neutrons while the nuclear radius $R = 7.2 \times 10^{-13}$ cm for ^{135}Xe and πR^2 is only 1.5 barns.

If $p_l > R$ there is no interaction. The critical l-value is thus given by $p_l = l_{\text{max}}\lambdabar = R$ or $l_{\text{max}} = R/\lambdabar = kR$ where $k = 2\pi/\lambda = 1/\lambdabar$ is the wave number of the incident particle.

The shadow region in fig. 6.2 suggests a total absorption cross section of πR^2. Using the wave–particle picture we get agreement if l_{max} is very large, i.e. a large number of partial waves are effective.

$$\sigma_r = \sum_l \sigma_T^l \leqq \pi\lambdabar^2 \sum_0^{R/\lambdabar} (2l+1) \leqq \pi(R+\lambdabar)^2. \tag{6.5}$$

This will be true for fairly high energy where l_{max} is very large and $\lambdabar$ fairly small.

2.2   Wave mechanical theory. Let us now consider the situation in terms of wave mechanics. If the particle approaches the nucleus parallel to the z-axis with velocity v we can represent it by a plane wave

$$\psi_{\text{inc}} = e^{ik(z-vt)} \tag{6.6}$$

where k is the wave number referring to the centre of mass system i.e. a particle of the *reduced mass* having channel energy, ε_α,

$$\varepsilon_\alpha = E_\alpha \frac{M_T}{M_T + M_p} \tag{6.7}$$

where α specifies the channel, T the target and p the bombarding particle.

The wave amplitude is put equal to unity so that there is only one particle per unit volume, and in the incident beam there are v particles per cm^2 per sec.

Since we cannot precisely specify the particle position with respect to the x,y-axes ($l\hbar$ may take on many values), we transform from plane waves into a superposition of spherical waves each of which represents a particle of definite angular momentum as follows

$$\psi_{\text{inc}} = e^{ikz} = e^{ikr\cos\theta}$$

$$\approx \frac{1}{kr}\sum_{0}^{\infty}(2l+1)i^{l}P_{l}(\cos\theta)\sin(kr-\tfrac{1}{2}l\pi)$$

$$= \frac{1}{kr}\sum_{0}^{\infty}(2l+1)i^{l}P_{l}(\cos\theta)\frac{e^{i(kr-\frac{1}{2}l\pi)}-e^{-i(kr-\frac{1}{2}l\pi)}}{2i} \tag{6.8}$$

where $P_{l}(\cos\theta)$ is the Legendre polynomial.

In other words we consider the plane wave to be the superposition of a series of spherical waves $e^{-i(kr-\frac{1}{2}l\pi)}$ converging on the nucleus along with a coherent superposition of waves $e^{i(kr-\frac{1}{2}l\pi)}$ diverging from the nucleus. Now if the interaction affects the outgoing waves *only in phase* but not in amplitude we call this *elastic* scattering. If in addition the outgoing waves are affected in *amplitude* as well as in phase (corresponding to 'loss' of original particles) we have *inelastic* processes. Thus we can write

$$\psi = \frac{1}{kr}\sum_{0}^{\infty}(2l+1)i^{l}P_{l}(\cos\theta)\frac{\eta_{l}e^{i(kr-\frac{1}{2}l\pi)}-e^{-i(kr-\frac{1}{2}l\pi)}}{2i} \tag{6.9}$$

where η_{l} is a complex constant which contains the effect of the scattering centre. Its *real* part gives the change in amplitude. Its *imaginary* part gives the change in phase.

We can also regard this as a superposition $\psi = \psi_{\text{inc}}+\psi_{\text{sc}}$ of a scattered wave ψ_{sc} upon the incident wave ψ_{inc}. The scattered wave is itself a superposition of partial waves and may be written

$$\psi_{\text{sc}} = \frac{f(\theta)e^{ikr}}{r} \tag{6.10}$$

where we define a *scattering amplitude* $f(\theta)$. Thus the number of particles crossing unit area per second in the incident plane wave is v and the number crossing unit area per second in the scattered beam is just

$$v|\psi_{\text{sc}}|^{2} = \frac{v|f(\theta)|^{2}}{r^{2}} = v|f(\theta)|^{2}\,d\Omega \tag{6.11}$$

where $d\Omega$ is an element of solid angle. Hence

$$d\sigma_{\text{sc}} = |f(\theta)|^{2}\,d\Omega = \sigma_{\text{el}}(\theta)\,d\Omega. \tag{6.12}$$

We can express $f(\theta)$ in terms of our previous parameter η_l as follows

$$f(\theta) = \sum_0^\infty f_l(\theta) = \frac{1}{2ik} \sum_0^\infty (\eta_l - 1)(2l+1)P_l(\cos\theta). \tag{6.13}$$

We get the total elastic scattering cross section if we integrate $d\sigma_{sc}$ over the angular range $0 \to \pi$ and using the above value for $f(\theta)$ we get

$$\sigma = \sum_0^{l_{max}} \sigma_l. \tag{6.14}$$

Now consider the different types of reaction.

1) *Elastic scattering.* There is no loss of incident particles, that is $|\eta_l|^2 = 1$. There is no change in amplitude between incident and scattered wave, that is, there is no change in the real part of η_l. So we can write $\eta_l = e^{2i\delta_l}$ where δ_l is a *real* quantity. Then

$$f(\theta) = \frac{1}{2ik} \sum_0^\infty (2l+1)(e^{2i\delta_l} - 1)P_l(\cos\theta) \tag{6.15}$$

and δ_l is the *phase shift* in the asymptotic form of the partial wave l. In other words, for $\eta_l = e^{2i\delta_l}$ our earlier expression

$$\frac{\eta_l e^{i(kr - \frac{1}{2}l\pi)} - e^{-i(kr - \frac{1}{2}l\pi)}}{2i}$$

becomes $e^{i\delta_l} \sin(kr - \frac{1}{2}l\pi + \delta_l)$ instead of $\sin(kr - \frac{1}{2}l\pi)$.

2) *Inelastic reactions.* Now $|\eta_l|^2 < 1$, i.e. there *is* a change in amplitude and δ_l is complex, $\delta_l = \alpha_l + i\beta_l$ where β_l is positive.

3   Cross sections

We now want to write down the cross section for *elastic* scattering.

$$d\sigma_{el}^l = \frac{1}{4k^2} |1 - \eta_l|^2 (2l+1)^2 [P_l(\cos\theta)]^2 \, d\Omega. \tag{6.16}$$

We use $\eta_l = e^{2i\delta_l}$ as before and also note that $\int [P_l(\cos\theta)]^2 d\Omega = 4\pi/(2l+1)$, then

$$\begin{aligned}
\sigma_{el}^l &= \frac{\pi}{k^2}(2l+1)|1 - \eta_l|^2 \\
&= 4\pi\lambdabar^2(2l+1)|e^{i\delta_l}\sin\delta_l|^2 \tag{6.17} \\
&= 4\pi\lambdabar^2(2l+1)\sin^2\delta_l.
\end{aligned}$$

The *inelastic cross section* can be written

$$\sigma^l_{\text{inel}} = \frac{\pi}{k^2}(2l+1)(1-|\eta_l|^2). \tag{6.18}$$

The *total cross section*

$$\sigma^l_{\text{T}} = \sigma^l_{\text{el}}+\sigma^l_{\text{inel}}$$

$$= \frac{\pi}{k^2}(2l+1)2(1-\text{Re }\eta_l). \tag{6.19}$$

But we know that $|\eta_l|$ must be ≤ 1 so we can deduce *maximum* values for the cross sections. For *elastic* interactions this occurs when $\eta_l=-1$. Then $\sigma^l_{\text{el}}(\text{max})=4\pi\lambda^2(2l+1)$; at this point $\sigma^l_{\text{inel}}=0$. For *inelastic* interactions this is a maximum when $\eta_l=0$. Then $\sigma^l_{\text{inel}}=\pi\lambda^2(2l+1)$; at this point $\sigma^l_{\text{el}}=\pi\lambda^2(2l+1)$ also and $\sigma_{\text{T}}=2\pi\lambda(2l+1)$. Of course the maximum value of σ_{T} is $4\pi\lambda^2(2l+1)=\sigma_{\text{el}}(\text{max})$. For $\eta_l=+1$ both $\sigma_{\text{el}}=\sigma_{\text{inel}}=0$.

Note that this tells us that we can have elastic scattering with no inelastic scattering but we cannot have inelastic scattering with no elastic scattering. In fig. 6.3 the possible values are seen to lie inside the curve.

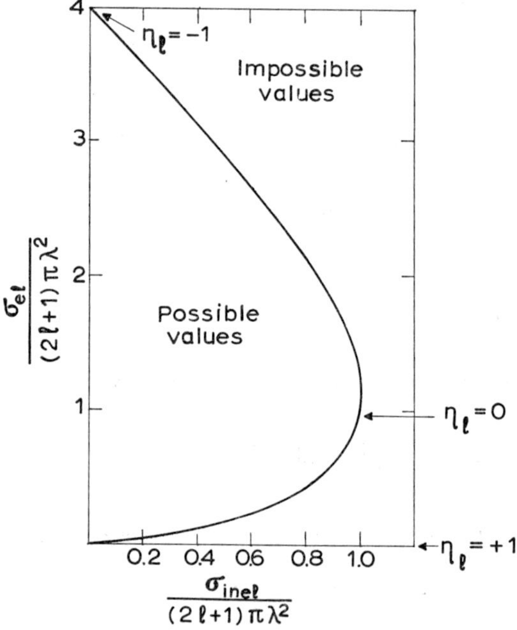

J. M. Blatt and V. F. Weisskopf, Theoretical Nuclear Physics (Wiley, New York 1952) p. 322

Fig. 6.3.   Possible values of elastic and inelastic scattering cross sections

Thus even for the case of a completely *absorbing* scatterer there is still an *elastic* scattering of up to $\pi\lambda^2(2l+1)$. This means that the total cross section is *twice* the geometrical cross section expected.

Also we want to understand why the maximum elastic cross section can be as high as four times the maximum absorption cross section. The reason for this is that in elastic scattering we have *coherence* between incident and scattered waves and these can interfere *constructively* as well as destructively. This does not occur for inelastic scattering as the incident and scattered waves are incoherent and we get no interference.

The reason we get elastic scattering even for a completely absorbing scatterer is the so-called *shadow scattering*. Suppose the wavelength of the incident particles is small compared to R, the radius of the scatterer. Then we expect $N\pi R^2$ particles to strike the nucleus per second and be absorbed. But in addition there is a shadow and because of diffraction of the waves, the shadow is not sharp and in fact is completely blurred at a distance R^2/λ just as in Fraunhofer diffraction in optics. This means that the particle waves are deviated from their straight paths, that is, are scattered. In the optical case R^2/λ is very large but in the nuclear case $R\approx\lambda$ and so R^2/λ is very small, of the order of R. So the number scattered in this way turns out to be just $N\pi R^2$ and hence the total is the sum of absorption plus shadow or diffraction scattering.

For an absorbing disc we have $\eta_l=0$ for $l\leq R/\lambda$ and $\eta_l=1$ for $l\geq R/\lambda$, whence for absorption

$$\sigma_{\text{inel}} \leqq \sum_0^{R/\lambda} \pi\lambda^2(2l+1) = \pi(R+\lambda)^2 \tag{6.20}$$

and for shadow scattering

$$\sigma_{\text{el}} \leqq \sum_0^{R/\lambda} \pi\lambda^2(2l+1) = \pi(R+\lambda)^2. \tag{6.21}$$

4 *Angular distributions*

The angular distribution for elastic scattering is given by the expression

$$\frac{d\sigma_{\text{el}}^l}{d\Omega} = \frac{1}{4k^2}|1-\eta_l|^2(2l+1)^2[P_l(\cos\theta)]^2. \tag{6.22}$$

Now

$$P_0(\cos\theta) = 1$$
$$P_1(\cos\theta) = \cos\theta$$
$$P_2(\cos\theta) = \tfrac{1}{2}(3\cos^2\theta-1)$$

that is, P_l will contain terms in $\cos \theta$ up to $\cos^l \theta$ and the angular distribution will contain powers of $\cos \theta$ up to $\cos^{2l} \theta$ *only*.

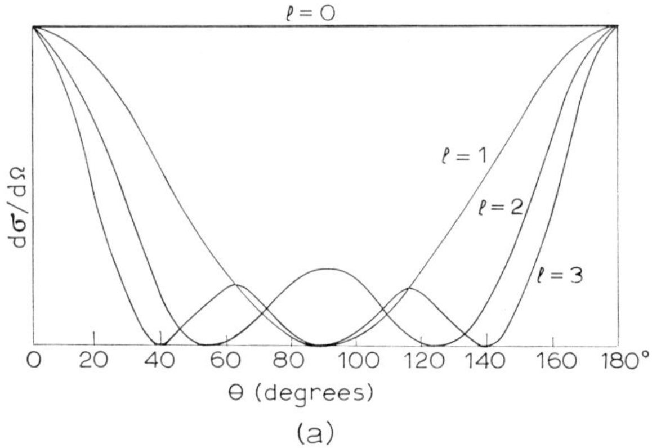

Fig. 6.4a.   The angular distribution for reaction scattering: the Legendre polynomial form

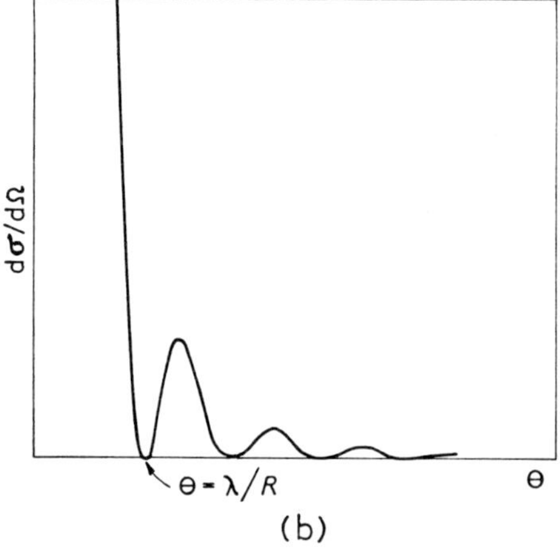

Fig. 6.4b.   The angular distribution for shadow scattering: the Bessel function form

For optical or shadow scattering or diffraction scattering, the angular distribution has the form as in optics

$$d\sigma_{\text{el}} = \frac{R^4}{\lambda^2} \left\{ \frac{J_1((2R/\lambda)\sin \tfrac{1}{2}\theta)}{(2R/\lambda)\sin \tfrac{1}{2}\theta} \right\}^2 d\Omega \tag{6.23}$$

where J_1 is a Bessel function.

The cross section has the form shown in fig. 6.4. We now consider the case where

$\lambda \gg R$ unlike the black disc approximation where $\lambda \ll R$. In this case we expect $l = 0$ only since $l_{max} = R/\lambda \to 0$ and the elastic scattering amplitude is

$$f_0(\theta) = \lambda\, e^{i\delta_0} \sin \delta_0; \tag{6.24}$$

the differential cross section is

$$d\sigma_{el}^0 = \lambda^2 \sin^2 \delta_0\, d\Omega. \tag{6.25}$$

The distribution is isotropic – no dependence on θ – since it depends solely on $l = 0$ waves.

If the scattering centre is an impenetrable sphere (hard sphere), the wave amplitude must vanish at the surface $r = R$ and

$$\eta_0 = e^{-2ikR} \tag{6.26}$$

hence $\delta_0 = -kR = -R/\lambda$ and the cross section

$$\sigma_{el}^0 = 4\pi\, d\sigma_{el}^0/d\Omega \approx 4\pi R^2 \quad \text{if} \quad \lambda \gg R. \tag{6.27}$$

This is expected to be the situation for *slow neutrons* where λ is large, s-wave interactions only are possible and we are away from resonances.

Often a different parameter than R is used, namely the *scattering length a* defined by $a = -f_0(\theta)$ thus

$$\sigma_{el}^0 = 4\pi a^2 \quad \text{(exactly)} \tag{6.28}$$

As we go to higher energies and more partial waves are of importance, more phase shifts will be required to specify the differential cross section. We have to assume details of the potential in order to predict them. In general for an attractive potential δ_l is positive and for a repulsive potential δ_l is negative.

5   Scattering of identical particles

There are special considerations when we are scattering particles from target particles of the same sort, for example p–p or α–α scattering.

The incident particles will scatter through an angle θ in the centre of mass system. The identical target particle will recoil through the angle $\pi - \theta$. In the laboratory system since $\theta = 2\psi$ the angles will be ψ and $\frac{1}{2}\pi - \psi$.

Then the total scattered intensity might be thought to be given by the sum of the intensities $|f(\theta)|^2 + |f(\pi - \theta)|^2$. However this is not correct since the wavefunction for a pair of identical particles must be either symmetric or antisymmetric for interchange of particles to satisfy the Pauli exclusion principle. Therefore

$$\psi_{sc} = [f(\theta) \pm f(\pi - \theta)]\, \frac{e^{ikr}}{r} \tag{6.29}$$

i.e. we must add *amplitudes* rather than *intensities* and the intensity at angle θ should be proportional to $|f(\theta)\pm f(\pi-\theta)|^2$.

For α–He scattering the spin of the α-particle is zero and the total wave function must be symmetric (bosons). Then we take the positive sign, and when $\theta=\pi-\theta$ $=90°$ or $\psi=45°$ we observe just twice the intensity predicted by the classical result.

For the case of identical particles with spin, a combination of symmetric and antisymmetric states is required and the results are more complicated.

6   *Transmission coefficients*

In all of this we have not allowed for any distortion or attenuation of the incident wave *before* it reaches the nucleus by, for example, the electric field or by reflection at the surface. That is, the limiting inelastic cross section should be written

$$\sigma_{\text{inel}} \lesssim \pi\lambda^2(2l+1)T_l \tag{6.30}$$

where T_l is the transmission coefficient for particles of angular momentum l. This is just the probability (≤ 1) for a particle incident on the barrier in fact passing the nuclear surface at $r=R$. The T_l is made up of two parts: 1) The barrier penetration factor, P_l, which is the probability of *reaching* the nuclear surface at $r=R$ in the presence of a potential barrier. 2) The potential discontinuity factor which gives the probability of passing the surface at $r=R$ and not being reflected. This arises because of the change from the momentum in free space (corresponding wave number k) to the momentum of the particle in the nuclear potential (corresponding wave number K).

The latter factor arises from the boundary conditions on the joining of the two waves at the surface (fig. 6.5). These conditions are that the value of $\psi(r)$ inside and outside must be equal *and* that the derivative $d\psi(r)/dr$ inside and outside must be equal so the slopes match. Suppose then just outside R we have

$$\psi(r) = A \exp(ikr) + B \exp(-ikr) \quad \text{for} \quad r > R \tag{6.31}$$
$$\underset{\text{(incident)}}{} \qquad \underset{\text{(reflected)}}{}$$

while just inside we have

$$\psi(r) = C \exp(iKr) \quad \text{for} \quad r < R. \tag{6.32}$$
$$\underset{\text{(transmitted)}}{}$$

Let us take $r=0$ at $r=R$. Then at $r=0$ $\psi(r)=A+B=C$ (1st condition) and $d\psi(r)/dr=ikA-ikB=iKC$ (2nd condition). Eliminating C we have

$$B = \left(\frac{k-K}{k+K}\right) A.$$

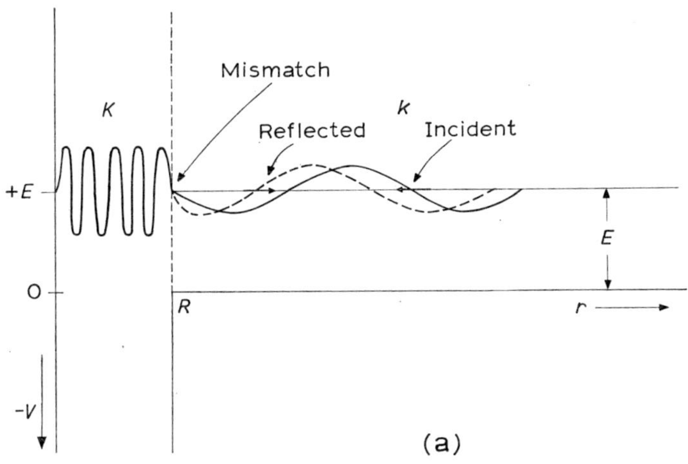

Fig. 6.5a.   The potential discontinuity factor e.g. for neutrons

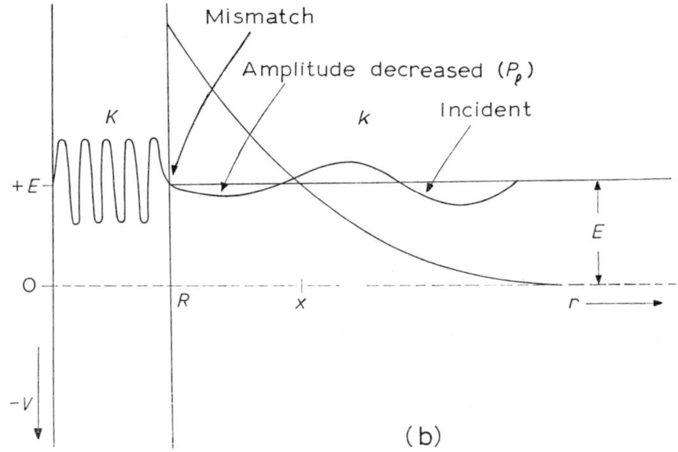

Fig. 6.5b.   Barrier penetration plus potential discontinuity e.g. for protons

But now our discontinuity factor is just

$$\frac{\text{intensity passing surface}}{\text{total incident intensity}} = \frac{|A|^2 - |B|^2}{|A|^2} = 1 - \left(\frac{k-K}{k+K}\right)^2$$

$$= \frac{4kK}{(k+K)^2} \approx 4k/K \quad \text{for} \quad K \gg k. \tag{6.33}$$

Thus for s-wave neutrons (low energy, k small)

$$T_0 \approx 4kP_0/K. \tag{6.34}$$

Now to evaluate P_l we must solve the wave equation for incident particles in the potential field and again find the ratio of particles reaching $r = R$ to particles incident on the barrier. The wave equation is

$$\frac{d^2\psi}{dr^2} + \frac{2M}{\hbar^2}\left[E - \frac{zZe^2}{r} - \frac{\hbar^2 l(l+1)}{2Mr^2}\right]\psi = 0. \tag{6.35}$$

The effective potential is made up of the electrostatic potential zZe^2/r and a second term $\hbar^2 l(l+1)/2Mr^2$ which can be considered as a centrifugal potential which keeps particles of high orbital momentum away from the nucleus. Its derivative with respect to r is just the centrifugal force. Thus it is often referred to as the centrifugal barrier. For neutrons therefore we have the potential discontinuity factor plus the centrifugal barrier. For charged particles we have the potential discontinuity plus the centrifugal barrier plus the coulomb barrier.

For the special case of slow neutrons $l=0$ so there is no centrifugal or coulomb barrier and $P_0 = 1$ hence $T_0 = 4k/K$.

For neutrons of $l \neq 0$ the centrifugal barrier must be included. In this case it is quite straightforward to evaluate P_l. Formulae are given in Blatt and Weisskopf's book* and graphs appear in *Nuclear Reactions* Vol. 1**.

For charged particles the solution is much more complicated. Again graphs appear in *Nuclear Reactions***. They are also collected in a Chalk River report[†].

For $l=0$ and high coulomb barrier we can write

$$P_0 = \frac{2\pi\eta}{e^{2\pi\eta} - 1} \quad \text{where} \quad \eta = \frac{zZe^2}{\hbar v} \tag{6.36}$$

(not connected with η_l for scattering. Weisskopf uses γ for this quantity).

For high barrier and low energy $\eta \gg 1$ and

$$P_0 \approx 2\pi\eta\, e^{-2\pi\eta} \approx \frac{1}{v}\, e^{-2\pi zZe^2/\hbar v}. \tag{6.37}$$

Then $T_0 = 4kP_0/K$ and

$$\sigma_{\text{inel}} \approx \pi\lambda^2 \frac{4k}{K}\frac{1}{v}\, e^{-2\pi zZe^2/\hbar v}$$

or

$$\sigma_{\text{inel}} \propto \frac{1}{v^2}\, e^{-2\pi zZe^2/\hbar v}. \tag{6.38}$$

* J. M. Blatt and V. F. Weisskopf, Theoretical Nuclear Physics (Wiley, New York 1952) p. 361.
** P. M. Endt and M. Demeur, eds., Nuclear Reactions 1 (North-Holland Publ. Co., Amsterdam 1959) p. 276, 273.
† W. T. Sharp, H. E. Gove and E. B. Paul, Graphs of Coulomb Functions. Chalk River Report TPI-70 (1955).

In this expression the exponential factor dominates over the E^{-1} dependence. The T_l-values for $R \approx 5 \times 10^{-13}$ cm and $Z \approx 20$ are shown in fig. 6.6. Note that at the top of the barrier the transmission for charged particles is still considerably less than unity.

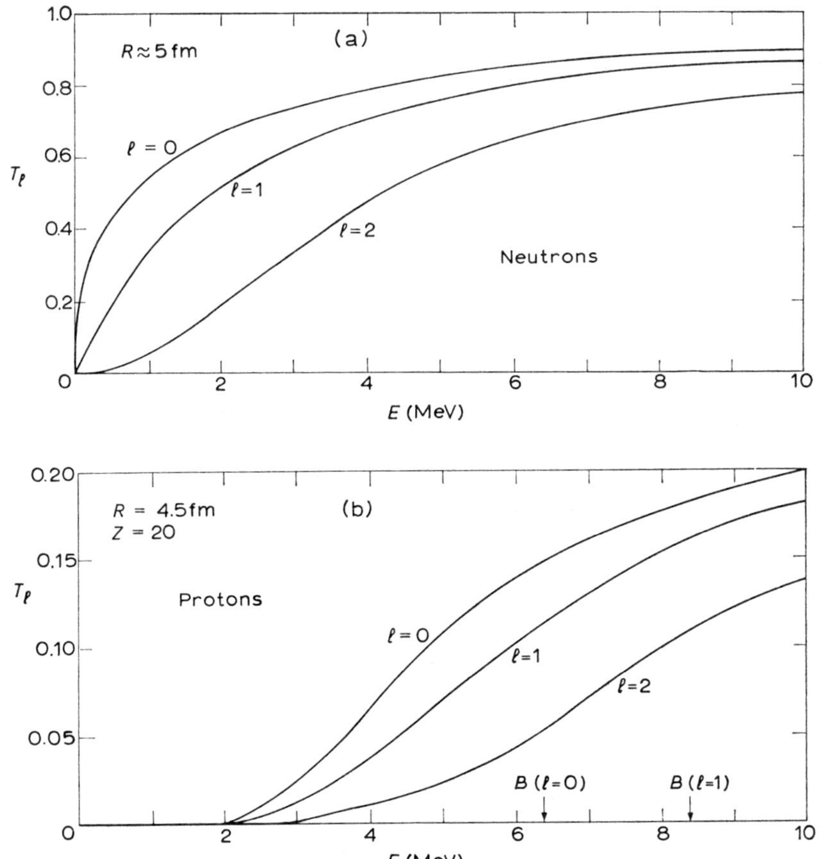

J. M. Blatt and V. F. Weisskopf, Theoretical Nuclear Physics (Wiley, New York 1952) p. 362, 363

Fig. 6.6.   Transmission coefficients for neutrons (a) and protons (b) of various angular momenta and energies. Barrier heights are indicated by arrows

In this region near the top of the barrier

$$\sigma_{\text{inel}} \approx \pi(R+\lambda)^2 \left[1 - \frac{zZe^2}{(R+\lambda)\varepsilon} \right] \tag{6.39}$$

where ε is the channel energy of the particle.

7

Reaction Mechanisms

1 Introduction

In the previous chapter we have outlined a framework based on general geometrical and wave mechanical ideas into which any more detailed nuclear reaction theory must fit. So far we can set certain limits on the possible size of cross sections and on the possible complexity of angular distributions but we cannot yet calculate these for a given nuclear case as this requires some model for the reaction mechanism.

In 1935 when nuclear reactions first began to be studied in detail using neutrons as projectiles it was soon determined that for high energy neutrons the total cross sections were of the order of πR^2 where R is the nuclear radius while for lower energy neutrons the cross sections became greater than πR^2 presumably approaching $\pi \lambdabar^2$ as expected. The first interpretation was that the incident neutron moved briefly in the potential well provided by the target nucleus. One would expect the probability of escaping with no change in energy, that is, of elastic scattering to be large and that the probability of not escaping, that is, of being captured, to be rather small. However the capture cross section should depend on the time spent near the target nucleus; it would vary as $1/v$ where v is the incident neutron velocity. On this basis for thermal energies one should expect capture and elastic scattering to be of comparable probability since each cross section approaches $\pi \lambdabar^2$ as a maximum value. We should expect to see resonances when the neutron wave could just be fitted inside the potential. These would then be spaced by 5–10 MeV (fig. 7.1a). They would have a width related to the time spent by the neutron inside the well which is about 10^{-21} sec that is $\Gamma = \hbar/\tau \approx 1$ MeV. Near thermal energy we should not expect the cross section to be much affected by these single particle resonances since the chance of having resonances with spacings of many MeV in the region of a few eV of the neutron binding energy is rather small.

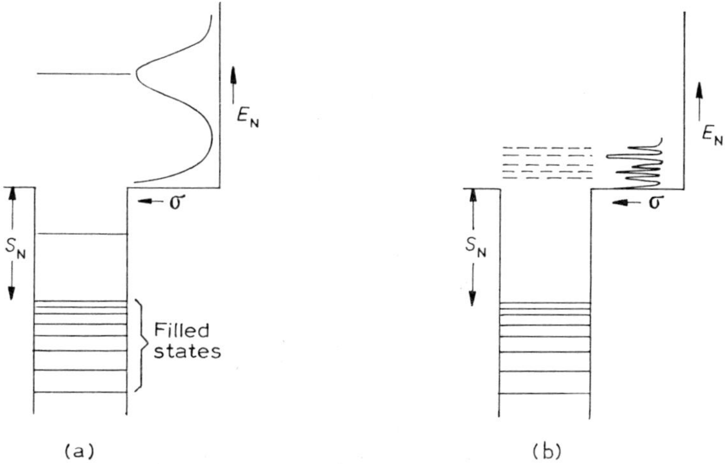

W. E. Burcham, Nuclear Physics (Longmans, London 1963) p. 528

Fig. 7.1.   a) Energy level diagram and cross section curve expected for single particle states above the binding energy S_N, b) The observed situation of many closely spaced levels above the binding energy

The observations with slow neutrons however were completely at variance with this picture. Many nuclei showed very large absorption cross sections for thermal energies while elastic scattering cross sections were generally very small. Furthermore the study of capture reactions such as $^{107}\mathrm{Ag} + \mathrm{n} \rightarrow {}^{108}\mathrm{Ag} + \gamma + 7.23$ MeV which could be studied by the subsequent β-decay of the $^{108}\mathrm{Ag}$ showed that the variation with neutron temperature corresponded to a variation proportional to $1/v$ as expected. However when the neutrons were passed through absorbers other effects showed up. When silver was used as an absorber a much reduced activation of another silver sample was observed. If equally thick absorbers of other materials were used no reduction in activation was seen. This is not consistent with a monotonic decrease in σ_{cap} with energy but instead we infer a number of closely spaced resonances at which σ_{cap} is very large (fig. 7.1b). It is found these resonances have widths of the order of a few tenths of an eV and spacings of a few eV.

2 The compound nucleus

To explain these observations Niels Bohr in 1936 proposed the compound nucleus hypothesis. In this picture we suppose that in the bombardment of nucleus X by particle x there is first formed a system in which X and x are amalgamated to form the compound nucleus C*. This is a system of strongly interacting particles and x thus has a very short mean free path for interaction with other nucleons. In these interactions the initial energy of x is very quickly shared out between all the other particles. A particle will not be remitted untill such time as sufficient energy is again associated with one particle so that it can emerge. If the original kinetic energy was small this may take a very long time. This means that the compound nucleus may have a decay lifetime of the order of 10^{-13}–10^{-15} sec, certainly very long compared to the traversal time of 10^{-21} sec. It is long enough for electromagnetic processes to compete very easily. Thus we have

$$x+X \rightarrow C^* \begin{array}{l} \longrightarrow X+x \quad \text{(unlikely)} \\ \longrightarrow C+\gamma \quad \text{(quite likely)} \end{array}$$

$$\underbrace{}_{\substack{\text{long time} \\ \text{delay}}}$$

This theory thus accounts for the predominance of radioactive capture over elastic scattering for slow neutrons. The many-body compound systems, being highly excited, will have many alternative modes of internal motion which will differ little in energy. Thus we account for the large number of closely spaced resonances observed. They are also very narrow resonances as this reflects the long times involved in the compound nucleus decay. Lifetimes of 10^{-13}–10^{-15} sec correspond to widths of 0.007–0.7 eV.

More generally we consider that a reaction takes place in two stages

$$X+x \rightarrow C^* \quad \text{formation of compound nucleus.}$$
$$C^* \rightarrow Y+y \quad \text{decay of compound nucleus}$$

Bohr proposed that the two stages are quite *independent* and are connected only by the properties of C*. The *decay* of C* depends only on the properties of C* and not upon how it was formed. The cross section for the reaction should then be

$$\sigma_{xy} = \sigma_x \Gamma_y / \Gamma \tag{7.1}$$

where σ_x is the cross section for formation of C* and Γ_y/Γ is simply the probability of C* in a given state breaking up into Y + y where Γ represents the total probability of decay of C* in all possible ways. This way of writing the cross section of course assumes that we are dealing with only one state of motion of C* – one resonance. This will be valid if other levels are far enough away in energy not to interfere. If $\bar{D}$ is the average level spacing we assume $\Gamma \ll \bar{D}$. On the other hand if $\Gamma \approx \bar{D}$ or

$\Gamma \gg \bar{D}$ we may have a few or many possible states of motion in the same energy interval. We consider first the isolated level case.

3   The single level formulation

An isolated level of the compound nucleus in a region where $\Gamma \ll \bar{D}$ will be characterised by 1) its excitation above the ground state, 2) its angular momentum and parity, 3) its partial widths Γ_i for decay into the ith channel of all those channels available. The excitation energy in the compound nucleus will be

$$E_0 = \varepsilon_y + S_y \tag{7.2}$$

where ε_y is the channel energy corresponding to the kinetic energy in the centre of mass system available for emission of particle y and S_y is the separation energy for

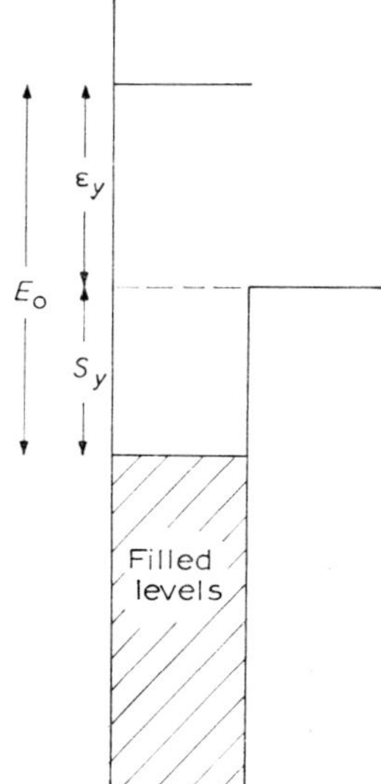

Fig. 7.2.   Compound nucleus energies

particle y in the compound nucleus (fig. 7.2). To derive the variation in cross section for excitation of this level with energy we use the analogy of a forced oscillation varying with frequency. There will be damping of such an oscillation through decay via the other open channels. We write the wavefunction of such a decaying state as

$$\psi_t = \psi_0 \, e^{-iE_0 t/\hbar} e^{-\Gamma t/2\hbar}. \tag{7.3}$$

This gives an exponential decrease of the intensity $|\psi_t|^2$ with time constant $\tau = \hbar/\Gamma$. This state can be built up as a superposition of states of slightly different energy

$$\psi_t = \int_{-\infty}^{+\infty} A(E)\, e^{-iEt/\hbar}\, dE \tag{7.4}$$

when $A(E)$ is just the amplitude of the state at energy E. By a Fourier transformation we have

$$|A(E)|^2 = \frac{|\psi_0|^2}{4\pi^2}\, \frac{1}{(E-E_0)^2 + \tfrac{1}{4}\Gamma^2}\,. \tag{7.5}$$

The cross section, proportional to the amplitude squared, is

$$\sigma = \text{const.} \times \frac{1}{(E-E_0)^2 + \tfrac{1}{4}\Gamma^2}\,. \tag{7.6}$$

We now determine the constant factor by considering that formation and decay take place in a volume Ω containing target nucleus X and bombarding particle x. The number of states of motion of the particle with momentum between p and $p+dp$ is $(4\pi p^2 dp/h^3)\Omega$. The effective collision area σ_x sweeps out a volume $\sigma_x v$ per second when v is the particle velocity. Therefore the probability of finding nucleus X in this volume is $\sigma_x v/\Omega$, and the probability of formation of the compound nucleus in a given channel per unit time is

$$\frac{\sigma_x v}{\Omega} \times \frac{4\pi p^2\, dp}{h^3}\, \Omega.$$

Integrating over the energy spectrum we have

$$\frac{4\pi}{h^3} \int_{-\infty}^{+\infty} v\sigma_x p^2 dp = \frac{4\pi}{h^3} \int_{-\infty}^{+\infty} \sigma_x p^2\, d\varepsilon_x = \frac{4\pi}{h} \int_{-\infty}^{+\infty} \frac{\sigma_x}{\lambda^2}\, d\varepsilon_x \tag{7.7}$$

where ε_x is the channel energy for the formation channel. Let us assume that over the small energy interval Γ the change in wavelength λ can be neglected. Then substituting the value of σ_x from eq. (7.6) we find the probability is $\text{const.}/\hbar\pi\lambda^2\Gamma$. We expect that the probability of formation via the channel $(x+X)_{\varepsilon_x}$ is just equal to the probability of decay via this same channel (elastic scattering). The decay probability we write $\Gamma_x/\hbar$. Thus the constant is equal to $\pi\lambda^2\Gamma\Gamma_x$ and the cross section for formation

$$\sigma_x = \pi\lambda^2\, \frac{\Gamma\Gamma_x}{(E-E_0)^2 + \tfrac{1}{4}\Gamma^2}\,. \tag{7.8}$$

We now have to consider the effect of angular momentum. For the present let us consider the spins of x, X to be zero. Then the spin of the compound nucleus state J_c is just equal to the orbital angular momentum l_x of particle x. The statistical weight of the compound state (the number of substates) is just $2J_c+1=2l_x+1$. Each of these substates can decay with equal probability and Γ_x is to be multiplied by $2l_x+1$.

$$\sigma_x = \pi\lambda^2(2l_x+1)\,\frac{\Gamma\Gamma_x}{(E-E_0)^2+\frac{1}{4}\Gamma^2}\,. \tag{7.9}$$

We now derive the cross section for the reaction $x+X\rightarrow Y+y$. This will be $\sigma_{xy}=\sigma_x\Gamma_y/\Gamma$, just the formation cross section times the probability of decaying via the $Y+y$ channel. Hence

$$\sigma_{xy} = \pi\lambda^2(2l_x+1)\,\frac{\Gamma_x\Gamma_y}{(E-E_0)^2+\frac{1}{4}\Gamma^2}\,. \tag{7.10}$$

This is just the single-level Breit–Wigner formula for the case of spinless particles.

For *elastic* scattering when no other processes are possible we have $\Gamma=\Gamma_x=\Gamma_y$ and at an energy equal to the resonance energy $E=E_0$ we have

$$\sigma_{el} = \sigma_{xx} = 4\pi\lambda^2(2l_x+1) \tag{7.11}$$

which is just the maximum elastic cross section derived in ch. 6. For non-elastic or total inelastic events we have $\Gamma_{inel}=\Gamma-\Gamma_x$ that is total less elastic, and at $E=E_0$ we have

$$\sigma_{inel} = \pi\lambda^2(2l_x+1)\,\frac{\Gamma_x(\Gamma-\Gamma_x)}{\frac{1}{4}\Gamma^2}\,. \tag{7.12}$$

This has a maximum value when $\Gamma_x=\frac{1}{2}\Gamma$, that is,

$$\sigma_{inel} = \pi\lambda^2(2l_x+1) \quad \text{and} \quad \sigma_{el} = \pi\lambda^2(2l_x+1) \tag{7.13}$$

again as derived before.

4   Particles with spin

In the foregoing we have assumed that the incoming particle and target nucleus were each of spin zero. This would apply for example to scattering or reaction of α-particles ($s=0$) with even–even nuclei ($I=0$). In general we consider particle x has spin s, the target nucleus has spin I and the orbital angular momentum of the incident particle is l. The spin of the compound nucleus level J is thus compounded $J=I+s+l$. The possible values of J range over all possible vector combinations of I, s, l. It is convenient to group target spin I and particle spin s into a channel spin $S = I + s$ and thus $J = S + l$ since the possible values of l may vary with energy. For example suppose we have protons ($s = \frac{1}{2}$) bombarding ^{27}Al($I = \frac{5}{2}$). The possible channel spins will be $S = \frac{5}{2} + \frac{1}{2} = 3$ or $S = \frac{5}{2} - \frac{1}{2} = 2$. For $l = 0$ the possible J-values for states in the compound nucleus, ^{28}Si will be $J = 2$ or 3 while for $l = 1$ the possible values are $J = 1, 2, 3$ ($S = 2$) or $2, 3, 4$ ($S = 3$). On the other hand for deuterons ($s = 1$) bombarding ^{27}Al the channel spins may be $\frac{3}{2}$, $\frac{5}{2}$ or $\frac{7}{2}$ and all half-integral J-values from $\frac{1}{2}$ to $\frac{9}{2}$ are possible.

In these strong interactions parity is also conserved and this will limit the possible states. We can make a table of allowed J^P-values for given combinations of S^P and l, table 7.1. Recall that the parity of a state having orbital angular momentum l is $(-1)^l$. Thus $(-1)^l P_S = P_J$. Then for the p$+^{27}$Al case the proton and ^{27}Al both have even parity. All integral values of J are allowed between the limits shown.

<table>
<tr><td colspan="3" align="center">TABLE 7.1</td><td colspan="2" align="center">TABLE 7.2</td></tr>
<tr><td>l</td><td>$S^P = 2^+$</td><td>$S^P = 3^+$</td><td>l</td><td>$S^P = \frac{3}{2}^-$</td></tr>
<tr><td>0</td><td>2^+</td><td>3^+</td><td>0</td><td>$\frac{3}{2}^-$</td></tr>
<tr><td>1</td><td>$1^-, 2^-, 3^-$</td><td>$2^-, 3^-, 4^-$</td><td>1</td><td>$\frac{1}{2}^+, \frac{3}{2}^+, \frac{5}{2}^+$</td></tr>
<tr><td>2</td><td>$0^+\text{–}4^+$</td><td>$1^+\text{–}5^+$</td><td>2</td><td>$\frac{1}{2}^-, \frac{3}{2}^-, \frac{5}{2}^-, \frac{7}{2}^-$</td></tr>
<tr><td>3</td><td>$1^-\text{–}5^-$</td><td>$0^-\text{–}6^-$</td><td>3</td><td>$\frac{3}{2}^+, \frac{5}{2}^+, \frac{7}{2}^+, \frac{9}{2}^+$</td></tr>
</table>

Another example is $\alpha + {}^7$Li. For the α-particle $s^P = 0^+$ and for ^{7}Li $I^P = \frac{3}{2}^-$. The allowed J^P-values are listed in table 7.2.

The probability of combining spin s and spin I to make channel spin S is just the statistical weight of channel spin S, $(2S+1)/(2s+1)(2I+1)$. For example for channel spins 2 and 3 the substates are $(2S+1)$ in number, that is 5 and 7 respectively. Their weights are $\frac{5}{12}$ and $\frac{7}{12}$. The probability of combining orbital angular momentum l and channel spin S to make a state of spin J is by similar arguments $(2J+1)/(2l+1)(2S+1)$. All of this assumes that there is no polarization involved of either particle or target, as this would imply a preferred combination and S would not arise statistically as above.

The cross section is thus to be multiplied by these statistical factors in addition to the $(2l+1)$ already occurring as a multiplying factor. The whole combination is termed g, the statistical factor

$$g = \frac{2S+1}{(2s+1)(2I+1)} \, \frac{2J+1}{(2l+1)(2S+1)} (2l+1) = \frac{2J+1}{(2s+1)(2I+1)} \tag{7.14}$$

and

$$\sigma_{xy} = \pi \lambda^2 g \, \frac{\Gamma_x \Gamma_y}{(E-E_0)^2 + \tfrac{1}{4}\Gamma^2} \, . \tag{7.15}$$

Note that the statistical factor is now (for $s \neq 0$, $I \neq 0$) independent of the orbital angular momentum involved. We assume the parity rule is also obeyed and do not show it explicitly.

We note also that in the absence of polarisation the states of different channel spins are incoherent and we can add contributions to the cross section from different channel spins. On the other hand the states of different orbital angular momentum combine coherently and we expect interference effects. An example of combined channel spins would be the $J=2^-$ state in the $^{27}\text{Al}+\text{p}$ example formed by $l=1$ with channel spin $S=2$ or 3. An example of combined orbital angular momenta would be the $J=2^+$ state formed by $l=0$ and $l=2$.

5   *Slow neutron resonances*

Let us now return to our starting point, the structure of slow neutron cross sections (cf. fig. 7.3). The measurements on many elements are shown as graphs in the document BNL-325.

The total widths Γ are mainly due to the capture process, since the elastic scattering is small. The neutron width is proportional to the density of states

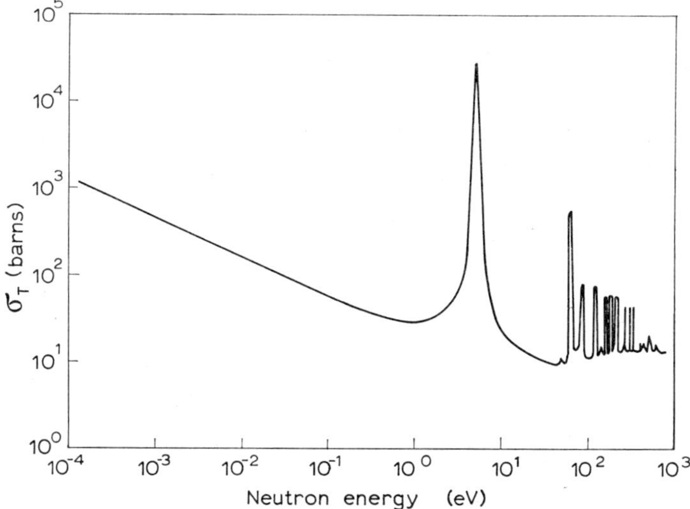

Fig. 7.3.   The slow neutron cross section of gold

of motion available for the particle per unit energy interval, which is just proportional to the velocity of the incident neutron:

$$\Gamma_n \propto \frac{4\pi p^2 \, dp}{\hbar^3 \, dE} \propto v. \qquad (7.16)$$

For very low energies the term in the resonance formula $(E-E_0)^2 + \tfrac{1}{4}\Gamma^2$ is approximately constant and

$$\sigma_{cap} \propto \lambda^2 \Gamma_n \Gamma_\gamma \qquad (7.17)$$

where Γ_n is the width for formation and Γ_γ is the width for radiative decay. The radiative width is not expected to vary appreciably over the energy range considered, so

$$\sigma_{cap} \propto \lambda^2 \Gamma_n \propto v^{-1} \qquad (7.18)$$

since $\Gamma_n \propto v$ and $\lambda^2 \propto v^{-2}$. This gives the shape of the cross section curve for capture from zero energy up to the velocity of the first resonance. Similarly

$$\sigma_{el} \propto \lambda^2 \Gamma_n^2 \propto v^2/v^2 \qquad (7.19)$$

is expected to be constant. Though the σ_{cap} curve will vary as v^{-1}, the magnitude of the cross section will be governed by $\Gamma_n \Gamma_\gamma$ of the compound state (or states) nearest to the neutron binding energy ($E_n = 0$). The controlling resonance may be just above or just below this point, that is, at a positive or negative corresponding neutron energy.

6   *Resonant and potential elastic scattering*

As we go to higher neutron energy, the elastic scattering will become more probable and eventually exceed the capture process. In the neighbourhood of a resonance

$$\sigma_{nn}(E) = \pi \lambda^2 g \, \frac{\Gamma_n^2}{(E-E_0)^2 + \tfrac{1}{4}\Gamma^2} . \qquad (7.20)$$

If elastic scattering is the dominant process then $\Gamma_n \approx \Gamma$ and

$$\sigma_{nn}(E) = \pi \lambda^2 g \, \frac{\Gamma^2}{(E-E_0)^2 + \tfrac{1}{4}\Gamma^2} . \qquad (7.21)$$

We have already derived an expression for elastic scattering (ch. 6) in terms of phase shifts (for no spin)

$$\sigma_{el}^l(E) = 4\pi \lambda^2 (2l+1) \sin^2 \delta_l . \qquad (7.22)$$

We can reproduce the Breit–Wigner form for this with g replacing $(2l+1)$ by making the substitution $\tan \delta_l = \Gamma/2(E_0-E)$. Thus far below a resonance $(E_0-E)\gg \Gamma$, δ_l is 0. At resonance $(E_0-E)=0$ and $\delta_l=\tfrac{1}{2}\pi$. Above the resonance δ_l approaches π. In other words at the resonance energy the phase shift passes through $\tfrac{1}{2}\pi$. Provided only elastic scattering is possible, this may be viewed as a definition of a resonance.

If we examine the elastic scattering cross section in a region of narrow isolated resonances $(\Gamma\ll D)$ we find two discrepancies with the theory outlined above. Firstly some of the resonances do not show the characteristic shape given by the Breit–Wigner formula and secondly the cross section between resonances does not reach the low value expected from the contributions from the two resonances. The compound nucleus cannot explain this but the model described previously of the single particle scattered by the nuclear potential can account for it. These effects are due to the potential scattering or shape elastic scattering. As we saw before this is expected to vary rather slowly with energy and we can respresent it by a phase shift ϕ_l for the scattering of partial wave l. Then our total phase shift δ_l is made up of two parts: $\delta_l=\beta_l-\phi_l$, where β_l is the phase shift associated with the resonance

$$\beta_l = \tan^{-1} \frac{\Gamma}{2(E_0-E)}.$$

The elastic scattering cross section including both resonant and potential scattering becomes

$$\sigma_{\text{el}}^l = 4\pi\lambda^2 g \left| \frac{e^{2i(\beta_l-\phi_l)}-1}{2i} \right|^2$$

$$= 4\pi\lambda^2 g \left| \frac{\tfrac{1}{2}\Gamma}{E_0-E-\tfrac{1}{2}i\Gamma} - e^{i\phi_l} \sin \phi_l \right|^2. \tag{7.23}$$

Far from a resonance, for $l=0$ the resonant term becomes small and $\sigma_{\text{el}}^l \to 4\pi\lambda^2 \sin^2\phi_0$ (fig. 7.4). This is just the expression we derived in ch. 6 for scattering from an impenetrable sphere by assuming $\phi_0 \approx -R/\lambda$. Thus the potential scattering is often referred to as the hard sphere scattering. Since the amplitudes for the two kinds of scattering are coherent we may expect interference. The resonant scattering is often called compound elastic to distinguish it from the shape elastic scattering.

In the case of elastic scattering of charged particles we must include also the coulomb scattering. It is also coherent with the other types of scattering. We have a total phase shift including a coulomb phase shift, a hard sphere phase shift and the resonant phase shift. This coherence of the incident and scattered waves is a characteristic of elastic scattering only. In the case of reactions the amplitude as well

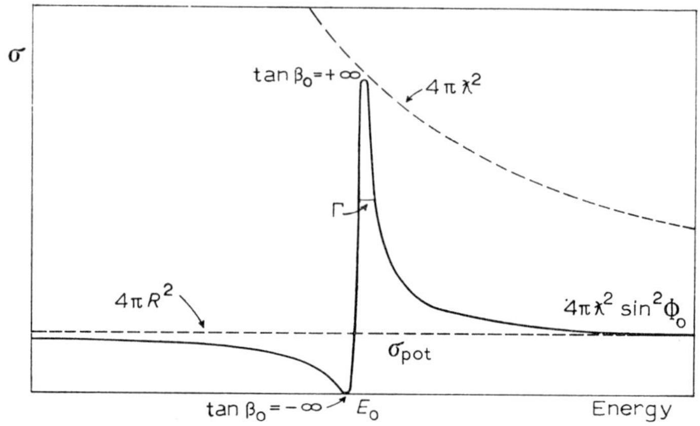

J. M. Blatt and V. F. Weisskopf, Theoretical Nuclear Physics (Wiley, New York 1952) p. 401

Fig. 7.4.   An elastic scattering resonance

as the phase is changed and ingoing and outgoing waves are incoherent. There is no interference involved except in the case of more than one resonance level.

7   *Resonances in charged particle reactions*

As an illustration of how the preceding ideas are applied in a practical case, consider a charged particle beam bombarding a target of thickness t cm and containing n distintegrable atoms per cm^3. The stopping power may be defined as

$$-\frac{\mathrm{d}T}{\mathrm{d}x} = \frac{4\pi z^2 e^4}{mv^2}\, nB = \varepsilon n \tag{7.24}$$

where B is proportional to the stopping power for the material in question. The target thickness in energy units is thus $\eta = nt\varepsilon$ and the yield observed from the target in terms of cross section is

$$Y = \int_{E-\eta}^{E} \frac{\sigma(E)}{\varepsilon}\, \mathrm{d}E \tag{7.25}$$

where E is the bombarding energy. In the vicinity of an isolated resonance we write the cross section

$$\sigma_{xy}(E) = \pi \lambdabar^2 g\, \frac{\Gamma_x \Gamma_y}{(E-E_0)^2 + \tfrac{1}{4}\Gamma^2} \tag{7.26}$$

and at the resonance

$$\sigma_{xy}(E_0) = \pi\lambda^2 g \frac{\Gamma_x \Gamma_y}{\frac{1}{4}\Gamma^2} = \sigma_0. \tag{7.27}$$

Therefore

$$\sigma_{xy}(E) = \sigma_0 \frac{\frac{1}{4}\Gamma^2}{(E-E_0)^2 + \frac{1}{4}\Gamma^2}. \tag{7.28}$$

Substituting, the yield is

$$Y(E) = \int_{E-\eta}^{E} \frac{1}{\varepsilon} \sigma_0 \frac{\frac{1}{4}\Gamma^2}{(E-E_0)^2 + \frac{1}{4}\Gamma^2} \, dE$$

$$= \frac{\sigma_0 \Gamma}{2\varepsilon} \left[\tan^{-1} \frac{(E-E_0)}{\frac{1}{2}\Gamma} - \tan^{-1} \frac{(E-E_0-\eta)}{\frac{1}{2}\Gamma} \right]. \tag{7.29}$$

Let us suppose now we have a thick target $\eta \gg \Gamma$

$$Y_\infty(E) = \frac{\sigma_0 \Gamma}{2\varepsilon} \left[\frac{1}{2}\pi + \tan^{-1} \frac{(E-E_0)}{\frac{1}{2}\Gamma} \right]. \tag{7.30}$$

At resonance $E = E_0$, the maximum yield for a target of thickness η is

$$Y_{max}(E_0, \eta) = (\sigma_0 \Gamma/\varepsilon) \tan^{-1} (\eta/\Gamma) \tag{7.31}$$

while for a thick target the step in the yield curve is just

$$Y_\infty(E_0) = \frac{1}{2}\pi \sigma_0 \Gamma/\varepsilon. \tag{7.32}$$

The ratio

$$\frac{Y_{max}(E_0, \eta)}{Y_\infty(E_0)} = \frac{2}{\pi} \tan^{-1} \frac{\eta}{\Gamma} \tag{7.33}$$

thus enables us to find the ratio of target thickness to resonance width. The area under the thin target curve is

$$A(E_0, \eta) = \int Y dE = \frac{\pi \sigma_0 \Gamma}{2\varepsilon} \eta = \eta Y_\infty. \tag{7.34}$$

A measurement of Y_∞ and $A(E_0, \eta)$ enables us to determine the thickness of the thin target.

We can also write

$$Y_\infty(E_0) = \frac{\pi\Gamma}{2\varepsilon} \pi\lambda^2 g \frac{\Gamma_x \Gamma_y}{\frac{1}{4}\Gamma^2} = \frac{\lambda^2}{2\varepsilon} g \frac{\Gamma_x \Gamma_y}{\Gamma}. \tag{7.35}$$

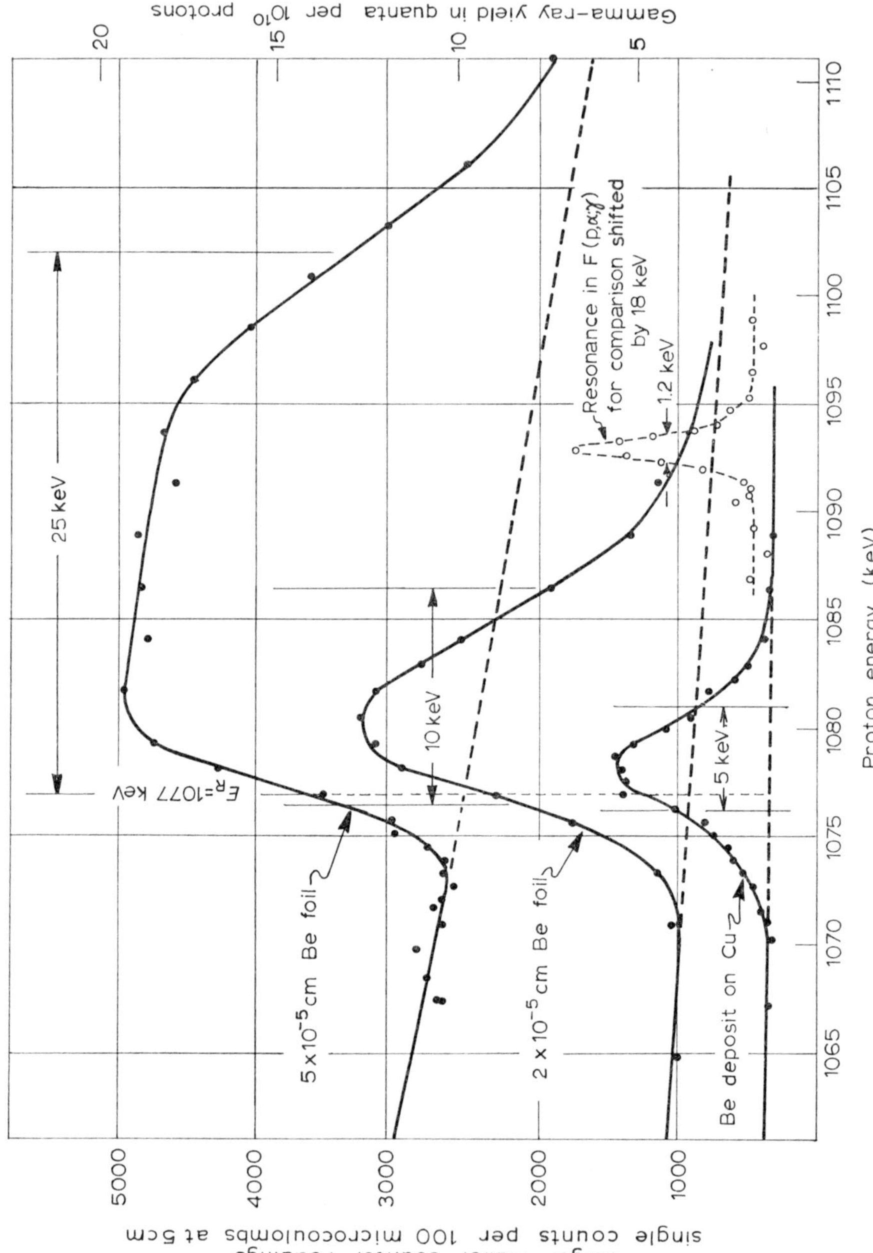

Fig. 7.5. The yield of the ^{9}Be(p, γ) reaction for different target thicknesses near the resonance at $E_p = 1077$ keV

Let us write $\xi = \Gamma_x \Gamma_y / \Gamma$ then

$$g\xi = 2\varepsilon Y_\infty(E_0)/\lambda^2. \tag{7.36}$$

The partials widths can then be written

$$\Gamma_x = \tfrac{1}{2}\Gamma \left[1 \pm \left(1 - \frac{4\xi}{\Gamma} \right)^{\frac{1}{2}} \right] \tag{7.37}$$

$$\Gamma_y = \tfrac{1}{2}\Gamma \left[1 \pm \left(1 - \frac{4\xi}{\Gamma} \right)^{\frac{1}{2}} \right]. \tag{7.38}$$

We are thus unable to tell which of Γ_x, Γ_y is the larger. If on other grounds we expect $\Gamma_x \gg \Gamma_y$ then $\Gamma_x \approx \Gamma$, $\xi \ll \Gamma$ and $\Gamma_y \approx \xi$. In other words the larger of the two fixes the total width while the smaller is of the order of $\Gamma_x \Gamma_y / \Gamma$. (We assume no other partial widths contribute appreciably.)

Comparison of thin and thick target yield curves (fig. 7.5) can thus enable us to determine target thickness, resonance width and something about the partial widths if only elastic scattering Γ_x and one other reaction channel Γ_y are possible.

8   Resonance widths

We have seen that the partial widths are related to the probability for decay of the compound state via the various open channels. To predict these decay probabilities we have to have some model of the structure of the compound state.

Suppose that one of the nucleons in the excited nucleus approaches the edge of the nucleus v times per second. The mean time for decay by emission of this nucleon would be just $\tau_0 = v^{-1}$. However there is in general a reduced probability for escape through the surface. For example if T_0 is the transmission coefficient for $l=0$ the lifetime would be $\tau = \tau_0/T_0$. The partial width for decay by emission of this kind of nucleon would be $\Gamma = \hbar/\tau = \hbar T_0/\tau_0$. To estimate v we can suppose the highly excited nucleus is oscillating with this frequency v. We then expect for a harmonic oscillator a set of equally spaced energy levels of spacing $D = hv = h/\tau_0$. Thus $\Gamma = \hbar T_0 D/h = D T_0/2\pi$. For s-wave neutrons $T_0 = 4k/K$. The partial width for emission of neutrons with $l=0$ would then be

$$\Gamma = \frac{4k}{K}\frac{D}{2\pi} = \frac{2Dk}{\pi K} \quad \text{or} \quad \frac{\Gamma}{D} = \frac{2k}{\pi K}. \tag{7.39}$$

In this expression the wave number k of the incident neutron is the only quantity depending on the experimental conditions since $k = (2mE)^{\frac{1}{2}}/\hbar$ depends on the incident energy. The other quantities D, K depend on internal features of the nuclear structure and are independent of the energy involved in the formation. In principle they are calculable if we know the structure of the compound state.

Accordingly we can write $\Gamma = 2kR\gamma^2$ for s-wave neutrons, where in γ^2, the *reduced width*, we lump together all the energy independent features of the internal structure. For neutrons of other orbital angular momentum l we define the reduced width by

$$\Gamma = 2kRP_l\gamma^2 \tag{7.40}$$

where P_l is the energy dependent barrier penetration factor. We now write an energy independent ratio $\gamma^2/D = 1/\pi KR$ whose value depends only on the internal structure. In the case of our simple model of equally spaced harmonic oscillator levels it is also almost independent of the excitation energy. However in more realistic models the structure may well depend on excitation energy. The ratio γ^2/D is called the strength function (for reasons we shall see later) and is an experimentally measurable quantity. For example for our s-wave neutron case we measure Γ for each resonance appearing at energy corresponding to k, the spacing between resonances D and the potential radius R is generally known from other experiments. The average value of the strength function $\langle\gamma^2/D\rangle$ can then be determined as a function of excitation energy. Its variations may reveal nuclear structure details.

A model based on the interaction of a single particle with a square potential well predicts a value for $\gamma_{sp}^2 = \hbar^2/MR^2$. This value is often used as a reference unit in discussing measured values of γ^2. Frequently the actual configuration is more complicated than a single particle configuration and measured reduced widths are smaller than γ_{sp}^2. On the other hand for some levels in which there are correlated effects in the internal motion such as rotations or vibrations, the reduced widths are larger than γ_{sp}^2 by a factor roughly corresponding to the number of cooperating particles. We have to describe such a state by a collective model.

9   *The statistical model*

9.1   Cross sections. As we go to higher excitation energies we might expect γ^2/D to remain roughly constant since as the complexity of the configurations increases γ^2 will decrease and as the level density increases D will decrease also. However the total width Γ will increase since it depends on k and P_l. P_l will increase exponentially and so will Γ on average. Eventually $\Gamma \gg D$ and rather than sharp isolated resonances we have many overlapping resonances contributing to the cross section at a given excitation.

Even for the case of $\Gamma \approx D$ with just a few levels overlapping it is no longer clear that the Bohr hypothesis of independence of formation and decay will be valid.

Since Γ is large the time scale is now considerably shortened. It turns out however that for $\Gamma \gg D$ we can make a different assumption. We can assume that the mode of formation is again forgotten, not now because of the time scale, but because any effect of formation will be averaged out over the many overlapping states. The mode of decay is again independent of the mode of formation because of the complexity of the linking process. This assumption is the *statistical assumption.*

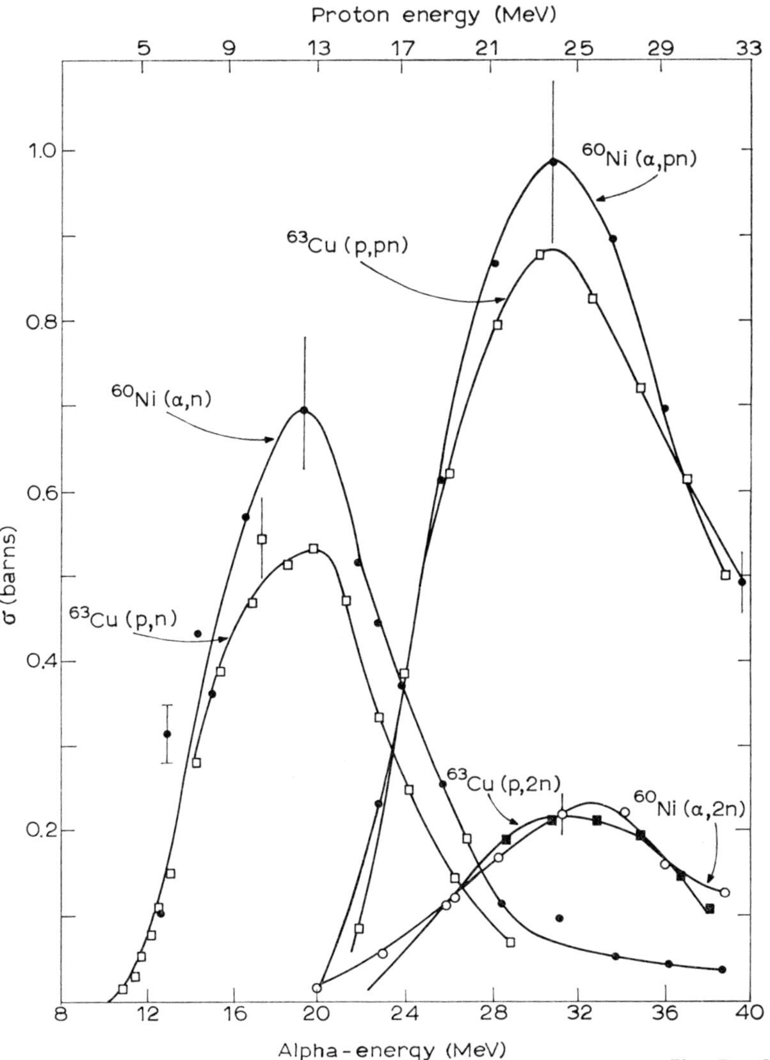

Fig. 7.6.   The Ghoshal experiment illustrating that compound nucleus decay is independent of the mode of formation

A classic experiment by Ghoshal* showed this assumption to be valid at least in certain cases (fig. 7.6). He studied first the reactions

$$^{60}\text{Ni} + \alpha \rightarrow {}^{64}\text{Zn} \rightarrow {}^{63}\text{Zn} + \text{n}$$
$$\rightarrow {}^{62}\text{Zn} + 2\text{n}$$
$$\rightarrow {}^{62}\text{Cu} + \text{p} + \text{n}$$

using α-energies between 8 and 40 MeV corresponding to excitation energies in the compound nucleus ^{64}Zn of 12 to 44 MeV. The α-binding energy in ^{64}Zn is 4 MeV. He compared yields from these reactions with yields from the reactions

$$^{63}\text{Cu} + \text{p} \rightarrow {}^{64}\text{Zn} \rightarrow {}^{63}\text{Zn} + \text{n}$$
$$\rightarrow {}^{62}\text{Zn} + 2\text{n}$$
$$\rightarrow {}^{62}\text{Cu} + \text{p} + \text{n}.$$

Since the proton binding energy in ^{64}Zn is 8 MeV he could use proton energies between 1 and 33 MeV and cover the identical range of excitation in ^{64}Zn. The results of the two sets of experiments plotted against excitation in ^{64}Zn are virtually identical thus showing the decay of ^{64}Zn is independent of how it is formed.

Consider a state in the compound nucleus C formed by X+x with kinetic energy of x, ε_x at an excitation energy $S_x + \varepsilon_x$ (fig. 7.7). This state decays by emission of

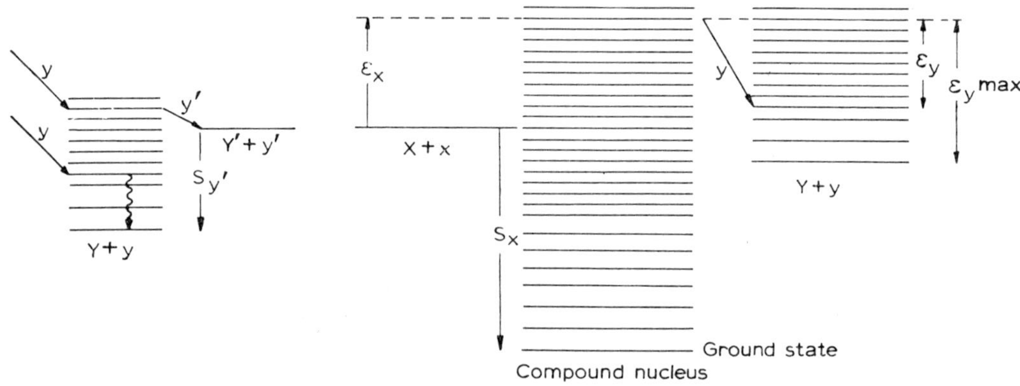

W. E. Burcham, Nuclear Physics (Longmans, London 1963) p. 545

Fig. 7.7.   Compound nucleus decay

particle y to an excited state of nucleus Y. This state may decay either by γ-emission to the ground state of nucleus Y or by further emission of another particle y' to nucleus Y' as is shown in the left section of fig. 7.7.

* S. N. Ghoshal, Phys. Rev. **80** (1950) 939

The maximum inelastic cross section for formation is

$$\sigma_x = \pi \lambda_x^2 (2l+1) T_l. \qquad (7.41)$$

But $\Gamma_y = D_c T_l/2\pi$ and for decay $T_l = 2\pi\Gamma_y/D_c$ where Γ_y is the partial width for decay of the compound state by channel $Y+y$ and D_c is the level spacing in the compound nucleus. Thus the decay cross section

$$\sigma_y = \pi \lambda_y^2 (2l+1) 2\pi \frac{\Gamma_y}{D_c} \qquad (7.42)$$

that is

$$\Gamma_y (2l+1) = \frac{D_c}{2\pi^2 \lambda_y^2} \sigma_y. \qquad (7.43)$$

This σ_y is just the cross section for *formation* of the compound state if we could bombard the excited state of Y by y with its energy at emission. However this refers to a specific energy ε_y for y. But y will be emitted with a range of possible energies from $0-\varepsilon_{y,\max}$ so we must integrate

$$\Gamma_y = \frac{D_c}{2\pi^2} \int_0^{\varepsilon_{y,\max}} \frac{\sigma_y}{\lambda_y^2} \omega_Y \, d\varepsilon_y \qquad (7.44)$$

where ω_Y is the level density in nucleus Y at excitation $E_y = \varepsilon_{y,\max} - \varepsilon_y$. Then $\omega_Y = D_Y^{-1}$ at this point. The formula for the decay width is often written

$$\Gamma_y = \frac{D_c M}{\pi^2 \hbar^2} f_y(\varepsilon_{y,\max}) \qquad (7.45)$$

where

$$f_y(\varepsilon_{y,\max}) = \int_0^{\varepsilon_{y,\max}} \varepsilon_y \sigma_y \omega_Y \, d\varepsilon_y.$$

We have to use approximate calculated expressions for σ_y and ω_Y and there are numerous published values for these. Using these we can calculate values of the widths for various reactions and thus the relative yields of the various reactions. In excitation energy regions where the statistical model might be expected to be correct, the agreement is reasonably good.

It is possible for Y to decay by further particle emission rather than by radiative decay, provided emission of y leaves Y with an excitation energy greater than the separation energy of y'. We then have

$$X+x \rightarrow C$$
$$C \rightarrow Y+y$$
$$Y \rightarrow Y'+y'.$$

A common example of this reaction is the (n, 2n) reaction. In this case the ratio of the (n, 2n) cross section to the (n, n′ γ) cross section is just

$$\frac{\sigma_{n,2n}}{\sigma_{n,n'\gamma}} = \frac{\int_0^{\varepsilon_{n',\max}-S_n} \varepsilon_{n'}\sigma_{n'}\omega_Y \, d\varepsilon_{n'}}{\int_{\varepsilon_{n',\max}-S_n}^{\varepsilon_{n',\max}} \varepsilon_{n'}\sigma_{n'}\omega_Y \, d\varepsilon_{n'}}. \tag{7.46}$$

That is, just the ratio of the integral for neutron emission leaving the residual nucleus excited above S_n the neutron binding energy in the residual nucleus to the integral for neutron emission leaving the residual nucleus at an excitation less than S_n. The operation of such a formula is clearly seen in multiple neutron reactions such

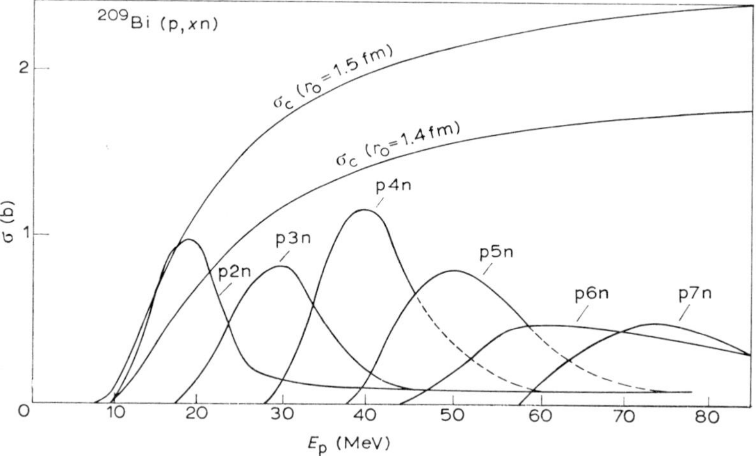

Can. J. Phys. **34** (1956) 745

Fig. 7.8.   Multiple neutron emission from compound nucleus decay. Formation cross sections σ_c are indicated for two assumed radius values

as (p, n), (p, 2n), (p, 3n) etc. A typical case is illustrated in fig. 7.8, where the results of a calculation of the formation cross section σ_c are shown. All decay cross sections must add up to this. Thus when a threshold is crossed and the next channel opens, the cross section via the channel still open must decrease.

9.2   Spectrum of emitted particles. The energy spectrum of the emitted particles is simply the quantity under the integral in the cross section expression $\varepsilon_y \sigma_y \omega_Y \, d\varepsilon_y$. If each level in Y were isolated and resolvable we might see a spectrum such as that in fig. 7.9a where E_y is the energy of the emitted particle and E_{ex} is the excitation in nucleus Y. The individual strengths are somewhat stronger at low excitation (γ^2, k larger), but the level density is greatly increased at high excitation. In practice the levels overlap and are not resolved. Accordingly for

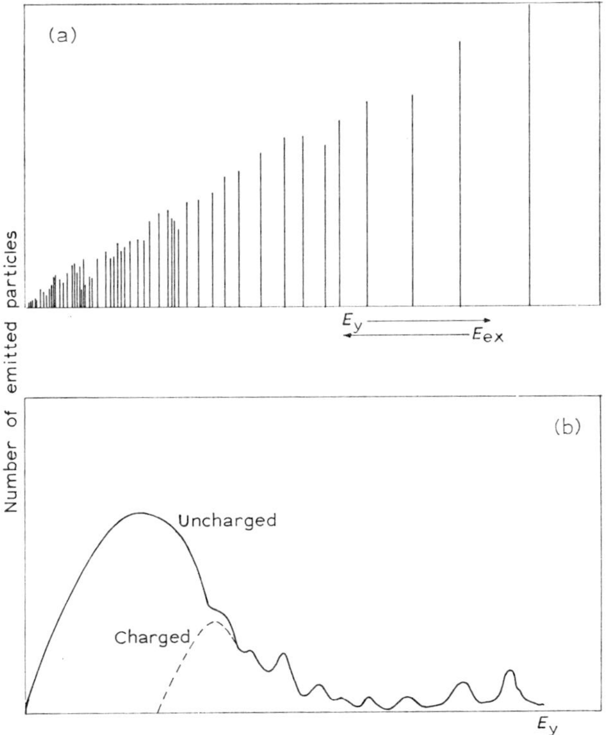

J. M. Blatt and V. F. Weisskopf, Theoretical Nuclear Physics (Wiley, New York 1952) p. 366

Fig. 7.9.   The energy spectrum of emitted particles; above, individual level strengths; below, overlapping levels as seen experimentally

uncharged particles we see the full curve while for charged particles the coulomb barrier prevents low energy particles from emerging as shown by the dotted curve in fig. 7.9b.

A convenient way to display level density is to plot $I(\varepsilon)/\varepsilon\sigma(\varepsilon)$ vs. ε where $I(\varepsilon) = \varepsilon_y \sigma_y \omega_Y$, that is, we plot level density as a function of excitation energy in nucleus Y, as in fig. 7.10. In practice we often see deviations from this spectrum in for example (n, p) or (p, p′) reactions. More particles are emitted with high energies and more with low energies than expected on the statistical model. The former are attributed to non-compound nucleus reactions, the so-called direct reactions. The latter are due to secondary reactions such as (n, np) or (p, np) which can occur when the neutron emission leaves the residual nucleus excited above the proton binding energy but below the neutron binding energy. Even though proton emission is hindered by the coulomb barrier it still is much more probable than radiative decay. It will only occur in those cases where $S_n > S_p$ for the residual nucleus.

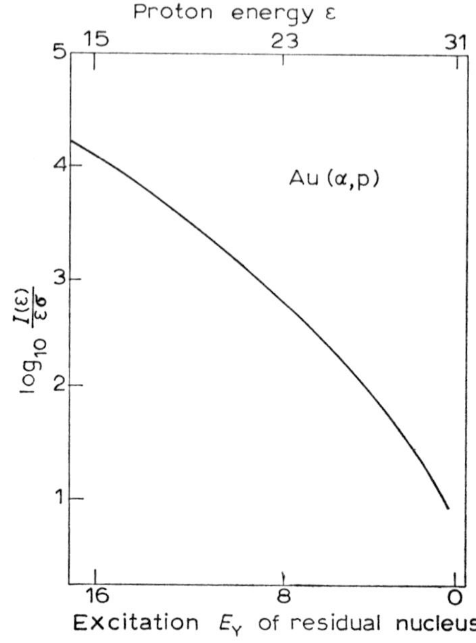

W. E. Burcham, Nuclear Physics (Longmans, London 1963) p. 551

Fig. 7.10.   Level density

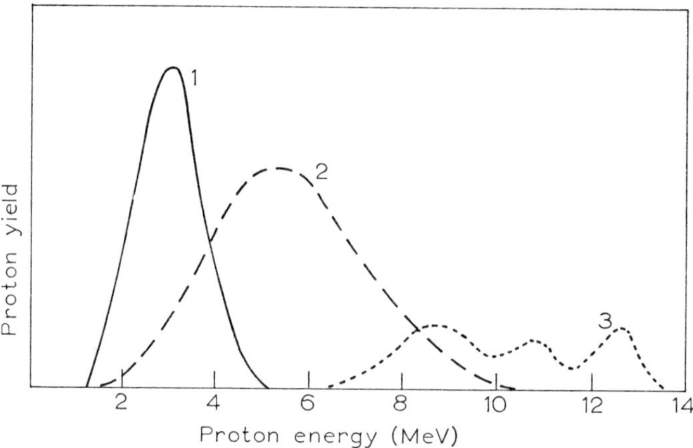

Proc. Phys. Soc. **70** (1957) 205

Fig. 7.11.   Proton spectrum from 14 MeV neutron bombardment of Fe. The yield is split into three curves: Curve 1, (n, np) reaction, compound nucleus. Curve 2, (n, pγ)+(n, pn) reactions, compound nucleus. Curve 3, (n, pγ) reaction, direct interaction

A spectrum showing all these effects is shown in fig. 7.11 for the case of iron bombarded by 14 MeV neutrons.

The angular distribution of particles emitted by a compound nucleus process in the region where $\Gamma \gg D$ is expected to be symmetric about $90°$ as a result of averaging over the phases of many interfering compound states. One often observes approximate isotropy but this is only expected to be the case when the level density in the residual nucleus is proportional to $(2I+1)$ for states of spin I. There is some evidence that this is often but not invariably so.

10   The evaporation model

The energy spectrum shown in fig. 7.11 resembles somewhat the distribution in energy amongst the molecules of a gas in thermal equilibrium, that is, a Maxwellian distribution. This suggests that highly excited nuclei with a fairly large number of nucleons can be considered as a thermodynamic system with a certain temperature. The emission of particles would then correspond to evaporation except that unlike the molecular case a rather large amount of energy is removed with the particle. We connect level density and entropy in the usual way

$$S(E) = \log \omega(E)$$

where $S(E)$ is the entropy. For our residual nucleus at excitation energy

$$E_{ex} = \varepsilon_{y,\,max} - \varepsilon_y$$

$$S(\varepsilon_{y,\,max} - \varepsilon_y) \approx S(\varepsilon_{y,\,max}) - \varepsilon_y (\partial S/\partial E)_{\varepsilon_{y,\,max}} \tag{7.47}$$

if we assume that $\varepsilon_y \ll \varepsilon_{y,\,max}$

$$= S(\varepsilon_{y,\,max}) - \varepsilon_y/T \tag{7.48}$$

where T is a temperature defined in the usual way

$$\frac{1}{T} = \frac{\partial S}{\partial E}$$

(we omit the Boltzmann constant k in defining nuclear temperature). T is then the temperature of the residual nucleus at maximum excitation ($\varepsilon_y = 0$). The energy spectrum is then given by

$$I(\varepsilon_y)d\varepsilon_y = \sigma_y \varepsilon_y \exp\left[S(\varepsilon_{y,\,max}) - \varepsilon_y/T\right]d\varepsilon_y$$

$$= \text{const.} \times \varepsilon_y \sigma_y [\exp - \varepsilon_y/T]d\varepsilon_y. \tag{7.49}$$

If σ_y varies slowly with energy this is just the Maxwellian distribution. T is the most probable energy of emission and the average energy of emission is $2T$.

For a Fermi gas we connect energy and temperature by

$$E = aT^2 \tag{7.50}$$

where a is a constant. The entropy S is then

$$S = \int \frac{\mathrm{d}E}{T} = 2aT + \mathrm{const.}$$
$$= 2(aE)^{\frac{1}{2}} + \mathrm{const.} \tag{7.51}$$

But since $S(E) = \log \omega(E)$ we have

$$\omega(E) = C \exp\left[2(aE)^{\frac{1}{2}}\right] \tag{7.52}$$

where C is another constant. By analysis of evaporation spectra, level spacings measured in excitation functions, etc., empirical values of C and a can be determined. These depend on the mass number A and *very approximately* have the following values.

$$a \approx (A/10.5)\ \mathrm{MeV}^{-1}$$
$$C \approx (10/A)\ \mathrm{MeV}^{-1} \text{ for even–even nuclei}$$
$$(20/A)\ \mathrm{MeV}^{-1} \text{ for odd-}A \text{ nuclei}$$
$$(40/A)\ \mathrm{MeV}^{-1} \text{ for odd–odd nuclei.}$$

The temperature T is determined from the slope of the spectral distribution curve since

$$\frac{1}{T} = -\frac{\partial}{\partial E} \log \frac{I(\varepsilon_y)}{\varepsilon_y \sigma_y}.$$

Typical values of T for the statistical region ($\Gamma \gg D$) are in the range 1.5–2.0 MeV. In fact even in this region of high excitation, effects due to details of the nuclear structure are apparent. The theory is thus only useful as a first approximation to the actual situation.

11   Direct interactions

11.1   General. We now consider the effects seen in some spectral distributions which could not be accounted for by compound nucleus effects. These effects show up in the following ways:

1) The emission of excess particles of *high* energy compared to the number expected by the evaporation theory. Since these lead to low-lying states of the residual nucleus they frequently appear as discrete groups. Emission of high energy particles leads to unusually large cross sections for (n, p) reactions in heavy nuclei compared

to those calculated on the statistical theory. There is a greater chance of their penetrating the high coulomb barriers for these nuclei.

2)   A forward peaked angular distribution instead of an angular distribution symmetric about 90° in the centre of mass system which is expected from evaporation or compound nucleus theory.

3)   A gradual monotonic dependence of the yield of a reaction upon bombarding energy. In a region where discrete levels appear, this may show up as a gradually varying background yield.

To explain these phenomena we assume that if we are far from a compound nucleus resonance and are at a sufficiently high energy (> 10 MeV) there is a good chance of a compound nucleus not being formed. Instead a single nucleon encounter takes place near the surface leaving the rest of the nucleus undisturbed.

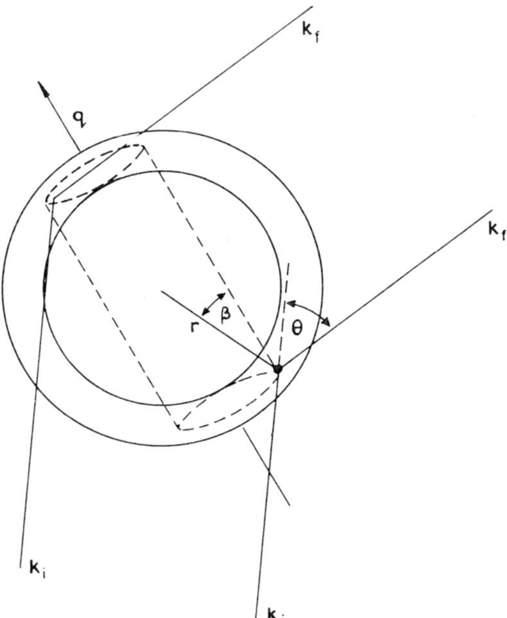

W. E. Burcham, Nuclear Physics (Longmans, London 1963) p. 557

Fig. 7.12.   Surface direct interaction

For clarity we confine our discussion to a surface encounter although a single nucleon interaction can occur inside the nucleus. In this case however the likelihood of reflection of the outgoing wave at the surface increases the chance of compound nucleus formation. Consider a shell of radius r (fig. 7.12). If k_i is the wave number of the incident wave and k_f of the final wave, the recoil momentum given the nucleus is q where

$$q\hbar = (k_i - k_f)\hbar$$

or

$$q = k_i - k_f, \qquad k = Mv/\hbar. \tag{7.53}$$

But now we expect to excite low lying states of the residual nucleus with high probability corresponding to large outgoing momenta i.e. $k_f \approx k_i$ and $q \to 0$. This means that the angle of emission $\theta \to 0°$ and we have an angular distribution peaked in the forward direction. Since we favour high energy particles in the spectrum of emitted particles this explains the large (n, p) cross sections. The extra cross section results from the extra particles which can penetrate the potential barrier.

However now our final state in the residual nucleus is a discrete state at low excitation with a definite angular momentum. Thus a definite l-value must have been transferred to the nucleus. We consider the angular momentum (fig. 7.13a)

$$l = (k_i - k_f) \times r\hbar$$
$$l = (q \times r)\hbar$$
$$l = qr \sin \beta \tag{7.54}$$

where β is the angle between r and q. Then for a given scattering angle θ,

$$q^2 = k_i^2 + k_f^2 - 2k_i k_f \cos \theta$$
$$= (k_i - k_f)^2 + 4k_i k_f \sin^2 \theta. \tag{7.55}$$

The scattering is thus restricted to the ends of a cylinder of radius l/q and axis in the q-direction. Again scattering from the sides of the cylinder in the nuclear interior is unlikely as mentioned above. The interference of the waves coming from the two ends of the cylinder leads to pronounced maxima and minima in the angular distribution. The rough relationship $l = qR \sin \beta$ where l is the change in angular momentum, q is the change in linear momentum, R is the interaction radius and β is the

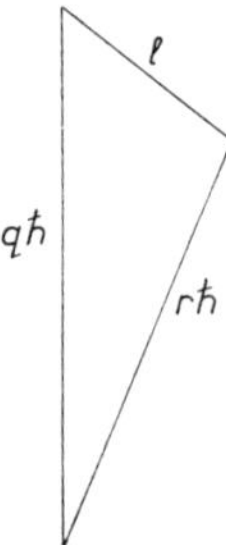

Fig. 7.13a.   Angular momenta in direct interaction

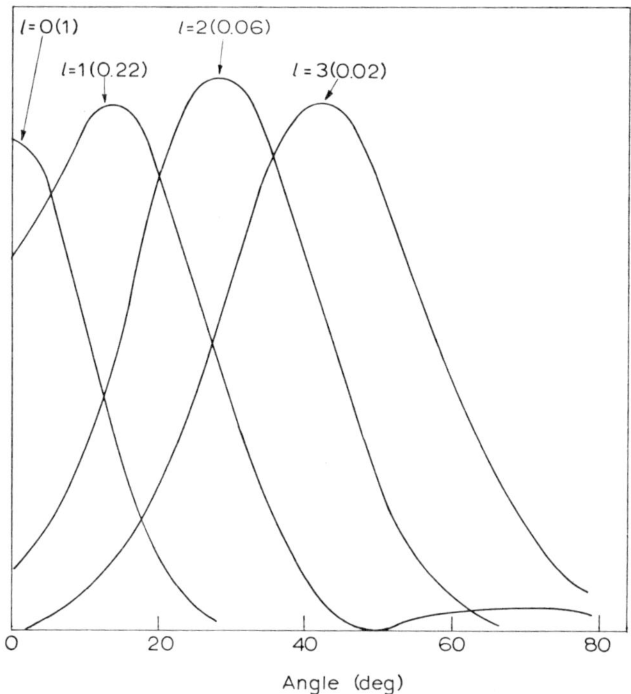

Fig. 7.13b.   Direct interaction angular distributions. Each curve is to be multiplied by the factor shown in brackets after the l-value

angle between q and r, tells us that the larger the l-value the greater is the angle at which the first peak occurs; for $l=0$ we expect a peak at $0°$ (cf. fig. 7.13b). In general the first peak occurs for $qR=l$. More detailed calculations show that the angular distribution is

$$\frac{d\sigma}{d\Omega} \propto [j_l(qR)]^2 \qquad (7.56)$$

where j_l is a spherical Bessel function which shows a maximum when $qR=l$. Such angular distributions are observed for high energy (n, p), (p, p') and (α, α') reactions. The l-value change can be deduced from the form of the angular distribution if a value of R is assumed.

11.2   Stripping and pickup reactions. The most important forms of direct interaction are the so-called stripping and pickup reactions. Early work on the (d, p) reaction e.g. $^{16}O+d\rightarrow^{17}O+p+1.9$ MeV showed strong forward peaks in the angular distribution of the protons, which were difficult to explain as a compound nucleus effect. For example it would require a Legendre polynomial series up to $P_{10}(\cos\theta)$

or higher to fit such a peak, implying l- and J-values up to 5 or more which were quite unreasonable.

The better explanation is that the proton is actually freed from the deuteron rath·

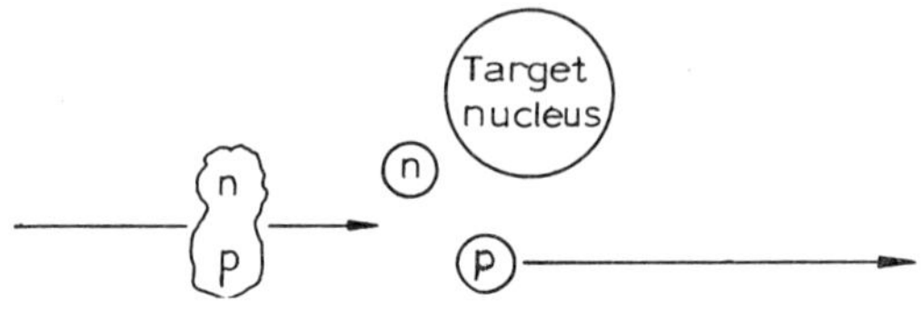

Ann. Rev. Nucl. Sci. **13** (1963) 204

Fig. 7.14a.   Schematic of deuteron stripping

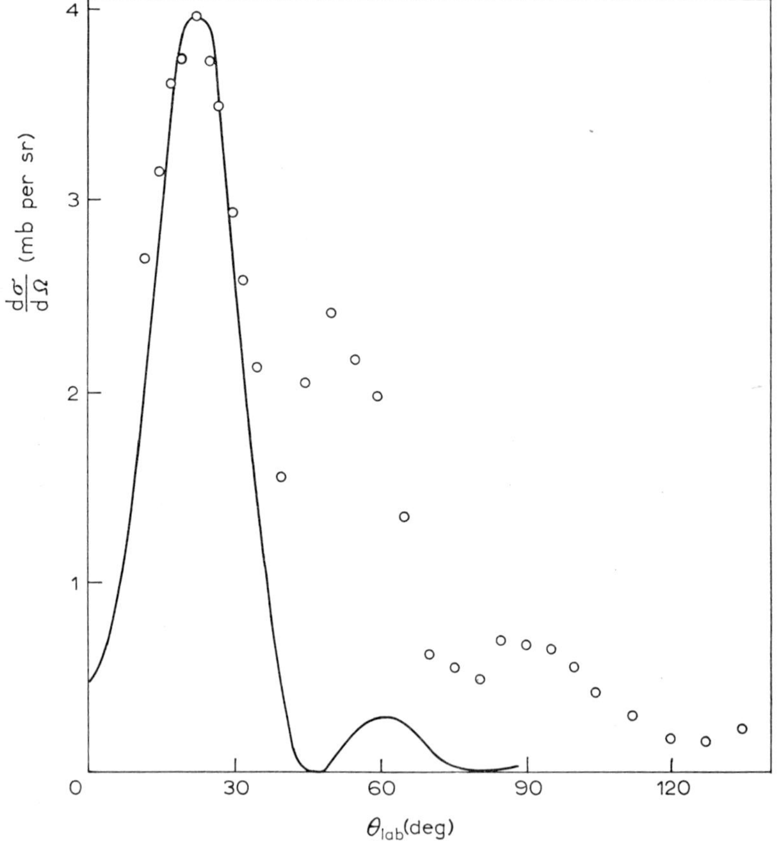

Ann. Rev. Nucl. Sci. **13** (1963) 204

Fig. 7.14b.   Angular distribution of protons from the reaction ^{90}Zr (d, p) ^{91}Zr (ground state). The curve drawn in the figure is the result of a stripping theory calculation with $l=2$

er far from the target nucleus. They can thus have large values of angular momentum and lead to sharp forward peaking. It was known that the deuteron is a fairly large loosely bound structure with binding energy 2.23 MeV whose constituent nucleons spend a large fraction of the time beyond the range of the nuclear forces between them. Thus we say that the neutron is absorbed at the nuclear surface while the proton continues on almost unaffected (fig. 7.14a). The neutron carries linear momentum $q\hbar$ into the nucleus where as before

$$q^2 = (k_d - k_p)^2 + 4k_p k_d \sin^2 \tfrac{1}{2}\theta \tag{7.57}$$

and the angular momentum change is

$$l = (q \times r)\hbar \tag{7.58}$$

where $|r| \approx R$, the interaction radius. Again we get interference for absorption of the neutron at different places on the nuclear surface and hence observe maxima and minima in the proton angular distribution (fig. 7.14b).

We note that the observation of a stripping pattern tells us at once the angular momentum carried into the target nucleus by the neutron. For example the reaction $^{16}\text{O} + \text{d} \rightarrow \text{p} + ^{17}\text{O}$ is equivalent to $^{16}\text{O} + \text{n} \rightarrow {}^{17}\text{O}$ with angular momenta

$$
\begin{array}{lll}
0^+ \quad \tfrac{1}{2}^+ \quad \tfrac{1}{2}^+ & \quad \text{for } l_n = 0 \\
\tfrac{1}{2}^-, \tfrac{3}{2}^- & \quad \text{for } l_n = 1 \\
\tfrac{3}{2}^+, \tfrac{5}{2}^+ & \quad \text{for } l_n = 2 \text{ etc.}
\end{array}
$$

Another important point is that although the l-values for neutron interactions could be determined directly by neutron scattering for excited states of ^{17}O above 4.1 MeV, the neutron binding energy, states below the binding energy (negative energy states) cannot be so studied. However the (d, p) stripping reaction can lead to all the states of ^{17}O including the ground state as can be seen from fig. 7.15. Thus l-values for neutron interactions can be determined for these bound states as well as for unbound states. The reduced neutron widths for the bound and unbound states can also be deduced from the magnitude of the stripping cross section in favourable cases.

The specification of the l-value of the captured particle does not uniquely fix the spin of the state. It does however uniquely define the parity of the state and limits the spin of the state, I_f to values such that

$$I_i + l + \tfrac{1}{2} \geqq I_f \geqq |I_i \pm l \pm \tfrac{1}{2}|_{\min} \tag{7.59}$$

where I_i is the target nucleus spin. So the stripping reaction angular distribution is sometimes called a 'parity gauge'.

In practice it is only in certain cases that simple angular distributions are observed in stripping and pickup reactions such that shape and absolute magnitude can be

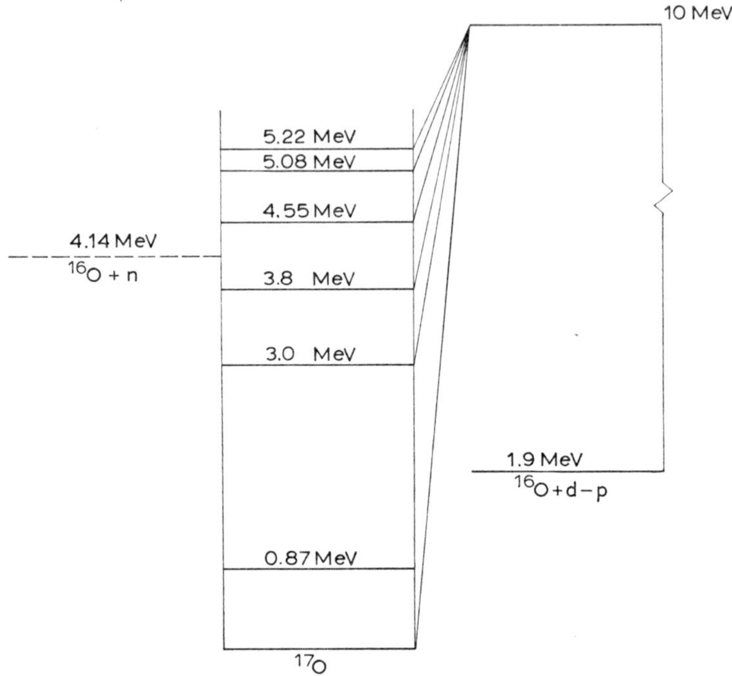

Fig. 7.15. ^{17}O level scheme

exactly fitted by the simple theory given above. More elaborate theories have been developed which take into account the distortion of initial and final wave functions by the target and the effects of the coulomb field. By such an analysis (distorted wave Born approximation, DWBA) reliable parameters can be extracted from the observations. The (d, n) reaction in which the proton is captured, the (p, d) reaction in which a neutron is picked up from the target, the (^{3}He, d) reaction in which a proton is stripped, the (^{3}He, α) reaction in which a neutron is picked up are other examples of such reactions which are extremely useful in nuclear spectroscopy.

12   The optical model

We now have considered two contrasting theories of nuclear interaction. The compound nucleus assumption is that there is independence of formation and decay and pictures the incident wave being heavily damped in the nuclear potential so that reemission of the same particle is improbable. In this model the incident particle is expected to have a short mean free path for interaction and the energy is quickly shared amongst the other particles. The direct interaction assumption is that the incident particle has a long mean free path for interaction so that elastic scattering is very probable and that essentially isolated nucleon encounters are also very probable at the nuclear surface. We now discuss a theory which combines the two extreme pictures.

Several experiments indicated how this could be done. We recall that the elastic scattering of neutrons is made up of a resonant part and a hard sphere or potential part away from resonances. At low energy this cross section has a value $4\pi a^2$ where a is the scattering length (eq. 6.28). Thus the cross section might be expected to vary as $R^2 \propto A^{\frac{2}{3}}$ and the scattering length to vary as $A^{\frac{1}{3}}$. However when scattering length measurements are plotted as a function of A (fig. 7.16) we note large deviations from the $A^{\frac{1}{3}}$ dependence. The spacing of the anomalies suggest single particle resonances in the potential well.

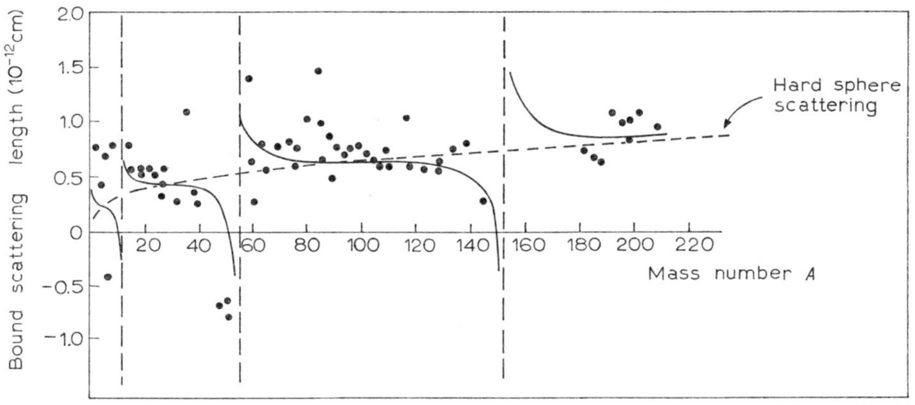

Phys. Rev. **79** (1950) 745

Fig. 7.16.   Scattering length for slow neutrons as a function of A. The vertical dashed lines indicate single particle states

Similar effects were observed in the total cross sections for fast neutrons of energy 1–3 MeV.* Maxima in the cross section at particular energies varied with A and again suggested potential well resonances. A peak showed up at low energy for $A \approx 60$, moved upwards to higher neutron energy as A increased to $A \approx 150$ when another peak appeared at low energy (fig. 7.17).

Finally when the averaged ratio of neutron width to level spacing was measured for slow neutron resonances, peaks were observed in the strength function $\langle \Gamma_n/D \rangle$

* H. H. Barschall, Phys. Rev. **86** (1952) 431; M. Walt and H. H. Barschall, Phys. Rev. **93** (1954) 1062.

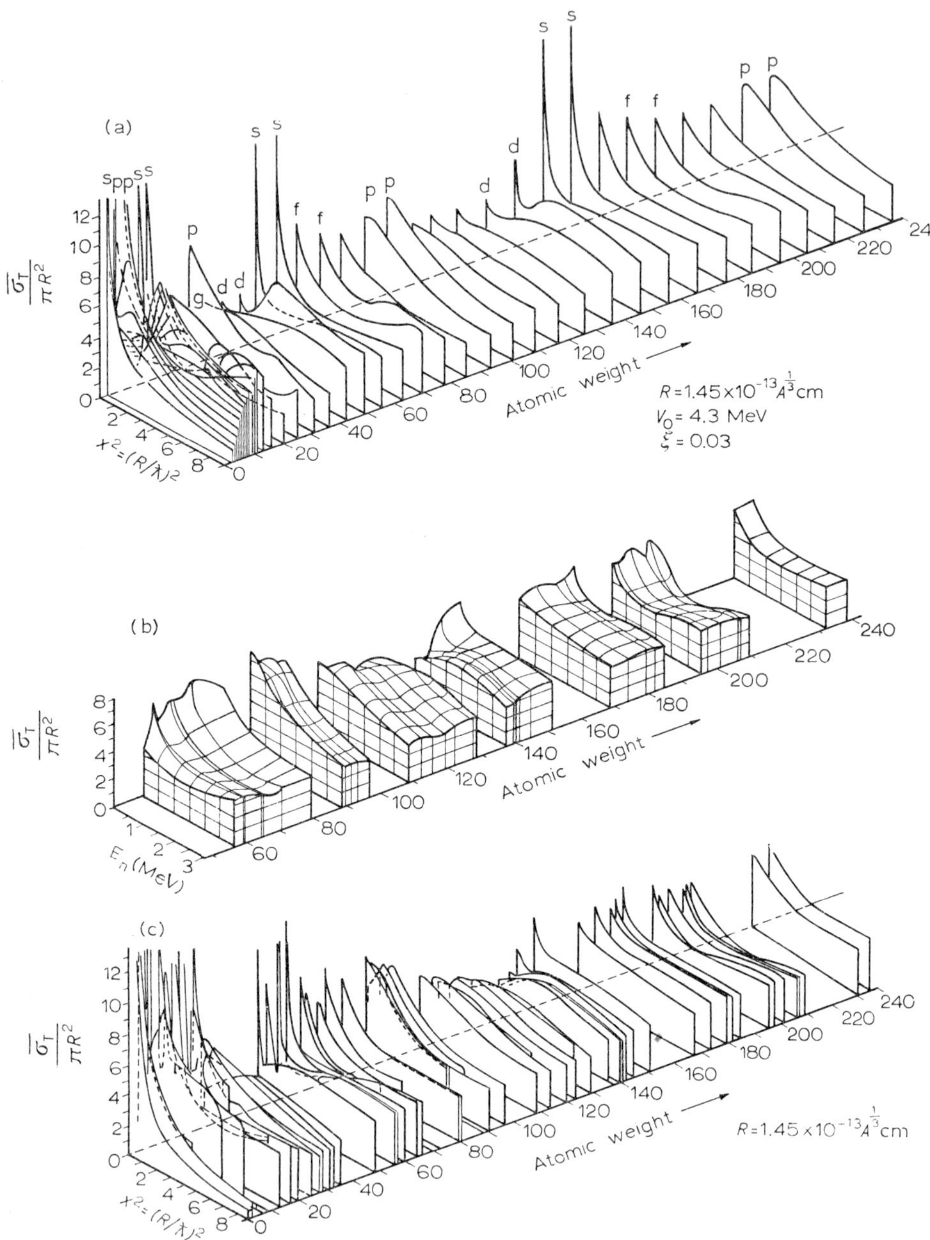

Phys. Rev. **96** (1954) 448

Fig. 7.17. a) Total neutron cross sections (calculated from the optical model) vs $x^2=(R/\lambda)^2$ and A. b) Measured total neutron cross sections averaged over resonances vs E and A. c) The data of (b) vs $(R/\lambda)^2$ and A

as shown in fig. 7.18. These are sometimes referred to as size resonances since the neutron wave can just fit into the well for certain values of the well radius.

These observations led Feshbach, Porter and Weisskopf* to develop a more general nuclear reaction model called the optical model. They supposed that the single

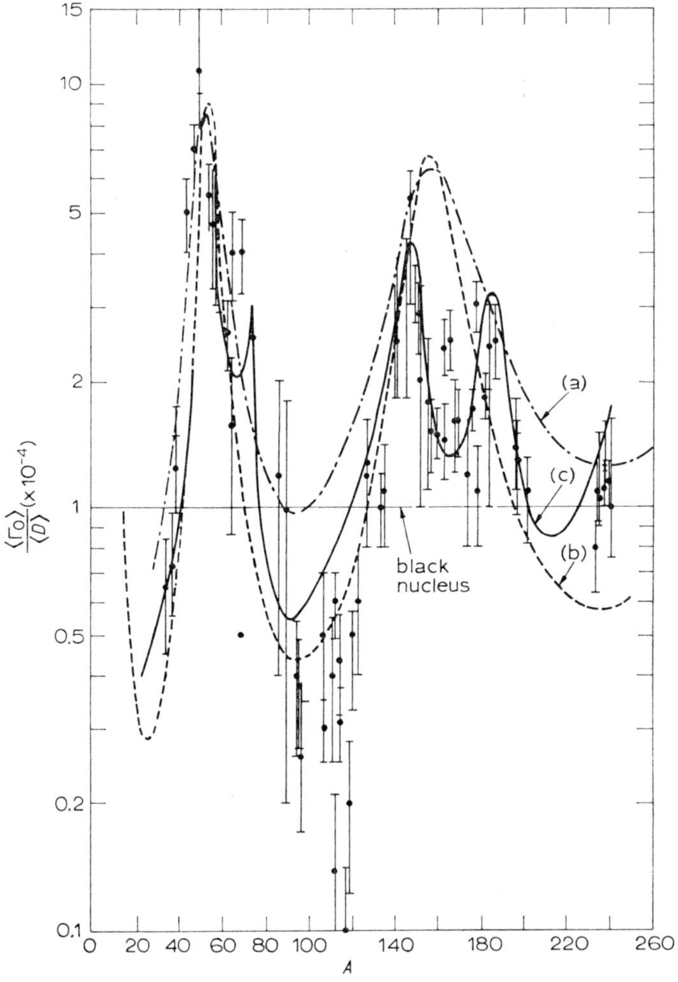

E. Segre, Nuclei and Particles (Benjamin, New York 1965) p. 469

Fig. 7.18.   Neutron strength function at zero energy vs A. The curves are calculated for a) optical model with Saxon potential, b) spherical nucleus, trapezoidal potential, surface absorption, c) deformed spheroidal nucleus

* H. Feshbach, C. E. Porter and V. F. Weisskopf, Phys. Rev. **96** (1954) 448.

particle sees a potential due to the target nucleus which is complex. We recall that in eq. (6.9) we introduced the scattering parameter η_l which was a complex number. The real part referred to elastic scattering and the imaginary part to inelastic processes. In the optical model the interaction of the incident particle wave with the nucleus is treated in analogy to the passage of light through a partially absorbing medium. The potential seen by the particle is written

$$U(r) = -[V(r)+iW(r)]. \tag{7.60}$$

We suppose that both the real and imaginary parts of the potential are energy dependent. The real part $V(r)$ is just the potential we assumed in our initial scattering model (§ 1) and at low energies describes the motion of a single particle. However we do also have effects due to short mean free path for the particle, that is, compound nucleus effects. These are controlled by $W(r)$, the imaginary part of the potential which attenuates the incident wave and thus controls the absorption of the particle. We can thus choose $V(r)$ so as to give the observed single-particle-like resonances seen in the scattering lengths and in the neutron total cross sections. We choose $W(r)$ to give the mean free path deduced for different energies. We also have to specify the radial variation of these potentials and if we use the form determined in ch. 1 § 6, we must specify at least 4 parameters, the magnitudes of $V(r)$ and $W(r)$, r_0 and a (if we assume the radial variation of V and W to be the same).

At an early stage in the development of the theory the complex potential was sometimes written

$$-V(r)(1+i\delta); \tag{7.61}$$

δ which is just W/V is then the absorption probability and turned out to be fairly small for neutrons of energy 1–5 MeV, of the order of 10 %. If we plot the magnitude of $V(r)$ and $W(r)$ as a function of energy as in fig. 7.19 we note a surprising feature. This is that at *low* energy the absorption is very small (W small) which means the mean free path is very large. Since we have assumed a very strong force between nucleons it seems unreasonable that a low energy neutron can cross and recross a nucleus without interacting with the densely packed nucleons. The answer (suggested by Weisskopf) recalls that the levels occupied by the target nucleons are all filled and the neutron is not allowed by the Pauli exclusion principle to occupy any of these states of motion. Thus unless its energy is large enough to have a chance of entering an unfilled state it cannot interact. This point is discussed more fully in ch. 15 § 8. For obvious reasons this model was at one time referred to as the 'cloudy crystal ball' model.

From experiments of various kinds we can empirically derive the optical model parameters, V, W and their radial dependence. We may expect that these will not vary very quickly from nucleus to nucleus or with energy. We must take care to

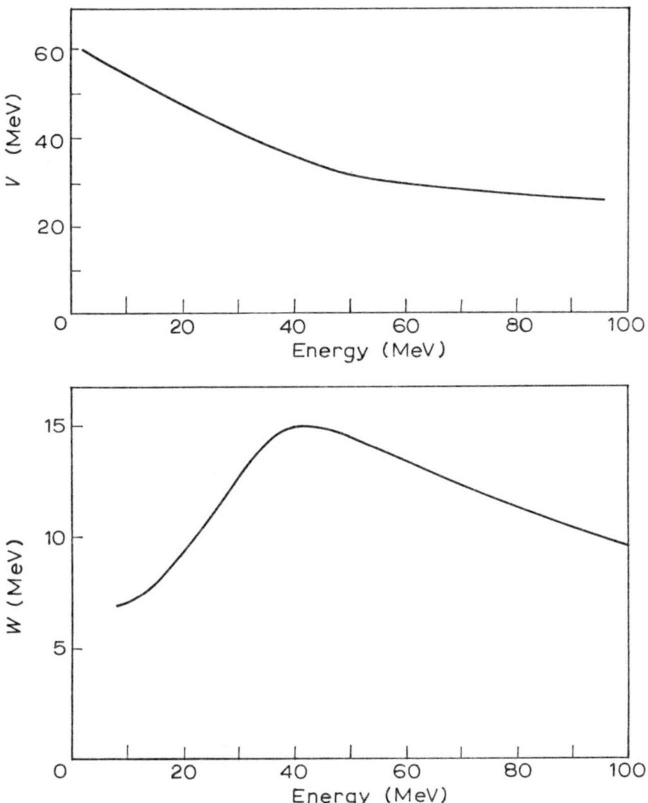

Fig. 7.19.   The variation of V and W with energy

ignore or rather to average over the effect of discrete compound nucleus resonances. This can sometimes be done experimentally by using poor energy resolution such that the fine structure is averaged out. Alternatively we can observe the fine structure with good resolution and perform the averaging afterwards. But this is often difficult as it requires some knowledge of the optical model trend which we are attempting to derive. At high energy where level density is high and level widths broad averaging is again achieved and the gross features show up. As a description of a large body of reaction data especially at high energies the optical model has been most valuable. If nuclear polarisation is observed in such experiments, a spin–orbit potential must be added to the main potential. The spin–orbit potential may itself be complex and has different radial dependence. In this case seven or eight parameters may have to be derived from the data.

In fig. 7.20 we show schematically the various modes of a nuclear interaction. Note that we cannot experimentally distinguish compound elastic scattering from shape elastic scattering The optical model cannot however predict compound elastic scattering and we can only interpret experiments where on other grounds we expect the compound elastic scattering to be small.

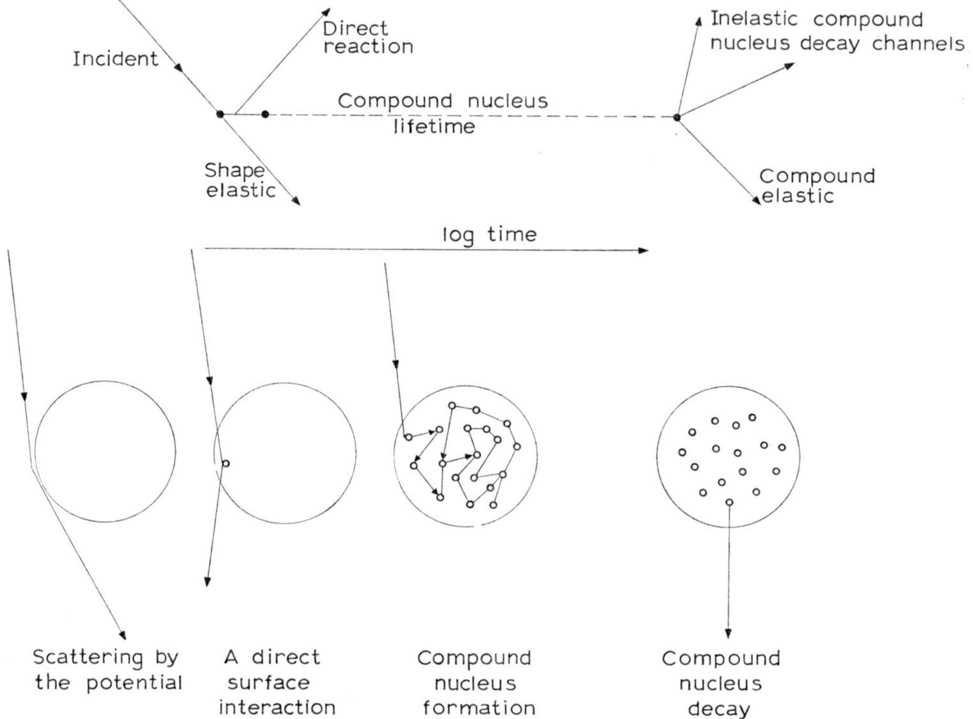

Fig. 7.20.   Schematic illustration of reaction mechanisms

For higher energies the optical model becomes more and more appropriate. In fact Serber formulated such a theory to explain high energy reactions before the cloudy crystal ball model was developed for low energy reactions.

13   General nuclear reaction theory

13.1   The S-matrix. Recall that reaction theory was formulated in terms of incoming and outgoing waves, so that our wavefunction

$$\psi(r) = \frac{1}{kr} \sum_l (2l+1) \mathrm{i}^l P_l(\cos\theta) \frac{\eta_l \mathrm{e}^{\mathrm{i}(kr-\frac{1}{2}l\pi)} - \mathrm{e}^{-\mathrm{i}(kr-\frac{1}{2}l\pi)}}{2\mathrm{i}}, \qquad (7.62)$$

where the constant η_l is complex and contains all the information about the scattering centre. If $|\eta_l|^2 = 1$ we have the possibility of a change in phase but no change in amplitude, corresponding to elastic scattering; while $|\eta_l|^2 < 1$ indicates a change in amplitude, i.e. loss of particles and change in phase corresponding to inelastic events.

For elastic scattering $\eta_l = e^{2i\delta_l}$ with δ_l real.

For reactions $\eta_l = e^{2i(\alpha_l + i\beta_l)}$ i.e. a complex phase.

Let us now generalise and label the possible channels a, b, c, etc. where a is the entrance channel.

Then in channel a for example

$$\psi_a = \psi(r_a)\chi_a$$

where $\psi(r_a)$ is a function of the relative coordinate of the particles in channel a and χ_a is a function describing the internal coordinates of the particles. Then if we assume r_a is *large* compared to the range of nuclear forces

$$\psi(r_a) = [A_a \exp(-ik_a r_a) + A'_a \exp(+ik_a r_a)]r_a^{-1}(4\pi v_a)^{-\frac{1}{2}}. \qquad (7.63)$$

The first term now represents the outgoing wave and the second an incoming wave. The wave number k_a and the relative velocity v_a in channel a obviously depend on the total energy of the system and of state a. The last factor normalises the wave to unit flux when $A_a = A'_a = 1$.

Now consider the case of an incoming wave present only in channel a. There will be outgoing waves in channel a and also in channels b, c etc. The first is the elastic scattering, the others are inelastic scattering and reactions. Again we look at the asymptotic behaviour of ψ_a in the region of exit channels b, c, to derive wavefunctions corresponding to reactions. Again the distances r_b, r_c are large compared to nuclear dimensions.

$$\psi_a = \psi_{ab}(r_b)\chi_b \qquad (7.64)$$

where

$$\psi_{ab}(r_b) = -S_{ab}(\exp ik_b r_b)(4\pi v_b)^{-\frac{1}{2}}r_b^{-1}. \qquad (7.65)$$

S_{ab} is a complex number function of energy and includes all the effects of scattering from channel a to channel b.

Suppose there are N channels altogether. Then there will be N^2 quantities like S_{ab} which form a matrix of which S_{ab} is an element. This is called the scattering or S-matrix or collision matrix or sometimes the U-matrix.

Such a matrix then includes all the information on the collision when we view it, as we must, from a distance large compared to nuclear force dimensions. The details of what goes on inside nuclear dimensions are all included and concealed in the χ_b function. The fact that these may be unknown does not affect our *description* of what happened in the collision as collected together in the S-matrix.

Although the direct calculation of the S-matrix is in general not possible we can make use of some important properties of the S-matrix which are based on very general principles.

1) The first called *unitarity* is expressed by the relation

$$\sum_n S_{an}^* S_{bn} = \delta_{ab} \tag{7.66}$$

where $\delta=1$ if subscripts are the same, $\delta=0$ if subscripts are not the same. In particular,

$$\sum_n |S_{an}|^2 = 1$$

means the sum of the probabilities of ending in one of the channels is 1. That is, we cannot create particles or destroy particles. If we have unit flux in, then when we add up all the outgoing particles we must also get unit flux.

2) The second property is that of time reversal invariance. This says that for the case of no spin the Hamiltonian is real and thus

$$H^* = H \quad \text{and} \quad S_{ab} = S_{ba}. \tag{7.67}$$

When spins are involved these must be reversed as well and the Hamiltonian is not real. But if we specify $-a$, $-b$ as the channels obtained by reversing all momenta *and* all spins then

$$S_{ab} = S_{-a-b}. \tag{7.68}$$

Obviously this relation has a close relation to the condition of detailed balance and has been checked experimentally to better than 1 %.

These two conditions then impose limitations on the N^2 elements of the S-matrix. Note that the diagonal elements of the S-matrix are $S_{aa}=\eta_l$ referred to before.

We can also express the cross section as

$$\sigma_{ab} = \pi \lambdabar_a^2 |\delta_{ab} - S_{ab}|^2. \tag{7.69}$$

Thus for $a=b$ we have the elastic scattering cross section

$$\sigma_{\text{el}} = \pi \lambdabar_a^2 |1 - S_{aa}|^2, \tag{7.70}$$

while for the reaction cross section we must sum over all channels $\neq a$

$$\sigma_{\text{r}} = \pi \lambdabar_a^2 \sum_{n \neq a} |S_{an}|^2. \tag{7.71}$$

But by unitarity

$$\sum_n |S_{an}|^2 = 1 = \sum_{n \neq a} |S_{an}|^2 + |S_{aa}|^2$$

so

$$\sigma_r = \pi \lambdabar_a^2 (1 - |S_{aa}|^2).$$

(7.72)

If now we consider values of S_{aa} between $+1$ and -1 we find

$$
\begin{aligned}
\sigma_{el,\,max} &= 4\pi \lambdabar_a^2 \quad \text{for} \quad S_{aa} = -1 \\
\sigma_{r,\,max} &= \pi \lambdabar_a^2 \quad \text{for} \quad S_{aa} = 0 \\
\sigma_{el} &= \pi \lambdabar_a^2 \quad \text{for} \quad S_{aa} = 0
\end{aligned}
$$

(7.73)

as we found before.

13.2 The R-matrix. So far this is a description in very general terms and has little specific physical content. It merely defines the asymptotic form of the wavefunction. But we have found previously that a specification of the continuity property (the matching conditions at the edge of the nuclear force region) will be sufficient to determine the wavefunction throughout the external region. That is, we need only specify the wavefunction and its derivative at the nuclear surface.

A very similar formulation has been made for the nuclear reaction problem. We define a quantity R_l as follows (for elastic scattering)

$$u_l(a) = R_l a \left(\frac{\partial u_l}{\partial r}\right)_{r=a}$$

(7.74)

connecting the value of the wavefunction with its derivative at the boundary $r=a$. We can easily relate this new function R with our S as

$$R_l = \left(\frac{u_l}{r\,\partial u_l/\partial r}\right)_{r=a} = \left[\frac{I_l - S_l O_l}{r(I_l' - S_l O_l')}\right]_{r=a}$$

(7.75)

where the prime indicates radial derivative and I is the incoming wave and O the outgoing wave and again we could derive cross sections in terms of R since we have derived them in terms of S.

But now we come to the new point of this method. We can express R in terms of certain states of the internal region. For example for those cases where the derivative of the inside wavefunction vanishes at $r=a$ we can call these *bound states* E_λ with corresponding eigenfunctions u_λ, $\lambda = 1, 2, \ldots$ etc.

Thus we connect our reaction theory description of what goes on outside (S-matrix) with what goes on inside which of course can tell us something about nuclear structure.

We specify our u_λ so that

$$(\partial u_\lambda/\partial r)_{r=a} = 0$$

(7.76)

and

$$\int_0^a u_\lambda u_{\lambda'}\, dr = \delta_{\lambda\lambda'}.$$

(7.77)

Suppose now we expand u_E, the wavefunction with any energy E

$$u_E = \sum_\lambda A_\lambda u_\lambda(r) \tag{7.78}$$

for a range $r = 0-a$. Since the u_λ are orthogonal

$$A_\lambda = \int_0^a u_\lambda u_E \, dr. \tag{7.79}$$

The wave equation for u is

$$\frac{d^2 u}{dr^2} + \frac{2M}{\hbar^2}\left(E - V - \frac{l(l+1)\hbar^2}{2Mr^2}\right) u = 0. \tag{7.80}$$

Let us multiply through by u_λ

$$u_\lambda \frac{d^2 u_E}{dr^2} + u_\lambda \frac{2M}{\hbar^2}\left(E - V - \frac{l(l+1)\hbar^2}{2Mr^2}\right) u_E = 0 \tag{7.81}$$

and multiply the equation for u_λ by u_E

$$u_E \frac{d^2 u_\lambda}{dr^2} + u_E \frac{2M}{\hbar^2}\left(E_\lambda - V - \frac{l(l+1)\hbar^2}{2Mr^2}\right) u_\lambda = 0. \tag{7.82}$$

Substract and integrate from 0 to a.

$$\int_0^a \left(u_\lambda \frac{d^2 u_E}{dr^2} - u_E \frac{d^2 u_\lambda}{dr^2}\right) dr + \frac{2M}{\hbar^2}(E - E_\lambda)\int_0^a u_E u_\lambda \, dr = 0. \tag{7.83}$$

At $r=0$, $u_\lambda = u_E = 0$ and at $r=a$, $du_\lambda/dr = 0$. Hence the first term is $u_\lambda(a)(du_E/dr)_a$ and the second term is $(2M/\hbar^2)(E - E_\lambda)A_\lambda$ or

$$A_\lambda = \frac{\hbar^2}{2M}(E_\lambda - E)^{-1} u_\lambda(a)\left(\frac{du_E}{dr}\right)_a \tag{7.84}$$

and

$$u_E(r) = \frac{\hbar^2}{2Ma} \sum_\lambda \frac{u_\lambda(r)u_\lambda(a)}{E_\lambda - E} \, a \left(\frac{du_E}{dr}\right)_a. \tag{7.85}$$

(Multiplication of top and bottom by a is to put the term in front in conventional form.) Putting this back into our definition of R we have

$$R = \frac{\hbar^2}{2Ma} \sum_\lambda \frac{u_\lambda^2(a)}{E_\lambda - E} = \sum_\lambda \frac{\gamma_\lambda^2}{E_\lambda - E} \tag{7.86}$$

where

$$\gamma_\lambda = \left(\frac{\hbar^2}{2Ma}\right)^{\frac{1}{2}} u_\lambda(a).$$

Obviously the form shows us that resonances may appear in scattering as a function of energy. The term γ_λ^2 is generally called the reduced width of the level in question. It will have a maximum value $\hbar^2/2Ma$, often called the Wigner limit or single particle reduced width.

This was all for elastic scattering i.e. one channel. We get an *R-matrix*

$$R_{an} = \sum_\lambda \frac{\gamma_{\lambda a}\gamma_{\lambda n}}{E_\lambda - E} = \sum_\lambda R_{an}^\lambda \tag{7.87}$$

when we allow all possible channels to interact.

It is not difficult to show that for a single λ the cross section formula reduces to just the single level Breit–Wigner formula we had before so the R-matrix formulation certainly includes the compound nucleus formation.

Now let us recall the details of the optical model and see if this is included too. You recall that if the single particle model were an exact description of a nucleus we should have only single particle states at fairly large separations (fig. 7.21a). But in fact there are residual interactions (corrections to the single particle Hamiltonian) so that a given state is not a pure and simple single particle state, but a mixture of such states. However since the single particle states form a complete set, we can always expand our actual state in terms of the single particle states. If *one* such state dominates the sum the actual state will approximate closely to a single particle state and this being the case at low energies we see agreement with the single particle model.

At higher energies the individual single particle states lose their importance and a given nuclear state contains many single particle configurations with more equal intensity. Or conversely we may say that a given single particle state u_p is distrib-

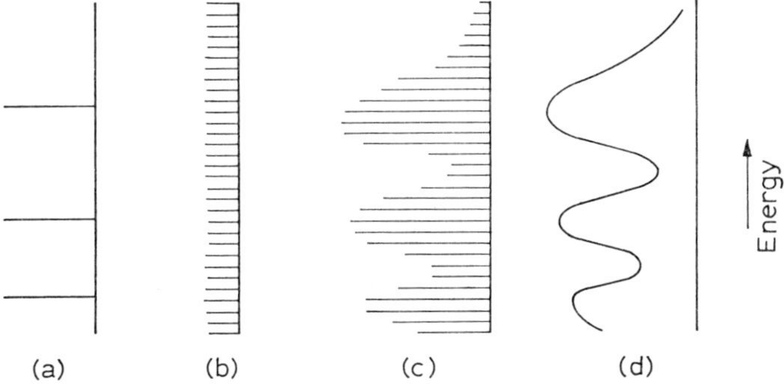

Fig. 7.21.   a) Single particle states, no residual interaction. b) Residual interaction very strong, complete smearing-out of individual states, the 'black ball' or uniform model. c) Residual interaction medium strong; some smearing-out of single particle states but not complete, the intermediate model. d) A poor resolution experiment showing resulting giant resonances

buted over many actual nuclear states. So if in our examination of the actual state we study some property characteristic of u_p we see it to the extent that this state contains u_p. The *strength* with which the process occurs at energy E depends on the admixture of u_p into the nuclear states at energy E. This is the basic idea of a *strength function* which we related to

$$\frac{\langle \gamma_{\lambda c}^2 \rangle}{\langle D \rangle} = \frac{\text{reduced width}}{\text{level spacing}} \text{ (average).}$$

Let us call the states of the system, nucleus plus incident particle, X_λ. The function $\psi_{\lambda c}(r)$ is a function of the radial coordinate of the scattered nucleon. We expand in terms of the eigenfunctions u_p of the single particle state of energy E_p. That is, the target nucleus in state c with energy E_c and extra nucleon in state p. Then

$$\chi_{cp} = \phi_c u_p(r). \tag{7.88}$$

ϕ_c describes the internal state of motion of the target nucleus. Since the χ_{cp} are a complete set

$$X_\lambda = \sum_{c,\,p} C_{\lambda,\,cp} \chi_{cp} \tag{7.89}$$

and we can write

$$\gamma_{\lambda c} = \left(\frac{\hbar^2}{2M_c a_c}\right)^{\frac{1}{2}} \sum_p C_{\lambda,\,cp} u_p(a_c). \tag{7.90}$$

at the boundary. Hence our *R*-function

$$R(E) = \sum_\lambda \frac{\gamma_\lambda^2}{E_\lambda - E}$$

$$= \frac{\hbar^2}{2Ma} \sum_{pp'} u_p(a)u_{p'}(a) \sum_\lambda C_{\lambda p}(E_\lambda - E)^{-1} C_{\lambda p'}$$

$$= \frac{\hbar^2}{2Ma} \sum_{pp'} u_p(a)u_{p'}(a) \langle \chi_p | (H-E)^{-1} | \chi_{p'} \rangle \tag{7.91}$$

where H is the total Hamiltonian and E the nucleon energy.

If now we take the intermediate picture we can assume that the state χ_p is found in appreciable intensity only in those states X_λ which have energy E_λ reasonably near energy of χ_p, i.e. E_p. So sums over p are dominated by a single term. Then

$$\gamma_\lambda^2 \approx \frac{\hbar^2}{2Ma} \sum_p u_p^2(a) C_{\lambda p}^2 \tag{7.92}$$

and

$$R(E) \approx \frac{\hbar^2}{2Ma} \sum_{\mathrm{p}} u_{\mathrm{p}}^2(a) \langle \chi_{\mathrm{p}} | (H-E)^{-1} | \chi_{\mathrm{p}'} \rangle. \tag{7.93}$$

For any E only one $C_{\lambda\mathrm{p}}$ is appreciable and we have only one term in this expression for the *average* of R over the interval E. Note that $\gamma_{\mathrm{p}}^2 = (\hbar^2/2Ma)u_{\mathrm{p}}^2(a)$ is the *single particle* reduced width. And since the X_λ form a complete set $\sum_\lambda C_{\lambda\mathrm{p}}^2 = 1$, we have $\sum_\lambda \gamma_\lambda^2 = \gamma_{\mathrm{p}}^2$, that is, the total reduced width of the single particle state, p, is completely distributed amongst the widths of the actual nuclear states, λ.

The R-function for a single particle potential we derived as

$$R_{\mathrm{sp}} = \frac{\hbar^2}{2Ma} \sum_{\mathrm{p}} \frac{u_{\mathrm{p}}^2(a)}{E_{\mathrm{p}} - E}. \tag{7.94}$$

The basis of the optical model is then that we can find empirically a complex potential $V(r) = V_0(r) + iW(r)$ which makes the expression for R_{sp} equal to the average value of R, $R(E)$, when we put such a modified potential into the Hamiltonian H. Note that this is approximately just the substitution in the denominator of R_{sp} of $E_{\mathrm{p}} - iW_{\mathrm{p}}(E)$ instead of E_{p}. This provides for the removal of particles via the complex part from the channel being viewed.

We can find the direct interaction effect also via R-matrix theory. To retrieve this we go back to the point where in the expression for $R(E)$ we picked out only the one term from the sum referring to the single particle state p and ignored all other terms. For example if we expand the expression $(H-E)^{-1}$ in powers of $H - H_0$ where H is the optical Hamiltonian and H_0 is the actual Hamiltonian,

$$\begin{aligned} H &= H_{\mathrm{A}} + T_0 + V_{\mathrm{opt}} \\ H_0 &= H_{\mathrm{A}} + T_0 + V_{\mathrm{actual}} \end{aligned} \tag{7.95}$$

(H_{A} is the target nucleus Hamiltonian) we get a series with some terms which refer to the contribution of a particular single particle state p as we did before and other terms referring to the contributions of all other states p′ which we essentially ignored in getting giant resonances by the assumption that the phases of their contributions would be random and hence cancel out. But this is not precisely correct and this second term turns out to have just the characteristics we require to describe direct interactions.

You recall that we described direct interactions phenomenologically as follows: a nucleon enters a nucleus and if absorption is small it has a large mean free path and can bounce around the nucleus until perhaps it undergoes a single nucleon encounter. We may originally have talked of this happening near the edge of the nucleus but the argument is formally the same. If the absorption is increased then compound nucleus formation becomes more and more probable, and in the 'black nucleus' approximation we would get no direct interaction. Neither do we get any optical model effects which were interpreted as spreading the effect of the single particle levels over a limited energy region of order of W.

So we begin to see the unifying effect of all this. The direct processes correspond to long range correlations of sign of the contributions of the $\gamma_{\lambda c}$ and the giant resonances correspond to one $\gamma_{\lambda c}$ predominating. If we apply a 'microscope' and

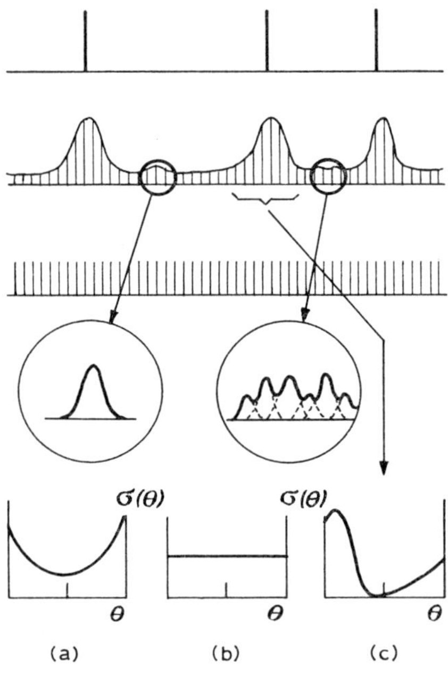

Fig. 7.22. Characteristic angular distributions: a) single isolated resonance, symmetric about 90°, b) overlapping resonances with random phases, isotropic approximately, c) phases not random when direct interaction or single particle effects are observed, forward peaked angular distribution

look at the fine structure compound nucleus states, we can see isolated levels at low energy corresponding to the Breit–Wigner form but at higher energies these will overlap and then if we look for correlations of the signs of the contributions of the *compound states* we do not see this and we have random signs as expected (fig. 7.22).

Since the widths associated with compound nucleus states are very narrow these correspond to a *long time*. But the widths associated with single particle effects are of order W, i.e. very wide, and characteristic times are very short. Thus by the R-matrix formulation we essentially connect our reaction observations far from the nucleus with the behaviour of the wavefunction at the nuclear boundary

$r=a$. We can in turn connect this directly with the wavefunction describing what happens inside (fig. 7.23). The average behaviour is dominated by the smeared-out effect of rather simple single particle motions. We get giant resonance effects if we are near a primeval simple state or direct interaction effects due to a number of distant ones. If we look with high energy resolution we see the compound nucleus states in all their complexity but by and large we cannot interpret them as this is a many-body problem involving a complicated force.

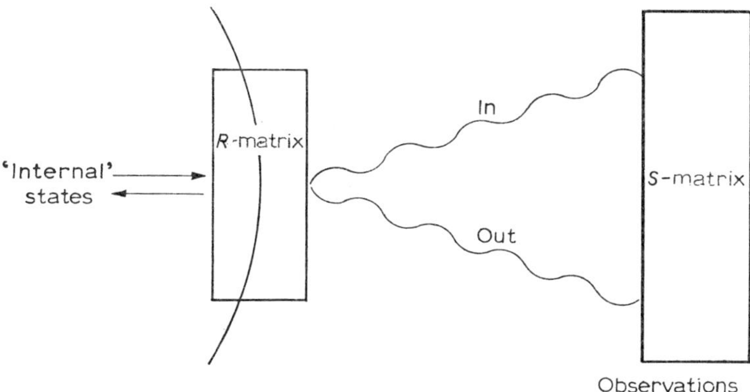

Fig. 7.23.   Relation of the S-matrix, R-matrix and internal states of the real nucleus

Of course we often prefer to interpret our observations in terms of some phenomenological formulation such as the optical model with empirically determined parameters describing the magnitude and radial dependence of the real and imaginary parts of the complex potential or the direct interaction model in which angular dependence and momentum transfer are simply related to the parity change. But it is satisfying that a basically much more rigorous formulation exists which in theory describes everything.

14   Nuclear fission

14.1   Introduction. We now consider briefly the phenomenon of fission which as we shall see is a rather special type of nuclear reaction. The fission reaction can be written using ^{235}U as an example,

$$^{235}U+n \rightarrow {}^{236}U \rightarrow A+B+Q$$

where A and B are radioactive nuclei of mass number between 70 and 170 and the energy release Q is of the order of 150 MeV. This is a very large energy release compared to other reactions. How it is made available can be understood by considering the semi-empirical binding energy formula. The binding energy per nucleon as a function of A is shown again in fig. 7.24. At $A \approx 120$ the B/A value is 8.5 MeV while at $A \approx 240$ it is 7.5 MeV. Thus division of a nucleus of $A = 240$ with total binding ≈ 1800 MeV into two nuclei of $A = 120$ with total binding $2 \times 120 \times 7.5 = 2040$ MeV results in an energy gain of 240 MeV.

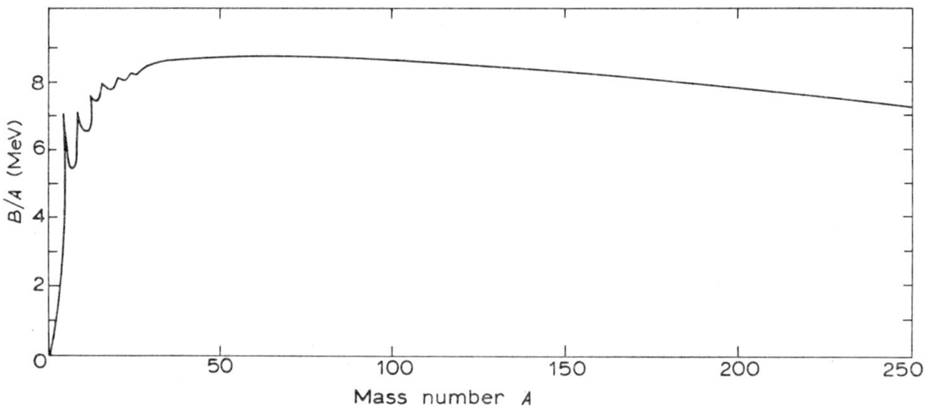

R. D. Evans, The Atomic Nucleus (McGraw-Hill, New York 1955) p. 299

Fig. 7.24.   Binding energy per nucleon vs A

We recall also that the neutron excess increases with A to counterbalance the coulomb repulsion of the increased number of protons. Thus when we divide a heavy nucleus with large neutron excess we create fission product nuclei with neutron excess much larger than for stable forms of these nuclei. The fission products will be β^--radioactive and will decay back to the stability line through a chain of successive β-decays.

The basic experimental facts concerning fission are as follows:

a) Uranium bombarded by neutrons breaks up as follows

$$_{92}U + n \rightarrow {}_{52}Te + {}_{40}Zr$$
$$\text{or } {}_{56}Ba + {}_{36}Kr \text{ etc.}$$

It was the positive chemical identification of barium by Hahn and Strassmann which led Meitner and Frisch to suggest that fission was taking place.

b) The fission fragments recoil in opposite directions with ranges which correspond to about 75 MeV per fragment. The range is very short as the fragment is heavy but the ionization is very intense because of the large energy deposition per unit path length.

c) The fragments are highly radioactive because of the neutron excess and give rise to chains of radioactive products such as

$$^{140}_{54}\text{Xe} \rightarrow \, ^{140}_{55}\text{Cs} \rightarrow \, ^{140}_{56}\text{Ba} \rightarrow \, ^{140}_{57}\text{La} \rightarrow \, ^{140}_{58}\text{Ce (stable)}.$$

d) A few fast neutrons and high energy γ-rays are emitted simultaneously with the fission fragments. If the extra neutrons can be arranged to induce more fissions, a chain reaction can be started.

Meitner, Hahn and Strassmann found also that when uranium was bombarded with neutrons in addition to fission a 23 minute β-activity was induced. The amount of this activity was such that it could not be due to the rare 0.7 % isotope ^{235}U but was due to neutron capture in the abundant isotope ^{238}U. The 23 minute decay was followed by a 2.3 day activity and the chain of events was eventually shown to be as follows:

$$^{238}\text{U} + \text{n} \rightarrow \, ^{239}\text{U} + \gamma$$
$$^{239}\text{U} \rightarrow \, ^{239}\text{Np} + \beta^- + \bar{\nu} \text{ (23 m)}$$
$$^{239}\text{Np} \rightarrow \, ^{239}\text{Pu} + \beta^- + \bar{\nu} \text{ (2.3 d)}.$$

Neptunium and plutonium were new transuranic nuclides.

14.2   Theory of fission. Bohr and Wheeler treated the fission process by assuming that the nucleus behaved like a liquid drop. This was the starting point we recall for the semi-empirical binding energy formula. The main term in the latter was the volume term proportional to A. This was reduced by a disruptive surface term proportional to $A^{\frac{2}{3}}$ and a disruptive coulomb term proportional to $Z^2/A^{\frac{1}{3}}$. We now compare the energy gain in splitting up a heavy nucleus (A, Z) into two nuclei each of $(\frac{1}{2}A, \frac{1}{2}Z)$. This is simply

$$Q = a_\text{s} A^{\frac{2}{3}}(1 - 2^{\frac{1}{3}}) + a_\text{c} Z^2 A^{-\frac{1}{3}}(1 - 2^{-\frac{2}{3}})$$

where a_s and a_c are the constants of the binding energy formula. For ^{236}U this gives about 169 MeV (cf. fig. 7.25). However a spherical nucleus must be deformed into a non-spherical shape with gain in potential energy for fission to occur. If we assume deformation into an ellipsoid of eccentricity ε, both the surface terms and the coulomb term will vary as follows:

$$E_\text{s} = a_\text{s} A^{\frac{2}{3}}(1 + \tfrac{2}{5}\varepsilon^2 + \ldots) \tag{7.96}$$

$$E_\text{c} = a_\text{c} Z^2 A^{-\frac{1}{3}}(1 - \tfrac{1}{5}\varepsilon^2 + \ldots). \tag{7.97}$$

The total change in energy as we go from sphere to ellipsoid (for ε small) is thus

$$\Delta E = \varepsilon^2 [\tfrac{2}{5} a_\text{s} A^{\frac{2}{3}} - \tfrac{1}{5} a_\text{c} Z^2 A^{-\frac{1}{3}}]. \tag{7.98}$$

Using values of the constants $a_\text{s} = 17.3$ MeV and $a_\text{c} = 0.70$ MeV we note that ΔE becomes negative for $Z^2/A > 49$. In this case the nucleus would spontaneously

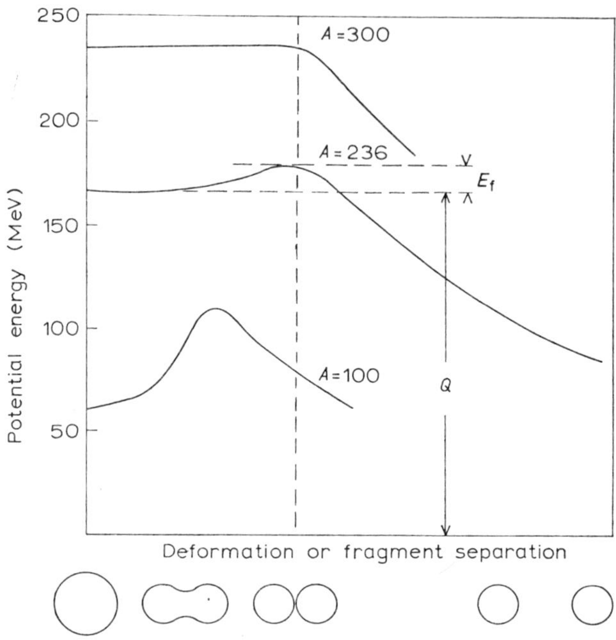

W. E. Burcham, Nuclear Physics (Longmans, London 1963) p. 700

Fig. 7.25.   The fission barrier

undergo fission in the characteristic nuclear time, 10^{-23} sec. Such a nucleus would have $Z \approx 115$ and $A \approx 270$. The heaviest known nucleus $^{257}_{103}$Lw has $Z^2/A = 41.4$. Its spontaneous fission half-life is expected to be a few minutes. Thus spontaneous fission effectively sets a limit on the size of nuclei which can be produced. Spontaneous fission can however occur for nuclei with $Z^2/A < 49$ by quantum mechanical penetration of the coulomb barrier $Z_1 Z_2 e^2/r$ when Z_1 and Z_2 are the fragment charges. Obviously the probability will show a very rapid variation with the value of Z^2/A as this is a measure of how far below the barrier the Q-value lies.

In the case of slow neutron fission the excitation of the compound nucleus is just the neutron separation energy S_n and for fission to take place this must exceed $E_c - Q$. If the neutron separation energy is less than $E_c - Q$ more energy will be required to induce fission and there will be a fission threshold. Appreciable fission will only occur when the neutron energy exceeds $E_c - Q - S_n$.

In table 7.3, note that all the thermally fissile nuclei (threshold <0) are odd-N nuclei so that the compound nucleus is even–even and the neutron separation energy is large. Shortly after the separation of the fragments prompt neutrons are emitted. These vary between 2 and 3 neutrons per fission differing somewhat between different nuclides and at different excitation energies. The neutrons have

TABLE 7.3

Neutron induced fission

Target	Compund nucleus	Z^2/A	S_n(MeV)	E_c-Q(MeV)	Effective threshold(MeV)
^{232}Th	^{233}Th	34.8	5.3	6.6	1.3
^{233}U	^{234}U	36.2	7.3	4.6	<0
^{234}U	^{235}U	36.0	5.7	6.1	0.4
^{235}U	^{236}U	35.9	6.9	5.3	<0
^{236}U	^{237}U	35.7	5.7	6.5	0.8
^{238}U	^{239}U	35.4	5.1	6.3	1.2
^{237}Np	^{238}Np	36.3	5.8	6.2	0.4
^{239}Pu	^{240}Pu	36.9	6.9	4.0	<0

an evaporation type spectrum with a peak energy of about 2 MeV and extending up to about 15 MeV. Prompt γ-rays are also emitted. The highly β-unstable fragments then start to decay via the chains mentioned above until a stable nuclide is reached. In some of the decay chains a highly unstable nuclide may β-decay to a level in the daughter nucleus above the neutron binding energy. This level at once breaks up with emission of a neutron. The neutrons will be observed to decay with the half-life of the parent β-decay. For ^{235}U fission the delayed neutrons with half-lives between 0.23 and 54 sec amount to 0.75 % of the prompt neutrons.

14.3   Thermally fissile nuclides. The three thermally fissile nuclides of importance for chain reactors since they can be produced in useful quantities are ^{235}U, ^{233}U and ^{239}Pu.

1) ^{235}U occurs naturally as $\frac{1}{140}$th of natural uranium and can be obtained in pure form by various isotope separation processes. It is an α-active nuclide with half-life 7×10^8 y.

2) ^{233}U can be produced in nuclear reactors by neutron capture in thorium

$$^{232}\text{Th}+\text{n} \rightarrow {}^{233}\text{Th}+\gamma$$
$$^{233}\text{Th} \rightarrow {}^{233}\text{Pa}+\beta^- +\bar{\nu} \text{ (23.6 min)}$$
$$^{233}\text{Pa} \rightarrow {}^{233}\text{U}+\beta^- +\bar{\nu} \text{ (27.4 days)}.$$

^{233}U is an α-active nuclide with half-life 1.6×10^5 y.

3) ^{239}Pu can be produced in nuclear reactors by neutron capture in ^{238}U, the abundant isotope of uranium, by the reactions mentioned in § 14.1. ^{239}Pu is also α-active with a half-life of 2.4×10^4 y.

The *average* energy division in fission is as shown in table 7.4. In a large assembly such as a reactor practically all except the antineutrinos will be

TABLE 7.4

Energy balance of an average fission

Kinetic energy of fragments	167 MeV
Kinetic energy of prompt neutrons (2.5)	5
Prompt γ-rays (5)	6
Fragment decay β-particles (7)	8
Fragment decay antineutrinos (7)	12
Fragment decay γ-rays (7)	6
	204 MeV

stopped and the energy converted into heat via ionization. Approximately 3.2×10^{10} fissions per second correspond to 1 watt.

In the foregoing crude treatment of fission theory we have assumed that the nucleus divided into two equal mass fragments. For low energy neutron fission and spontaneous fission such symmetric fission occurs rather infrequently but

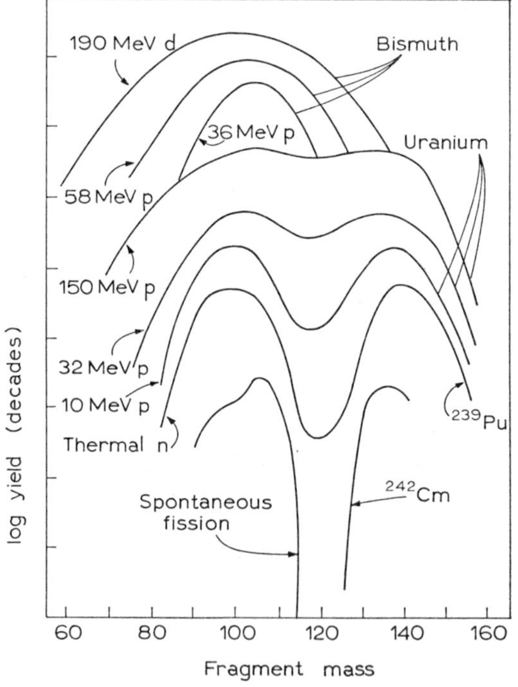

Ann. Rev. Nucl. Sci. 9 (1959) 311

Fig. 7.26.   The mass distribution in fission

instead shows a mass distribution with two peaks at $A=90$ and $A=145$ (for ^{235}U thermal neutron fission). The ratio of asymmetric to symmetric fission is in this case several hundred. As the excitation energy is increased symmetric fission becomes more probable and eventually dominates. Symmetric fission is also more probable for nuclides lighter than thorium. The preference for division into unequal fragments at low energies is doubtless connected with shell structure stability in the fragments but no completely satisfactory theory is yet available. Fig. 7.26 shows a variety of mass distribution curve shapes.

The cross section for neutron fission shows similar characteristics at low energy to the capture reaction. At thermal energy the cross section varies as v^{-1} while at higher energy resonances in the fission cross section appear. The main competing reaction is the (n, γ) capture.

14.4 Nuclear reactors. In order to set up a self-sustaining chain reaction, the fast prompt neutrons emitted in a fission process must be slowed down to near thermal energies where the fission cross section is large. This is accomplished by mixing with the fissile fuel, a moderator to do the slowing down. Light elements such as D, Be and C are used which have small neutron absorption cross sections. A neutron elastically scattered from a light element can lose a large fraction of its energy. For the reaction to be self-sustaining there must be sufficient neutrons available to compensate for losses due to absorption by the (n, γ) process in fuel and moderator, absorption in structural materials and leakage out of the reactor. If natural uranium is used as fuel, absorption, in particular in the resonance region in ^{238}U, is another loss mechanism.

If in each generation the neutron flux is multiplied by k the flux will increase as

$$1+k+k^2+k^3+ \ldots = \frac{1}{1-k} \qquad (k < 1).$$

That is, if $k>1$ the flux diverges while if $k<1$ the series converges and the reactor is not self-sustaining. Let us consider an infinite system. Suppose that we start with N thermal neutrons. The probability for absorption in the fissile material is f (the thermal utilisation factor) while $1-f$ is the probability for absorption in the moderator or structural materials. The Nf neutrons absorbed in the fuel may produce fission or be captured. If the fission cross section is σ_f and the capture cross section is σ_a we define $\alpha=\sigma_a/\sigma_f$. Then the probability of producing a fission is just $1/(1+\alpha)$ and $Nf/(1+\alpha)$ fissions take place. If now each fission results in v prompt fast neutrons, the ratio of fast neutrons produced to slow neutrons absorbed we call $\eta=v/(1+\alpha)$ and we have $Nf\eta$ fast neutrons. Some of these neutrons may produce fissions before being slowed down. This will increase the fast neutrons by a factor ε (the fast fission factor) and $Nf\eta\varepsilon$ are available for slowing down. During the slowing down time some of these neutrons will undergo resonant capture in the fuel and in ^{238}U if present. Suppose the probability of a neutron reaching thermal energy is p (the resonance escape probability) then the new generation of neutrons number $Nf\eta\varepsilon p$ and the reproduction factor $k=f\eta\varepsilon p$. This is the 'four factor' formula for an infinite system. The factor f depends on the composition of the reactor, η depends on the nuclear properties of the fissile nuclide, ε and p depend on both the composition and the geometrical arrangement of fuel and moderator. The standard technique of distributing the fuel on a lattice embedded in the moderator has several effects. 1) It increases ε. 2) Since moderation is carried out in a fuel-free region it reduces the chance of resonant capture in ^{238}U and increases p. 3) Since the resonant capture cross sections are very large capture will occur only in the surface of the fuel lumps and to maximize p the lumps should be large to minimize surface to volume. On the other hand increasing the volume decreases f since the neutron flux density is lower in or near the lumps, and the chance of absorption in the moderator is increased. Thus there is an optimum geometry to yield optimum values of p and f.

The four factor formula gives a value for k in an infinite medium and depends only on the composition and lattice arrangement. For a finite reactor neutrons will leak out of the reactor and the effective multiplication factor $k_{eff}=kP$ when P is the probability that a neutron will not leak out. Reactors are frequently provided with a reflector usually made up of one of the moderating materials, so there is an increased chance of the neutron re-entering the reactor core and k_{eff} is increased.

Typical values for the factors of a natural uranium fuelled reactor are $\alpha=0.88$, $\eta=1.31$, $f=0.88$, $\varepsilon=1.02$, $p=0.89$ and $k=1.05$ for an infinite system. To calculate k_{eff} and hence the critical size of a reactor so that k_{eff} exceeds 1 is very complex and usually must be checked by experiment. Suppose our reactor is spherical and that M, the migration length represents the average distance travelled by a neutron between emission as a prompt fission neutron and absorption as a thermal neutron. Suppose that all neutrons originating in a shell of thickness M in fact escape. Then the ratio of the volume of the shell to the total volume is $\approx 3M/R$ (for $M \ll R$) and if we assume that the neutron density increases linearly toward the centre the shell density/sphere density $\approx M/R$. The probability for escape is thus $3M^2/R^2$. Hence

$$k_{eff} = k_\infty(1-3M^2/R^2). \tag{7.99}$$

The critical radius at which $k_{eff}=1$ is thus

$$R_c^2 = \frac{3M^2 k_\infty}{k_\infty-1} \approx \frac{3M^2}{k_\infty-1}. \tag{7.100}$$

M is about 50 cm for graphite moderator and thus in our example above $R_c=385$ cm.

As the neutron energy increases from thermal σ_a decreases and v increases hence k_∞ is larger and the dimensions of a *fast* reactor can be made smaller.

14.5   Reactor control. A nuclear reactor which is critical must be controllable; the operator must be able to control k_{eff}. This is done by inserting control rods containing materials such as Cd or B which have large neutron absorption cross sections. These are withdrawn from a reactor of critical size until k_{eff} exceeds unity. The neutron density, ρ, then increases with time according to the equation

$$\frac{d\rho}{dt} = \frac{\rho(k-1)}{\tau_0} \qquad (7.101)$$

where τ_0 is the average neutron lifetime in the reactor, ≈ 1.4 milliseconds. Most of this time is spent in the slowing down process. The neutron density rises exponentially with a time constant

$$T = \tau_0/(k-1). \qquad (7.102)$$

Thus even a small excess reactivity say $k=1.01$ leads to a doubling time for the reactor power of 0.1 second. This is inconveniently short and practical control would be impossible. However the presence of the delayed neutrons in the fission product decay chains makes control possible. Suppose β is the ratio of the number of delayed neutrons per fission to the number of prompt neutrons per fission v. For ^{235}U, $\beta=0.0064$. The reproduction factor k is thus made up of a factor due to the prompt neutrons k_p and a factor due to the delayed neutrons, $k_d=\beta k$. Thus $k=k_p/(1-\beta)$. If $k_{\text{eff}}>1$ or $k_{p\ \text{eff}}>1/(1-\beta)$ $(=1.007$ for ^{235}U) the time constant will be little affected by the delayed neutrons and the reactor is prompt critical. However if the reactor is subcritical on prompt neutrons but supercritical on prompt plus delayed neutrons, the time constant is given by the equation

$$\frac{\tau_0}{k^2 T} + \frac{1}{k}\sum_i \frac{\beta_i \tau_i}{T+\tau_i} = (k_{\text{eff}}-1) \qquad (7.103)$$

where β_i, τ_i are number of delayed neutrons and their decay lifetimes. The latter range from 0.23 to 54 sec. When the excess reactivity $(k_{\text{eff}}-1)=2.6\times10^{-5}$ the reactor period is 1 hour. This is frequently used as a unit of excess reactivity called the inhour (inverse hour).

14.6   Breeding in reactors. We have seen that the absorption of neutrons in a thermal reactor by ^{238}U or ^{232}Th can lead to the production of further fissile nuclides ^{239}Pu or ^{233}U. For this reason ^{238}U and ^{232}Th are called fertile nuclides. Some of the ^{235}U fuel used to keep the reactor operating is converted into ^{239}Pu which is fresh fuel. The ratio of new fuel created to old fuel destroyed is called the conversion gain C while $C-1=G$ is termed the breeding ratio. We note that

$$\begin{aligned} C &= \eta-1-(1-P) \\ G &= \eta-2-(1-P) \end{aligned} \qquad (7.104)$$

where $(1-P)$ is the probability for escape of a neutron. From table 7.5 we note that η is only just greater than 2 for all thermally fissile nuclei except ^{233}U for which $\eta = 2.28$. Hence thermal breeding reactors will have to be based on this fissile material. We noted above however that η increases with mean neutron energy. Thus

TABLE 7.5

Values of nuclear constants of importance for reactors

	Natural U	^{233}U	^{235}U	^{239}Pu
σ_a (b)	7.68	588	694	1025
σ_f (b)	4.18	532	582	738
ν	2.47	2.52	2.47	2.91
η	1.34	2.28	2.07	2.09

at 1 MeV η is about 2.45 for ^{233}U, 2.3 for ^{235}U and 2.7 for ^{239}Pu. Thus fast reactors are favourable candidates for breeder reactors. Fast reactors are however more difficult to build from the engineering standpoint as the reactor core is much smaller and the heat produced per unit volume is difficult to remove quickly enough. Liquid metal coolants must be used. The intrinsic reactor period τ_0 may now be a thousand times shorter than for a thermal reactor and control is more difficult.

8

Nuclear Models:

Shell Model

1 Introduction

We now want to discuss the structure of the nucleus in more detail and consider the various models which have been proposed to describe and predict nuclear properties. Such properties associated with the structure are the angular momentum, parity and moments of the ground state, that is the lowest stable configuration, as well as the energies of metastable configurations or excited states, their widths or lifetimes and their angular momenta and parities.

There have been several approaches to this problem, most of which are based on analogies with models which have proven useful of atomic or thermodynamic systems. The *Fermi Gas Model* is an analogy with the electron theory of solids. We assume the nucleons attract in pairs and are confined to a volume of radius R. Then many features of the thermodynamics of a perfect gas are applicable. The wavelength of the nucleons $\lambda = \hbar/p$ must be $\leq R/A^{\frac{1}{3}}$ hence p, the momentum, $\approx \hbar A^{\frac{1}{3}}/R$. Thus the kinetic energy per particle is $p^2/2M = \hbar^2 A^{\frac{2}{3}}/2MR^2$. The total kinetic energy for A nucleons is thus $\hbar^2 A^{\frac{5}{3}}/2MR^2$. However the potential energy must be proportional to the number of interacting pairs, $\frac{1}{2}A(A-1)$. The potential energy thus varies as A^2 while the kinetic energy varies as $A^{\frac{5}{3}}/R^2 \propto A$. Thus as A increases the nucleus would collapse if we believe the forces are attractive up to $r = 0$. This does not occur of course but this model is useful in describing phenomena which are sensitive to the high momentum part of the nucleon spectrum. For example in high energy scattering the shape of the momentum spectrum gives the energy spread in the scattered particles. In this case a momentum spectrum based

255

on this model fits quite well. For tightly bound states however it fails.

The *Liquid Drop Model* has already been used in discussing the theory of fission (Bohr and Wheeler) and in parameterising the semi-empirical mass formula. It was applied to nuclear structure by Bohr and Kalckar in 1937 but as it gave rise to very closely space energy levels it seemed unsuited to describe low lying excited states which are quite widely spaced. In this model we assume a chunk of nuclear matter with strong attractions between the constituents as in a liquid. In such a many-body system many configurations differing in energy by a small amount are expected. Such close spacing is indeed seen in neutron capture cross sections, that is in nuclear levels near the neutron binding energy (≈ 8 MeV). The fact that the semi-empirical binding energy formula based on the average behaviour of such a system works so well indicates that the basic idea of a strong short range interaction must be correct in some respects. We return to modern versions of the liquid drop model when we discuss collective models in ch. 9.

The *Shell Model* is suggested in analogy with the electronic structure of the atom and we suppose that by fixing our attention on the last bound particle or particles we will see properties analogous to those imparted to atoms by the valence electrons. Unlike the other two models, this one will be very sensitive to nucleon angular momentum and if valid should enable us to predict total angular momentum and parity. We should expect to see periodicity in properties as we do in atoms as the various nucleon shells are filled up according to the Pauli principle. We shall have to arrange that the rest of the nucleons do not play a detailed role other than to supply an average spherically symmetric potential for our valence nucleons to move in. This is not a problem in atomic structure as the electron interaction between shells is weak. It is however a central problem in nuclei where we know the forces are strong.

2 *Experimental evidence for shell effects*

We review now the experimental evidence which lends support to a single particle orbit picture. The first clue to atomic electron orbits were the periodical effects observed in chemical behaviour as expressed in the Mendeleev table and in X-ray characteristic frequencies as observed by Moseley. Thus we seek similar periodic effects in nuclear properties. Such effects are indeed observed and are summarised below.

a) Discontinuities are observed in the neutron binding energy as a function of A. These are determined by measurement of (n, γ) and (d, p) Q-values. Similar effects are seen in proton binding energies. When a neutron (or proton) is added to a nucleus already containing a number of neutrons (or protons) equal to 2, 8, 20,

TABLE 8.1

Nucleus	Neutron number	S_n(MeV)	
^{15}O	7	13.2	
^{16}O	8	15.7	
^{17}O	9	4.14	Neutron bound to 8 neutrons very weakly
^{18}O	10	8.05	
^{40}Ca	20	15.7	
^{41}Ca	21	8.4	Neutron bound to 20 neutrons weakly
^{42}Ca	22	11.4	
^{43}Ca	23	7.9	
^{44}Ca	24	11.1	
^{45}Ca	25	7.4	
^{46}Ca	26	10.4	
^{47}Ca	27	7.3	
^{48}Ca	28	9.9	
^{49}Ca	29	5.1	Neutron bound to 28 neutrons weakly

28, 50, 82 or 126, the binding energy of the added particle is anomalously low. Some examples are shown in table 8.1.

Those numbers of neutrons (or protons) which appear particularly stable are often referred to as the 'magic numbers'. Note that nuclei such as ^{16}O, ^{40}Ca or ^{208}Pb ($Z=82$, $N=126$) contain magic numbers of both neutrons and protons and are said to be doubly magic.

b)   Anomalies in the abundance of elements as a function of N or Z are observed. We discuss this point later (ch. 15 § 1) but it will be shown that in the formation of the elements, those containing a magic number are preferentially formed. If we look at the elements containing more than one stable isotope oxygen ($Z=8$) is the first to have 3 stable isotopes, calcium ($Z=20$) is the first to have 5 stable isotopes and tin ($Z=50$) is the first to have 9 stable isotopes.

c)   If we look at the excitation energy of the first excited state of even-Z, even-N nuclei we find an anomalously high value for those nuclei with N or Z magic (fig. 8.1). Thus it requires more energy to disturb a magic number configuration.

d)   If we look at trends in α- and β-decay energies we find in general decays to magic number nuclei have the highest energies. That is, these nuclei are more tightly bound than the average. Half-lives which depend on the energy available also show the periodic effects.

e)   One of the first clear indications of magic numbers was a measurement of the absorption cross section (fig. 8.2) for all the elements for neutrons emerging from a

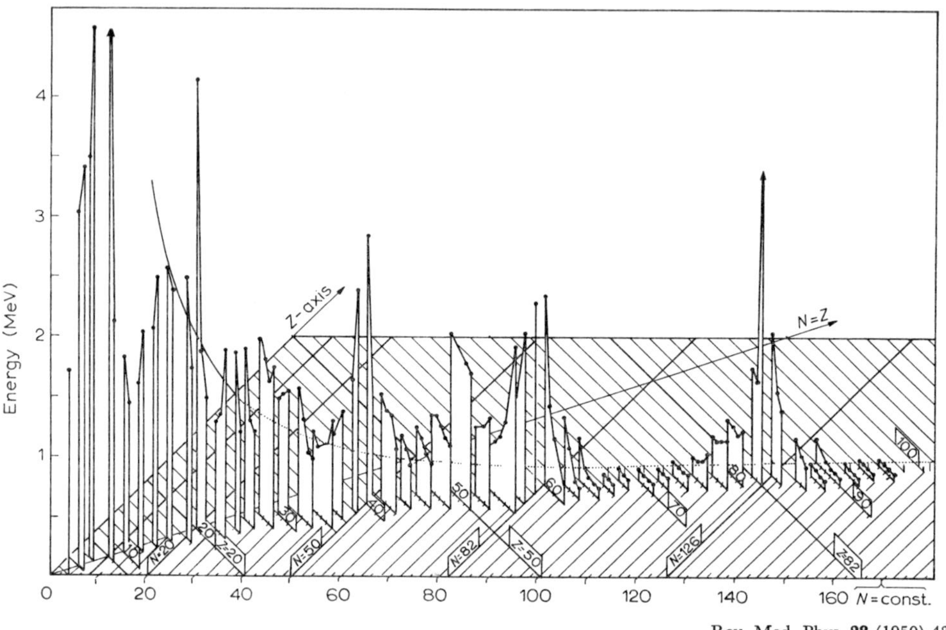

Rev. Mod. Phys. **28** (1950) 432

Fig. 8.1.   Energies of the first excited 2^+ state as a function of N, Z

nuclear reactor, i.e. a fission spectrum of a few MeV mean energy This indicated
a low level density for magic nuclei. They must therefore have a more symmetrical
structure with fewer variations in configuration possible.

f)   If we look at the angular momentum and parity of odd-A nuclei in which we
have an odd number of either neutrons or protons we see a pattern emerging in
which series of nuclei with the same spin and parity show up.

g)   The moments of nuclear ground states also show certain trends: the quadrupole
moments show minima at magic nuclei indicating that these nuclei are most nearly
spherically symmetric. Also the magnetic moments are related to the angular momentum of the last odd particle.

h)   Nuclear isomers are nuclear excited states having long lifetimes of the order
of seconds or longer. These occur when we have high spin difference between
ground state and excited state. We note clusters or 'islands' of isomers occurring at
particular places through the periodic table, again connected with the sudden
changes in orbit of the outer valence particle or particles.

So to summarise we seem to have particular stability associated with neutron
number or proton number equal to 2, 8, 20, 28, 50, 82, 126. When both N and Z
are 'magic' the resulting configurations are some of the most stable nuclei known.

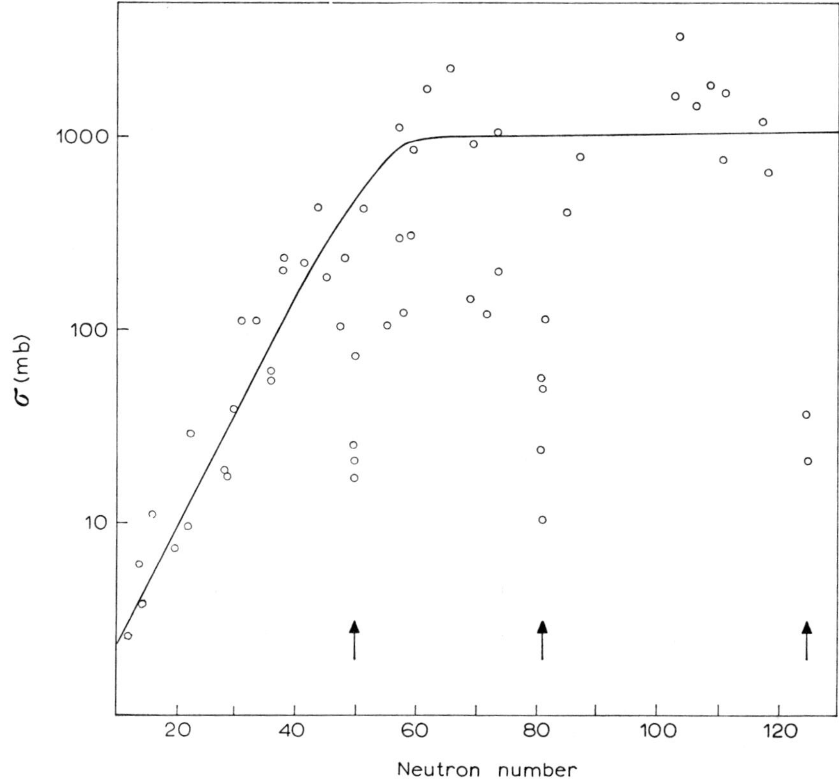

D. J. Hughes, Pile Neutron Research (Addison-Wesley, Reading 1953) p. 113

Fig. 8.2.   Slow neutron capture cross sections as a function of N

3   The single particle shell model

3.1   General. This model is based on the assumptions that we can ignore the detailed interactions between nucleons and that we can consider each particle to move in a state of definite orbital angular momentum independently of all other particles. It moves in the field of force which is the average smoothed-out interaction with all the other particles.

Suppose the actual short range interaction potential between two nucleons i, j is $v(r_{ij})$. We replace this with its average acting on each particle $V_i = \langle \sum_j v(r_{ij}) \rangle$. This procedure is analogous to the self-consistent Hartree field approximation of atomic physics.

We can write the actual Hamiltonian for the system

$$H = \sum_i T_i + \sum_{ij} v(r_{ij}) \tag{8.1}$$

in the following form

$$H' = \sum_i [T_i + V(r_i)] + \lambda [\sum_{ij} v(r_{ij}) - \sum_i V(r_i)]. \tag{8.2}$$

For $\lambda = 1$ we have $H = H'$, just the real Hamiltonian. The shell model assumption is that $\lambda = 0$ or is small enough to be neglected. In other words the average central interaction given by the first term is so much larger than the *residual interactions* given by the second term that we can neglect the latter term.

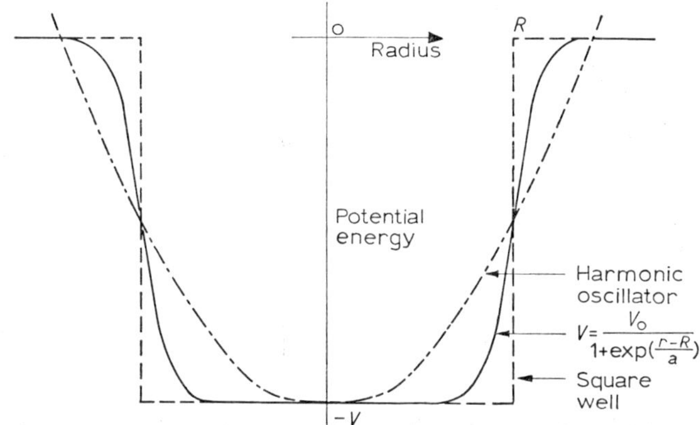

Fig. 8.3.   Various forms of potential wells

We now have to choose a form for the average potential $V(r_i)$ which is a *central* potential because it depends only on r_i, the distance of the particle from the centre. $V(r_i)$ is made up of the superposition of the short range internucleon potentials

$$V(r_i) = \int v(|r - r'|)\varrho(r')\,dr' \tag{8.3}$$

where $\varrho(r')$ is the density distribution in the nucleus. Now if we represent the internucleon potential as a delta function $v(r_{ij}) = -V_0\delta(r_{ij})$ then

$$V(r_i) = V_0\varrho(r). \tag{8.4}$$

Thus if the $v(r_{ij})$ potential is of short range confined within the nuclear radius R we expect $V(r)$ to resemble the density distribution which has been experimentally measured (ch. 1). In such a potential there will be a number of bound levels given by solutions of the Schrödinger equation $(T + V)\psi(r) = E\psi(r)$.

It is interesting to consider first some approximations (fig. 8.3) to this potential well shape to appreciate the sensitivity of the level structure in the well to this shape.

3.2   The square well potential. We can write the Schrödinger equation for our problem as follows

$$\frac{d^2}{dr^2}(rR_{nl}) + \frac{2M}{\hbar^2}\left\{E_{nl} - V(r) - \frac{l(l+1)\hbar^2}{2Mr^2}\right\}(rR_{nl}) = 0 \qquad (8.5)$$

for a nucleon of mass M moving in a spherically symmetric potential $V(r)$ with angular momentum $l\hbar$. This is the radial part of the wave equation, R_{nl} is the radial eigenfunction and E_{nl} is the eigenvalue, both depending on total quantum number n and orbital angular momentum l.

For the square well we find a solution

$$R_{nl} = \frac{A}{(Kr)^{\frac{1}{2}}} J_{l+\frac{1}{2}}(Kr) \qquad (8.6)$$

where A is some constant, K is the wave number of the nucleon,

$$K^2 = \frac{2M}{\hbar^2}(E_{nl} + V),$$

$J_{l+\frac{1}{2}}(Kr)$ is a spherical Bessel function and $-V$ is the well depth. If we measure from the bottom of the well

$$K^2 = \frac{2M}{\hbar^2} E'_{nl}$$

where E'_{nl} is positive.

Now at the boundary of the potential well we must arrange that the wavefunction vanish; $R_{nl}(R) = 0$ for $r = R$. That is, we must arrange for a zero of the Bessel function to be reached at $r = R$. This may be the first zero, the second zero etc. (fig. 8.4), and we label these allowed wavefunctions by $n = 1, 2, 3, 4 \ldots$ depending on which node is at $r = R$. The more oscillations we fit in for given l, the larger is n, the larger is K and hence the greater is the energy

$$E'_{nl} = \frac{K^2\hbar^2}{2M} = \frac{(KR)_0^2\hbar^2}{2MR^2} \qquad (8.7)$$

where $(KR)_0$ is the value of Kr which gives a zero at $r = R$. When we calculate these energy levels (cf. table 8.2) we get a level order for levels characterised by

$$1s, \ 1p, \ 1d, \ 2s, \ 1f, \ 2p, \ 1g, \ 2d, \ \text{etc.}$$

where 2p means the second level of orbital angular momentum 1 corresponding to the second zero in the Bessel function $J_{\frac{3}{2}}$. Each level characterised by l has $2l+1$

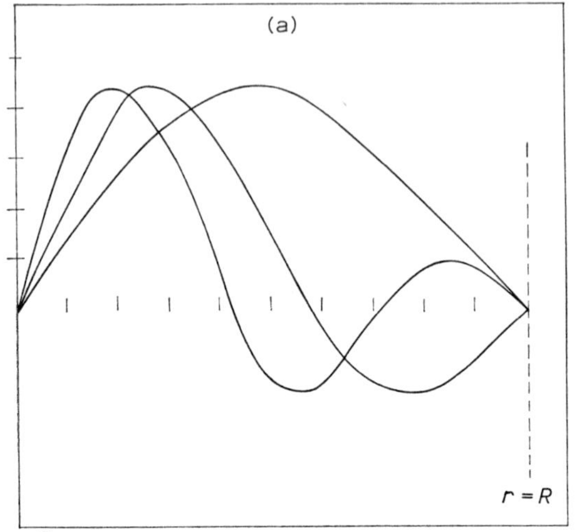

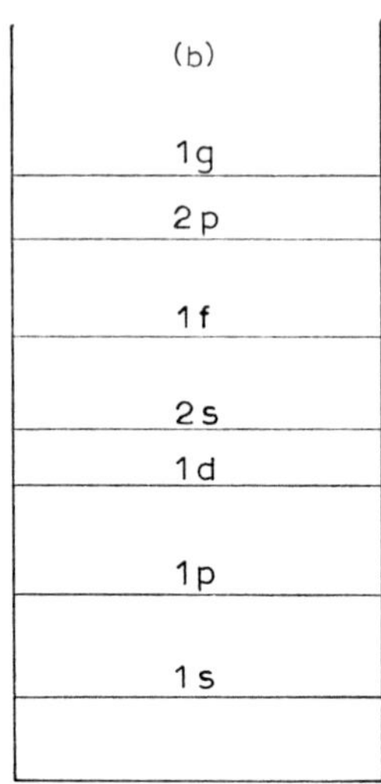

Fig. 8.4.   a) The Bessel function $J_{\frac{3}{2}}(Kr)$ fitted into the square well for $l=1$ and $n=1$, 2 and 3. b) The level order in the square well

substates. Each substate can be occupied by two identical nucleons with spins opposed. Hence each state above is occupied by $2(2l+1)$ identical particles.

TABLE 8.2

Level	Occupation	Total	E'_{nl} (units of $\hbar^2/2MR^2$)
1s	2	2	9.87
1p	6	8	20.14
1d	10	18	33.21
2s	2	20	39.48
1f	14	34	48.83
2p	6	40	59.68
1g	18	58	66.96

The clearest gaps in the energy spectrum which might show up as shell closure are at 2, 8, 20, 40, 58, 92, 138. These do not correspond to the magic numbers we adduced from experiments. We might guess that the choice of potential is at fault. The rise at the edge of the square well certainly does not correspond to the density distribution at the edge of the nucleus, observed experimentally. We now go to another extreme in which the potential changes slowly all the way from the origin.

3.3   The harmonic oscillator potential. This potential rises all the way from $r=0$ to $r=R$. It can be represented as

$$V(r) = -U + \tfrac{1}{2}M\omega^2 r^2 \tag{8.8}$$

(cf. fig. 8.3) where U is the well depth and ω is the frequency of the simple harmonic oscillation of the particle. As before we substitute in the Schrödinger equation, and the solutions now contain Hermite polynomials rather than Bessel functions. In the one-dimensional case $E'_n = (n+\tfrac{1}{2})\hbar\omega$ while in the three-dimensional case

$$E'_{n_1 n_2 n_3} = (n_1 + n_2 + n_3 + \tfrac{3}{2})\hbar\omega$$

or

$$E'_N = (N+\tfrac{3}{2})\hbar\omega \tag{8.9}$$

where $N = n_1 + n_2 + n_3$, the n_i are all integers specifying the wavefunction and N is called the oscillator quantum number.

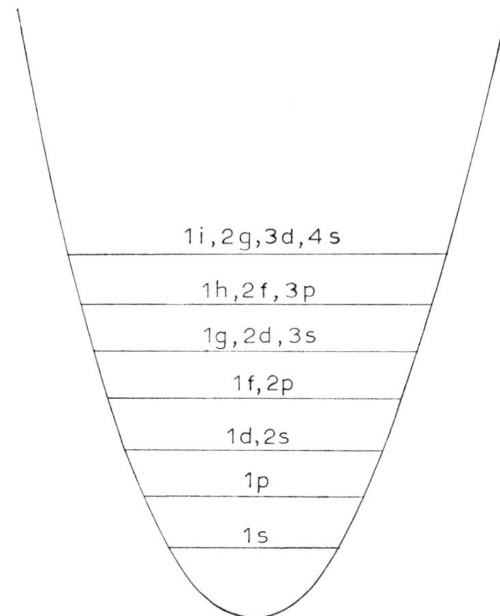

Fig. 8.5.   The level order in the harmonic oscillator well

We now compute E'_N as before and find that for each value of N there is a degenerate group of levels (all with the same energy) with different values of l, such that $l \leq N$ and even l corresponds to even N, odd l to odd N (cf. fig. 8.5). Thus

$$N=0 \text{ contains 1s}$$
$$N=1 \text{ contains 1p}$$
$$N=2 \text{ contains 1d 2s}$$
$$N=3 \text{ contains 1f 2p}$$
$$N=4 \text{ contains 1g 2d 3s etc.}$$

These degenerate sets of levels are obviously equally spaced in terms of $(N+\frac{3}{2})\hbar\omega$. Each set of levels can be occupied by $(N+1)(N+2)$ identical nucleons (table 8.3).

TABLE 8.3

N	Allowed l-values	E'_N	Occupation $(N+1)(N+2)$	Totals
0	0	$\frac{3}{2}\hbar\omega$	2	2
1	1	$\frac{5}{2}h\omega$	6	8
2	2, 0	$\frac{7}{2}\hbar\omega$	12	20
3	3,1	$\frac{9}{2}\hbar\omega$	20	40
4	4, 2, 0	$\frac{11}{2}\hbar\omega$	30	70
5	5, 3, 1	$\frac{13}{2}\hbar\omega$	42	112

Again this does not correspond to the shell closures indicated by experiment. Since the actual nuclear potential has a shape somewhere between these two extremes it is unlikely that any further adjustment of the potential can help us to get the correct magic numbers.

3.4 The central plus spin–orbit potential. Although such single particle models had been discussed since 1934, the year in which Rutherford looked forward to a future Mendeleev who would find a 'Natural Order of Nuclei', it was fifteen years before the problem was solved. The solution was suggested independently by Maria Mayer and by Haxel, Jensen and Suess, in 1949.* They suggested that in addition to the central static force there should be a non-central component as well. They suggested that this should have a magnitude which depends on the relative orientation of the orbital angular momentum vector, and the spin vector of the nucleon. The nucleon spin has hitherto not played a role except to restrict the orbit filling to 2 nucleons per orbital substate. This force will remove the degeneracy in levels of the same l and split them up into substates with total angular momentum $|j|=|l+s|=l\pm\frac{1}{2}$. Furthermore the sign of the force must be such that the level with the higher j-value $(l+\frac{1}{2})$ lies lower i.e. is more tightly bound. That is, when l and s are aligned, the force is stronger. This is opposite in sign to the atomic electron case where spin–orbit coupling is also seen in heavy atoms.

The levels of a harmonic oscillator are split as shown in fig. 8.6. We note each state of given j can contain $2j+1$ identical particles. Note also that the strength of the spin–orbit force can be made such that the $2p_{\frac{3}{2}}$, $1f_{\frac{5}{2}}$ and $2p_{\frac{1}{2}}$ levels with $N=3$

---

* M. G. Mayer, Phys. Rev. **78** (1950) 16.
O. Haxel, J. H. D. Jensen and H. E. Suess, Z. Physik **128** (1950) 301.

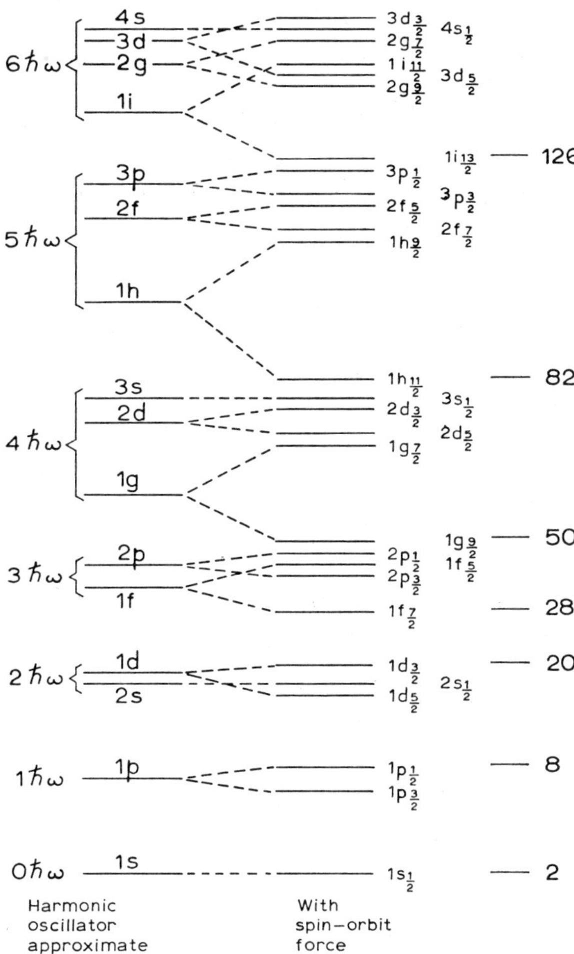

Fig. 8.6. The change in level order when a spin–orbit potential is added to a central force harmonic oscillator potential

are grouped with the $1g_{9/2}$ level with $N=4$. With similar shifts we can produce the observed magic number closures 2, 8, 20, 28, 50, 82, 126.

In fact we are now able to adopt a more realistic potential intermediate between the harmonic oscillator and the square well for the central potential and adjust the spin–orbit potential strength to reproduce the observed shell closures.

In the present discussion and historically, the spin–orbit force was introduced in order to explain the observed closures. But we shall see later that there is other experimental evidence for the existence of a spin–orbit force. We write it $V_s(r)l \cdot s$ where l, s are vectors. The energy change introduced is thus proportional to l.

The potential has the *form* resulting from the effect of interaction between the *magnetic fields* due to the spin and to the orbital angular motion. But the size of this effect turns out to be much smaller than that required. So the spin–orbit potential must be more intimately connected with the nuclear force.

The occupation of each of the levels has always been in terms of the number of *identical* particles. Thus in each level we can put $2j+1$ protons and also $2j+1$ neutrons. So we can write down shell model orbit assignments as in table 8.4.

TABLE 8.4

Nucleus	Orbit assignments		Nuclear spin
	Protons	Neutrons	
^{4}He	$s_{\frac{1}{2}}^2$	$s_{\frac{1}{2}}^2$	0^+
^{6}He	$s_{\frac{1}{2}}^2$	$s_{\frac{1}{2}}^2 p_{\frac{3}{2}}^2$	2^+
^{7}Li	$s_{\frac{1}{2}}^2 (p_{\frac{3}{2}}^1)$	$s_{\frac{1}{2}}^2 p_{\frac{3}{2}}^2$	$\frac{3}{2}^-$
^{11}B	$s_{\frac{1}{2}}^2 (p_{\frac{3}{2}}^3)$	$s_{\frac{1}{2}}^2 p_{\frac{3}{2}}^4$	$\frac{3}{2}^-$
^{12}C	$s_{\frac{1}{2}}^2 p_{\frac{3}{2}}^4$	$s_{\frac{1}{2}}^2 p_{\frac{3}{2}}^4$	0^+ (closed subshell)
^{15}N	$s_{\frac{1}{2}}^2 p_{\frac{3}{2}}^4 (p_{\frac{1}{2}}^1)$	$s_{\frac{1}{2}}^2 p_{\frac{3}{2}}^4 p_{\frac{1}{2}}^2$	$\frac{1}{2}^-$
^{16}O	$s_{\frac{1}{2}}^2 p_{\frac{3}{2}}^4 p_{\frac{1}{2}}^2$	$s_{\frac{1}{2}}^2 p_{\frac{3}{2}}^4 p_{\frac{1}{2}}^2$	0^+ (closed major shell in p and n)
^{17}O	$s_{\frac{1}{2}}^2 p_{\frac{3}{2}}^4 p_{\frac{1}{2}}^2$	$s_{\frac{1}{2}}^2 p_{\frac{3}{2}}^4 p_{\frac{1}{2}}^2 (d_{\frac{5}{2}}^1)$	$\frac{5}{2}^+$

In each odd-A case the configuration with the odd particle is bracketed. In the single particle shell model this should give the nuclear spin. In these examples this is the case.

In general we find that the spins and the parities of the ground states of odd-A nuclei are well represented by a shell model representation. The discrepancies are generally for nuclei far from closed shells. The first two ground state spins at variance with the single particle model are for ^{19}F $(\frac{1}{2}^+)$ and ^{23}Na $(\frac{3}{2}^+)$ both of which might be expected to have $\frac{5}{2}^+$.

The spins and parities of the first few excited states of odd-A nuclei are also in agreement in many cases especially when there are only one or three particles or holes associated with a closed shell core.

4   *Magnetic moments of odd-A nuclei*

We have already considered the magnetic moments expected for a single particle of $j = l + s$ moving in a potential. The results were, for an *odd neutron*:

$$j = l+\tfrac{1}{2}, \qquad \mu_j = \mu_n$$

$$j = l-\tfrac{1}{2}, \qquad \mu_j = \frac{-j}{j+1}\,\mu_n$$

where μ_n is the free neutron magnetic moment, -1.913 nuclear magnetons, and for an *odd proton*:

$$j = l+\tfrac{1}{2}, \qquad \mu_j = (j-\tfrac{1}{2})+\mu_p \text{ n.m}$$

$$j = l-\tfrac{1}{2}, \qquad \mu_j = \frac{j}{j+1}\,[(j+\tfrac{3}{2})-\mu_p]\ \text{n.m}$$

where μ_p is the free proton moment, $+2.79$ n.m.

Our single particle shell model assumption for odd-*A* nuclei is that the last odd particle gives its angular momentum and magnetic moment to the nucleus as a whole i.e. $j = I$, $\mu_j = \mu_I$. A plot of μ_I vs I then gives a set of four lines, two for an odd proton with $j = l \pm \tfrac{1}{2}$ and two for an odd neutron with $j = l \pm \tfrac{1}{2}$, as shown in fig. 8.7.

If we now plot on this graph the measured values of magnetic moments for odd-*A* nuclei we find that the values with one or two exceptions lie inside the lines (generally called the Schmidt lines). Those nuclei nearest closed shells lie nearest the lines but others act as if they were a mixture of $l+\tfrac{1}{2}$ and $l-\tfrac{1}{2}$ configurations with a tendency to be closer to the line we should have predicted.

If we plot the deviation from the Schmidt line vs N or Z we can see the shell effect more clearly as shown in fig. 8.8.

The tendency to group near the lines sometimes enables us to confirm the configuration involved in a nuclear ground state. For example ^{7}Li with 3 protons and 4 neutrons has $I=\tfrac{3}{2}$. The single particle prediction is $\mu_{sp}=3.79$ n.m for a p-orbit $(l+\tfrac{1}{2})$ or $\mu_{sp}=0.12$ n.m for a d-orbit $(l-\tfrac{1}{2})$. The observed value is 3.26 n.m. So we assume the p-orbit assignment is correct. For ^{23}Na on the other hand which has a spin $\tfrac{3}{2}$ at variance with the shell model prediction of $\tfrac{5}{2}$, the magnetic moment is 2.22 n.m, almost halfway between the two lines indicating severe mixing of states.

When we cross a shell we observe the expected sharp change in the magnetic moment from that corresponding to $l-\tfrac{1}{2}$ to that corresponding to $l+\tfrac{1}{2}$. For example

$$^{39}_{19}\text{K}_{20} \qquad I = \tfrac{3}{2}, \quad \mu_{obs} = 0.22; \quad \mu_{sp} \ \text{for} \ \text{d-orbit}\,(l-\tfrac{1}{2}) = 0.12.$$

$$^{45}_{21}\text{Sc}_{24} \qquad I = \tfrac{7}{2}, \quad \mu_{obs} = 4.76; \quad \mu_{sp} \ \text{for} \ \text{f-orbit}\,(l+\tfrac{1}{2}) = 5.79.$$

Thus the single particle shell model predicts some trends in and limits to the values of magnetic moments but in general fails to predict the value with any precision.

For even-even nuclei the magnetic moment is always zero so this tells us nothing about shell structure. For odd-odd nuclei the combination of proton and neutron orbits is usually too complex to make comparisons useful.

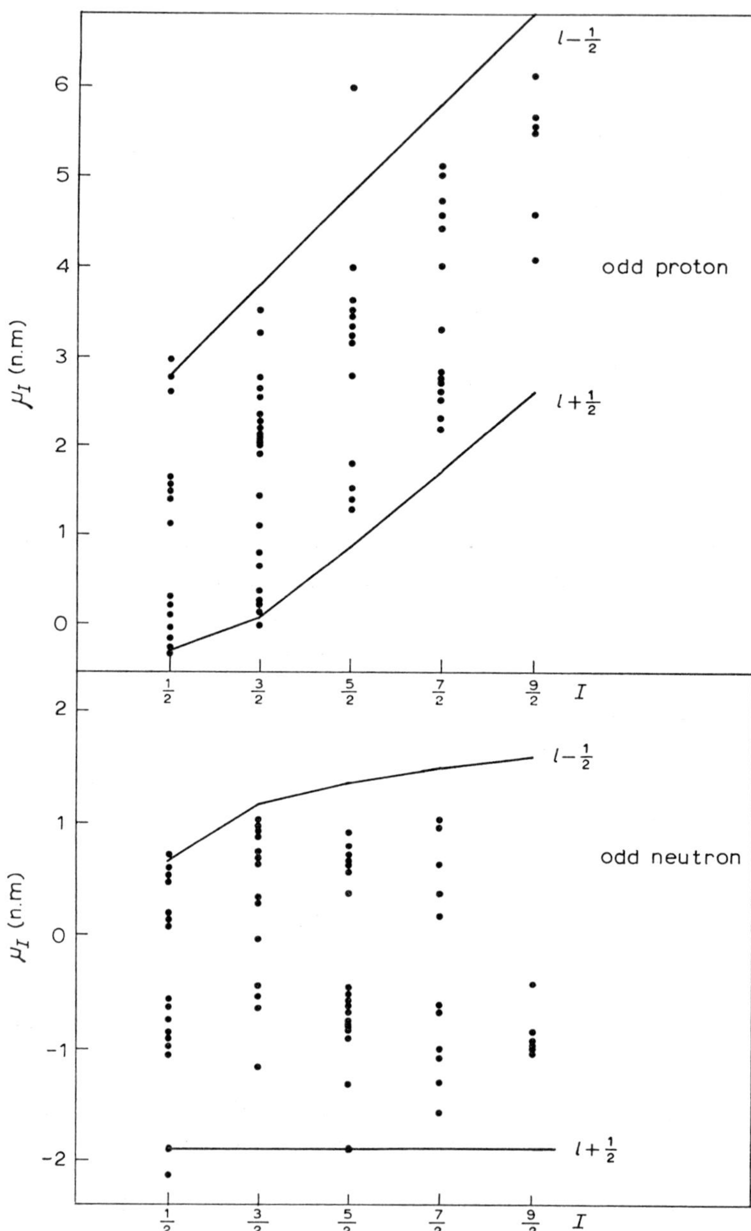

Rev. Mod. Phys. **28** (1956) 86

Fig. 8.7.   Observed magnetic moments for odd proton and odd neutron nuclei as a function of nuclear spin I. The lines are the single particle values (Schmidt lines)

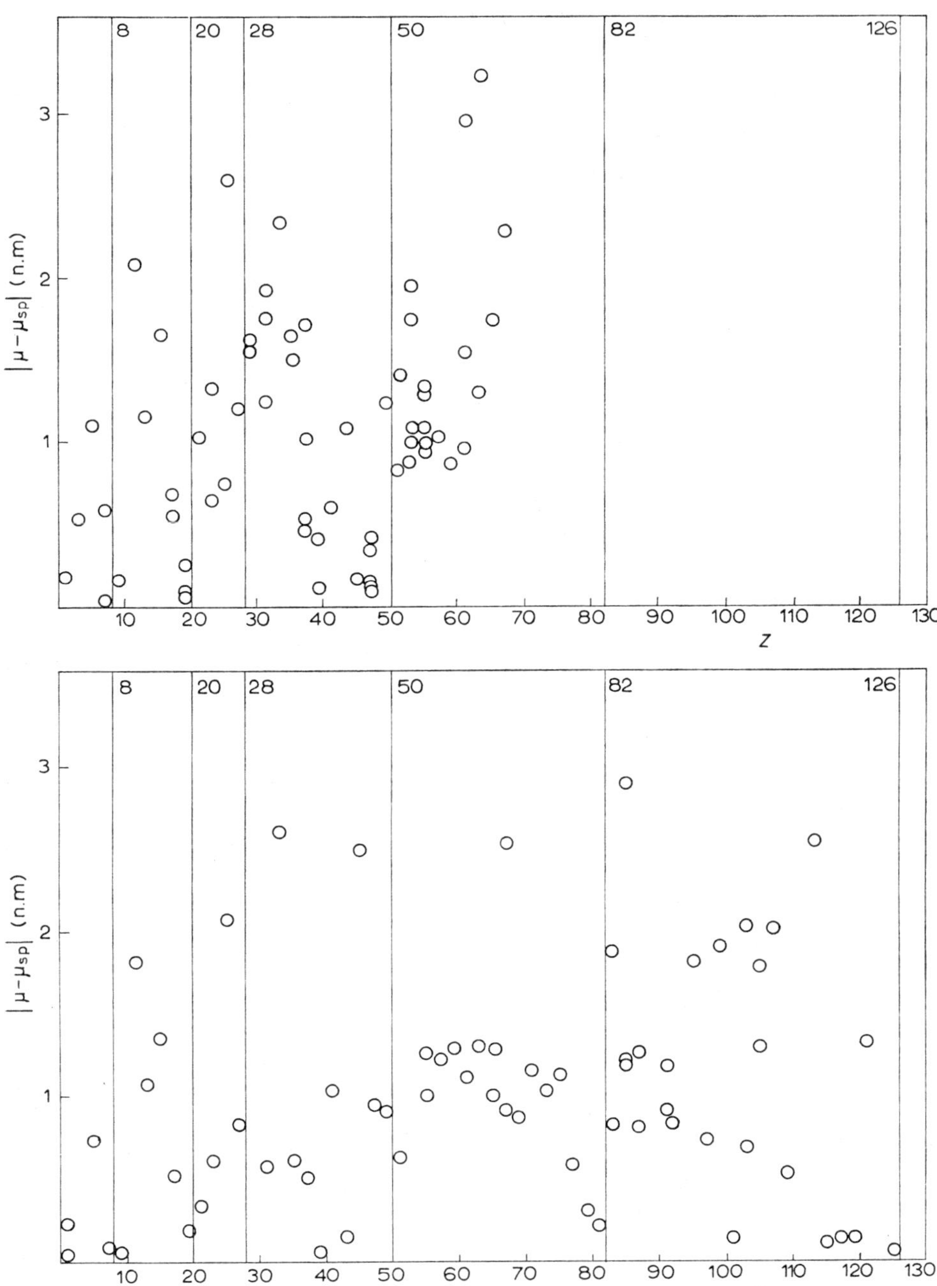

Fig. 8.8.   The deviation of the observed magnetic moment from the single particle value as a function of Z and N

5 *Quadrupole moments of odd-A nuclei*

We turn now to the other type of static moment of importance, the electric quadrupole moment. We have already discussed this in ch. 1 and when we considered the effect of a single proton we derived

$$Q_{sp}(\text{proton}) = -\overline{r^2}\,\frac{2j-1}{2(j+1)}$$

where $\overline{r^2} \leq \frac{1}{2}R^2$ is the mean square radius. For a single neutron we might expect $Q=0$ but there is an effect on the distribution of the other protons by a shift of the centre of mass which leads to

$$Q_{sp}(\text{neutron}) = \frac{Z}{A^2}\,Q_{sp}(\text{proton}).$$

We assume in all this that the remainder of the nucleus, the core, is spherically symmetric and thus does not contribute to the moment. In general if we have more than one loose particle outside a closed shell, all in orbit j, we get a resultant Q for n loose particles.

$$Q_n = -\overline{r^2}\,\frac{2j-1}{2(j+1)}\left[1 - \frac{2(n-1)}{2j-1}\right]. \tag{8.10}$$

That is, it varies linearly with n.

We turn now to the observed quadrupole moments for odd-A nuclei. Near closed shells the agreement with the single particle picture is quite good and the sign agrees, that is, negative for an odd particle, positive for an odd hole. For example

$$^{63}_{29}\text{Cu}\,(\text{p}_{\frac{3}{2}})^1, \qquad ^{123}_{51}\text{Sb}\,(\text{g}_{\frac{7}{2}})^1, \qquad ^{129}_{53}I\,(\text{g}_{\frac{7}{2}})^3, \qquad ^{209}_{83}\text{Bi}\,(\text{h}_{\frac{9}{2}})^1$$

all have negative quadrupole moments, while

$$^{27}_{13}\text{Al}\,(\text{d}_{\frac{5}{2}})^{-1}, \qquad ^{55}_{25}\text{Mn}\,(\text{f}_{\frac{7}{2}})^{-3}, \qquad ^{69}_{31}\text{Ga}\,(\text{p}_{\frac{3}{2}})^{-1}, \qquad ^{113}_{49}\text{In}\,(\text{g}_{\frac{9}{2}})^{-1}$$

all have positive quadrupole moments. However we expect the magnitudes to be of the order of R^2 i.e. 10^{-24}–10^{-25} cm^2 since $\overline{r^2}$ for a uniformly charged sphere is $\frac{3}{5}R^2$. But in certain regions of the periodic table in particular near $Z=60-80$, $A \approx 150$ (the rare earth nuclei) we observe extremely large values up to the order of $10R^2$. Also the odd neutron nuclei do not show small values as expected from the factor Z/A^2 but instead we see quadrupole moments as large as the odd proton moments. In fact the largest measured quadrupole moment is for $^{167}_{68}\text{Er}_{99}$, an odd neutron nucleus where $Q \approx 30R^2$. All these large quadrupole moments are seen far from closed shells. Near to closed shells the single particle values are more nearly in agreement.

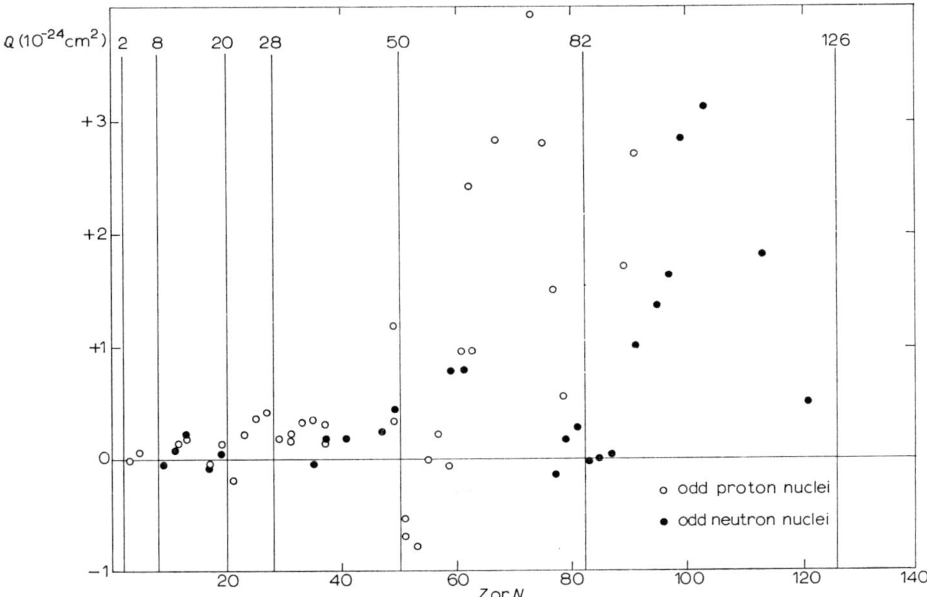

Fig. 8.9.   Observed quadrupole moments for odd proton and odd neutron nuclei as a function of proton or neutron number

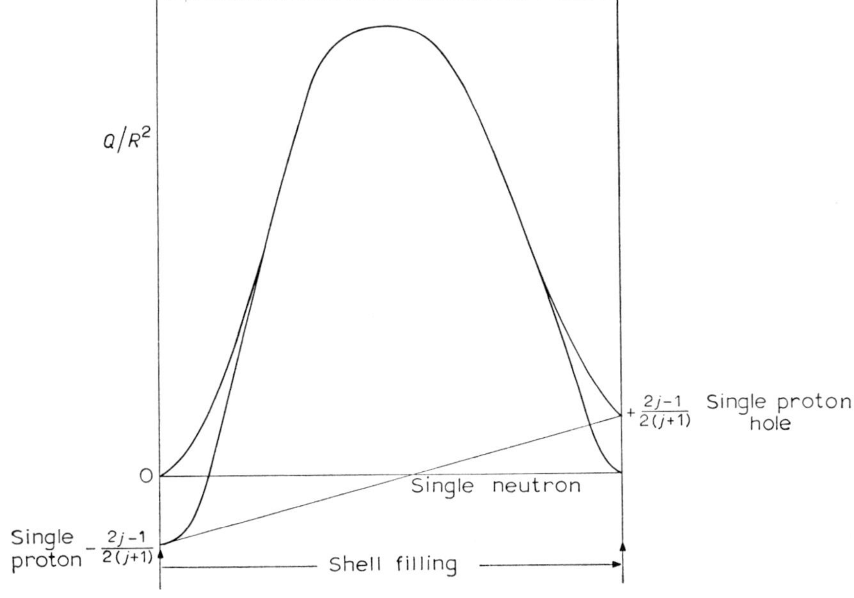

Fig. 8.10.   The behaviour of the quadrupole moment whilst a shell is being filled

If we plot Q vs N or Z we can see the shell effects as in fig. 8.9. If we plot Q/R^2 from one closed shell to the next we will obtain a variation as shown in fig. 8.10.

Thus we have seen that the single particle shell model is remarkably successful in predicting periodicities, angular momentum and parity of ground states and general trends in magnetic moments. Far from closed shells the discrepancies become very marked however, particularly in the magnitude of quadrupole and magnetic moments.

6   Isobaric spin

6.1   Charge independence. We now need to introduce a new concept which describes the fact that there are two kinds of nucleons, neutrons and protons. In our discussion of nuclear radius determination we referred to *mirror nuclei*. These were nuclei with the same value of A and having A odd which differed only in that the odd nucleon was in one case a neutron, in the other a proton. For example if we add to ^{12}C a proton we get $^{13}_{7}$N$_6$ while if we add a neutron we get $^{13}_{6}$C$_7$. These two nuclei have different binding energies since ^{13}N decays by β^+-emission to stable ^{13}C with an energy release of 2.2 MeV. However if we attribute this difference to the coulomb energy due to the extra proton in ^{13}N less the neutron–proton mass difference we find the binding energies remaining to be nearly identical. The coulomb energy due to the extra proton is (assuming a uniformly charged sphere)

$$\Delta E_c = \frac{3e^2}{5R}[Z^2 - (Z-1)^2] \tag{8.11}$$

then the energy release turns out to be equal to

$$\Delta E_c - (M_n - M_p)c^2.$$

That is, the nuclear force between p and ^{12}C and n and ^{12}C is the same, and the nuclear force is said to be *charge symmetric*. In other words if we change from a charged to an uncharged nucleon nothing is changed in the nuclear part of the interaction.

In the diagrams of fig. 8.11 the horizontal lines represent spatial states or given orbits. Thus each line can contain at most four particles by the exclusion principle proton spin up, proton spin down, neutron spin up, neutron spin down. The symmetry between ^{13}N and ^{13}C due to charge exchange is clearly seen.

It could still be that the nuclear force is charge dependent since we have only proven from mirror nuclei that the p–p forces are identical to the n–n forces. To compare the n–p forces we have to study nuclei which differ in the last *pair* of nucleons, so we consider even-A nuclei having pp, np and nn as this last pair. If

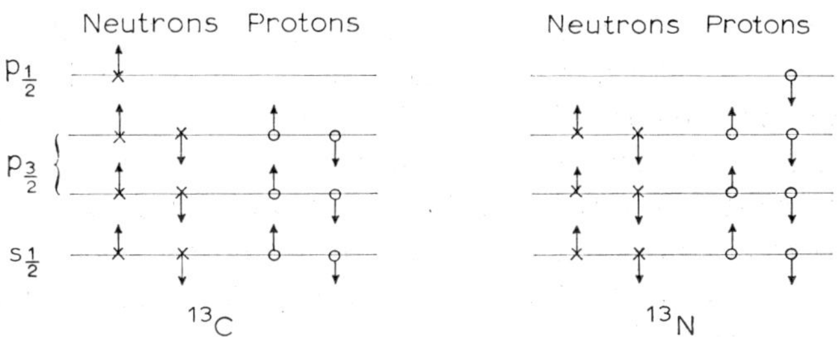

Fig. 8.11.   Bar diagram for mass 13 mirror nuclei

we add these to $^{12}_6C_6$ we get $^{14}_8O_6$, $^{14}_7N_7$, and $^{14}_6C_8$. Again we correct the observed binding energies for the different coulomb energies and neutron–proton mass differences, and compare the remaining binding energies.

We find that ^{14}C and ^{14}O come out the same as we expect from charge symmetry [fig. 8.12] and that there is an excited state at 2.31 MeV excitation in ^{14}N which also corresponds exactly to these. However it is not the ground state of ^{14}N, which lies lower and has different spin. The fact that there are three states which correspond

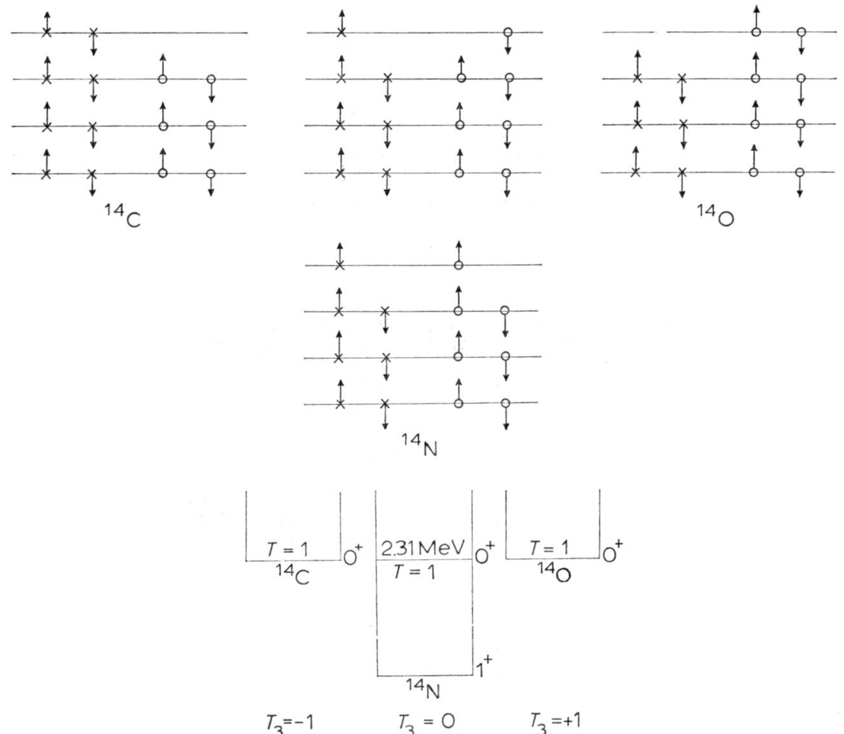

Fig. 8.12.   Mass 14 isobars

tells us that the nuclear force is indeed *charge independent* and that the p–p, n–p and n–n forces are identical. Why is it however that ^{14}N, that is, the n–p system has another more tightly bound state? This is a consequence of another symmetry.

We note that the ^{14}N ground state is more symmetric for the exchange of *spin* than is the excited state at 2.31 MeV. This added symmetry results in extra binding energy of 2–3 MeV. Furthermore we cannot make such a spin-symmetric state with ^{14}C or ^{14}O as we cannot have two identical particles of the same spin in the same spatial state by the Pauli principle.

This was an example of nuclei with $A = 4n + 2$. For nuclei with $A = 4n$ we get a different situation. Let us consider mass 12 similarly corrected for coulomb energy and n–p mass difference (fig. 8.13). Here we find the state in ^{12}C corresponding to

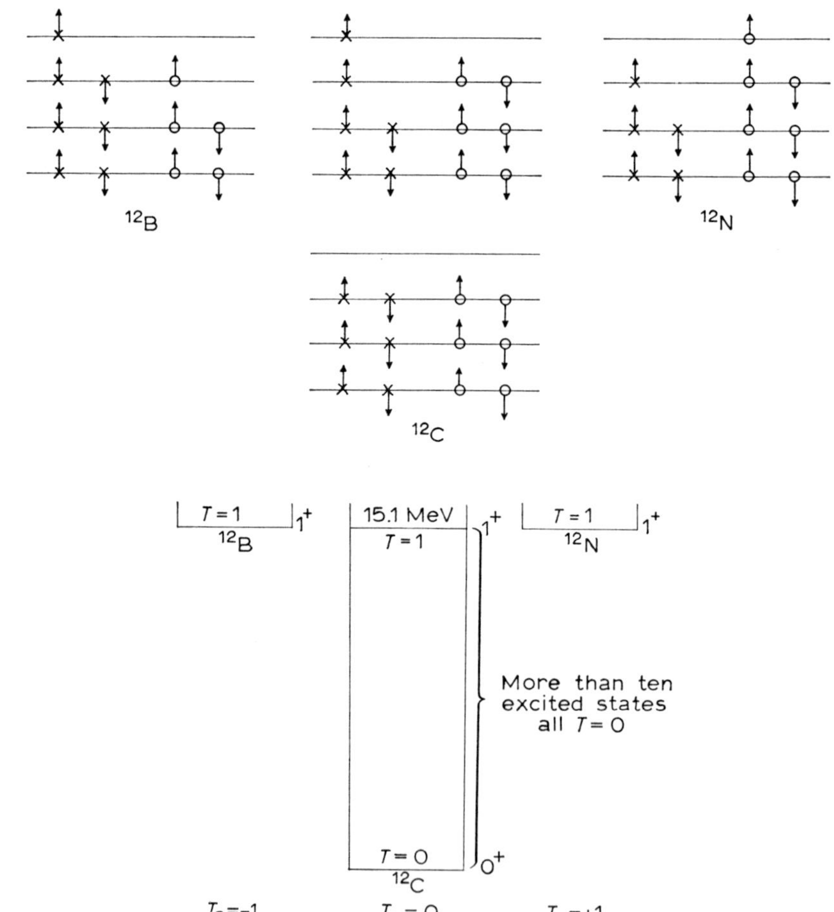

Fig. 8.13.   Mass 12 isobars

^{12}B and ^{12}N much higher in energy. In this case the greater spatial symmetry of the ^{12}C ground state causes it to be bound 15.1 MeV more tightly. Again the ^{12}B or ^{12}N cannot achieve this symmetry because of the Pauli principle.

We say the common system of states in such nuclei constitutes a charge triplet whereas the more symmetric ground state of the central nucleus is a charge singlet. Similarly the mirror nuclei would be a charge doublet. This suggests assigning quantum numbers to describe the charge symmetry by analogy with the assignment of angular momentum quantum numbers to describe space or spin symmetry. We call the quantum number isobaric spin or simply isospin. The charge triplet would have isospin $T=1$ with $2T+1$ components. These components which for angular momentum are the allowed projections along the z spatial axis, are taken to be projections of the isospin along some axis in a hypothetical 'isospin space'. The projections T_z or $T_3 = -1$, 0, $+1$ for $T=1$. Similarly the singlet will be $T=0$, $T_3 = 0$ and the doublet will be $T=\frac{1}{2}$, $T_3 = \pm\frac{1}{2}$. In the approximation we have been using, that the coulomb electrical force is absent or negligible, these substates are degenerate in energy, that is they have the same binding energy. When we take into account the coulomb force it splits the levels just as an external magnetic field splits angular momentum substates. If they are split by a small amount compared to the *total* binding then they are still approximately degenerate and the T-formalism has some validity. If they are split by a large amount, T will cease to have meaning and we expect this to happen with increasing nuclear charge Z.

Thus in ^{14}C and ^{14}O all the states have $T=1$, $T_3 = \pm 1$, while in ^{14}N the lowest state has $T=0$, $T_3 = 0$, while some of the excited states have $T=1$, $T_3 = 0$. The mirror nuclei ^{13}C, ^{13}N have $T=\frac{1}{2}$ $T_3 = \pm\frac{1}{2}$. The signs of T_3 are given by the following convention

$$T_3 = \tfrac{1}{2}(Z-N) = \tfrac{1}{2}(A-2N) = -\tfrac{1}{2}(A-2Z). \tag{8.12}$$

We can also assign isobaric spin to the two nucleons which are an isospin doublet with $T_3 = +\frac{1}{2}$ for the proton and $T_3 = -\frac{1}{2}$ for the neutron. (This is the sign convention generally used in high energy and low energy nuclear physics nowadays. In earlier low energy work the opposite convention is sometimes used.)

All the states of ^{13}C–^{13}N up to 15 MeV have $T=\frac{1}{2}$, $T_3 = \pm\frac{1}{2}$. At 15.07 MeV in ^{13}N the analog state is seen corresponding to the ground states of ^{13}B and ^{13}O (fig. 8.14). Above this energy in ^{13}C–^{13}N some states will have $T=\frac{1}{2}$ and others (analogs of excited states of ^{13}B, ^{13}O) will have $T=\frac{3}{2}$.

Selection rules involving conservation of isospin will be observed if T is a good quantum number. For example consider

$$^{10}\text{B}+{}^2\text{H}\rightarrow{}^{10}\text{B*}+{}^2\text{H}-Q.$$
$$T= \qquad 0 \quad 0 \quad\ \ 0 \quad\ \ 0$$
$$\text{or } 1$$

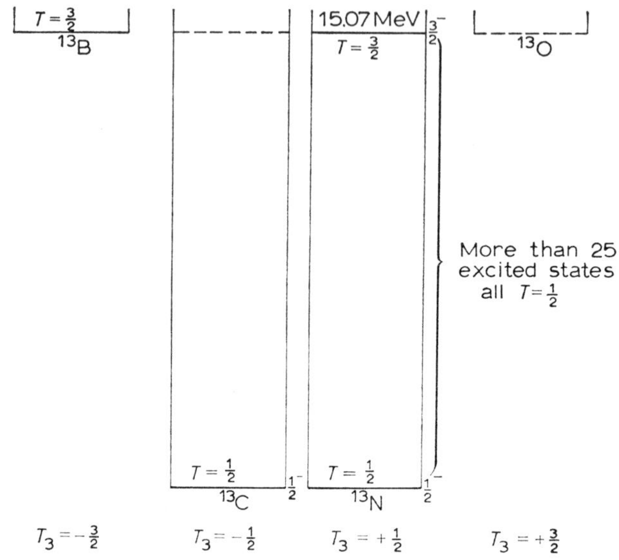

Fig. 8.14.   Mass 13 isobars

It is clear that we cannot produce $T=1$ states in ^{10}B by inelastic deuteron scattering if T is conserved. However by inelastic proton scattering we can do so

$$^{10}\text{B}+{}^{1}\text{H}\rightarrow{}^{10}\text{B}^{*}+{}^{1}\text{H}-Q.$$

$$T=\qquad 0\qquad \tfrac{1}{2}\qquad 0\qquad \tfrac{1}{2}$$
$$\text{or } 1$$

The fact that the 1.74 MeV state of ^{10}B could not be excited by the (d,d′) reaction but could by the (p, p′) reaction confirmed that this was the first $T=1$ state of ^{10}B corresponding to the ground states of ^{10}Be and ^{10}C.

Thus isobaric spin does seem to be conserved in strong nuclear interactions. In electromagnetic interactions it is obviously not conserved. In weak interactions the situation is more complex and will be discussed later. However note that in *all* circumstances T_3 is conserved as this is just charge conservation since the charge is $T_3+\tfrac{1}{2}A$.

6.2 Isobaric spin in radiative transitions. Suppose that the transition involves the change in state of a *single* nucleon with $T=\tfrac{1}{2}$. Then we can see that a change in T between initial and final states can be at most 1. So the first rule is that radiation is forbidden except for $\varDelta T=0,\ \pm1$. This is of little practical interest since states with $\varDelta T=2$ are generally far above the particle binding energy where radiation cannot compete with particle emission anyhow.

Of more interest is a rule for electric dipole radiation in self-conjugate nuclei

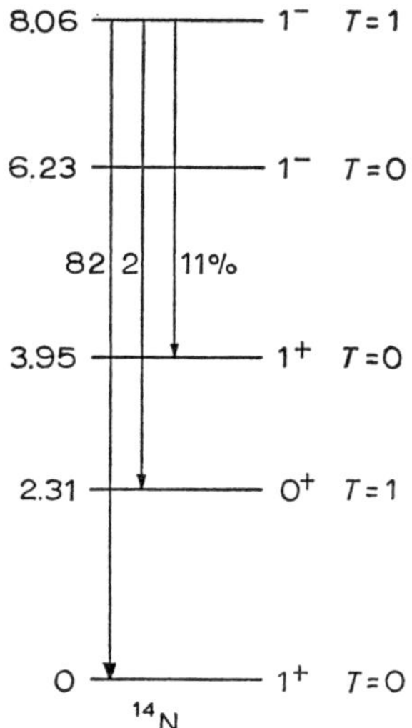

Fig. 8.15.   Decay of the 8.06 MeV state of ^{14}N

i.e. with $Z=N$. The rule states that $\Delta T=\pm 1$. Since for $T=0\to T=0$ transitions
there is no way to distinguish neutrons and protons, each has effective charge $\frac{1}{2}e$
and no dipole moment can be induced, only a movement of the centre of mass
which does not result in radiation. Thus in nuclei with $Z=N$ the isospin must change
for E1 radiation. An example of this behaviour is shown in fig. 8.15.

In ^{16}O the ΔT rule inhibits emission of E1 by a factor of 10^5 for the 7.12 MeV

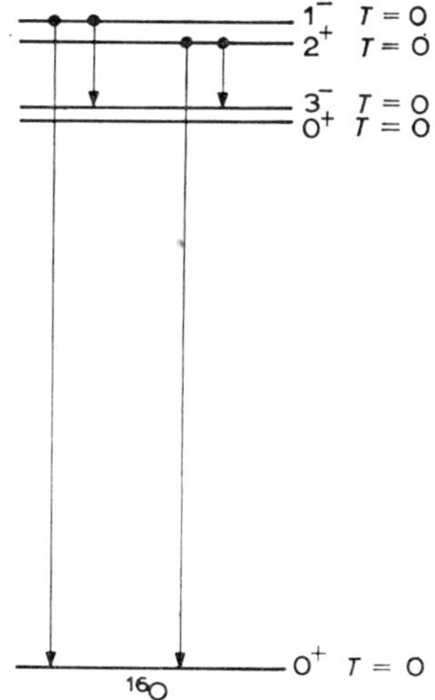

Fig. 8.16.   Decay of ^{16}O states

(fourth excited state) transitions and by more than a factor of 10^3 for the 6.91 MeV (third excited state) transitions (see fig. 8.16). In the case of the 7.12 MeV state we can attribute the E1 we *do* see to impurity in isospin of the 7.12 MeV level. That is, for a small fraction of the time this state has $J^P=1^-$, $T=1$. The amount of the admixture will depend on the strength of the coulomb interaction which breaks down isospin and the energy separation from the nearest $J^P=1^-$, $T=1$ state which could share with it. For example there is such a state at 13.10 MeV and it may be this which is mixing with the 7.12 MeV state. We can use this rule to estimate isospin impurity. However note that for the 6.91 MeV state something else seems to be inhibiting the E1 emission as well. In this case basic differences in the two states are probably the reason.

6.3   Isobaric analog states in heavy nuclei. We noted that on general grounds we should expect isobaric spin to be a less good quantum number as Z increases and

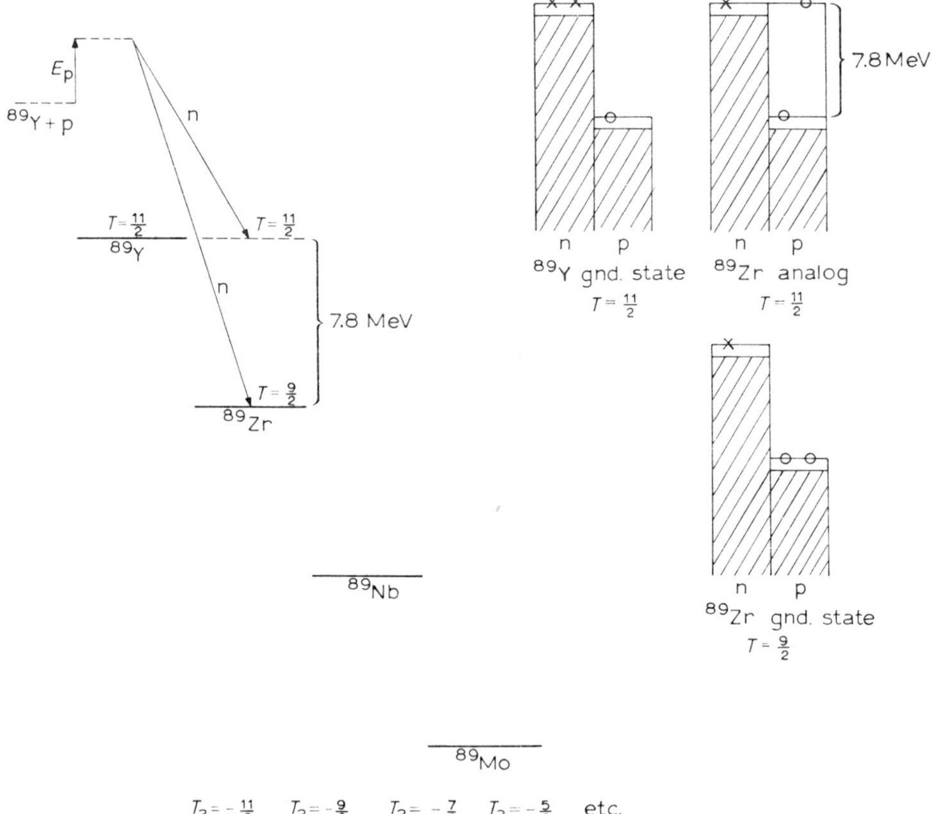

Fig. 8.17.   Analog states in ^{89}Zr. In the upper right is shown the bar diagram for the states involved

the coulomb energy becomes larger. Thus the observation of isobaric spin analog states in nuclei around mass 90 was at first sight somewhat surprising. In such heavy nuclei the neutron excess is already quite large. That is, the T-values involved are large. For example let us take the case of ^{89}Y + p which has been much studied. ^{89}Y has 39 protons and 50 neutrons. Thus $N - Z = 11$ and $T_3 = -\frac{11}{2}$. Suppose now we add a proton and remove a neutron. We can do this if we perform a (p, n) reaction at sufficiently high energy so that a direct charge exchange reaction takes place with no intermediate state involved. The product is $^{89}_{40}$Zr$_{49}$ with $N - Z = 9$ and $T_3 = -\frac{9}{2}$,

$$^{89}\text{Y} + \text{p} \rightarrow {}^{89}\text{Zr} + \text{n}.$$

Now if we look at the level diagram arranged as before with coulomb energy correction we would expect the analog state at an excitation of 7.8 MeV in ^{89}Zr (fig. 8.17).

Thus if we observe the neutron spectrum leading to states in ^{89}Zr (fig. 8.18) we expect a peak at 7.8 MeV below the maximum neutron energy (leading to the ^{89}Zr ground state). This just corresponds to how high we have to go in excitation before we can exchange neutron and proton in the same spatial state. The peak observed actually is so wide that it must consist of a number of individual levels. The strength of the narrow ^{89}Y ground state is distributed over a relatively limited region of levels as shown in fig. 8.19. Together they form a peak in the neutron spectrum. Also shown in fig. 8.18 is the neutron spectrum from the ^{51}V(p, n) reaction leading to the isobaric analog state in ^{51}Cr.

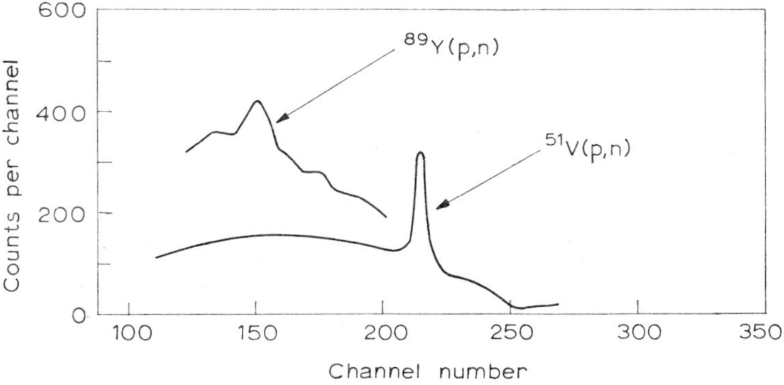

Phys. Rev. **126** (1962) 2170

Fig. 8.18.   Neutron spectra showing states analogous to the ground state of the target

Thus the observation of an analog 'state' is not inconsistent with great impurity in isospin of the fine structure states.

Another type of isobaric analog state has been observed in the same mass region.

We bombard ^{89}Y again with protons but at lower energy such that compound nucleus formation is more likely. Then if we look at the *excitation function* for ^{89}Y + p, for example, the elastic scattering, then at a certain excitation in the compound nucleus ^{90}Zr, we will see the first $T=6$ state, the analog of the ground state of ^{90}Y. Again we see a large proton width distributed over a number of compound nucleus states in a limited region.

If we look at the excitation function for the ^{89}Y(p, n) reaction in the same region of proton energy we also see peaks in the same positions. This is less expected if we consider the isospins involved. For the proton scattering

$$^{89}\text{Y} + \text{p} \rightarrow {}^{90}\text{Zr} \rightarrow {}^{89}\text{Y} + \text{p}$$
$$T = \quad \tfrac{11}{2} \ \tfrac{1}{2} \quad 6 \quad \tfrac{11}{2} \ \tfrac{1}{2}$$
$$\text{or } 5$$

both $T=6$ and $T=5$ states are formed and can decay via elastic scattering. Now for the (p, n) reaction

$$^{89}\text{Y} + \text{p} \rightarrow {}^{90}\text{Zr} \rightarrow {}^{89}\text{Zr} + \text{n}$$
$$T = \quad \tfrac{11}{2} \ \tfrac{1}{2} \quad 6 \ \rightarrow \ \tfrac{9}{2} \ \tfrac{1}{2} \quad \text{forbidden}$$
$$\text{or } 5 \ \rightarrow \ \tfrac{9}{2} \ \tfrac{1}{2} \quad \text{allowed.}$$

That is, only the $T=5$ states can decay to ^{89}Zr + n while the $T=6$ states cannot. However this argument assumes that T is a good quantum number, yet we know the analog 'state' consists of heavily mixed fine structure states. The *proton* widths for these states are anomalously large. The *neutron* widths are actually small since the neutron decay can only proceed via the small $T=5$ component.

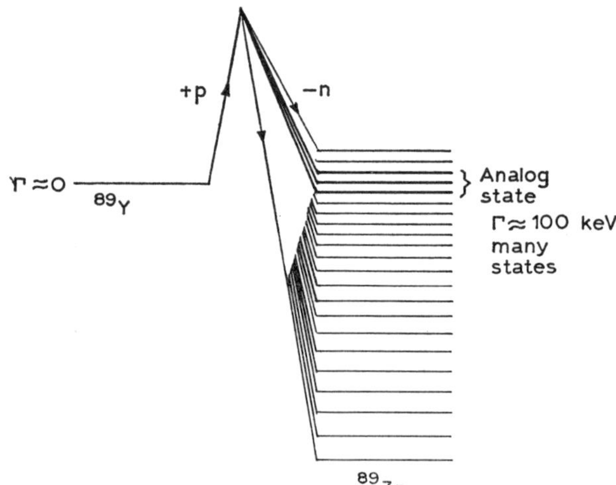

Fig. 8.19. The analog state spread amongst the 'normal' states

Suppose that at all the normal $T=5$ states $\Gamma_n \approx 10\Gamma_p$ and let $\Gamma_p=1$ then

$$\sigma_{pn} \propto \frac{\Gamma_p \Gamma_n}{\Gamma} \approx \frac{\Gamma_p \Gamma_n}{\Gamma_n} = \Gamma_p = 1$$

and

$$\sigma_{pp} \propto \frac{\Gamma_p \Gamma_p}{\Gamma} \approx \frac{\Gamma_p \Gamma_p}{\Gamma_n} = 0.1.$$

For the analog $T=6$ states suppose Γ'_p is larger by a factor of 10 but Γ'_n is unchanged. Then $\Gamma'_p \approx \Gamma'_n$ and

$$\sigma_{pn}(\text{analog}) \propto \frac{\Gamma'_p \Gamma'_n}{\Gamma} \approx \frac{\Gamma'_p \Gamma'_n}{\Gamma'_p + \Gamma'_n} = 5$$

$$\sigma_{pp}(\text{analog}) \propto \frac{\Gamma'_p \Gamma'_p}{\Gamma} \approx \frac{\Gamma'_p \Gamma'_p}{\Gamma'_p + \Gamma'_n} = 5.$$

Both cross sections are enhanced in the region of the analog states.

Studies of these analog states can yield much information about the isobaric spin mixing which is indeed as large as expected from the magnitude of the coulomb energy.

7   Intermediate coupling shell model

In our shell model so far we have assumed that all the nucleons, except for the last odd particle in odd-A nuclei, formed an inert core with zero angular momentum, even parity, spherical symmetry etc. This core served only to form the average central field of force in which the odd nucleon moved. We assumed the core was formed because the pairing energy, that is the binding between two identical particles with opposite spins in the same orbital state was large enough to exceed the disruptive effect of the single particle.

If this latter is not a good assumption the next less uncertain assumption is to consider a core composed of the last closed shell. The extra binding of this is of course the intershell energy, in general somewhat larger than the pairing energy. We then assume that the nuclear properties are given by the behaviour of the 'loose' particles combined in all possible ways. This is the *independent particle model*. Thus ^{7}Li treated as $(p_{\frac{3}{2}})^1$ on the single particle model would now be $(p_{\frac{3}{2}})^3$ on the independent particle model. Similarly ^{19}F which did not agree with the single particle model prediction of $(d_{\frac{5}{2}})^1$ would now be $(d_{\frac{5}{2}})^3$ or more precisely $(d_{\frac{5}{2}}, s_{\frac{1}{2}}, d_{\frac{3}{2}})^3$ where the configurations inside the bracket indicate *possible* orbits for the shell that is filling.

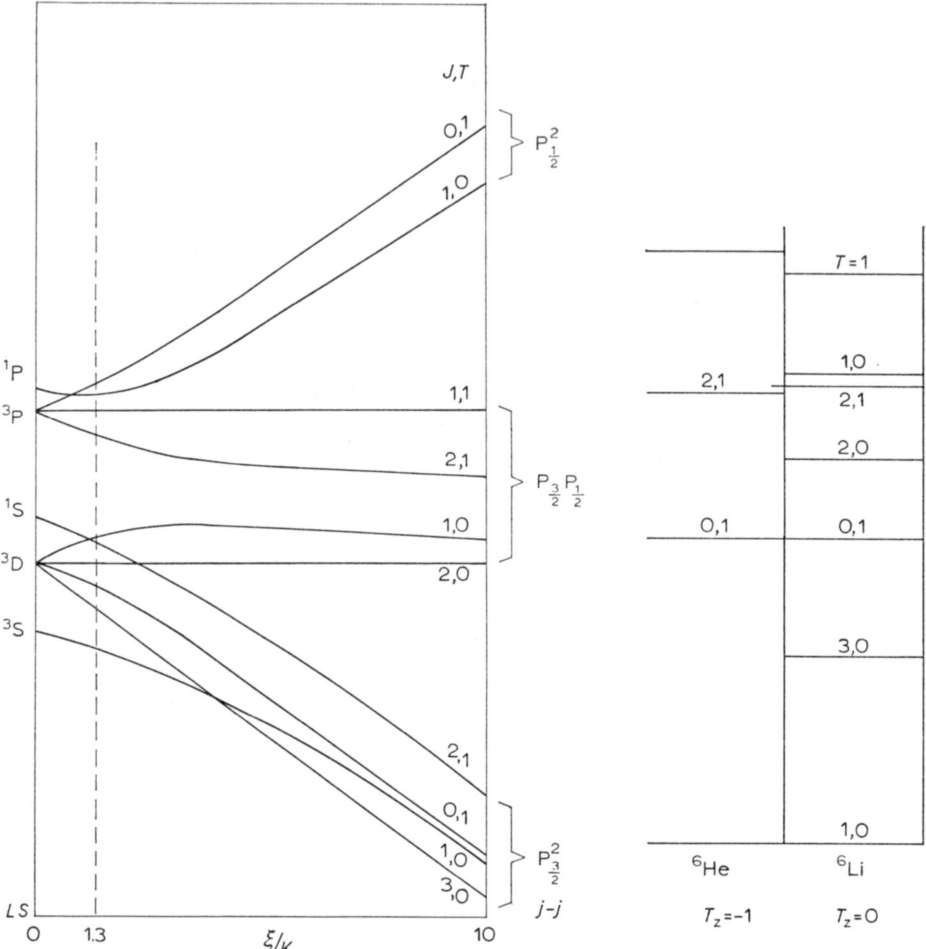

Rev. Mod. Phys. **25** (1953) 410

Fig. 8.20.   Intermediate coupling for mass 6

We assume that there is a spin–orbit force between nucleons as suggested by the single particle model and that it has strength $\xi(\boldsymbol{l}\cdot\boldsymbol{s})$ per nucleon pair. This leads us to a classification of states according to j–j coupling as before. Particles 1, 2, 3 combine as follows.

$$\begin{aligned}
\boldsymbol{l}_1+\boldsymbol{s}_1 &= \boldsymbol{j}_1 \\
\boldsymbol{l}_2+\boldsymbol{s}_2 &= \boldsymbol{j}_2 \qquad \boldsymbol{j}_1+\boldsymbol{j}_2+\boldsymbol{j}_3 = I,\ \text{the nuclear spin.} \\
\boldsymbol{l}_3+\boldsymbol{s}_3 &= \boldsymbol{j}_3
\end{aligned}$$

But if there were no spin–orbit force we should get a different classification of states due to angular momentum dependent parts of the *central force*. We have already recognised effects of spin exchange and spatial exchange. The latter, the main part of the exchange force, leads to coupling of the nuclear angular momenta in a different way.

$$l_1 + l_2 + l_3 = L$$
$$s_1 + s_2 + s_3 = S \qquad L + S = I.$$

This is just *L–S* or Russell–Saunders coupling familiar in atomic structure. We suppose this central exchange force is of strength K.

The level ordering will then depend on the ratio of the strengths of the two couplings ξ/K. Calculations of structure based on these ideas are called *intermediate coupling shell model* calculations. We can start with the *L–S* states that can be made up from our loose particles, taking account of the Pauli principle and sorting out the state according to energy, spin and isospin. This corresponds to $\xi/K = 0$, a pure central potential with exchange component. Then we turn on the residual interaction allowing ξ/K to increase and note the effect on the energies of the states. The residual interaction splits the states (removes degeneracy) and some of them can cross over. By plotting the energy spectrum vs ξ/K as in fig. 8.20, we can find a value of ξ/K which gives best agreement with the observed level spectrum.

A simple example is ^{6}He which has two loose neutrons in configuration p^2, $T = 1$. The lowest *L–S* states (all $T = 1$) are

$$
\begin{array}{lll}
l = 1 - 1 & L = 0 & \\
s = \tfrac{1}{2} - \tfrac{1}{2} & S = 0 & {}^1S_0 \\[2mm]
l = 1 + 1 & L = 2 & \\
s = \tfrac{1}{2} - \tfrac{1}{2} & S = 0 & {}^1D_2 \\[2mm]
l = 1 + 1 & L = 1 & \\
s = \tfrac{1}{2} + \tfrac{1}{2} & S = 1 & {}^3P_{0,\,1,\,2}
\end{array}
$$

The lowest *j–j* states are

$$
\begin{array}{lll}
p^2_{\frac{3}{2}}, & j_1 + j_2 = \tfrac{3}{2} + \tfrac{3}{2}, & I = 0, 2 \\[2mm]
p_{\frac{3}{2}}, p_{\frac{1}{2}}, & j_1 + j_2 = \tfrac{3}{2} + \tfrac{1}{2}, & I = 1, 2 \\[2mm]
p^2_{\frac{1}{2}}, & j_1 + j_2 = \tfrac{1}{2} + \tfrac{1}{2}, & I = 0
\end{array}
$$

The Pauli principle does not allow $I = 1, 3$ for $p^2_{\frac{3}{2}}$ or $I = 1$ for $p^2_{\frac{1}{2}}$, as these are identical particles. The calculation then indicates how the corresponding states are to be joined up. The dotted line in fig. 8.20 shows the value of ξ/K which gives the best fit to the mass 6 level spectrum.

It can easily be shown that the states for some intermediate coupling can be expressed as linear combinations of *L–S* states suitably weighted or alternatively as linear combinations of j–j states. This configuration mixture was just what was indicated by the observed values of magnetic moments intermediate between the Schmidt single particle values. When magnetic moments are calculated from such intermediate coupling wavefunctions the agreement with experiment is excellent. Somewhat better agreement with observed quadrupole moments is also obtained when the 'loose' nucleons are taken into account in this way. Thus 'odd neutron' nuclei may have quadrupole moments produced by the 'loose' protons.

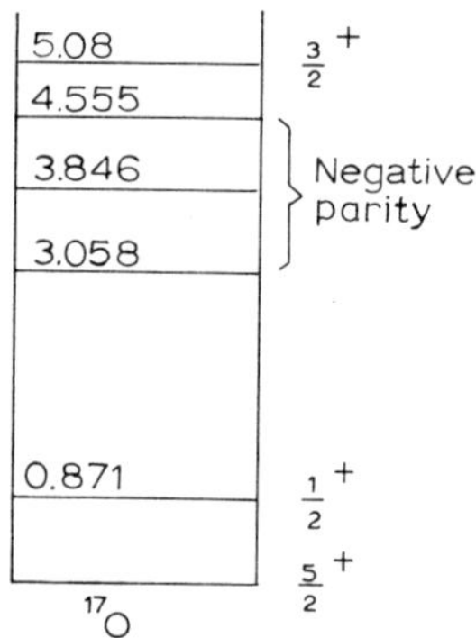

Fig. 8.21. ^{17}O states

One of the first dramatic successes of this model was for ^{19}F. The calculation of Elliott and Flowers* for the three loose particles required a large computer. They chose to fix the spin–orbit part of the force from the level spacings in ^{17}O (fig. 8.21) with one loose neutron outside the doubly closed shell of ^{16}O, hence a single particle nucleus on both approximations. This gave them the separation of the $d_{\frac{5}{2}}$, $s_{\frac{1}{2}}$ and $d_{\frac{3}{2}}$ orbits of the (sd) shell. The level order and other properties were then studied as a function of V_0, the central part of the force (cf. fig. 8.22). This was equivalent to plotting as a function of K.

We note that for $V_0 > 25$ MeV the $\frac{1}{2}^+$ state lies lower than the $\frac{5}{2}$ state. At about 40–50 MeV the observed level ordering is reproduced. This value for the depth of the central potential is about correct and can be arrived at in other ways. Many other properties of the ^{19}F states could be predicted and were mostly in good agreement with observation. The mixed configuration of the state is illustrated by the $\frac{1}{2}^+$ ground state which had a composition for the three particles (per cent) d^3 12, d^{2}s 59, ds^2 0, s^3 29.

Note that we are only attempting to fit positive parity states of ^{19}F which can be made by (sd) combinations. Negative parity states are also seen and these must arise from either p-shell configurations produced by breaking into the ^{16}O core or by

* J. P. Elliott and B. H. Flowers, Proc. Roy. Soc. **A229** (1955) 536.

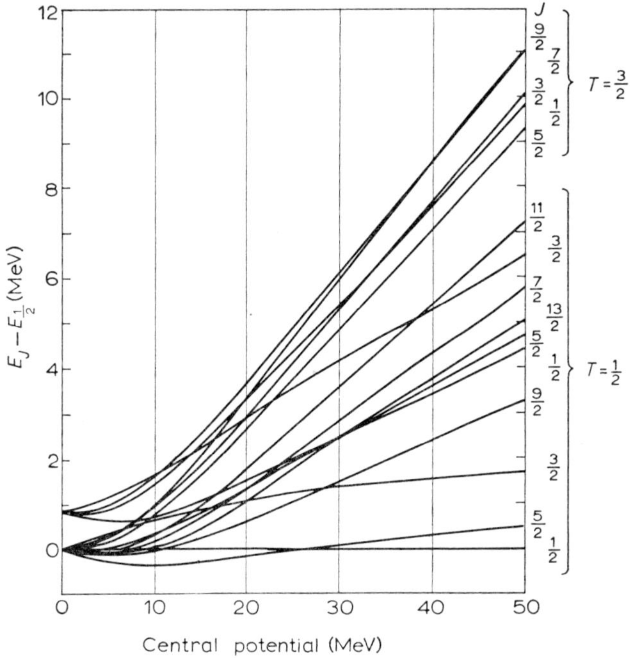

Proc. Roy. Soc. **A229** (1955) 551

Fig. 8.22.   Intermediate coupling for ^{19}F

f-shell configurations from the next higher shell. Some doubt is cast on the assumptions of the model by the fact that the first negative parity state of $\frac{1}{2}^-$ is seen at 0.1 MeV as the first excited state! If the ^{16}O core can be broken into so easily how can we consider it as an inert core? We return to this point later.

Many such calculations have been carried out for up to 3 particles outside a closed shell with generally excellent agreement except for quadrupole moments and E2 transition strengths. For more than 3 particles the calculation begins to be too large even for modern computers.

8   *Electromagnetic decay of shell model states*

In addition to the level spectrum and the static moments, the decay probabilities of the states can also be predicted. For electromagnetic decay the transition probability $T(L)$ can be written

$$T(L) = \frac{8\pi(L+1)}{L[(2l+1)!!]^2} \frac{1}{\hbar} \left(\frac{E_\gamma}{\hbar c}\right)^{2L+1} B(L) \text{ sec}^{-1}, \tag{8.13}$$

where E_γ is the γ-ray energy and $(2L+1)!! = 1 \cdot 3 \cdot 5 \cdots (2L+1)$. $B(L)$ is the *reduced* transition probability that is with the energy dependence removed.

For a single particle transition in which we assume that the radiation arose from the transition of a single nucleon from one orbital state to another, Weisskopf showed that for $l=0$ in the final state

$$B(\mathrm{E}L) = \frac{e^2}{4\pi}\left(\frac{3R^L}{L+3}\right)^2 \qquad \text{for electric multipoles} \qquad (8.14)$$

and

$$B(\mathrm{M}L) = 10\left(\frac{\hbar}{M_\mathrm{p}cR}\right)^2 B(\mathrm{E}L) \qquad \text{for magnetic multipoles} \qquad (8.15)$$

where R is of the order of the nuclear radius and M_p is the proton mass.

We can derive other formulae for final states with $l \neq 0$ by addition of vector addition coefficients to the above expressions.

It is convenient to use these single particle estimates of $B(L)$ as units in which to quote observed $B(L)$-values. Thus we can quote the observed transition probability as a fraction or multiple of the single particle value or the 'Weisskopf unit'.

$$|M|^2 = \frac{\Gamma_\gamma(\mathrm{obs})}{\Gamma_\gamma(\mathrm{Weisskopf})} \qquad (8.16)$$

where

$$\begin{aligned}
\Gamma_\gamma(\mathrm{E1}) &= 0.07 E_\gamma^3 A^{\frac{2}{3}} \\
\Gamma_\gamma(\mathrm{M1}) &= 0.021 E_\gamma^3 \\
\Gamma_\gamma(\mathrm{E2}) &= 4.9 \times 10^{-8} E_\gamma^5 A^{\frac{4}{3}}
\end{aligned}$$

are the Weisskopf units with E_γ in MeV.

In addition to the angular momentum selection rules for radiation there are additional ones involving the isospin. The most important is that in nuclei with $N=Z$ ($T=0$) E1 radiation is allowed only for $\Delta T=0$. In fact E1 transitions for $\Delta T \neq 0$ are observed in such nulei but the $B(\mathrm{E1})$ is some 20–50 times smaller than for $\Delta T=0$ transitions.

The general trends in observed radiative widths can be summarised as follows.
1)   For M1 and E1 all observed values of $|M|^2$ are <1 that is, are slower than the single particle estimate. This is to be expected from mixed configurations.
2)   For light nuclei E1 transitions have $|M|^2 \approx 3\%$ for isospin-allowed transitions.
3)   For light nuclei M1 transitions have $|M|^2 \approx 15\%$.
4)   E2 transitions for both light and heavy nuclei in general have $|M|^2 > 1$ and are frequently 10–100 times as large as single particle values, especially in those regions of the periodic table where large ground state quadrupole moments are observed.

5)   For transitions of higher L there is less spread amongst the values of $|M|^2$ than for dipole or quadrupole and they are not often enhanced except for a few E3 transitions.

6)   There is no clear difference in transition strengths in single proton and single neutron odd-A nuclei.

The general trends of the transition probabilities are similar to those observed in the static moments; magnetic moments and transitions indicating mixed configurations and quadrupole moments and B(E2) values being much larger than the single particle estimates.

9   Summary

We have seen that many of the periodic features observed in nuclear properties can be quite well reproduced by a single particle shell model. We suppose that the odd particle moves in the average central field which is a result of averaging the effects of all the individual two-body interactions. We found however that at least one portion of the residual interaction not included in the average central field had to be included to produce the observed shell ordering, that is that ordering which fitted the observed periodicities in properties. This was the spin–orbit force which binds a particle more tightly when its spin vector and its orbital angular momentum vector are aligned, than when these two are in opposite directions. Thus we wrote our potential $V(r) = V_0(r) + V_s(r)\mathbf{l}\cdot\mathbf{s}$ and using a harmonic oscillator radial dependence obtained considerable agreement for the case of one particle or one hole outside a closed shell.

The next step was to improve the agreement when more than one particle or one hole was outside. The single particle assumption that all but one particle paired off did not work very well. Thus all possible configurations of 2, 3 or 4 particles were allowed to mix and some parameter controlling the mixing was allowed to vary until one or more observed level spacings were fitted. This technique improved the agreement with experiment considerably. However computational difficulties prevented it being applied very widely.

The outstanding parameters which even the independent particle shell model still failed to account for were the large quadrupole moments and large E2 transition probabilities observed in certain regions of the periodic table.

9

Nuclear Models:

Collective Model

1 Introduction

A most significant failure of the shell model approach to nuclear structure is its inability to explain the very large quadrupole moments which are observed for nuclei in certain regions of the periodic table. The large $B(E2)$ values indicate also large quadrupole effects. If we observe a Q which is n times Q_{sp} we can say roughly that at least $2n$ particles must have been involved since the neutrons cannot directly contribute to the moment. Thus we are led to consider the collective motion of large numbers of nucleons in order to explain the large observed moments far from closed shells. The features of such a model which could explain the moments were first given by Rainwater[*] and a complete description combining single particle and collective features into a unified model was elaborated by Bohr and Mottelson[**] and their collaborators.

2 Collective quadrupole moments

Suppose that we assume a constant volume deformation of a uniformly charged sphere of radius R and total charge Ze. Suppose that it is an ellipsoid of revolution (fig. 9.1) of eccentricity $\varepsilon = (a-b)/R$ where a, b are the major and minor semi-axes.

[*] J. Rainwater, Phys. Rev. **79** (1950) 432.
[**] A. Bohr and B. Mottelson, Kgl. Dan. Vid. Selsk. Mat.-Fys. Medd. **27**, No. 16 (1953).

Then we can write in polar coordinates for the surface shape

$$R(\theta) = R[1+\tfrac{2}{3}\varepsilon P_2(\cos\theta)] = R[1+\tfrac{1}{3}\varepsilon(\cos^2\theta-1)] \tag{9.1}$$

so that $R(0)=a$ and $R(\tfrac{1}{2}\pi)=b$ where $P_2(\cos\theta)$ is the Legendre polynomial function of order 2. Now we let the volume $\tfrac{4}{3}\pi R^3$ represent a potential well of depth $-V_0$

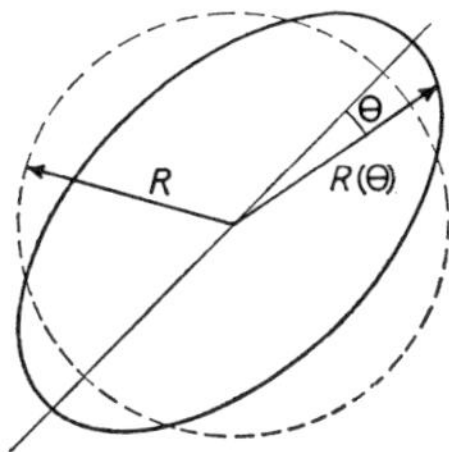

Fig. 9.1.   Radius vector in ellipsoid of revolution

which rises abruptly at the surface $R(\theta)$ to $V=0$. This is the potential seen by a single nucleon. We assume the potential and the nuclear matter deform in the same way. Then the first order change in the single particle energy due to the deformation is

$$\Delta E_{\mathrm p} = \int \tilde{V}|\psi_{\mathrm p}|^2\,\mathrm d\tau$$

where $\psi_{\mathrm p}$ is the unperturbed wavefunction of the odd particle and $\tilde{V}$ is the difference between the deformed and undeformed potentials. Thus for $R(\theta) > R$, $\tilde{V}=-V_0$

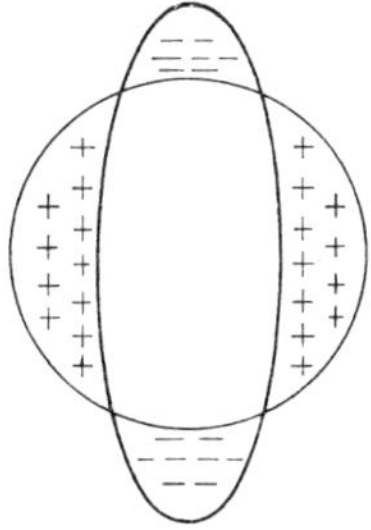

Fig. 9.2.   Regions of positive and negative potential energy

and for $R(\theta)<R$, $\tilde{V}=+V_0$. Fig. 9.2 indicates the sign of the potential in various regions of its orbit. Now let $\psi_{\mathrm p}=u(r)Y(\theta,\phi)$ where $u(r)$ is the radial part of the wavefunction and $Y(\theta,\phi)$ is a spherical harmonic function. To first order we can neglect the variation of $u(r)$ through the deformed region so we replace it with $u(R)$. Then

$$\Delta E_{\mathrm{p}} = -V_0 |u(R)|^2 \int \mathrm{d}\Omega |Y(\theta, \phi)|^2 \int_R^{R(\theta)} r^2 \,\mathrm{d}r$$

$$= -\tfrac{1}{3}\varepsilon V_0 R^3 |u(R)|^2 \int \mathrm{d}\Omega (3\cos^2\theta - 1)|Y(\theta, \phi)|^2$$

$$= -\tfrac{1}{3}\varepsilon V_0 R^3 |u(R)|^2 Q_{\mathrm{sp}}/\overline{r^2} \tag{9.2}$$

where Q_{sp} is the single particle unperturbed proton quadrupole moment irrespective of whether the nucleon in our problem is a proton or a neutron.

Now the surface energy is changed by an amount proportional to the eccentricity squared, ε^2, thus

$$\Delta E_{\mathrm{def}} = -\varepsilon U Q_{\mathrm{sp}}/\overline{r^2} + \tfrac{1}{2}S\varepsilon^2 \tag{9.3}$$

where $U = \tfrac{1}{3}V_0 R^3 |u(R)|^2$ and S is a constant representing both the surface energy and a second order term in ΔE_{p} which is also positive.

Now we note that ΔE_{def} has a minimum at

$$\varepsilon = 2U Q_{\mathrm{sp}}/S\overline{r^2}.$$

The corresponding value of the quadrupole moment is

$$Q/R^2 = \tfrac{4}{5}\varepsilon Z = \tfrac{4}{5}Z U Q_{\mathrm{sp}}/S\overline{r^2}. \tag{9.4}$$

The collective quadrupole moment thus depends on the total number of protons Z and it is easy to see that $Q_{\mathrm{obs}}/Q_{\mathrm{sp}}$ might easily be 10–20. This is just the collective *part* of the moment. If it is also an odd proton nucleus we add to it Q_{sp} due to the proton motion.

3   Collective spectra

3.1   Introduction. Since this simple model offers an explanation for the large quadrupole moments by assuming a permanent deformation for the nucleus we must now examine other consequences of such deformation. For example, what excited states are expected due to motions of the deformed structure itself such as rotations or vibrations and what changes are expected in the excited states due to single particle motion if the potential is now deformed rather than spherical? We consider first the rotational and vibrational spectra. We expect to see these most clearly in the even–even nuclei where single particle effects are absent. If we look at the states of even–even nuclei we find all the ground states to be 0^+. The first excited states are with a few exceptions 2^+. The exceptions are almost always closed shell nuclei. In fig. 9.3 are plotted the excitation energies of the 2^+ states of even–even nuclei vs N or Z.

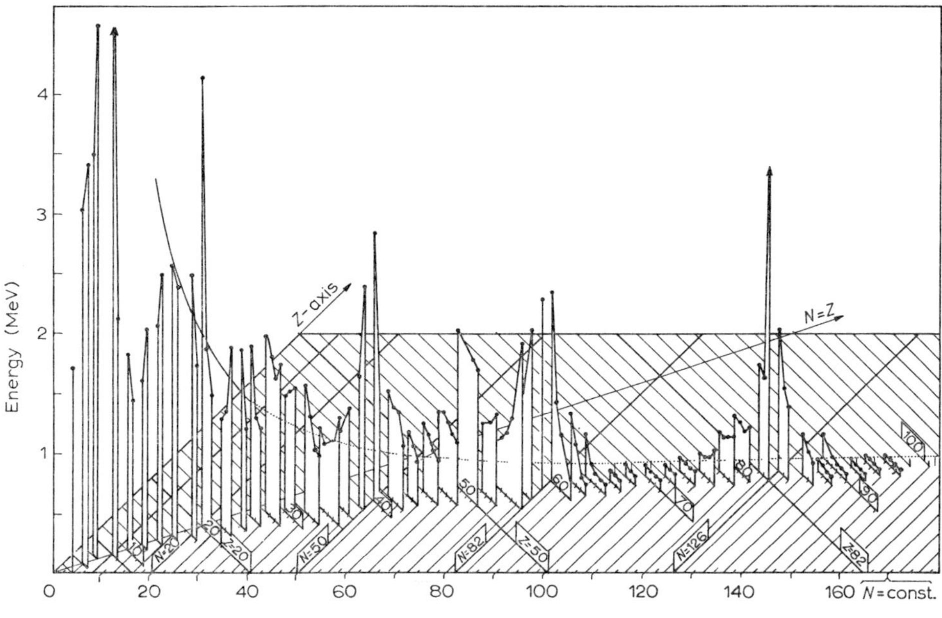

Rev. Mod Phys. **18** (1950) 432

Fig. 9.3.   Energies of the first excited 2$^+$ state as a function of N, Z

In the regions far from closed shells where the quadrupole moments are largest we find the energies of the 2$^+$ states are very low and show a very smooth variation with N or Z. When we recall the complexity of individual particle orbits it seems difficult to associate the 2$^+$ state with combinations of such motions. Rather it suggests that the 2$^+$ states have a collective origin. Furthermore when we look at the levels of even–even nuclei in the rare earth region we find a characteristic system of low lying levels. The spectrum of $^{180}_{72}\text{Hf}_{108}$ is shown in fig. 9.4. The spacings are close to being proportional to I^2 which suggests an analogy with the energy levels of a symmetric top

$$E_I = \frac{\hbar^2}{2\mathscr{I}} I(I+1) \tag{9.5}$$

where $\mathscr{I}$ is the moment of inertia of the top. Suppose we fix the value of $\hbar^2/2\mathscr{I}$ from the 0$^+$–2$^+$ spacing of 93 keV. Then

$$\frac{\hbar^2}{2\mathscr{I}} = \frac{93}{6-0} = 15.5 \text{ keV}.$$

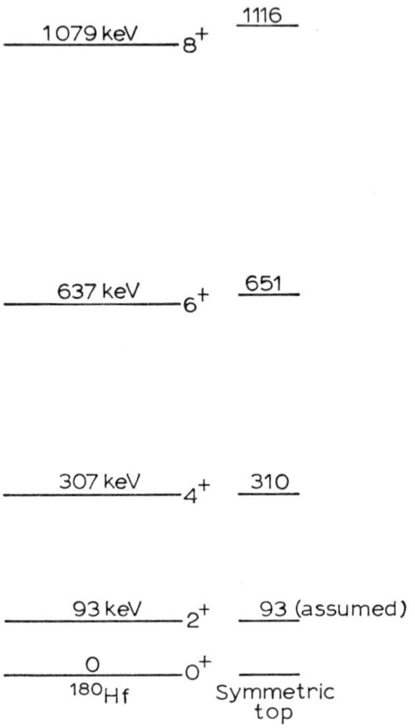

Fig. 9.4.   Rotational states in ^{180}Hf

Then we can compute the energies of the higher states

The 4^+ state $E_4 - E_0 = 15.5(20-0) = 310$ keV
The 6^+ state $E_6 - E_0 = 15.5(42-0) = 651$ keV
The 8^+ state $E_8 - E_0 = 15.5(72-0) = 1116$ keV

which are in remarkable agreement with the observed energies.

Many such bands of levels have been observed. Sometimes several bands are seen in the same nucleus and we assume they are based on different intrinsic states of motion, rotation of each configuration giving rise to levels with the $I(I+1)$ spacing.

When we look at levels in even–even nuclei nearer to closed shells we see a different level system. The energy of the first excited 2^+ state is now higher and the second excited state instead of being $\frac{10}{3}$ times the spacing of the first excited state and spin 4^+, is now closer to twice the first 2^+ spacing and has spin 2^+, 4^+ or 0^+ (fig. 9.5). Frequently two states with two of these spins are seen close together. Again this is a simpler effect than would be expected from single particle effects and is also ascribed to collective motions. The equal spacing suggests simple harmonic vibrations of the structure about an equilibrium *spherical* shape. The quadrupole phonons $\hbar\omega$ each carry angular momentum 2^+. Then the ground state represents no phonons, spherical shape, the first excited state one phonon 2^+, the second excited state two phonons each of 2^+, giving a degenerate triplet of states 0^+, 2^+, 4^+ and so on.

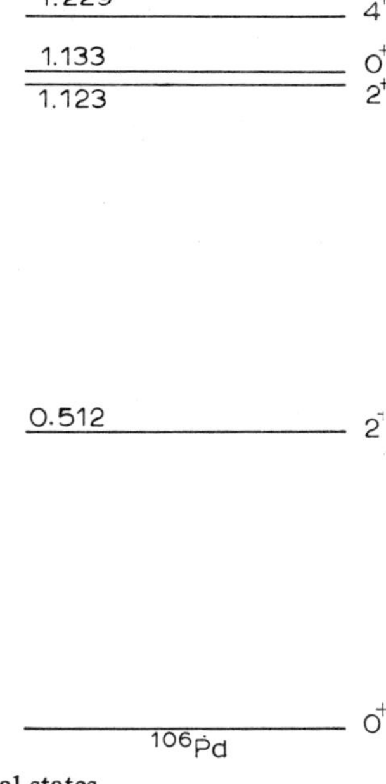

Fig. 9.5.   Quadrupole vibrational states

Thus in even–even nuclei we see trends emerging in which, far from closed shells, the loose nucleons lead to large permanent deformations characterised by rotational spectra; nearer closed shells the equilibrium shape is spherical but surface phonon vibrations lead to characteristic vibrational spectra; at closed shells excited states can only be produced by break-up of the core to give new particle states with no uniformity in angular momenta.

3.2 Vibrational spectra. We consider first the surface vibrations. We can describe these ellipsoidal vibrations by

$$R(\theta, \phi) = R_0\left[1 + \sum_\mu \alpha_\mu Y_\mu^{(2)}(\theta, \phi)\right]. \tag{9.6}$$

The surface shape is then described by the five parameters α_μ which specify the oscillation as a function of time and are our dynamical variables. For α small, the Hamiltonian is

$$H = T + V = \tfrac{1}{2}\sum_\mu B|\dot{\alpha}_\mu|^2 + \tfrac{1}{2}\sum_\mu C|\alpha_\mu|^2 \tag{9.7}$$

where B, the inertial parameter, and C, the force or restoring potential, both depend on details of the structure.

If we consider the nuclear matter to be an incompressible, irrotational uniformly charged fluid we can derive the parameters from the semi-empirical mass formula. We have

$$B = \tfrac{1}{2}\varrho R_0^5 = \frac{3}{8\pi} AMR_0^2 \qquad (9.8)$$

$$C = 4R_0^2 S - \frac{3}{10\pi} \frac{Z^2 e^2}{R_0} \qquad (9.9)$$

where ϱ is the density, A is the mass number, M the nucleon mass, Z the nuclear charge and S the surface tension parameter. The restoring force is the surface tension opposed by the electrostatic repulsion.

From the Hamiltonian we thus derive harmonic oscillations of frequency $\omega = (C/B)^{\frac{1}{2}}$. We say the surface can contain 0, 1, 2, ... phonons of surface energy, each of energy $\hbar\omega$. Fig. 9.6 shows the spacings and spin associated with vibrational levels.

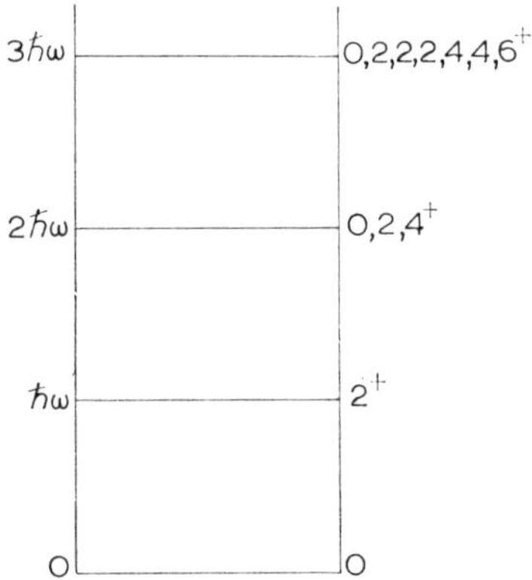

Fig. 9.6.   Types of vibrational motion

Each phonon carries angular momentum 2^+ and can be considered a boson, that is, it obeys Bose–Einstein statistics. We are led to a surface vibration spectrum.

We expect this crude model to be obeyed only roughly and that anharmonic terms we have neglected will remove the degeneracies in the 0, 2, 4^+ states for example. We do, however, see frequent examples of a state at twice the excitation of the 2^+ state which has spin 2^+, 4^+ less often and 0^+ rather rarely.

This is just one type of vibration but it is assumed that the surface wave vibration will require least energy and hence will lie lowest. The collective dipole oscil-

lation shown in fig. 9.7 we shall discuss in more detail with photonuclear phenomena. It lies at 12–20 MeV excitation.

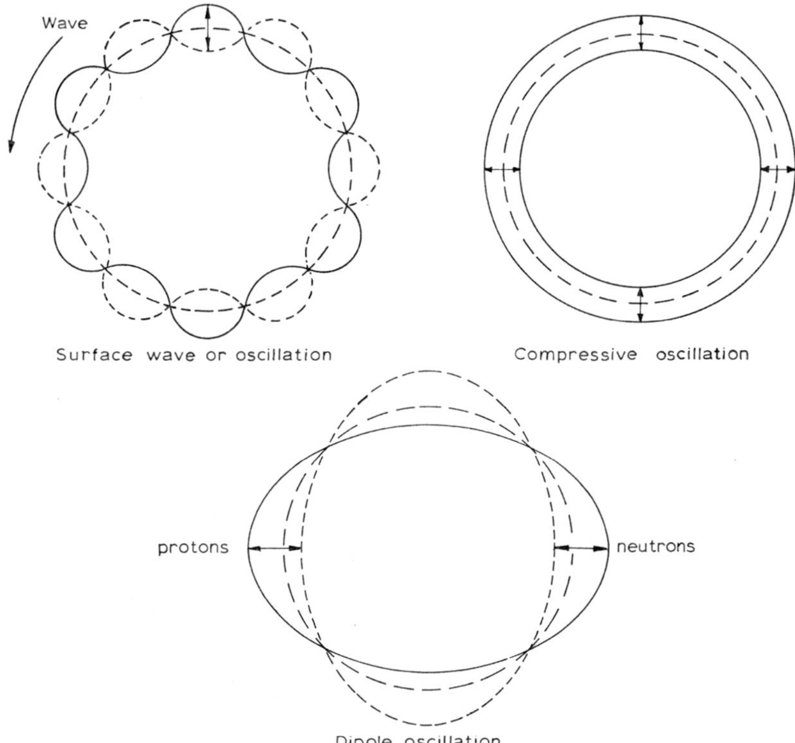

Fig. 9.7.   The quadrupole phonon model

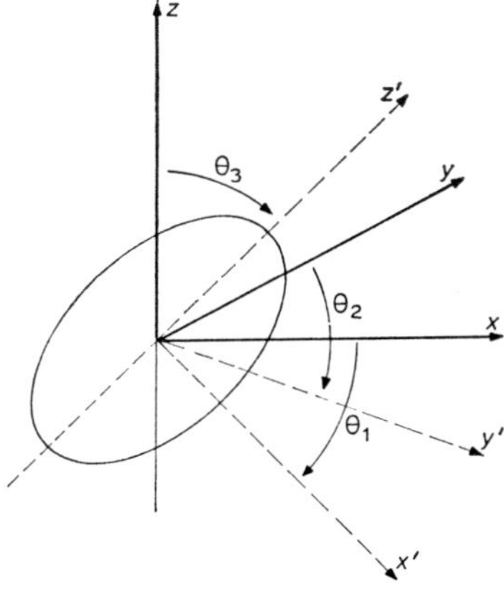

Fig. 9.8.   Change from space coordinates to coordinates relative to the symmetry axis

3.3   Rotational spectra. We consider now the characteristic spectrum like that of a symmetric top which we see in even–even nuclei with many nucleons outside the closed shell. It is convenient to change our variables α_μ to ones of more direct physical significance. We transform from a system of laboratory fixed coordinates (x, y, z) to the coordinate system corresponding to the fixed axes of the ellipsoid (x', y', z'). We must specify the Euler angles θ_1, θ_2, θ_3 (cf. fig. 9.8) which fix the orientation of the $x'y'z'$ system relative to the xyz system. We have two remaining variables to specify the magnitude and the shape of the distortion.

We associate the three principal eccentricities

$$\varepsilon_1 = \frac{2}{R_0}(\delta R_2 - \delta R_3)$$

$$\varepsilon_2 = \frac{2}{R_0}(\delta R_3 - \delta R_1) \tag{9.10}$$

$$\varepsilon_3 = \frac{2}{R_0}(\delta R_1 - \delta R_2)$$

with new variables β, γ so that

$$\varepsilon_\kappa = (15/\pi)^{\frac{1}{2}}\beta \sin(\gamma - \tfrac{2}{3}\pi\kappa). \tag{9.11}$$

Here $\beta^2 = \sum_\mu |\alpha_\mu|^2$ is the magnitude of the deformation while the value of γ gives the shape of the deformation. For $\gamma = 0$ or $\gamma = \frac{1}{3}n\pi$ with n integral we have an ellipsoid of revolution such as we have been discussing. Our five parameters α_μ are replaced by β, γ and $\theta_{1,2,3}$.

The potential energy is $V = \frac{1}{2}C\beta^2$; the kinetic energy is $T = T_{\text{vib}} + T_{\text{rot}}$. T_{vib} can be written

$$T_{\text{vib}} = \tfrac{1}{2}B(\dot{\beta}^2 + \beta^2\dot{\gamma}^2) \tag{9.12}$$

describing the oscillations in magnitude and shape of the distortion independent of orientation, i.e. of θ_i. Thus as before the vibrations are eigenfunctions of

$$H_{\text{vib}} = T_{\text{vib}} + V(\beta)$$

where

$$V(\beta) = \tfrac{1}{2}C\beta^2 \qquad \text{(for } \beta \text{ small)}. \tag{9.13}$$

T_{rot} does not involve changes in β or γ but changes in θ_i, that is, it is a function of $\dot{\theta}_i$, β, γ but not of $\dot{\beta}$ or $\dot{\gamma}$. It describes the free rotation of the deformed nucleus without change in magnitude of the deformation or its shape. It is identical with the Hamiltonian of a free rotor which gives

$$T_{\text{rot}} = \tfrac{1}{2}\sum_\kappa \mathscr{I}_\kappa \omega_\kappa^2 \tag{9.14}$$

where $\mathscr{I}_\kappa = \frac{4}{15}\pi B\varepsilon_\kappa^2$; ω_κ is the angular velocity about the κth principal axis and $\mathscr{I}_\kappa$ is the corresponding moment of inertia. For our ellipsoid of revolution two of the ε^κ are equal and thus two of the $\mathscr{I}_\kappa$ are equal. The $\mathscr{I}_3$ about the symmetry axis is zero. Thus the one value of $\mathscr{I}$ defines the motion of the symmetric top.

We now consider the angular momentum involved in such a deformed nucleus with axial symmetry. We consider the motion of the loose nucleons. The total angular momentum of the nucleus is $I(\equiv \hbar[I(I+1)]^{\frac{1}{2}})$ with component along the space axis $M\hbar$ and component along the symmetry axis $K\hbar$ (fig. 9.9).

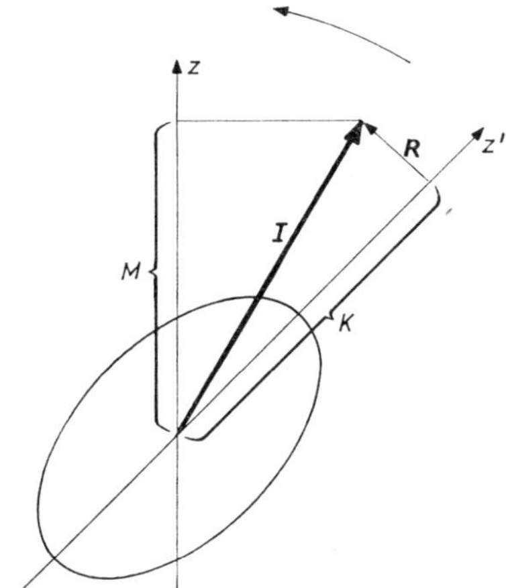

Fig. 9.9.   Angular momenta in the ellipsoid

Each individual particle of angular momentum j has a projection on the symmetry axis Ω. Since $\sum j = I$ then $\sum \Omega = K$. The angular momentum of the collective motion is R (rotation perpendicular to the symmetry axis). We thus obtain

$$|R|^2 = [I(I+1)-K^2]\hbar^2. \tag{9.15}$$

We can write the rotational energy

$$E_{\text{rot}} = \tfrac{1}{2}\mathscr{I}\omega^2 \tag{9.16}$$

where $\mathscr{I}$ is the moment of inertia. We now wish to estimate $\mathscr{I}$.

If the nucleus behaves as a rigid body the moment of inertia is simply $\mathscr{I}_{\text{rig}} = \frac{2}{5}MAR_0^2$, just the rotation of a mass MA. Alternatively the nuclear rotation could be considered as a wave travelling around the nuclear surface, leading to irrotational flow of the loose nucleons (fig. 9.10). That is, oscillations of the particles give rise to rotation of the *shape* but not of the particles. In this case the moment of inertia is given by

$$\mathscr{I}_{\text{irrot}} = \tfrac{2}{5}MA(\varDelta R)^2. \tag{9.17}$$

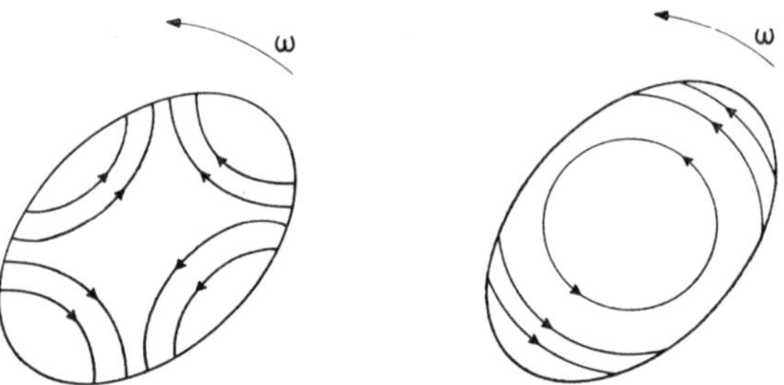

Fig. 9.10.   Irrotational (left) and rigid rotation

It is clear that the moment of inertia depends on the deformation

$$\mathscr{I}_{\text{irrot}} = \mathscr{I}_{\text{rig}}(\Delta R/R_0)^2 \tag{9.18}$$

and $\Delta R/R_0 = \varepsilon$ is close to the β-parameter previously defined,

$$\beta = \tfrac{4}{3}(\tfrac{1}{5}\pi)^{\frac{1}{2}}\Delta R/R_0 = 1.055\Delta R/R_0 = 1.055\delta \tag{9.19}$$

(δ is often used to stand for $\Delta R/R_0$).

In fact we find the observed values of $\mathscr{I}$ lie intermediate between $\mathscr{I}_{\text{irrot}}$ and $\mathscr{I}_{\text{rig}}$ being about 30–50 % of the latter (fig. 9.11). Obviously for a strict independent particle model we should expect $\mathscr{I}_{\text{rig}}$ while the complete breakdown of shell effects by the residual interactions would lead to $\mathscr{I}_{\text{irrot}}$. We obviously have an intermediate situation, i.e. partial breakdown of the shell model.

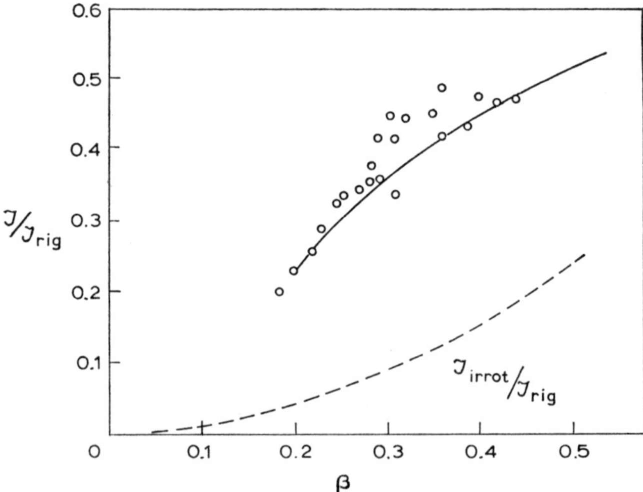

P. M. Endt and M. Demeur (eds.), Nuclear Reactions 1 (North-Holland, Amsterdam 1959) p. 480

Fig. 9.11.   Moments of inertia relative to the rigid rotation prediction as a function of deformation. The irrotational prediction is shown dashed

If we take for the present the irrotational value of $\mathscr{I}$ and since the rotational angular momentum $|\boldsymbol{R}|$ is $\mathscr{I}\omega$ then

$$|\boldsymbol{R}|^2 = \mathscr{I}^2\omega^2$$

and

$$E_{\text{rot}} = \tfrac{1}{2}\mathscr{I}\omega^2$$
$$|\boldsymbol{R}|^2 = 2\mathscr{I}E_{\text{rot}}$$
$$2\mathscr{I}E_{\text{rot}} = [I(I+1)-K^2]\hbar^2$$
$$E_{\text{rot}} = \frac{\hbar^2}{2\mathscr{I}}[I(I+1)-K^2] \tag{9.20}$$

just the energy levels of a symmetric top. (The constant term $\hbar^2 K^2/2\mathscr{I}$ does not affect the level *spacings*.)

For even–even nuclei we have pairs of identical particles which add successively Ω and $-\Omega$, hence $K=0$. There is symmetry about a plane perpendicular to the symmetry axis and the wavefunction must be invariant upon reflection through 180°. The angular momentum values in this case are restricted to $I=0^+, 2^+, 4^+, 6^+$ etc. as in the rotation of a homopolar molecule. This is just the lowest level sequence observed in many even–even nuclei in the strongly deformed regions, $A\approx 24$, $A=150$–190 and $A>200$. Using the energy formula we see that the spacings should be

$$E_{4^+}/E_{2^+} = \tfrac{20}{6}, \qquad E_{6^+}/E_{2^+} = \tfrac{42}{6}, \qquad E_{8^+}/E_{2^+} = \tfrac{72}{6}, \text{ etc.}$$

For odd-A nuclei in the deformed nucleus regions, the odd particle contributes $\Omega = K$ as projection on the symmetry axis since the remaining loose nucleons pair off to $\sum \Omega = 0$. Hence (for $K\neq\tfrac{1}{2}$) we get levels in the following order $I=K, K+1$, $K+2, K+3$ etc. all half-integral and of the same parity as the odd particle configuration. For $K=\tfrac{1}{2}$ we get a special situation wherein there is a partial decoupling of the intrinsic particle spin from the rotational motion. For $K=\tfrac{1}{2}$ we have

$$E_{\text{rot}} = \frac{\hbar^2}{2\mathscr{I}}\left[I(I+1)+a(-1)^{I+\frac{1}{2}}(I+\tfrac{1}{2})-K^2\right] \tag{9.21}$$

where the parameter a, called the decoupling parameter, has the value

$$a = \sum_j (-1)^{j-\frac{1}{2}}(j+\tfrac{1}{2})|c_j|^2 \tag{9.22}$$

where $|c_j|^2$ is the probability that the last odd particle has angular momentum j. This term can vary the level order so that we do not always get the lowest state with $I=K$. For example if we assume *pure j–j* coupling states the level orders are

j	Order of states	a
$\frac{1}{2}$	$\frac{1}{2}$, $(\frac{3}{2}\ \frac{5}{2})$, $(\frac{7}{2}\ \frac{9}{2})$	$+1$
$\frac{3}{2}$	$\frac{3}{2}$, $\frac{1}{2}$, $\frac{7}{2}$, $\frac{5}{2}$	-2
$\frac{5}{2}$	$\frac{1}{2}$, $\frac{5}{2}$, $\frac{3}{2}$, $\frac{9}{2}$	$+3$
$\frac{7}{2}$	$\frac{3}{2}$, $\frac{7}{2}$, $\frac{1}{2}$, $\frac{11}{2}$	-4

For mixed j configurations we expect intermediate values of a.

For odd–odd nuclei we expect the resultant K to be $\Omega_p \pm \Omega_n$. Thus there will be two almost degenerate states made up of proton and neutron aligned and of proton and neutron opposed and a rotational band will be built on each. There is some experimental evidence for this.

4   Nuclear moments of strongly deformed nuclei

This deformed nucleus model would lead to a value of the intrinsic quadrupole moment of

$$Q_0 = \tfrac{4}{5}ZR_0^2 \frac{\Delta R}{R_0} \tag{9.23}$$

with respect to the symmetry axis. However the quadrupole moment which we measure is related to the asymmetry in the *averaged* charge distribution. Thus in our strong coupling model

$$Q_{\text{obs}} = \frac{I(2I-1)}{(I+1)(2I+3)} Q_0 . \tag{9.24}$$

Thus in a deformed even–even nucleus the ground state has $I=0$ and $Q_{\text{obs}}=0$ even though Q_0 has a large value. The averaging effect of the quantum fluctuations in the axis direction ensures that the *average* charge distribution is spherically symmetric. For $I\neq 0$, Q_{obs} is not zero but is reduced below Q_0 by the factor shown. This factor is called the spectroscopic factor.

In the case of the magnetic moments we get two effects which cause the value to differ from the single particle values. One is due to the sharing of angular momentum between particle and collective motion, the other is due to the change in the particle orbit as the potential is non-spherical. For strongly deformed odd-A nuclei for $I=K\neq\frac{1}{2}$ we have

$$\mu_I = \frac{I^2}{I+1} g_\Omega + \frac{I}{I+1} g_R . \tag{9.25}$$

Here g_Ω is the g-factor of the last odd particle in the deformed potential and g_R is the g-factor for the collective rotational motion. The latter has the value Z/A for the incompressible model.

The reduced radiative transition probabilities may also be compared with predictions of this model. These values of, in particular, B(E2) may be determined by measurement of lifetimes of excited states. However in many cases if the B(E2) is very large the lifetime is very short. Another method has proven particularly suited to studying these enhanced E2 transitions in rotational spectra. This is coulomb excitation.

5   Coulomb excitation

Suppose that a charged particle passes near a nucleus with an energy considerably less than the coulomb barrier so that a nuclear interaction is improbable. It will follow a classical trajectory and be deflected according to the Rutherford scattering law. The electric field at the nucleus due to the charged particle will accordingly vary with time. This time dependent field can be represented quantum mechanically as a process of emission and reabsorption of virtual photons. The photon spectrum will be continuous, that is, all energies up to the particle energy will be present. Hence if the nucleus has a low lying excited state there will be some photons of exactly the right energy to be absorbed and excite this level (fig. 9.12). The cross section for such coulomb excitation will depend on the trajectory of the particle and on

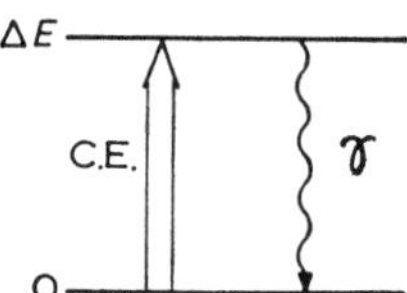

Fig. 9.12.   Coulomb excitation

the radiation width of the level as this will determine the probability of absorption of energy from the electromagnetic field. Thus wide levels (short lifetime) will have large coulomb excitation cross sections and we can relate the cross section to the value of $B(L)$ for the transition.

$$\sigma_L = f(E, Z, \Delta E)B(L) \tag{9.26}$$

where E is the incident particle energy, Ze is the charge of the bombarded nucleus, $\Delta E = E_\gamma$ is the exitation energy of the level and $f(E, Z, \Delta E)$ is a function which has been evaluated.* We *observe* either the subsequent de-excitation of the level by

* K. Alder, A. Bohr, T. Huus, B. Mottelson and A. Winther, Rev. Mod. Phys. **30** (1958) 353.

γ-ray emission or internal conversion, or the energy loss in the scattered particles.

Background effects of two kinds may interfere with coulomb excitation measurements. The charged particle will radiate bremsstrahlung as it passes the nucleus. For this reason heavy ions are frequently used in this work as they are travelling more slowly for the same energy and the bremsstrahlung yield is lower. Also X-rays resulting from atomic excitation of the target atom may be seen but since these occur as discrete lines they can be distinguished from the de-excitation γ-rays.

In even–even nuclei, states in the ground state, $K=0$, band differ by 2 units of angular momentum. Hence pure E2 transitions between them may be observed as a γ-ray cascade and successive levels may be excited by multiple coulomb excitation (fig. 9.13).

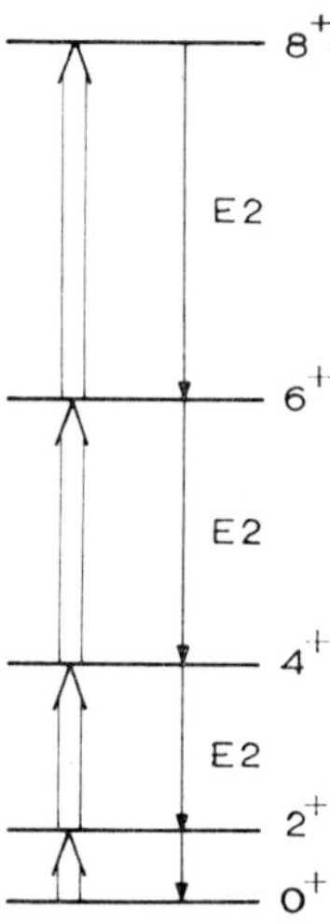

Fig. 9.13.   Multiple coulomb excitation

Our strongly deformed nucleus model predicts that within a band

$$B(\text{E2}) = \frac{5}{16\pi}\, e^2 Q_0^2 (I_i 2K0|I_f K) \tag{9.27}$$

where $Q_0 = \frac{4}{5}ZR_0^2(\Delta R/R_0)$ is the intrinsic quadrupole moment, I_i and I_f are initial and final state spins (for the γ-decay) and $(I_i 2K0|I_f K)$ is a vector combination coefficient called a Clebsch–Gordan coefficient.

In all strongly deformed nuclei even with $K \neq 0$ we expect strongly enhanced E2 transitions corresponding to the large intrinsic quadrupole moments. However in odd-A nuclei for example the de-excitation γ-ray may contain M1 as well as E2 multipoles.

6   Single particle states in a deformed potential

We now consider the motion of a single particle in a field of force which is no longer spherically symmetric. We expect that this will have the effect of varying the
level order from that we deduced for example for a harmonic oscillator central
potential plus a spin–orbit force with *spherical* symmetry. Such a calculation was
first made by Nilsson*. A nucleon with angular momentum l_j in the spherical case
($\beta = 0$) can now occupy $2j+1$ states with component angular momenta $\Omega_p = K_p = j$,
$j-1, j-2, \ldots -(j-1), -j$ along the symmetry axis. States of K_p and $-K_p$ are

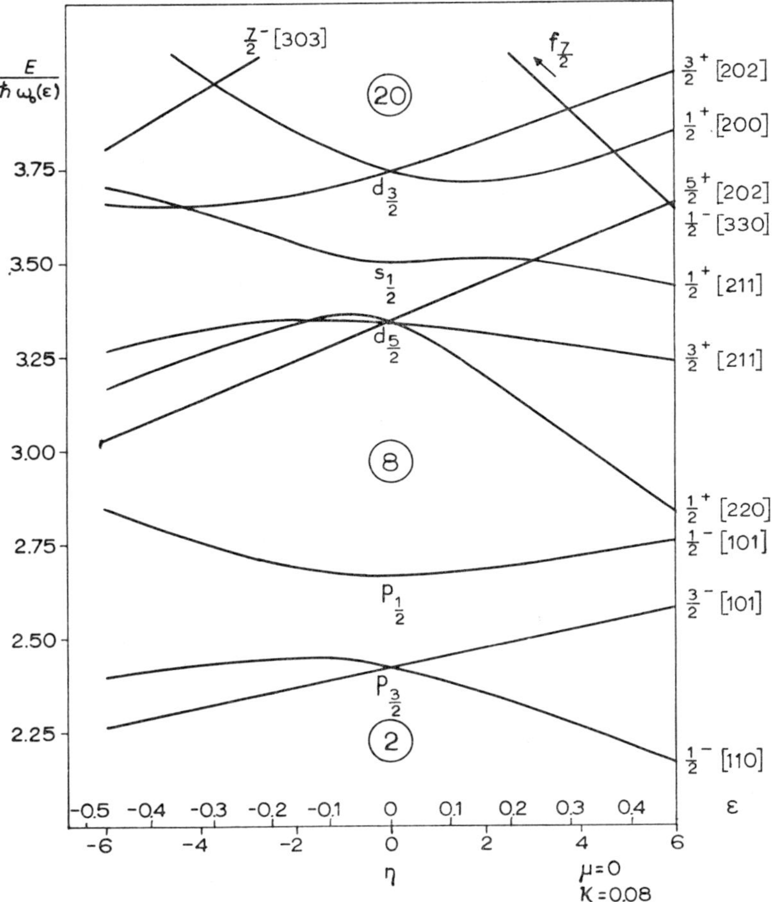

Fig. 9.14. Single particle energy levels as a function of deformation according to the Nilsson model

* S. G. Nilsson, Kgl. Dan. Vid. Selsk. Mat.-Fys. Medd. **29**, No. 16 (1955).

degenerate so we find that each shell model state characterised by l_j splits into $\frac{1}{2}(2j+1)$ states labelled by K (or Ω) and parity. Thus

$$
\begin{aligned}
1s_{\frac{1}{2}} \qquad & K = \tfrac{1}{2}^{+} \\
1p_{\frac{3}{2}} \qquad & K = \tfrac{1}{2}^{-}, \tfrac{3}{2}^{-} \\
1p_{\frac{1}{2}} \qquad & K = \tfrac{1}{2}^{-} \\
1d_{\frac{5}{2}} \qquad & K = \tfrac{1}{2}^{+}, \tfrac{3}{2}^{+}, \tfrac{5}{2}^{+} \\
2s_{\frac{1}{2}} \qquad & K = \tfrac{1}{2}^{+} \\
1d_{\frac{3}{2}} \qquad & K = \tfrac{1}{2}^{+}, \tfrac{3}{2}^{+} \\
1f_{\frac{7}{2}} \qquad & K = \tfrac{1}{2}^{-}, \tfrac{3}{2}^{-}, \tfrac{5}{2}^{-}, \tfrac{7}{2}^{-} \ \text{etc.}
\end{aligned}
$$

In fig. 9.14 the positions of the states are seen as a function of deformation parameter β. The states are filled pairwise with two protons and two neutrons. These are *intrinsic* particle states and on each of these states we might expect a rotational band to be 'built'.

Using the wavefunctions of Nilsson we can predict many properties of odd-A nuclei as a function of deformation such as magnetic and quadrupole moments, $B(L)$, reduced transition probabilities, reduced particle widths etc. If K can be considered a good quantum number certain K-selection rules will apply. Multipole radiation of order L will be forbidden between states for which $|K_f - K_i| > L$.

The observed $B(\text{E2})$ and static quadrupole moment will be compounded of a part due to the single particle motion in the non-spherical potential and a part due to the direct collective effect discussed above. For light nuclei the two effects may be comparable. For heavier nuclei the collective effect proportional to Z will dominate completely.

7   *Vibrations of strongly deformed nuclei*

If we plot the potential energy as a function of the deformation for different numbers of particles outside the closed shell we get a series of curves as shown in fig. 9.15.

For curve (a) near the closed shell the pairing forces favour grouping the nucleons to give a spherical equilibrium shape. The first vibrational state will lie at a high excitation. As more particles are added (curve b) the long range deforming forces begin to counteract the pairing forces, the nucleus becomes 'softer' and the frequency of the spheroidal vibrations becomes less. Finally as the long range forces dominate, the spheroidal shape becomes unstable and the nucleus acquires a permanent deformation. We can now expect rotations of the deformed nucleus with β, γ constant and at a slightly higher energy vibrations about the stable β-value (β-vibrations) and also oscillations in the *shape* with constant β (γ-vibrations).

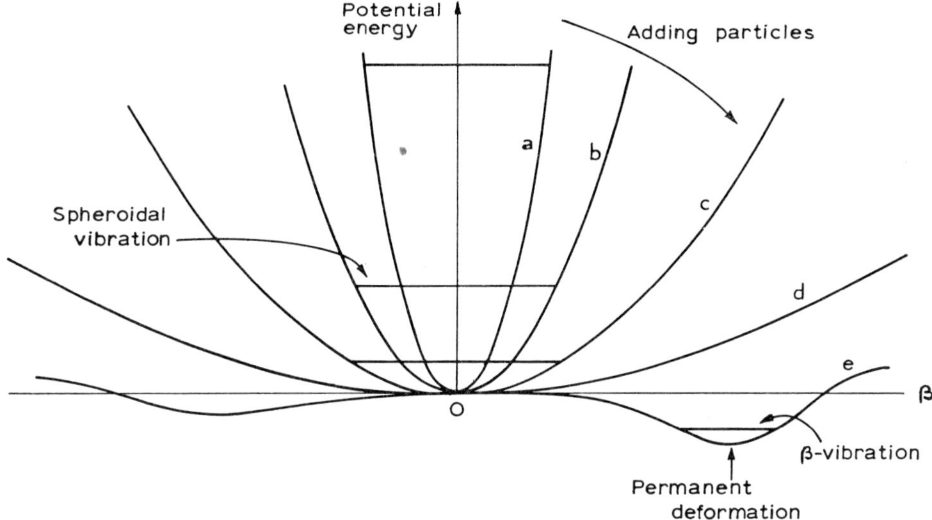

Fig. 9.15.   Change in potential energy of spheroidal nucleus as particles are added to closed shell (curve a). Low lying spheroidal vibrational states are seen as in curves b, c. Finally permanent deformation appears as in curve e. Beta-vibrational states may be observed

We see the effect of these vibrations both indirectly as a modification of the rotational spectrum and in the heaviest deformed region above Pb we can see these vibrational states directly.

The indirect effect is to add a term proportional to $[I(I+1)]^2$ to the rotational state energy formula. This is just the form seen from rotation–vibration coupling in molecules. It has the effect of depressing the higher spin states of a sequence below their expected value. We can note this effect in the spectrum of ^{180}Hf (fig. 9.4) where the higher spin levels lie somewhat below the predictions. We can get an *exact* fit to these levels by adding a term $-5 \times 10^{-4} \hbar^2 [I(I+1)]^2 / 2\mathscr{I}$. The rotation–vibration interaction transfers energy from the state in the rotational band to the state in the vibrational band of the same spin thus splitting the two states.

The β-vibration of lowest order in even–even nuclei is expected to have $K=0$, even parity like the ground state band while the γ-vibration of lowest order will have $K=2$, even parity. We see these states directly in ^{240}Pu (fig. 9.16).

These lowest β- and γ-vibrations can be seen in many of the heavy nuclei from Po upward. In some cases the rotational levels based on them can also be identified.

We also see in ^{240}Pu and other deformed nuclei the relatively low lying negative parity states with 1^-, 3^-. What is their origin? We have assumed all along that quadrupole deformations only are important since our first clue was the large quadrupole moments. However higher order deformations may also occur such as octupole. This corresponds to a pear-shaped nucleus rather than an ellipsoid. Ro-

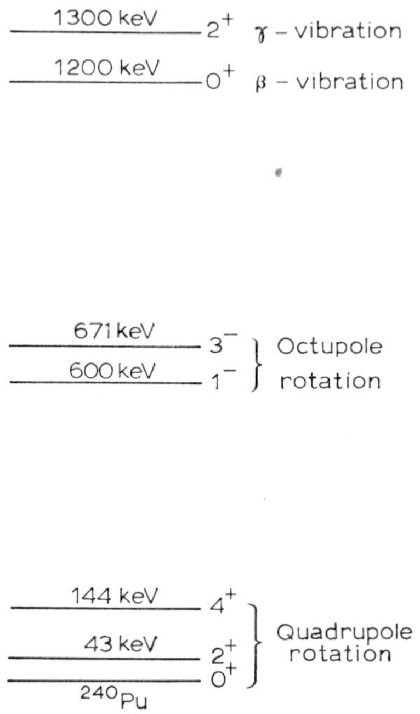

$$\underline{\quad 1300\ \text{keV} \quad} \quad 2^{+} \quad \gamma-\text{vibration}$$

$$\underline{\quad 1200\ \text{keV} \quad} \quad 0^{+} \quad \beta-\text{vibration}$$

$$\underline{\quad 671\ \text{keV} \quad} \quad 3^{-} \quad \left.\right\} \text{Octupole}$$

$$\underline{\quad 600\ \text{keV} \quad} \quad 1^{-} \quad \left.\right\} \text{rotation}$$

$$\underline{\quad 144\ \text{keV} \quad} \quad 4^{+}$$

$$\underline{\quad 43\ \text{keV} \quad} \quad 2^{+} \quad \left.\right\} \begin{array}{c}\text{Quadrupole}\\ \text{rotation}\end{array}$$

$$\underline{\qquad\qquad} \quad 0^{+}$$

$$^{240}\text{Pu}$$

Fig. 9.16. Rotational and vibrational states in ^{240}Pu

tations of such an object lead again to a $K=0$ band but reflection symmetry demands negative parity and suppression of even I-values. So we expect for $K=0$, $I=1^{-}, 3^{-}, 5^{-}$ etc. Note that in ^{240}Pu, $\hbar^{2}/2\mathscr{I}$ is 7.2 keV for the quadrupole rotational band and is 7.1 keV for the octupole rotational band.

8 *Rotation–particle coupling*

We may also expect in nuclei with permanent deformation that states of the same spin in different rotational bands may be mixed by the rotation–particle coupling. Energy is removed from the lower state and added to the upper state via the particle motion. The states will appear to be split or to repel each other. This effect is normally only of importance for bands differing by $\Delta K=1$. Classically it corresponds to the Coriolis interaction and is the same effect which contributed directly to the energies of $K=\frac{1}{2}$ bands. If two intrinsic particle states lie fairly close together so that their energy difference is comparable to the rotational energy, states of the same I in the two bands will be mixed. This represents a breakdown in the K quantum number. Many examples of this are observed. This rotation–particle coupling is just the effect of the Coriolis force due to the particle motion.

9 *Spheroidal vibrations*

Returning to the spheroidal vibrations with equilibrium value of $\beta=0$, we expected on the quadrupole phonon model equally spaced states in even–even nuclei with spins 0^+, 2^+ and a degenerate triplet 0^+, 2^+, 4^+. In fact such states are

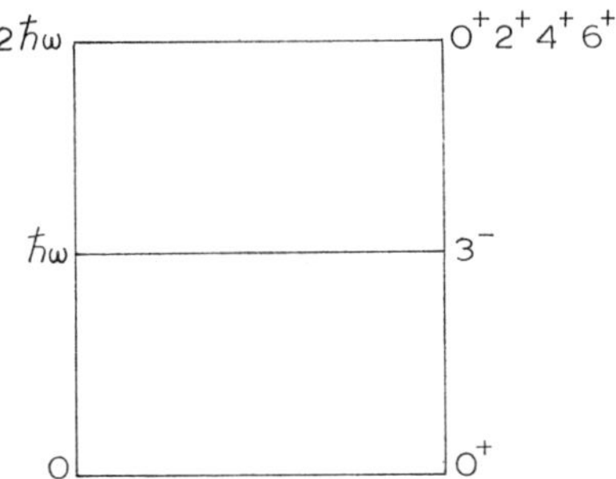

Fig. 9.17.   States of an octupole phonon model

observed but only rarely are all three states of the triplet seen. The 2^+ member is seen most often. Moreover the radiative decay of these two phonon states does not seem to be in agreement with predictions based on this model. It would appear the model is much too simple.

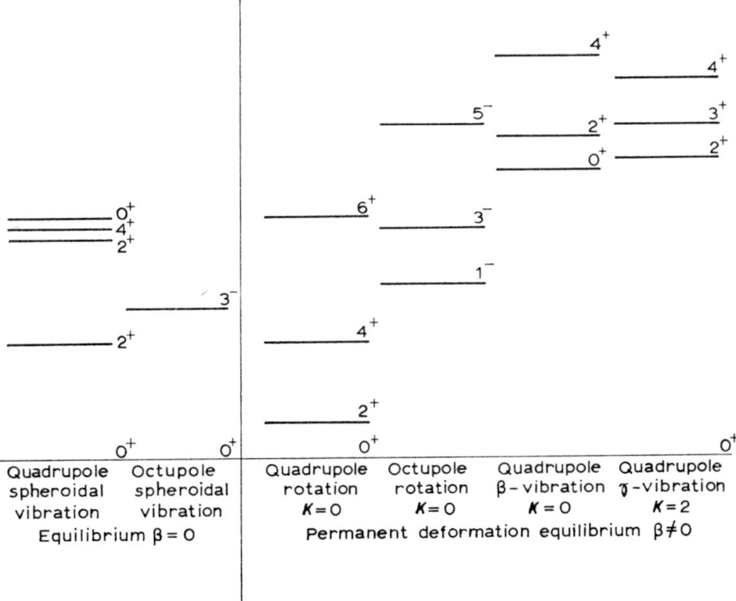

Fig. 9.18.   Types of collective states seen in even–even nuclei

We also see evidence of collective states which could correspond to octupole spheroidal vibrations. In this case the expected sequence is as shown in fig. 9.17 and there are many examples of a 3^- state at 2–3 MeV which show an enhanced E3 transition to the ground state and hence must be of collective origin. To summarise, the lowest collective states expected in even–even nuclei are shown in fig. 9.18.

10   Unified models

We thus seem to be approaching a *unified* model in which all the features of the individual particle motion and of the cooperative motion can be logically combined.

In fact since in some cases we have two models which give equivalent agreement with experiment we guess that the two pictures are in fact identical basically. This was dramatised some time ago for the case of ^{19}F. Shortly after the intermediate coupling shell model calculations were made for this nucleus Paul[*] pointed out that the ^{19}F states could also be described using the Nilsson model if mixing of K bands was also assumed. The results of the two calculations agreed with each other more closely than they agreed with experiment. This led Elliott[**] to examine the overlap more carefully. The introduction of the residual forces to the single particle shell model mixes configurations in a certain way so as to agree with experiment. The non-spherical potential shape similarly mixes single particle states and the end result must be identical. Elliott discovered a group theoretical recipe for mixing the configurations of the four particles in ^{20}Ne and ended up with rotational bands! Kurath[†] also studied the p-shell and found the wave functions derived from the two methods to be identical.

There is a third method which should be mentioned which has also been applied to these and lighter nuclei at least in a qualitative way. In this case ^{19}F is considered to be ^{16}O+triton. One then uses the triton wavefunction to describe the loose particles. The correlations so introduced turn out to give wavefunctions identical to those derived from the intermediate coupling shell model or the Nilsson model.

So we now rather think of the three *models* as three *methods* for arriving at a configuration-mixed model for deformed nuclei of odd A and we use the method which is most convenient. This is frequently the Nilsson method, hence we have described it in more detail.

Now this is a unified model in the sense that in certain regions where there is a

[*] E. B. Paul, Phil. Mag. **2** (1957) 311.
[**] J. P. Elliott, Proc. Phys. Soc. **A245** (1958) 128, 562.
[†] D. Kurath and L. Picman, Nucl. Phys. **10** (1959) 313.

permanent deformation the interplay of the single particle motions in the deformed potential and the cooperative effects due to the deformation induced by the loose particles is described.

However it does not pretend to a simple description of the transitional regions between region of deformed nuclei and region of spherical stable nuclei, that is, the regions where we see some evidence for spheroidal vibrations about a spherical shape of E2 and E3 character.

Also it gives no real help when we try to explain the higher states in even–even nuclei. If we try to explain the levels seen as merely the promotion of one of the paired particles to another orbit we do not get agreement.

We now describe some recent attempts to put the whole theory on a somewhat more general basis. We start once again with the single particle shell model with a spin–orbit force. We can say that 1) the main part of the forces between nucleons can be considered to coalesce into an average spherical potential and 2) there is a part of the nucleon–nucleon force 'left over', the residual interaction, and we must treat this separately.

We note a similarity with the theory of conduction electrons in a metal. These electrons are in an average potential well of depth equal to the Fermi energy plus

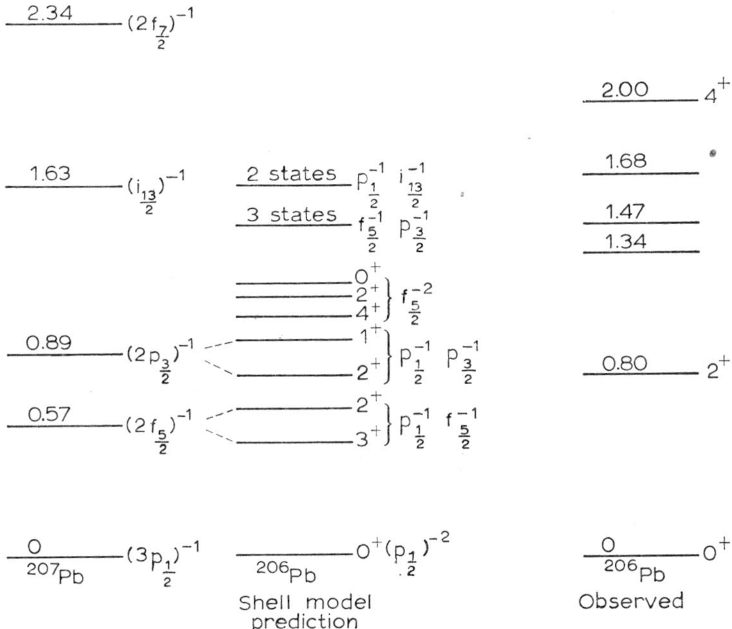

Fig. 9.19.   States of ^{206}Pb predicted via the simple shell model from states of ^{207}Pb and the ^{206}Pb states actually observed

the thermionic work function. However the electrons also interact with one another and near absolute zero these interactions become important and lead to the phenomenon of superconductivity. So we can treat the residual interactions in nuclei in a very similar way.

We know from observation of even-even nuclei that like particles pair up to give $I = 0$. However these pairing effects were not included in the shell model as such except to ensure that the core would be inert to the single particle motion, and this neglect of the pairing force leads to disagreement with observed spectra in the following way. Suppose we compare levels of ^{207}Pb and ^{206}Pb which are both magic in protons (82). ^{207}Pb has one neutron hole and its spectrum as shown in fig. 9.19 is just that to be expected as a neutron is lifted successively from inner levels in the $N = 126$ shell. If we now consider ^{206}Pb with two neutron holes the spectrum generated by similar promotion of neutrons from inner levels contains many states. The actual observed level scheme of ^{206}Pb does not agree with our prediction at all. The expected low lying levels are not observed (except for the 2^+ state) and a considerable energy is required before the numerous states show up. This phenomenon is called the energy gap. It comes about through having to break a neutron pair in the ground state of ^{206}Pb and this takes extra energy, more than the shell model accounts for. Similarly the pairing of electrons is the basis of superconductivity and a similar energy gap is seen.

We try then to introduce this short range pairing force into our shell model ideas. The action of the pairing between nucleons in different shells is to break down the definite j-dependence. The pairing force tends to destroy a pair in one shell in order to create a pair in another shell state.

In other words, a particle may be found partly in one shell, partly in another. This is just the *configuration mixing* we have spoken of before. This means shells are partially occupied. If we plot occupancy vs level energy and assume a very large number of levels so their energy E_j is a continuous function we get a curve as shown in fig. 9.20.

In the single particle shell model we assume occupancy up to a certain maximum energy λ which we can call the chemical potential, determined by the density of particles with account taken of the exclusion principle. For energies larger than λ, there is no occupancy. The pairing interaction mixes configurations so shell states near λ are only partially occupied. The Fermi surface is no longer sharp but has a transition region of thickness 2Δ.

We can compute λ and Δ from the shell model if we assume a pairing force of strength G. Δ will depend on the magnitude of G. For $G = 0$, no pairing, we get just the single particle shell model. The curve shown in fig 9.20 refers to the ground state of an even–even nucleus. For excited states the occupation probability must change. In the single particle shell model it corresponds to creation of particles in previously

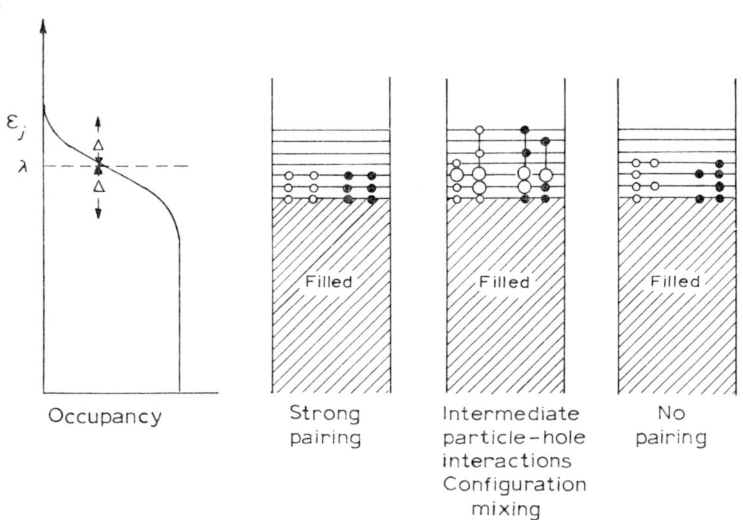

Fig. 9.20. Quasi-particles and particle–hole interactions

empty shells and of holes in full shells. Now however we have excitations in the
region of *partial* occupancy, so we are creating less than whole particles and less
than whole holes. These are called quasi-particles.

We say that a quasi-particle in the j-shell contributes an excitation energy

$$E_j = [(\varepsilon_j - \lambda)^2 + \Delta^2]^{\frac{1}{2}}. \tag{9.28}$$

For shells far from the Fermi surface $|\varepsilon_j - \lambda| \gg \Delta$ and the excitation energy is just
$\varepsilon_j - \lambda$, which is identical with the independent particle shell model. But note that
the excitation energy can never be *smaller* than Δ. Hence Δ is just the magnitude of
the energy gap.

Even–even ground states have no quasi-particles since all the particles are paired.
It is the quasi-particle vacuum and we can only create excited states by formation of
quasi-particle pairs. Since each quasi-particle requires for formation an energy
of at least Δ, the energy gap seen in even–even nuclei has a magnitude of at least 2Δ.

Odd-A nuclei have one quasi-particle i.e. unpaired particle in the ground state
and this can be excited to a higher energy in the normal way without any *extra* energy needed for its creation. So no gap is observed in odd-A nuclei.

We have considered the inclusion of a so-called pairing force which arose from
the attraction of like particles with opposing spin. This could not have been included in the original shell model with a central force and spin–orbit force and so is
a new part of the residual interaction to be put in. Its effects could be thought of in
terms of the configuration mixing near the top of the Fermi sea leading to partially

filled and partially empty states. Breaking a pair in this picture then required an extra amount of energy corresponding to the thickness of this region of partial occupancy. It is convenient to consider the partially filled state in terms of quasi-particles so that a shell model based on quasi-particles and quasi-holes had built into the definitions of the quasi-particles the correction for the pairing interaction. In the absence of a pairing force, real particles obey a shell model picture. In presence of a pairing force quasi-particles obey a shell model picture.

We are then able to explain features of the even–even nuclear spectra whilst retaining the shell model appearance of odd-A spectra. Note that odd–odd nuclei are to be described as two independent quasi-particle systems as we do not assume an n–p pairing force. This is a disturbing drawback of the theory.

We should also like to include in the model some account of the collective excitation which can be considered as coherent scattering of quasi-particles. This long range force is introduced in the following way.

Consider two particles 1, 2 at distances r_1 and r_2 from the centre of the nucleus. A potential acting between them of scalar form and spin independent can be expanded in terms of Legendre polynomials $P_l(\cos \theta_{12})$ where

$$\cos \theta_{12} = \frac{r_1 \cdot r_2}{|r_1||r_2|} .$$

The first term in P_0 can be neglected as it is just the spherical well already included corresponding to volume changing vibrations which are not observed. The second term in P_1 must vanish as it just acts to move the centre of mass of the nuclear matter with respect to the centre of the well. The third term in P_2 is the first significant term and can lead to spheroidal vibrations and quadrupole moments as we have seen. So we introduce a force $-\chi F(r_1, r_2)P_2(\cos \theta_{12})$ where χ is a constant and F is a function of $|r_1 - r_2|$. The P_2 function favours angles of 0 and π. So if F is of long enough range, the particles tend to align to produce the quadrupole distortion observed in the nucleus as a whole.

So finally our Hamiltonian will contain three terms

1)   the shell model part; spherical harmonic oscillator $+$ spin–orbit,
2)   the pairing part, short range, tending to stabilise the spherical shape by pairing to $J=0$,
3)   the $P_2(\cos \theta)$ part, long range, tending to align nucleons into a stable deformed shape.

This sort of treatment has been fairly successful for nuclei with a closed shell in one particle, e.g. Sn with $Z=50$ and in certain vibrational nuclei.*

* L. S. Kisslinger and R. A. Sorensen, Kgl. Dan. Vid. Selsk. Mat.-Fys. Medd. **32**, No. 9 (1960).

Our picture is thus as follows: the basic 'shell model' nucleus described as a spherical harmonic oscillator with spin–orbit interaction is in reality subject to two different effects whose relative importance depends on how many nucleons are outside the closed shell (fig. 9.21). The two-particle pairing interaction tends to

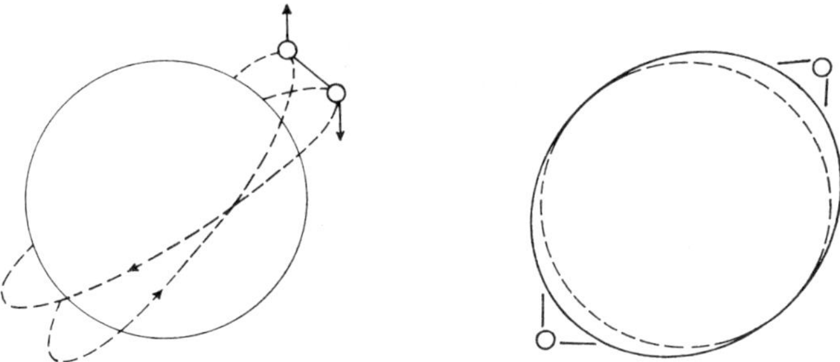

Fig. 9.21.   The pairing force and the long range aligning force

stabilise the spherical shape and the fluctuations of each pair from spherical symmetry will result in overall fluctuations of the nucleus from spherical symmetry. In this circumstance we observe spheroidal vibrations. The interactions of nucleons on opposite sides of the nucleus can result in a force of the $P_2(\cos\theta)$ form of apparently long range. This can result, if there are enough loose nucleons, in a clustering of nucleons on opposite sides of the nucleus and thus in a permanent nuclear deformation. This deformation gives rise to somewhat different single particle orbits as well as allowing energy states to result from rotations and vibrations of the nucleus as a whole.

In principle then, our understanding of nuclear structure in terms of this unified model is very good. However the problems of calculating nuclear properties with sufficient accuracy on the basis of such a complex situation are still far from solution.

10

Nuclear Orientation

1 Introduction

We have referred frequently to nuclear spin and in ch. 1 outlined some atomic phenomena which enable us to deduce its value. We now consider some further direct consequences of the spin in nuclear reactions.

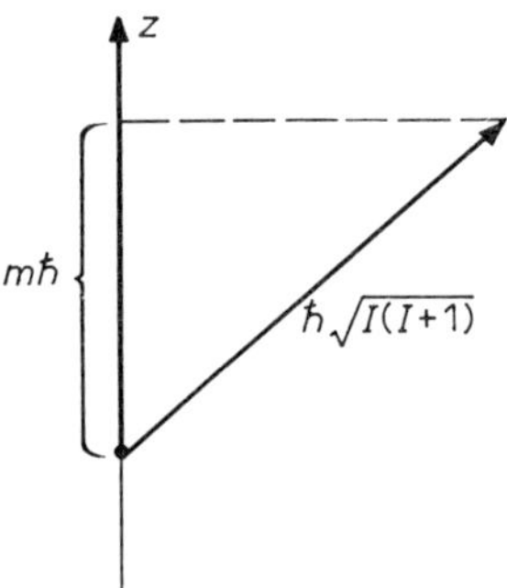

Fig. 10.1. Angular momentum and its projection

We recall the definitions of the angular momentum vector I and the associated magnetic moment μ_I (fig. 10.1). The absolute value of I is $\hbar[I(I+1)]^{\frac{1}{2}}$ and its projections on an externally defined space axis are $m\hbar$ where m takes on the $2I+1$ values

$$I, I-1, I-2, \ldots, -(I-2), -(I-1), -I.$$

The magnetic moment is measured in nuclear magnetons and again μ_I is the maximum value of its projection on a space axis. The nuclear gyromagnetic ratio is

$$\gamma_I = \frac{\mu_I}{I} = \frac{\mu_I}{I\hbar}.\tag{10.1}$$

The nuclear g-factor is

$$g_I = \gamma_I \frac{2M_\mathrm{p}c}{e}$$

where $e\hbar/2M_\mathrm{p}c$ is the value of the nuclear magneton.

If we now observe particles or quanta from a radioactive source in a field-free region we find them distributed isotropically in space. For one particular nucleus the emission may not be isotropic but as the nuclear directions are quite random we observe isotropy.

If we define an external space axis O_z then for the whole assembly of nuclei, there are equal numbers having projections $[I, I-1, \ldots, -(I-1), -I]$ on O_z. Thus for a single nucleus the probability $W(m)$ of the occurrence of any particular value, $m\hbar$, of the projected angular momentum is independent of m. We say the system is *unoriented* (fig. 10.2a). In the absence of a magnetic field the energies associated with the various m-values are identical – they are degenerate. An external magnetic field can split these states, hence they are referred to as magnetic substates and in the unoriented case we say the substates are equally populated.

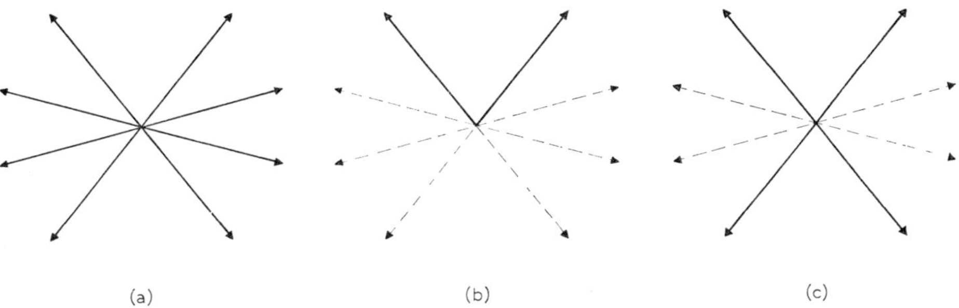

Fig. 10.2.   Orientation definitions: a) unoriented, b) polarisation, c) alignment

If $W(m)$ does depend on m, that is, if the substate populations are unequal, the system is said to be *oriented*. We distinguish two possibilities. If the spin direction is more likely to be in one direction, say $+O_z$, than in the other, $-O_z$, then $W(+m) \neq W(-m)$ and we term the system *polarised* (fig. 10.2b). On the other hand if equal numbers point in each direction but the populations of different $|m|$ substates are different we term the system *aligned* and $W(m) \propto m^2$ (fig. 10.2c). In this case the spin must be $\frac{3}{2}$ or greater since for spin $\frac{1}{2}$ we have only one substate of either sign.

Polarisation is defined as $P = I^{-1}\sum mW(m)$ which is simply the excess of spins pointing in one direction. For spin $\frac{1}{2}$, $m = \pm\frac{1}{2}$ corresponding to up and down. Suppose N_+ are pointing up in state $m = +\frac{1}{2}$ and N_- are pointing down in state $m = -\frac{1}{2}$. Then

$$W(+\tfrac{1}{2}) = \frac{N_+}{N_+ + N_-} \quad \text{and} \quad W(-\tfrac{1}{2}) = \frac{N_-}{N_+ + N_-}$$

and

$$P = 2\left(\frac{1}{2}\frac{N_+}{N_+ + N_-} - \frac{1}{2}\frac{N_-}{N_+ + N_-}\right) = \frac{N_+ - N_-}{N_+ + N_-}. \tag{10.2}$$

P can take on any value between $+1$ and -1.

For the deuteron with spin 1 the situation is more complicated as we have three substates $m = +1, 0, -1$. If we ignore the population of the $m = 0$ substate we can define the *vector polarisation* as before

$$P_3 = \frac{N_+ + N_-}{N_+ + N_0 + N_-}. \tag{10.3}$$

But we also must define another polarisation to describe the $m = 0$ population N_0. This is the *tensor polarisation*.

$$P_{33} = \frac{N_+ - N_0 + N_- - N_0}{N_+ + N_0 + N_-} = \frac{1 - 3N_0}{N_+ + N_0 + N_-}. \tag{10.4}$$

This can take on values from $+1$ to -2.

2 Detection of nuclear orientation

A common method to check whether nuclear orientation has taken place is to observe the emission of γ-radiation as a function of angle. A γ-ray of multipolarity (L, M) emitted in a transition between an initial state of spin (I_i, m_i) and a final state (I_f, m_f) has an angular distribution $F_L^M(\theta)$ with respect to the quantisation axis O_z. The selection rules for the transition are

$$I_i + I_f \geq L \geq |I_i - I_f|$$
$$m_i - m_f = M \quad \text{where} \quad M \leq L.$$

These limit the number of components which are permitted. In optical spectroscopy the substates can rather easily be split apart as shown in fig. 10.3. In nuclear

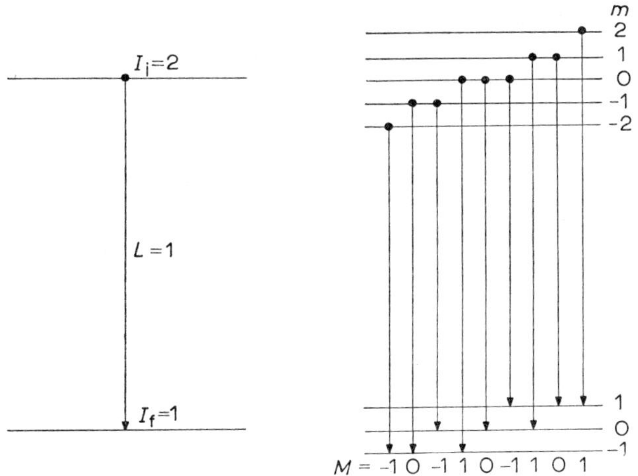

Fig. 10.3.   Radiative decay and its components

spectroscopy with available magnetic fields the separation is only about 10^{-8} eV, and the splitting is impossible to observe (except in a special case, the Mössbauer effect). So what we observe are the separate components combined incoherently. We have to include the probability of the vectors I_i and I_f with certain magnetic quantum numbers combining to give the vector L with a particular magnetic quantum number. This probability is given by what is variously termed a transformation amplitude, a vector addition coefficient, a Wigner coefficient or when we are combining 2 vectors only, a Clebsch–Gordan coefficient. This is written as $(I_i I_f m_i m_f | LM)$ and may be evaluated using algebraic formulae or by consulting tables.

The angular distribution of radiation from a given substate m_i is

$$W(\theta) = \sum_M (I_i I_f m_i m_f | LM)^2 F_L^M(\theta) \tag{10.5}$$

and if the inital substates have weights $W(m_i)$ which depend on the value of m, the total angular distribution is

$$W(\theta) = \sum_{M, m_i} W(m_i)(I_i I_f m_i m_f | LM)^2 F_L^M(\theta). \tag{10.6}$$

If $W(m)$ is independent of m, $W(\theta)$ must be a constant, that is the distribution is isotropic. For example if $I_i = 1$, $I_f = 0$ and $L = 1$ then $F_1^0 \propto \sin^2\theta$ and $F_1^{\pm 1} \propto \frac{1}{2}(1 + \cos^2\theta)$. The Clebsch–Gordan coefficients are $(1\ 0 \pm 1\ 0 | 1 \pm 1)^2 = 1$ and $(1\ 0\ 0\ 0 | 1\ 0)^2 = 1$. Thus

$$W(\theta) = (\sin^2 \theta) \times 1 + \tfrac{1}{2}(1 + \cos^2 \theta) \times 1 + \tfrac{1}{2}(1 + \cos^2 \theta) \times 1$$
$$= (1 - \cos^2 \theta) + (1 + \cos^2 \theta)$$

which is constant, as expected.

The function F_L^M contains only Legendre polynomials of even order for pure multipoles so we can write the angular distribution as

$$W(\theta) = a_0 + a_2 \cos^2 \theta + a_4 \cos^4 \theta + \ldots a_{2L} \cos^{2L} \theta. \tag{10.7}$$

No term in $\cos \theta$ of power higher than $2L$ can appear. Also since only even powers of $\cos \theta$ appear our distribution must be symmetric about 90°. Accordingly the observation of an anisotropic angular distribution is evidence for nuclear orientation of the decaying state.

3   Production of nuclear orientation

3.1   Low temperature methods. The whole idea of low temperature orientation is, to be left with only a very few possible energy levels in the solid and to use the thermal equilibrium distribution of energies to populate these with different weights. If the nuclei are differently oriented depending on which state they are in, one gets a net orientation. Most systems of nuclei with spin exhibit a weak paramagnetism in a magnetic field (to separate the levels), because of excess population of the lower levels through the Boltzmann distribution factor. In a field H_0 there is then a nuclear polarisation which is directly proportional to the paramagnetic susceptibility. For example for a nucleus of spin $\tfrac{1}{2}$ and moment μ_I there are two levels separated by an energy $\mu_I H_0 / I = 2\mu_I H_0$ (fig. 10.4). Then if the temperature is such that $kT \approx \mu_I H_0$ we will get unequal populations. For $\mu_I = 1$ nuclear magneton $H_0 / T = 2.8 \times 10^7$ gauss/°K. The largest practical external fields are about 5×10^4 gauss so we must go to temperatures of the order of 0.002 °K. In fact polarisation of about 2 % has been observed in ^{115}In at 0.05 °K.

Instead of an external field we can sometimes make use of the very large internal magnetic fields in paramagnetic *ions* due to the electron spin. Because of the very

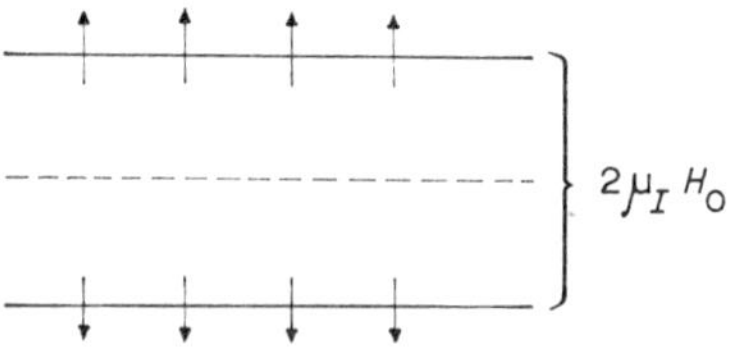

Fig. 10.4.   A level split in an external field

high electronic moment we can use modest external fields of a few hundred gauss to orient the *ion*. The nuclear spin will follow the electronic spin because of the high internal fields of 10^5–10^6 gauss, and again we get polarisation in thermal equilibrium.

Suppose we have a system in which the internal field is due to a single electron of spin $s=\frac{1}{2}$. This acts on a nucleus of spin $I=\frac{1}{2}$. The interaction energy is $Am_I m_s$ where $m_i = \pm\frac{1}{2}, m_s = \pm\frac{1}{2}$ with respect to the axis of quantisation fixed by an external field. The field splits the F levels into substates m_F. In the top two states the electron spin is up while in the bottom two states the electron spin is down. On fig. 10.7 is shown the behaviour of the energy splittings of the magnetic substates by a strong magnetic field.

In one method suggested by Gorter and Rose the sample is brought to a low temperature, ≈ 0.01 °K and then $kT \approx \frac{1}{2}A$ in a residual field H_0. The lowest state d is most heavily populated and we get a polarisation of nuclei *up*.

Bleaney and Pound suggested another technique in which a single crystal is used in which anisotropic magnetic fields exist. The crystal field now provides the axis of quantisation and no external field is required. In this case the energies of states with $m_I = \pm m$ are the same and we get alignment. So this method only orients nuclei with $I \geq 1$.

Since a γ-ray distribution contains only even powers of $\cos \theta$ an aligned system and a polarised system give the same pattern. However only a polarised system could show *asymmetry* such as was observed in β-decay with parity violation.

Experimentally some radioactive atoms to be studied are incorporated into a single paramagnetic crystal which has suitable internal fields. This is then brought to a temperature of ≈ 0.01 °K by adiabatic demagnetisation. For alignment we demagnetise to zero field but for polarisation we must be left with a residual field and so have to start with a higher field. The angular distribution of γ-rays is then observed with respect to the axis of the crystal or of the field. This is often

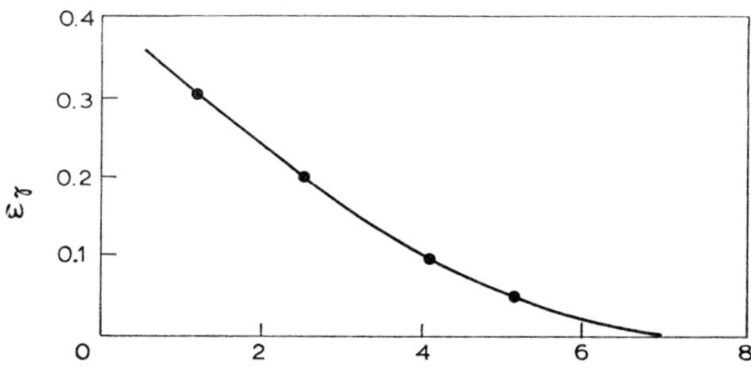

Fig. 10.5.   Anisotropy of γ-emission vs temperature (which rises with time)

expressed as an anisotropy

$$\varepsilon_\gamma = \frac{W(90°) - W(0°)}{W(90°)} .$$

(10.8)

We then measure this as a function of temperature as the crystal warms up (fig. 10.5). Analysis of this result can give information on multipolarity or mixture of multipoles if such exists. This in turn can lead to spin assignments for the levels involved.

3.2   Optical pumping. Another method again uses the fact that if we can orient the electronic spin the hyperfine coupling will cause the nuclear spin to be oriented too. If we have free atoms (not bound in a solid) they can be oriented by absorption of resonance radiation. This leads to a change in the population of the levels. For example suppose we have an atom with $F(=I+S)=\frac{3}{2}$ in the excited state and $F=\frac{1}{2}$ in the ground state. If we illuminate only with right circularly polarised light we populate only the substates $+\frac{3}{2}$ and $+\frac{1}{2}$ of the upper level (fig. 10.6). These however can decay by either kind of transition. So by this optical pumping we increase the population of the $+\frac{1}{2}$ substate of the ground state with respect to the $-\frac{1}{2}$ substate. This can go on at room temperature and the nuclear spin will follow, hence we get nuclear orientation.

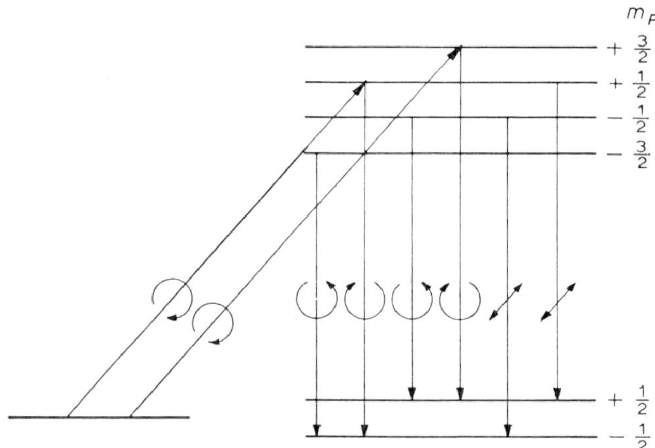

Fig. 10.6.   Optical pumping

We must use gas or vapour at rather low pressure to minimise collisions which cause disorientation. We must also avoid stray magnetic fields. However this method can provide polarised targets of ^{3}He, for example.

3.3   Polarised beams. We have so far considered methods for orienting target nuclei. We now consider oriented beams of nuclei. We restrict our discussion to hydrogen since this is most used and is simplest. It was proposed by Clausnitzer et al.* to perform a Stern–Gerlach experiment on a beam of neutral hydrogen atoms. If we pass such a beam through a magnetic field the substates will be split as before into four components, and all levels will be equally populated as in fig. 10.7.

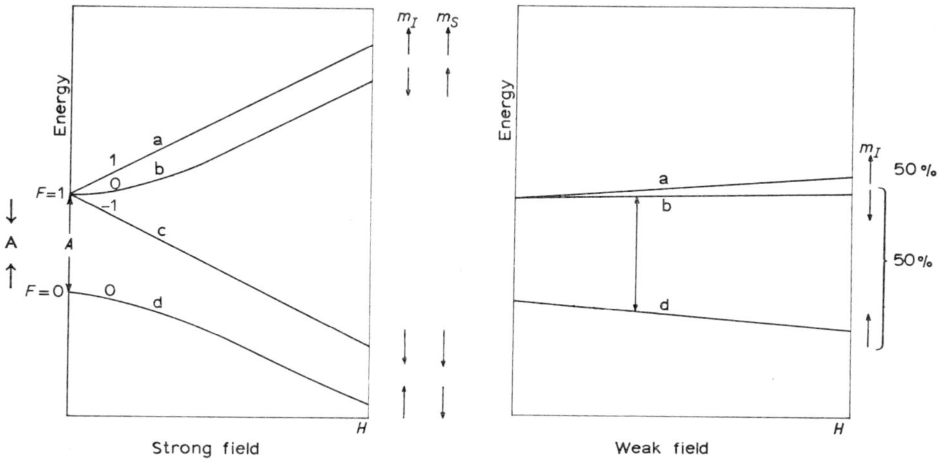

Fig. 10.7. Polarisation of an atomic beam. On the left is shown the energy splittings of the magnetic substates in a strong field.

But in the two upper states (a, b) the electron spin is up and in the two lower states (c, d) it is down, and if the beam is sent into a non-uniform magnetic field the two sets of states will be oppositely deflected. Suppose we thus get rid of the lower states c, d so we have left the two upper states a, b. We still have no polarisation as a = ↑↑ and b = ↓↑. But now we let the beam pass out of the high magnetic field into a weak field region. Now the hyperfine coupling takes over and state b↓↑ is mixed with state d↑↓ each of $m_F = 0$ with no net polarisation but state a ($m_F = +1$) remains pure and is polarised 100 %. So we can get at most 50 % polarisation. If in addition we can apply a radio frequency field to induce transitions between states b and d, we can deplete b and increase the net polarisation.

Practically the source consists of a beam of neutral atoms which passes through a sextupole magnet which focuses only the desired states a and b. The beam then enters a weak field region in which the neutral atoms are ionized. First trials on a linear accelerator gave $\approx 10^{-5}$ μA with 30 % polarisation at 30 MeV.

3.4 Polarisation by scattering. We know that to produce the shell model states observed we required a spin–orbit force, that is, the energy of the nucleon state depended on the orientation of its spin and orbital momenta. Suppose the bombarding nucleons are polarised. Then in passing to right or left of a scattering centre

* G. Clausnitzer, R. Fleishmann and H. Schopper, Z. Physik **144** (1956) 336.

they generate orbital momentum vectors of opposite sign (fig. 10.8). Now suppose there is a spin–orbit force, that is, the scattering depends on the scalar product $l \cdot s$. Then these will be unequal on the left and right sides with respect to the beam direction. This leads to an asymmetric angular distribution, if it now interferes with some background scattering.

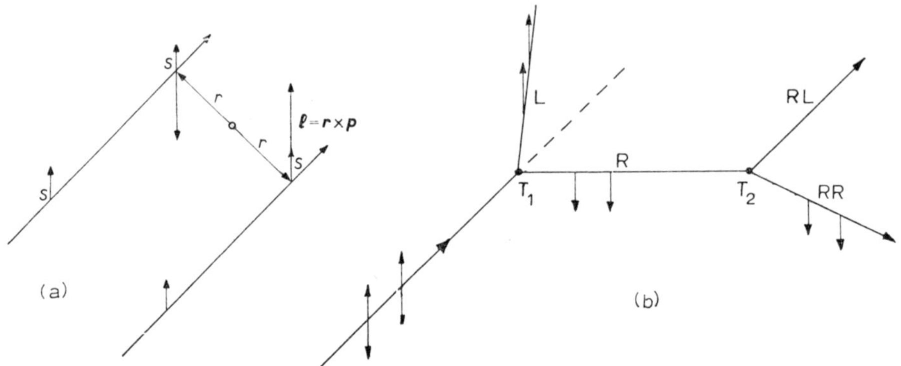

Fig. 10.8.   a) Polarisation by scattering. b) Asymmetry observed by second scattering of polarised beam. The left (L) and right (R) scattered beams are of equal intensity but opposite polarisation. After the second scattering, the intensities of the left (RL) and right (RR) scattered beams are different.

If we start with an unpolarised beam it is easy to see that the beam scattered to left and to right will be of equal intensity but one will be polarised in one sense and the other in the opposite sense. This can be proved by a further scattering.

An example much used is the scattering of protons by helium or the inverse. This forms a compound nucleus ^{5}Li which has a ground state $P_{\frac{3}{2}}$ ($j=l+s$) and an excited state $P_{\frac{1}{2}}$ ($j=l-s$). If the energy is such that one of these states dominates, the interaction is stronger for one value of $l \cdot s$ than the other and polarisation of the proton results. This interferes with the non-resonant s-wave background and leads to large polarisation, of the order of 100 %. This method is used both as a source of polarised protons and as a polarisation analyser.

4   Angular distributions

These are often divided into two types.
1) When a nuclear reaction takes place the differential cross section is measured as a function of the angle between the emitted particle and the incident beam. The latter provides a reference axis.
2) In the decay of a radioactive nucleus, two kinds of radiation may be emitted, both isotropically since there is no specified direction, e.g. a β-decay followed by a

γ-decay or two γ-decays in cascade. If these are detected in coincidence, the co-incidence rate as a function of the angle between the counters is recorded. This is sometimes called an angular correlation.

If the reaction is $$X + a \rightarrow C^* \rightarrow Y + b$$

then the angular distribution will depend on 1) the spin of target nucleus X, 2) the orbital angular momentum of particle a, 3) the angular momentum J of the compound nucleus, 4) the orbital angular momentum of particle b, 5) the spin of the residual nucleus Y. Since the magnetic substates of the level J in the compound nucleus are unequally populated (oriented) this leads to an angular dependence of the intensity with respect to the direction of the incident beam of particles. The distribution will depend on the ways that the angular momenta of the constituent particles couple up in the reaction. Therefore we can learn from angular distribution measurements about preferred ways of coupling which are pertinent to the nuclear structure problem.

In an excited nucleus we frequently find γ-ray cascades especially if I_2 differs much in spin from I_0. Then if we observe γ_a in a particular direction it picks out a set of nuclei in states I_1 with a non-isotropic set of spin directions. If this set of

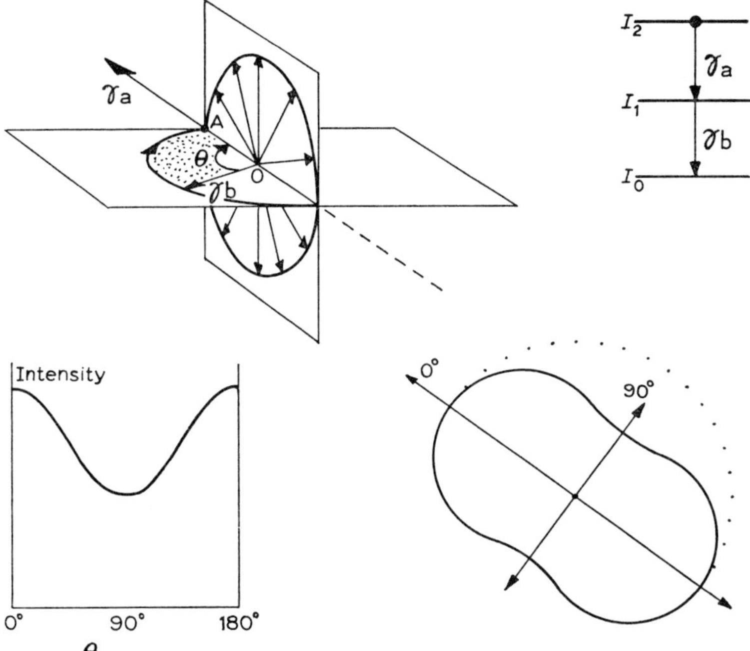

Fig. 10.9. Angular correlation of gamma-ray γ_b with respect to gamma-ray γ_a

nuclei almost immediately emit γ_b then γ_b is emitted from a system oriented with respect to the γ_a direction.

Suppose γ_a, γ_b are both dipole and we can neglect intrinsic photon spin. Let $I_1 = 1\hbar$, $I_2 = I_0 = 0$. If γ_a is observed (fig. 10.9) in direction OA then the spin I_1 will be aligned perpendicular to OA (since dipole radiation is emitted at right angles to the dipole). The photon γ_b carries away this spin and hence is emitted in a plane perpendicular to it with equal probability for all angles θ in this plane. But the aligned dipole will as before give an angular distribution as viewed by counter b of the form $a_0 + a_1 \cos^2\theta$ in the plane of $\gamma_a - \gamma_b$ and hence as the angle between γ_a and γ_b approaches 0 or 180° we will approach a maximum i.e. it is most probable to see γ_b coming out in the same direction as γ_a or 180° away. In the general case we have for the relative population of the substates of the intermediate level I_1

$$W(m_1) = \sum_{m_2} (I_1 I_2 m_1 m_2 | L_a M_a)^2 F_{L_a}^{M_a}(\theta) \quad \text{when} \quad M_a = m_2 - m_1. \quad (10.9)$$

If we choose the axis of quantisation along the direction of γ_a we get a simplification since the orbital angular momentum of the photon has no component along this axis and the intrinsic angular momentum has components $\pm 1\hbar$. Then only the functions $F_{L_a}^1$ and $F_{L_a}^{-1}$ appear for $\theta = 0°$. Then the correlation is (when the γ_a direction is taken as $0°$)

$$W(\theta) = \sum_{m_0 m_1 m_2} (I_1 I_2 m_1 m_2 | L_a 1)^2 F_{L_a}^{\pm 1}(0) \times (I_0 I_1 m_0 m_1 | L_b M_b)^2 F_{L_b}^{M_b}(\theta). \quad (10.10)$$

For our example $I_2 = I_0 = 0$ and $I_1 = 1$ and the only transitions permitted are $(I, m) = 0, 0 \rightleftarrows 1, \pm 1$. The appropriate Clebsch–Gordan coefficients are all equal to one and

$$W(\theta) = F_1^{\pm 1}(0) \times [F_1^1(\theta) + F_1^{-1}(\theta)].$$

But $F_1^{\pm 1}(\theta) = \frac{1}{2}(1 + \cos^2\theta)$, therefore

$$\begin{aligned} W(\theta) &= (\tfrac{1}{2} + \tfrac{1}{2}) \times [\tfrac{1}{2}(1 + \cos^2\theta) + \tfrac{1}{2}(1 + \cos^2\theta)] \\ &= 1 + \cos^2\theta \end{aligned}$$

as expected.

There have been many ways devised to divide up the coefficients to be summed. I choose one described by Sharp et al.* to illustrate the quantum numbers used. Here the angular functions are the Legendre polynomials and the combinatorial coefficients are summed into terms called Z-coefficients. There are two kinds of such coefficients, Z-coefficients for particles and Z_1-coefficients for photons (which have a different polarisation).

---

* W. T. Sharp, J. M. Kennedy, B. J. Sears and M. G. Hoyle, Tables of Coefficients for Angular Distribution Analysis. Chalk River Report CRT-556 (1953).

The γ–γ correlation then becomes

$$W(\theta) = \sum_k (-1)^{I_1 - I_2} Z_1(L_1\,jL_1'\,j,\,I_1\,k)Z_1(L_2\,jL_2'\,j,\,I_2\,k)P_k(\cos\theta); \qquad (10.11)$$

k will run over even integer values 0, 2, 4, . . . up to a maximum value set as before by $2n$ where n is the smallest of L_1, j, or L_2.

The particle in – particle out correlation is

$$W(\theta) = \sum_k (-1)^{S_1 - S_2} Z(l_1\,jl_1',\,S_1\,k)Z(l_2\,jl_2',\,S_2\,k)P_k(\cos\theta) \qquad (10.12)$$

where $S_1 = I_T + s_{in}$, $S_2 = I_F + s_{out}$, I_T is target spin, I_F is final nucleus spin, s_{in} and s_{out} are particle spins.

The particle in – γ-ray out correlation is

$$W(\theta) = \sum_k (-1)^{S-I}\, Z(ljl'j,\,Sk)Z_1(LjL'j\,,\,Ik)P_k(\cos\theta). \qquad (10.13)$$

Values of the coefficients required for common cases are tabulated in the report mentioned above.

All of the above correlations are double correlations in which one incoming particle or photon and one outgoing particle or photon is involved. The more complex case of a triple correlation in which three radiations are recorded can often give more unambiguous information but in general requires use of a computer to analyse the result. A few systems for doing this are referred to at the end of this chapter.

In all of the above cases it is assumed that the detector is insensitive to polarisation and that the incoming radiation is unpolarised. If this is not the case, the analysis is still more difficult but in favourable cases even more information is in principle available.

5   Gamma-ray polarisation

5.1   Introduction. In our discussion of nuclear orientation and angular distributions we remarked on various measurements of the angular distribution of γ-rays arising from nuclei aligned or polarised by nuclear reaction mechanisms. We want now to consider methods for studying the polarisation of the γ-rays themselves.

A study of the angular distribution patterns of radiation emitted from nuclei aligned in various ways can be related very directly to the multipolarity, dipole, quadrupole etc. of the radiation. The angular distribution in the case shown in fig. 10.10 is given by a vector spherical harmonic $X_{J_1}^{M_1}$.

These arise from combinations of the eigenstates of spin 1 for the γ-ray with the

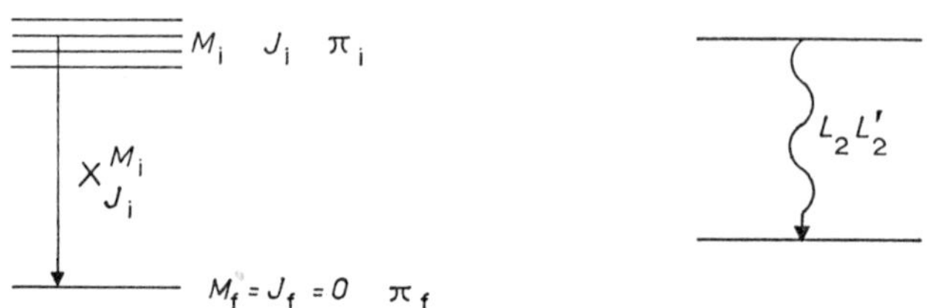

Fig. 10.10.   a) Radiative transition, b) Interfering multipoles

eigenfunctions of orbital angular momentum $Y_l^M(\theta, \phi)$. According to whether there is parity change or not $(-1)^{J_i} = \pm\pi_i\pi_f = \pm\pi$ and the quantum emitted is EJ_i or MJ_i, electric or magnetic.

But we do not usually separate the substates so the distribution is derived by summing over all substates

$$W(\theta) = \sum_{k \text{ even}} A_k(1)A_k(2)P_k \cos \theta \qquad (10.14)$$

where the A_i are appropriate Z- or Z_1-coefficients for example.

But in general this does not define parity. In other words, dipole gives the same distribution whether it is electric or magnetic and quadrupole gives the same distribution whether it is electric or magnetic. But if the linear polarisation is known it tells us at once the parity of the radiation. In some cases the radiation is circularly polarised and this gives additional information.

Let us summarise various situations in table 10.1.

TABLE 10.1

Initial particle	Initial condition	Type of orientation	Type of correlation
p, n, etc.	direction	aligned	p–γ direction–polarisation (linear)
p, n, etc.	polarised	polarised	p–γ polarisation–circular
β	direction	polarised	β–γ direction–circular
γ	direction	aligned	γ–γ direction–polarisation (linear)
γ	polarised	aligned	γ–γ polarisation–polarisation (linear)
γ	circular	polarised	γ–γ circular–circular
–	low temperature [a]	aligned	alignment–polarisation (linear)
–	low temperature [b]	polarised	polarisation–circular

[a] Bleaney and Pound

[b] Rose and Gorter

5.2   Linear polarisation. Let us consider for example the direction–polarisation case. To get the polarisation distribution we change the distribution function as follows. For each term characterised by $L_2 L_2'$ and k we replace $P_k(\cos\theta)$ by

$$P_k(\cos\theta)\,(\pm)_{L_2}\,(\cos 2\gamma)[K_k(L_2 L_2')]P_k^{(2)}(\cos\theta). \tag{10.15}$$

γ is the angle between the polarisation vector and the reaction plane. $K_k(L_2 L_2')$ is a numerical factor containing the quotient of two Clebsch–Gordan coefficients involving only the multipolarity, and the $P_k^{(2)}(\cos\theta)$ are the associated Legendre polynomials. Then the plus or minus sign applies according as the 2^{L_2}-pole radiation is electric or magnetic. In this way we can determine the parity.

The polarisation is often separated out as follows, since $\gamma = 0°$ or $90°$ is used where $\cos 2\gamma = \pm 1$. If the intensities of polarised radiation at $0°$ and $90°$ are called J_0 and J_{90} then

$$P_1 W(\theta) = J_0 - J_{90} = \sum (\pm)_{L_2'} A_k K_k(L_2 L_2')P_k^2(\cos\theta) \tag{10.16}$$

where P_1 is the degree of linear polarisation. For example consider the reaction

$$^{19}\text{F}(p,\alpha)^{16}\text{O*} \ (6.13 \text{ MeV state})$$
$$\downarrow$$
$$^{16}\text{O} + \gamma \ (\text{octupole, } 6.13 \text{ MeV}).$$

At $E_p = 330\,\text{keV}$ the proton capture is via an s-state i.e. the ^{20}Ne state is unoriented so the α–γ emission is an independent process. The direction–direction correlation $(\alpha$–$\gamma)$ is

$$W(\theta) = 1 + 111 \cos^2\theta - 305 \cos^4\theta + 225 \cos^6\theta.$$

The direction–polarisation correlation is

$$P_1 W(\theta) = \pm[-1 + 131 \cos^2\theta - 355 \cos^4\theta + 225 \cos^6\theta].$$

Thus at $\theta = 118°$ or $62°$ the polarisation is complete i.e. $J_0/J_{90} = 0$ or ∞, depending on whether the radiation is electric or magnetic. Of course it is electric so $J_0 = 0$.

5.3   Circular polarisation. The γ–γ circular–circular polarisation can be written

$$W(\theta, p_1, p_2) = \sum_k (-p_1)^k A_k(1)(-p_2)^k A_k(2)P_k(\cos\theta) \tag{10.17}$$

where k runs over odd and even integers and $p = \pm 1$ according as the circular polarisation is right- or left-handed. Right-handed circular polarisation has its spin in the direction of its momentum.

If we sum the above over p_1 or p_2 we get just $W(\theta)$, the direction–direction correlation. Hence there are no direction–circular or linear–circular correlations.

If as before we define P_3, the excess of right-handed over left-handed,

$$P_3 W(\theta) = \sum_{k \text{ odd}} A_k(1)A_k(2)P_k(\cos\theta). \tag{10.18}$$

In the case of nuclear polarisation by the low temperature method $A_k(1)$ is replaced by an orientation parameter. The sign of this parameter depends on the sign of the nuclear magnetic moment; hence this can be derived.

Circularly polarised radiation can be produced following capture or emission of a polarised particle, e.g. by capture of polarised thermal neutrons. In this case at $\theta = 0°$ the direction of neutron polarisation, $P_3 = +1$ and the γ-rays are completely circularly polarised (right-handed). Beta-particles are also polarised resulting from helicity conservation and hence β–γ correlations lead to circularly polarised γ-rays.

6   Detection of γ-ray polarisation

6.1   The Compton effect. The cross section for this effect, the Klein–Nishina formula, can be written

$$\frac{\mathrm{d}\sigma_\phi}{\mathrm{d}\Omega} = \frac{r_0^2}{2} \frac{k^2}{k_0^2} \left[\frac{k_0}{k} + \frac{k}{k_0} - 2 \sin^2 \theta \cos^2 \phi \right]$$

where r_0 is the classical electron radius $e^2/m_0 c^2$, θ is the angle through which the photon is scattered (fig. 10.11), ϕ is the angle between the electric vector of the

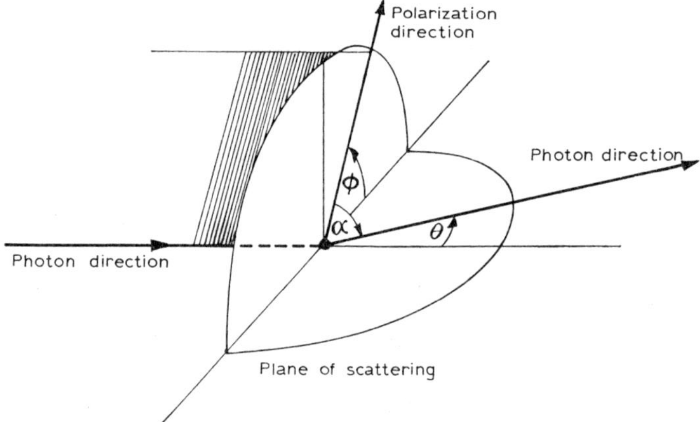

Fig. 10.11.   Compton scattering

incident photon and the plane of scattering, k_0 and k are energies of incident and scattered photons. These are related, you recall, by

$$k = \frac{k_0}{1 + k_0(1 - \cos \theta)/m_0 c^2}$$

The $\cos^2\phi$ term tells us that the scattering is a maximum in the plane normal to the electric vector of the incident photon, i.e., for $\phi=90°$. Thus in the limit of zero energies the first two terms approach 2 and at $\theta=90°$ the differential cross section is zero for $\phi=0°$ and maximum for $\phi=90°$.

At higher energies the asymmetry ratio, the ratio of the intensity at 90° to the intensity at 0°, decreases. It is about 3 at 1 MeV, 2 at 2 MeV and 1.5 at 4 MeV. Also, of course, the cross section decreases with increasing energy. Hence the method is useful for γ-rays up to several MeV energy. An experimental arrangement is shown in fig. 10.12.

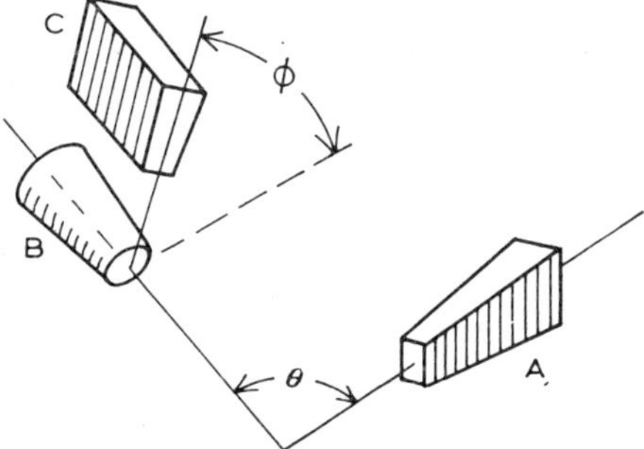

Phys. Rev. **78** (1950) 551

Fig. 10.12.   Polarisation detection by Compton scattering. A, Monitor counter. B, Recoil electron detector. C, Scattered photon detector

6.2   Deuteron photodisintegration. For higher energy γ-rays the linear polarisation can be detected via the photodisintegration of the deuteron. Of course it cannot work below the threshold at 2.2 MeV, but it is satisfactory up to 20 MeV. The electric disintegration is the one that is polarisation sensitive and the cross section can be written

$$\frac{d\sigma(\text{E1})}{d\Omega} = 2\left(\frac{e^2}{\hbar c}\right)\left[\frac{\gamma k^3}{(\gamma^2+k^2)^3}\right]\left(\frac{1}{1-\gamma r_{0t}}\right)\cos^2\alpha \qquad (10.19)$$

where α is the angle between the direction of polarisation of the γ-ray and the direction of motion of the proton or neutron, r_{0t} is the effective range in the triplet state, k is given by

$$E = h\nu - W_1 = \frac{h^2 k^2}{M}$$

and γ is given by

$$W_1 = \frac{h^2\gamma^2}{M}$$

where hv is the γ-ray energy, W_1 is the deuteron binding energy and M is the nucleon mass.

A more useful form in terms of θ and ϕ as before is

$$\frac{d\sigma_\phi}{d\Omega} = 2 \left(\frac{e^2}{\hbar c}\right)\left[\frac{\gamma k^3}{(\gamma^2 + k^2)^3}\right]\left(\frac{1}{1-\gamma r_{0\iota}}\right) \sin^2\theta \cos^2\phi. \tag{10.20}$$

Of course correction must be made for the proportion of photomagnetic disintegration which occurs, especially at energies near threshold. This is not sensitive to polarisation and dilutes the effect. For further discussion of this effect see ch. 11 § 6.

The atomic photoelectric effect also has a polarisation sensitive term. However, it is only useful for energies well below 0.5 MeV. Also the low energy electron is rather hard to detect.

6.3 Circular polarisation. The Compton effect can also be used for detecting this by studying the interaction with polarised electrons in magnetised iron. In magnetised iron about 2 electrons out of the 26 are polarised so the effect is small but useable, of the order of a few per cent.

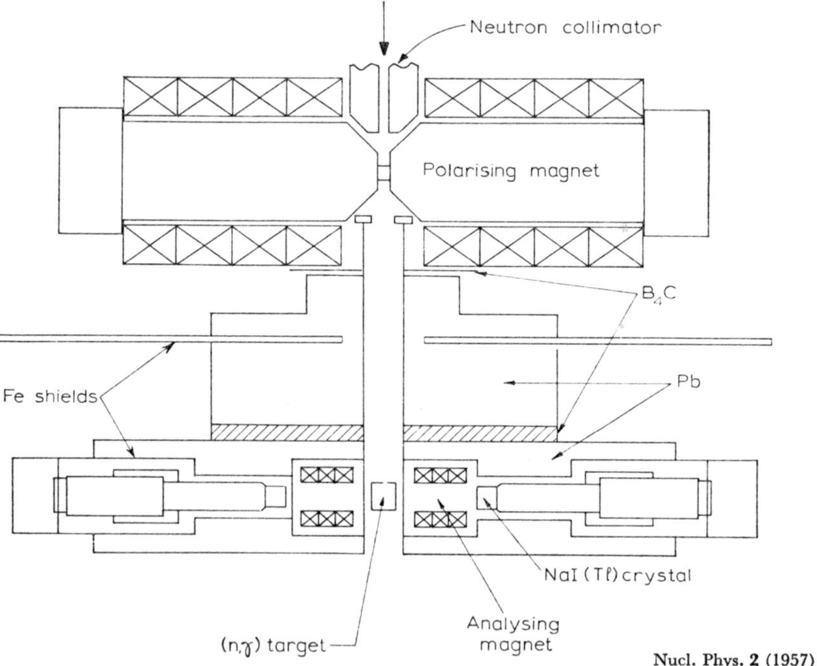

Fig. 10.13. Detection of polarised γ-radiation from polarised neutron capture. Slow neutrons from a reactor are polarised by the magnet at the top. Gamma-ray circular polarisation can be detected by the analysing magnet–scintillation counters arrangement

The main interest has been in the study of γ-rays from nuclei polarised by the low temperature method, and in the study of γ-rays following β-decay which, because the β-particle comes off with a definite helicity (relation between polarisation and momentum), the γ-rays are circularly polarised. A study of this direction–polarisation correlation is then equivalent to a study of the polarisation–direction correlation of the β-particle from low temperature polarised nuclei as in the original experiment on parity violation (ch. 2 § 6.5).

We also mentioned the circular polarisation produced by capture of polarised neutrons. In this case the apparatus was as shown in fig. 10.13.

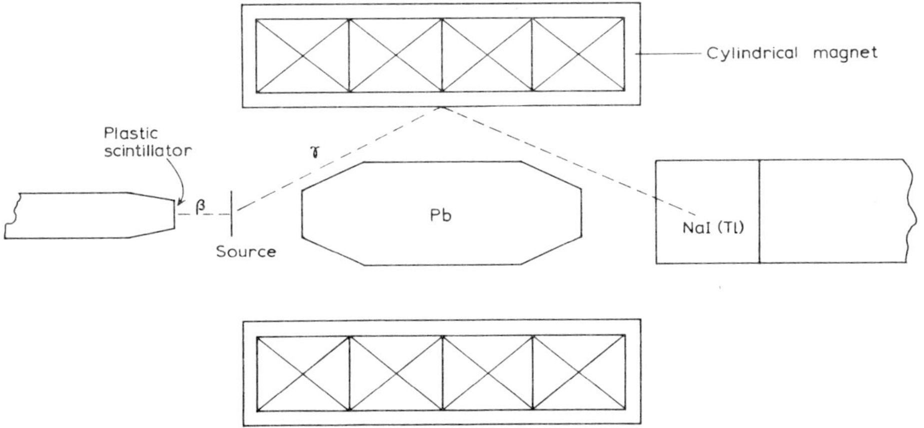

Fig. 10.14.   Beta–gamma circular polarisation detection

Many β–γ direction–circular polarisation correlations have been measured as shown in fig. 10.14. The γ-rays emitted at almost 180° to the β-particles are observed. Schopper* studied the decays

$$^{22}\text{Na} \rightarrow {}^{22}\text{Ne} + \beta^{+} + \nu$$
$$^{60}\text{Co} \rightarrow {}^{60}\text{Ni} + \beta^{-} + \bar{\nu}.$$

He measured $P_3 = (v/c)A \cos \theta$, θ being the angle between the β-ray and the γ-ray. He found $A = +0.39$ for ^{22}Na (right-handed positrons) and $A = -0.41$ for ^{60}Co (left-handed electrons). The polarisations are opposite in sense and show that the two neutrinos have opposite helicities.

In the experiment of Goldhaber, Grodzins and Sunyar (ch. 2 § 6.5) a measurement of circular polarisation of the γ-ray was used in the determination of the neutrino helicity.

---

* H. Schopper, Phil. Mag. **2** (1957) 710.

11

Nuclear Forces

1   *Introduction*

In our discussion of nuclear properties and nuclear models we have assembled a number of facts about the nuclear force. The binding energy of nuclei suggested certain properties the force must have such as short range and saturation. In addition it must be several orders of magnitude stronger than the coulomb force. The nuclear models were based on various simplified versions of the force which in practice described observed properties. Often the success of a model indicated some new component of the nuclear force such as the spin–orbit part required by the shell model.

From studies of the nucleon–nucleon interaction either from scattering experiments or from the properties of the bound two-nucleon system, the deuteron, we derive many more details of the force. However we shall see that the force as derived from the properties of complex nuclei and the force derived from the nucleon–nucleon interaction differ in significant ways. We must thus in addition have a theory of nuclear matter to connect the two.

2   *The two-nucleon system*

We have one stable nucleus made up of two nucleons, the deuteron. It has a binding energy of 2.23 MeV and has no bound excited states. The deuteron has angular momentum 1^+, a magnetic moment equal to 0.857393 n.m and a quadrupole moment equal to 2.82×10^{-27} cm^2. It is stable. The other two possible systems of two nucleons, the di-neutron and the di-proton, are not observed and are thus unstable against immediate break-up into their constituents.

The magnetic moment of the free proton is 2.79270 n.m and of the free neutron is -1.91316 n.m. The sum, 0.87954 n.m, is so close to the observed moment of the deuteron that we can assume the two nucleons are in a state of zero orbital angular momentum, an S state. The observed spin of the deuteron is 1 so the two spin-$\frac{1}{2}$ nucleons must be aligned and in a $^3S_1$ state, a triplet state. Since no other bound state is seen it must be that the state in which the two spins are antiparallel, the $^1S_0$ singlet state, is at least 2.23 MeV less bound. We thus infer a spin dependence of the force by which nucleons with spins parallel are at least 2.23 MeV more tightly bound than with spins antiparallel.

The quadrupole moment is very small. For example the radius of the deuteron is about 2.15×10^{-13} cm and $\pi R^2 = 140 \times 10^{-27}$ cm^2. As the observed moment is about 50 times smaller we infer that the deuteron is close to being spherically symmetric.

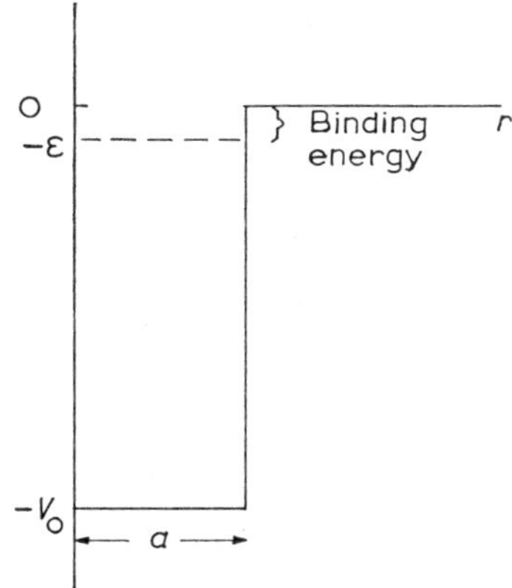

Fig. 11.1.   Nuclear potential well

Suppose that we assume the inter-nucleon potential to have the form of a square well of radius equal to the range of the force, a, as in fig. 11.1. Whenever the proton and neutron approach within a distance a, they feel the strong attractive potential $-V_0$. The wavefunction for the relative motion can in general be written

$$\psi = R(r)Y_l^m(\theta, \phi). \tag{11.1}$$

However in our case of a spherically symmetric S state there is no angular dependence and

$$\psi = R(r) = \frac{u(r)}{r} \tag{11.2}$$

which is just the probability of finding a separation r between the nucleons. The Schrödinger equation in terms of the radial function $u(r)$ is just

$$\frac{d^2u(r)}{dr^2} + \frac{2M}{\hbar^2}[E - V(r)]u(r) = 0 \tag{11.3}$$

where M is the mass of the nucleon moving in a potential $V(r)$ and E is the energy of the stationary state. In the deuteron the masses of neutron and proton are approximately equal and they move about their common centre of mass. Thus in the equation we use the reduced mass, $\frac{1}{2}M$. $[E - V(r)]$ is just the kinetic energy of the two particles for $r > a$. For the deuteron this is negative, $-\varepsilon$, where ε is the binding energy.

We study the solution of the Schrödinger equation in two regions.

For $r > a$, $V(r) = 0$ and $E = -\varepsilon$. The equation becomes

$$\frac{d^2u}{dr^2} - \frac{M}{\hbar^2}\varepsilon u = 0 \tag{11.4}$$

which has the solution $u = Ae^{-\alpha r}$ where $\alpha^2 = M\varepsilon/\hbar^2$ provided we assume that $u(r) \to 0$ as $r \to 0$ which it must do if $|\psi|^2$, that is, the particle density, is to remain finite. The shape of the wavefunction outside $r = a$ is thus an exponential.

For $r < a$, $V(r) = -V_0$, $E = -\varepsilon$. The equation becomes

$$\frac{d^2u}{dr^2} + \frac{M}{\hbar^2}(V_0 - \varepsilon)u = 0 \tag{11.5}$$

which has the solution $u = B\sin Kr$ where $K = M(V_0 - \varepsilon)/\hbar^2$ with the same assumption about $u(r) = 0$ at $r = 0$. $(V_0 - \varepsilon)$ is simply the internal kinetic energy of the bound nucleon and K is therefore the wave number of the internal motion for $r < a$.

We must now choose the constants A and B in the two solutions such that the two parts of the wavefunction join smoothly at the edge of the well. As usual we require ψ and $d\psi/dr$, that is u and du/dr, to be continuous at $r = a$. For u to be continuous we have

$$Ae^{-\alpha a} = B\sin Ka$$

while for du/dr to be continuous we have

$$-A\alpha\, e^{-\alpha a} = KB\cos Ka.$$

Dividing we eliminate A and B and have

$$-\alpha = K\cot Ka \quad \text{or} \quad \alpha^2 = K^2\cot^2 Ka.$$

But

$$\alpha^2 = M\varepsilon/\hbar^2 = K^2\cot^2 Ka$$

or

$$\cot^2 Ka = \frac{\varepsilon}{V_0 - \varepsilon} .$$

However V_0 is considerably larger than the binding energy ε. Hence $\cot Ka \approx 0$ or $Ka \approx \tfrac{1}{2}\pi$.

The quantity $1/\alpha$ is sometimes referred to as the radius of the deuteron. More precisely, it corresponds to the point at which the outside wavefunction has fallen to $1/e$ of its value at $r = a$. It is the decay length in the potential well for the relative motion of the two nucleons and has a value about 4.3×10^{-13} cm. The average radius of the neutron–proton system is one half of this, 2.15×10^{-13} cm, and more properly defines the deuteron size. This tells us that there is a large probability

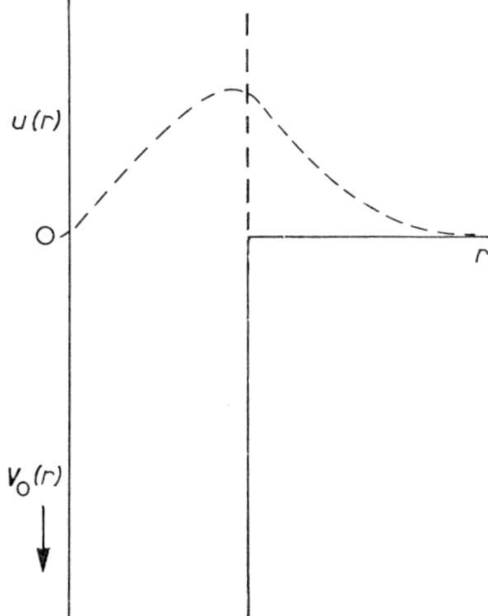

Fig. 11.2.   The wavefunction superimposed on the potential well illustrating joining at the nuclear surface

that the two nucleons are further apart than the range of the force. Since $Ka \approx \tfrac{1}{2}\pi$, $a \approx \tfrac{1}{2}\pi \times \lambda/2\pi = \tfrac{1}{4}\lambda$, that is, the range is about one quarter of the wavelength of the relative motion. The wave function for $r < a$ just turns over at $r = a$ to meet the rising exponential part for $r > a$. This is shown superimposed on the potential diagram in fig. 11.2. Also since

$$a^2 \approx \frac{1}{K^2} (\tfrac{1}{2}\pi)^2 \approx \frac{\hbar^2}{M} \frac{1}{(V_0 - \varepsilon)} (\tfrac{1}{2}\pi)^2$$

we have

$$V_0 a^2 \approx \frac{\hbar^2}{M} (\tfrac{1}{2}\pi)^2 \tag{11.6}$$

for $\varepsilon=0$ (just zero binding energy). For a well depth V_0 of 30 MeV and $a=1.9\times 10^{-13}$ cm, the kinetic energy for separation is $30-\varepsilon\approx28$ MeV.

All of this has been for the chosen potential shape. Similar solutions can be derived with other potential shapes. Those most commonly used are

1) the square well
$$V(r) = -V_0 \quad \text{for} \quad r < a,$$
$$V(r) = 0 \quad\quad \text{for} \quad r > a,$$

2) the gaussian well
$$V(r) = -V_0 e^{-(r/a)^2},$$

3) the exponential well
$$V(r) = -V_0 e^{-2(r/a)},$$

4) the Yukawa well
$$V(r) = -V_0 e^{-r/a}/a.$$

The last has great basic significance and we return to it at the end of this chapter.

3   Scattering of slow neutrons by protons

Having considered the bound system we now turn to the interaction of neutrons with protons. At low energy we might expect that the theory of the deuteron binding would also be applicable. The slow neutron may be captured by the proton, the binding energy released as a 2.23 MeV γ-ray and a bound deuteron formed. Classically the neutron could not then enter the region $r>a$ as its wave number would become imaginary (kinetic energy negative). However, quantum mechanically we have seen that the exponential part of the wavefunction could extend into the region $r>a$.

If capture does not take place, the neutron is elastically scattered by the proton, and emerges back into region $r>a$ with its original energy E_n and wave number k. Inside the region $r<a$ the wave number increased because of the attractive potential and became equal approximately (for $E_n\rightarrow0$) to K, the internal wave number which equals $\hbar^{-1}[M(V_0-\varepsilon)]^{\frac{1}{2}}$. We can use the method described in ch. 6 to calculate the elastic scattering cross section. The wavefunction can be written

$$\psi = \frac{1}{kr}\sum_0^\infty (2l+1)P_l(\cos\theta)\frac{\eta_l e^{i(kr-\frac{1}{2}l\pi)}-e^{-i(kr-\frac{1}{2}l\pi)}}{2i} \qquad (11.7)$$

where the effect of the scattering centre is included in η_l. We assume s-wave scattering, $l=0$, so that $(2l+1)P_l(\cos\theta)=1$. The wavelength of the relative motion (for 1 MeV neutrons, $\lambdabar=55\times10^{-13}$ cm) is so much larger than the range of the force that we can also make the approximation of $a=0$, the *zero range approximation*. In other words we can assume that the scattering is completely defined by the form of the external exponential part of the wavefunction. There is no loss of particles,

$|\eta_l|^2=1$ and $\eta_0=e^{2i\delta_0}$ where δ_0 has the significance of the phase shift between incident and scattered waves. The last factor in the expression for ψ becomes $e^{i\delta_0}\sin(kr+\delta_0)$ and

$$u(r) = r\psi = \frac{e^{i\delta_0}}{k}\sin(kr+\delta_0) \tag{11.8}$$

where $k^2=\tfrac{1}{2}ME_n/\hbar^2$, $\tfrac{1}{2}M_n$ is the reduced mass and E_n the neutron energy. We can now try to join this wavefunction to the inside wavefunction derived for the deuteron. The inside wavefunction was $u(r)=B\sin Kr$ and for scattering the external wavefunction is $u(r)=C\sin(kr+\delta_0)$. The joining conditions at $r=a$ are

$$B\sin Ka = C\sin(ka+\delta_0)$$
$$KB\cos Ka = kC\cos(ka+\delta_0).$$

Dividing we have

$$K\cot Ka = k\cot(ka+\delta_0).$$

But $K\cot Ka=-\alpha=-\hbar^{-1}M^{\frac{1}{2}}\varepsilon^{\frac{1}{2}}$. Hence

$$\delta_0 = \cot^{-1}(-\alpha/k)-ka. \tag{11.9}$$

Since k and a are both small $\delta_0\approx\cot^{-1}(-\alpha/k)$. We can then derive the cross section for elastic scattering for $l=0$:

$$\sigma_{el}^0 = \frac{4\pi}{k^2}\sin^2\delta_0 = \frac{4\pi}{k^2(1+\cot^2\delta_0)}$$

$$= \frac{4\pi}{k^2+\alpha^2} = \frac{4\pi}{(ME_n/2\hbar^2)+(M\varepsilon/\hbar^2)}$$

$$= \frac{4\pi\hbar^2}{M}\frac{1}{\varepsilon+\tfrac{1}{2}E_n} = 5.25\frac{1}{\varepsilon+\tfrac{1}{2}E_n}\ \text{barns for } \varepsilon, E \text{ in MeV.} \tag{11.10}$$

Obviously for low energies, $E_n\ll2.23$ MeV, we should expect an approximately constant cross section of about $5.25/2.23=2.35$ barns while as E_n becomes comparable to ε the cross section will decrease.

If we now compare our prediction for the scattering cross section derived from the deuteron properties with the measurement (fig. 11.3) we find a serious discrepancy. The flat region is about a factor of 10 higher than we calculated although the theoretical and experimental curves are in agreement at higher energy where $E_n\gg\varepsilon$. The mistake is that we have omitted to include in our calculation the unbound $^1S_0$ state of the deuteron. The two states will affect the scattering cross section in proportion to the number of substates $2I+1$ in each state. There are three times as many triplet interactions via the $^3S_1$ state as singlet interactions via

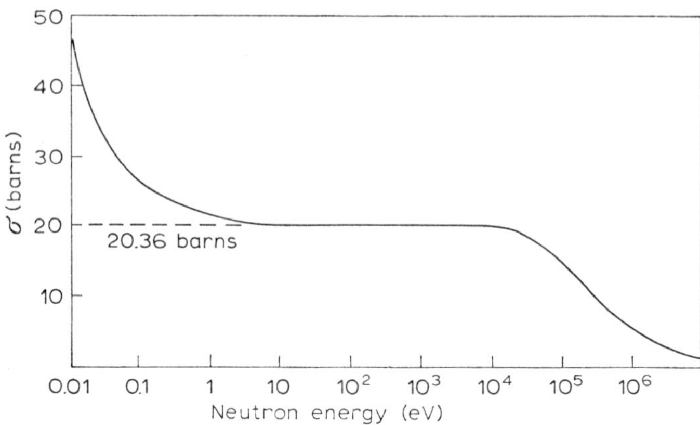

W. E. Burcham, Nuclear Physics (Longmans, London 1963) p. 665

Fig. 11.3.   The n–p total cross section

the $^1S_0$ state. Hence we write the cross section

$$\sigma_{el}^0 = \frac{4\pi}{k^2}\left(\tfrac{3}{4}\sin^2\delta_{0t} + \tfrac{1}{4}\sin^2\delta_{0s}\right)$$

$$= \frac{4\pi\hbar^2}{M}\frac{1}{4}\left(\frac{3}{\varepsilon + \tfrac{1}{2}E_n} + \frac{1}{|\varepsilon'| + \tfrac{1}{2}E_n}\right)$$

$$= 1.28\left(\frac{3}{\varepsilon + \tfrac{1}{2}E_n} + \frac{1}{|\varepsilon'| + \tfrac{1}{2}E_n}\right) \text{ barns (for energies in MeV)} \quad (11.11)$$

where δ_{0t} is the triplet and δ_{0s} the singlet phase shift, and ε' is the binding of the singlet state, which we know to be negative. If we choose ε' to give the observed thermal scattering cross section of 20.36 barns we find $|\varepsilon'| = 70$ keV. Thus we find the $^1S_0$ state to be unbound by 70 keV or at an excitation of $2.23 + 0.07$ MeV.

For elastic scattering the scattering amplitude $f(\theta)$ for s-wave scattering could be written

$$f(\theta) = \frac{e^{2i\delta_0} - 1}{2ik} = \frac{e^{i\delta_0}\sin\delta_0}{k}. \quad (11.12)$$

As k approaches zero, that is, for low energy, δ_0 must also approach zero so that $f(\theta)$ remains finite. In the limit $e^{i\delta_0} \to 1$ and $\sin\delta_0 \to \delta_0$ so that $f(\theta) \to \delta_0/k$ as $k \to 0$. This ratio has the dimensions of a length and is termed the *scattering length* a_s,

$$a_s = -\delta_0/k \quad (11.13)$$

where the negative sign is a convention. It has the following meaning. Consider the outside wave function

$$u(r) = \frac{e^{i\delta_0}}{k} \sin(kr + \delta_0).$$

As $k \to 0$,

$$u(r) \to \frac{1}{k}(kr + \delta_0) = r + \frac{\delta_0}{k} = r - a_s$$

which is just the equation of a straight line cutting the r-axis at $r = a_s$ (fig. 11.4).

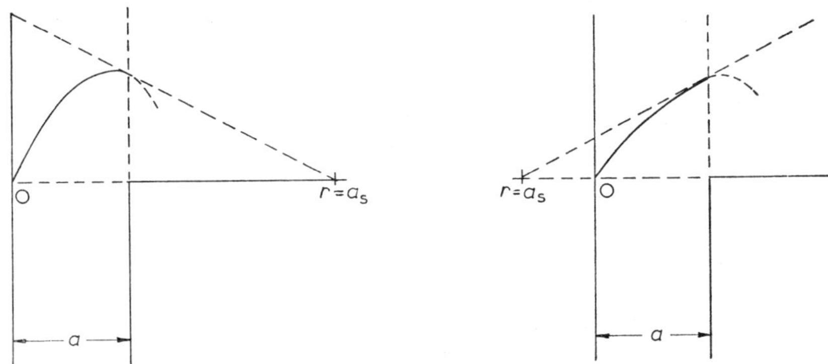

Fig. 11.4.   Illustration of a) a positive scattering length (bound state) and b) a negative scattering length (unbound state)

A positive value of a_s implies a bound state in which $u(r)$ inside has turned over to join at $r = a$. A negative value of a_s implies an unbound state in which $u(r)$ has failed to turn over at $r = a$.

We shall now formulate the low energy nucleon–nucleon interaction theory so that it is less dependent on the exact shape of the potential well. It can be shown that for any reasonable well shape

$$k \cot \delta_0 = -\frac{1}{a_s} + k^2 \int_0^\infty (vv_0 - uu_0)\,\mathrm{d}r \tag{11.14}$$

where:

v is the modified wavefunction, $r\psi$, outside the range of the nuclear potential,

v_0 is the modified wavefunction for zero incident energy $E_n = 0$,

u is the modified wavefunction $r\psi$ inside the range of the potential,

u_0 is the modified wavefunction for zero energy.

Since $V(r) \gg E_n$ we can approximate $v = v_0$ and $u = u_0$. The resulting integral $2 \int_0^\infty (v_0^2 - u_0^2)\,\mathrm{d}r$ is called the effective range r_0 where the factor 2 is an arbitrary normalisation. Therefore

$$k \cot \delta_0 = -\frac{1}{a_s} + \tfrac{1}{2}k^2 r_0. \tag{11.15}$$

The next term in the expansion can be written $-P_0 r_0^3 k^4$ where P_0 is a coefficient including details of a particular potential shape. If this term is neglected we speak of the *shape independent approximation*. For the deuteron the effective range expression is a definition of $\alpha = M\varepsilon/\hbar^2$,

$$\alpha = \frac{1}{a_{st}} + \tfrac{1}{2}\alpha^2 r_{0t} \tag{11.16}$$

thus connecting the binding energy with r_0 and a_s. Approximately $|\varepsilon| = \hbar^2/Ma_s^2$ and $\sigma_{el}^0 \to 4\pi a_s^2$ as $k \to 0$. Note that the cross section measurement gives the magnitude but not the sign of a_s. Thus a scattering cross section measurement cannot distinguish between bound and unbound states.

We now return to a further unexplained feature of the observed n–p cross section which is the steep increase in the cross section for neutron energies below about 0.1 eV. Our theoretical treatment referred to free target protons. We were then justified in taking $\tfrac{1}{2}M_n$ for the reduced mass in the centre of mass system. However typically our target hydrogen atoms will be bound into molecules and when the neutron energy becomes comparable to or less than the chemical binding, the reduced mass increases and the cross section and scattering length increases. In the one extreme of a free proton we can define a coherent scattering length

$$a_s(\text{free}) = \tfrac{3}{4}a_{st} + \tfrac{1}{4}a_{ss} \tag{11.17}$$

while in the other extreme of a tightly bound proton in a crystal lattice for example we find the reduced mass doubled and hence

$$a_s(\text{bound}) = 2a_s(\text{free}) = \tfrac{1}{2}(3a_{st} + a_{ss}). \tag{11.18}$$

4　Measurement of the scattering lengths

It is clear that measurements of the coherent scattering lengths together with the elastic cross section measurement can yield both magnitude and sign of the two scattering lengths. The classic method for measuring the coherent scattering length is by a study of the scattering from protons bound into the two types of hydrogen molecule. These differ in the relative orientation of the two proton spins. In *orthohydrogen* the spins are parallel while in *parahydrogen* the spins are anti-parallel. Nearly pure samples of either of these types of hydrogen molecule can be prepared at low temperatures ($\approx 20\ °\text{K}$) using activated charcoal as a catalyst. If we now scatter neutrons from each of these forms and if there is, as we expect, a spin dependence of the scattering forces, we should see different cross sections. This will be so if there is coherence between the scattered waves from the two protons. The separation of the two protons is 0.78×10^{-8} cm and if we use neutrons whose wavelength is long compared to this, the coherence condition will be satisfied. Accordingly we use 'cold' neutrons corresponding to temperatures of 10–30 °K which correspond to energies of 8–25×10^{-4} eV or wavelengths of 10–5.7×10^{-8} cm.

For parahydrogen with spins opposed the two protons will appear uncorrelated and the scattering length for the molecule should be just double that for a single proton.

$$a_s(\text{para}) = 2a_s(\text{free}) = 2(\tfrac{3}{4}a_{st}+\tfrac{1}{4}a_{ss}). \tag{11.19}$$

The scattering cross section will be

$$\sigma_{\text{para}} = \tfrac{16}{9}\, 4\pi[a_s(\text{para})]^2 \tag{11.20}$$

where the factor $\tfrac{16}{9}$ is a correction for the new reduced mass which is $\tfrac{2}{3}M_n$ rather than $\tfrac{1}{2}M_n$.

For orthohydrogen with parallel proton spins the scattering length will include interference between the two kinds of spin interaction

$$a_s(\text{ortho}) = 2[(\tfrac{3}{4}a_{st}+\tfrac{1}{4}a_{ss})^2+2(\tfrac{1}{4}a_{st}-\tfrac{1}{4}a_{ss})^2]^{\frac{1}{2}} \tag{11.21}$$

where the weighting factor 2 in front of the second term is just $F(F+1)$ where F is the molecular spin. In this case for orthohydrogen, $F=1$. For the parahydrogen case $F=0$ and there is no interference term. The cross section for scattering from orthohydrogen is thus

$$\sigma_{\text{ortho}} = \tfrac{16}{9}\, 4\pi[a_s(\text{ortho})]^2. \tag{11.22}$$

The ratio of the scattering cross sections from the two molecules is

$$\frac{\sigma_{\text{ortho}}}{\sigma_{\text{para}}} = 1+2\left[\frac{a_{st}-a_{ss}}{3a_{st}+a_{ss}}\right]^2. \tag{11.23}$$

The experimental values for a neutron energy of 0.002 eV are $\sigma_{\text{ortho}}=120$ barns, $\sigma_{\text{para}}=4$ barns, $\sigma_{\text{ortho}}/\sigma_{\text{para}}=30$. From these measurements we can conclude the following:

1) The n–p force is spin dependent. If it were not we should expect the two cross sections to be the same.

2) The singlet state of the deuteron is virtual, that is, is unbound. If this is so ε' is negative and a_{ss} is negative. From the elastic scattering cross section we deduce that $a_{st}=+5.38\times 10^{-13}$ cm and $a_{ss}=\pm24\times 10^{-13}$ cm. If the singlet state were bound, a_{ss} positive, we should get $\sigma_{\text{ortho}}/\sigma_{\text{para}}=1.4$. If it is unbound the predicted

ratio is 40. The observed ratio of 30 thus confirms that a_{ss} is negative and the state is unbound.

3) These results also serve to confirm the spin of the neutron as $\frac{1}{2}$. Higher values of the spin such as $\frac{3}{2}$ would lead to ratios of about 2 or less.

4) These results can also be used to evaluate an effective range r_0 for the triplet and singlet interactions. It is found that r_{0t} is smaller than r_{0s}.

The analysis of the experiments has been much simplified in the foregoing discussion. Inelastic scattering of the neutron can occur in which orthohydrogen is converted to parahydrogen with an increase in energy of the scattered neutron. If the neutron energy is too high (>0.02 eV) we can also get conversion of para- to orthohydrogen. All such effects must be corrected for.

Another method for studying the coherent scattering length is by study of crystal diffraction of slow neutrons. Typically a monochromatic beam of neutrons is produced by scattering by a crystal (NaCl for example) of a beam of slow neutrons from a reactor. This monochromatic beam then strikes a polycrystalline sample containing hydrogen (NaH for example) and the powder diffraction pattern recorded. The amplitudes for scattering from each scattering centre depend on the scattering lengths for each of the individual nuclei involved. The intensity of the Bragg reflections can be related to these scattering lengths. The intensity will be proportional for example to $(a_{sH}-a_{sNa})^2$. The sodium scattering length can be determined by another experiment and a_{sH} can be deduced.

A third method is to study the total reflection of a neutron beam from a mirror surface. In analogy to optical diffraction we can derive an index of refraction

$$n = \frac{1-N\lambda^2 a_s(\text{bound})}{2\pi} \tag{11.24}$$

where N is the number of scattering centres per unit volume, λ is the neutron wavelength and $a_s(\text{bound})$ is the bound coherent scattering length. If a_s is positive we get external total reflection ($n<1$) while if a_s is negative we get internal total reflection ($n>1$). The critical angle $\theta_c=n$ is about 10 minutes of arc in a typical case. If a liquid hydrogen mirror were used where a_s is negative we get total internal reflection which is inconvenient to observe. In practice a liquid hydrocarbon mirror is used. The carbon scattering converts the total reflection into an external effect which is easier to observe. From a knowledge of the carbon scattering length, the hydrogen scattering length can be derived.

The results of these various techniques are summarised in tables 11.1 and 11.2.

TABLE 11.1

Values of the coherent scattering length for hydrogen

Method	
Ortho-para scattering	$-3.80 \pm 0.05 \times 10^{-13}$ cm
Crystal diffraction	$-3.96 \pm 0.02 \times 10^{-13}$ cm
Mirror total reflection	$-3.78 \pm 0.02 \times 10^{-13}$ cm

TABLE 11.2

Values of the free neutron–proton scattering lengths and effective ranges

$$a_{st} = +\ 5.378 \times 10^{-13} \text{ cm}$$
$$r_{0t} = +\ 1.70\ \times 10^{-13} \text{ cm}$$
$$a_{ss} = -23.69\ \times 10^{-13} \text{ cm}$$
$$r_{0s} = +\ 2.7\ \ \times 10^{-13} \text{ cm}$$

5 *Capture of neutrons by protons*

We now turn to the alternative process to elastic scattering in the n–p interaction. In 1935 Fermi found that neutrons exist for only about 10^{-4} sec in a block of paraffin wax or in a tank of water and that their disappearance is due to the capture of the neutrons by protons:

$$^{1}\text{n} + {}^{1}\text{H} \rightarrow {}^{2}\text{H} + \gamma + 2.23 \text{ MeV.}$$

The measurement of the photon energy is the most accurate way of measuring the deuteron binding energy*. The cross section is however small compared to elastic scattering at thermal energies, being about 0.3 barns compared to 20 barns for scattering.

Up to about 10 MeV neutron energy, the n–p interaction is entirely in the S state as evidenced by the isotropy of the scattering in the centre of mass system. For higher orbital momenta, the two particles are too far apart to interact. The probability for capture will vary inversely as the neutron velocity, that is, depend on the time the neutron spends near the proton. The interaction must be via either $^{1}\text{S}_0$ or $^{3}\text{S}_1$ states and the radiative transitions can only be $^{1}\text{S}_0 \rightarrow {}^{3}\text{S}_1$ or $^{3}\text{S}_1 \rightarrow {}^{3}\text{S}_1$. In both cases the radiation would be magnetic dipole. However in the triplet–triplet transition there is no change in either spin or orbital angular momentum and M1 is forbidden since the one unit of angular momentum required for dipole

* R. E. Bell and L. G. Elliott, Phys. Rev. **79** (1950) 282.

radiation cannot be supplied. So the only possibility is the $^1S_0 \rightarrow {}^3S_1$ transition in which the angular momentum of the dipole is supplied by flipping the spin of one of the particles.

The matrix element for spin-flip M1 radiation can be written

$$M_{10}(\text{i, f}) = \frac{e\hbar}{2Mc} \left(\frac{3}{4\pi}\right)^{\frac{1}{2}} \sum_{k=1}^{2} \mu_k \int \phi_f^* \sigma_{kz} \phi_i \, dr \tag{11.25}$$

where σ_{kz} is the z-component of the spin vector of the kth particle and μ_k is its magnetic moment, ϕ_i and ϕ_f are wavefunctions of the initial and final states. The operator in the above expression can be written

$$\mu_n \sigma_n + \mu_p \sigma_p = \tfrac{1}{2}(\mu_n + \mu_p)(\sigma_n + \sigma_p) + \tfrac{1}{2}(\mu_n - \mu_p)(\sigma_n - \sigma_p). \tag{11.26}$$

The factor in the first term $(\sigma_n + \sigma_p)$ is zero for a singlet state and thus only the second term need be considered. The final state is the triplet $^3S_1$ ground state of the deuteron. We write its wavefunction in slightly different form

$$\phi_f = \frac{1}{(4\pi)^{\frac{1}{2}}} \frac{u_{0t}(r)}{r} \chi_{1,0}$$

where $\chi_{1,0}$ is the triplet spin eigenfunction and $u_{0t}(r)$ is the radial wavefunction which in the limit of large r becomes $u_{0t} \rightarrow A e^{-\alpha r}$ and is normalised so that $\int_0^\infty u_{0t}^2(r) dr = 1$. The initial state is the singlet $^1S_0$ state in the continuum.

$$\phi_i = \frac{u_{0s}(r)}{kr} \chi_0$$

where χ_0 is the singlet spin eigenfunction, k is the wave number of the relative motion, and $u_{0s}(r)$ in the limit of large r becomes

$$u_{0s}(r) \rightarrow \sin (kr + \delta_{0s}).$$

The matrix element then becomes

$$M_{10}(\text{i, f}) = \frac{\sqrt{3}}{2} \frac{e\hbar}{Mck} (\mu_n - \mu_p) \int_0^\infty u_{0s}(r) u_{0t}(r) dr. \tag{11.27}$$

The cross section is then

$$\sigma_{\text{cap}}^M = \frac{2e^2}{3\hbar c} \left(\frac{\hbar}{Mc}\right)^2 \left(\frac{k^2 + \alpha^2}{k^2}\right) (\mu_n - \mu_p)^2 I^2 \tag{11.28}$$

where $I = \int_0^\infty u_{0s}(r) u_{0t}(r) dr$. To evaluate I we use the zero range approximation. From the asymptotic form of $u_{0t}(r)$ we find $A = (2\alpha)^{\frac{1}{2}}$ and also using the asymptotic form of $u_{0s}(r)$ we obtain

$$I = (2\alpha)^{\frac{1}{2}} \frac{k \cos \delta_{0s} + \alpha \sin \delta_{0s}}{k^2 + \alpha^2}. \tag{11.29}$$

As we are interested in capture near zero energy we can neglect k^2 compared to α^2 and use the relation $k \cot \delta_{0s} = -a_{ss}^{-1}$. We introduce also the binding energy $\varepsilon = \hbar^2 \alpha^2 / M$ and the laboratory neutron energy $E_n = 2\hbar^2 k^2 / M$. We also have to multiply the cross section by a statistical factor of $\frac{3}{4}$ since the probability of finding neutron and proton in a singlet state is $\frac{1}{4}$ and there are three transitions possible to the three substates of the triplet ground state. The final cross section expression in the zero range approximation is

$$\sigma_{cap}^M = \frac{\pi e^2}{\hbar c} \left(\frac{\hbar}{Mc}\right)^2 \frac{\varepsilon}{Mc^2} (\mu_n - \mu_p)^2 (1 - \alpha a_{ss})^2 \left(\frac{2\varepsilon}{E_n}\right)^{\frac{1}{2}}. \tag{11.30}$$

For thermal neutrons this leads to a value of 0.3 barn in good agreement with the observed value of 0.332 barn.

This formulation tells us several things:

1) the neutron and proton magnetic moments must be different, otherwise we should see no radiation.

2) the cross section is sensitive to the sign of a_{ss}. If the $^1S_0$ state were bound we should expect a ratio $\sigma_{el}/\sigma_{cap} \approx 118$ while if it were unbound we expect a ratio ≈ 71. The observed ratio is $20.36/0.332 = 61$. Thus the fact that the singlet state is unbound was first discovered in 1935 from the magnitude of the capture cross section.

6   *Photodisintegration of the deuteron*

The inverse process to n–p capture can also be studied, $^2H + \gamma \rightarrow {}^1H + {}^1n - 2.23$ MeV. The process can proceed either by a transition from the $^3S_1$ ground state to the $^1S_0$ state in the continuum, exactly the inverse of the capture process for slow neutrons or by a transition from the $^3S_1$ ground state to a ^3P state in the continuum. The former will be absorption of magnetic dipole radiation, photomagnetic disintegration, while the latter will be absorption of electric dipole radiation, photoelectric disintegration analogous to the photoelectric effect in the atom. Obviously we must have γ-rays of energy in excess of 2.23 MeV before any photodisintegration can take place.

We consider first the photomagnetic disintegration. It will have a closely similar form to the capture cross section formula.

$$\sigma_{dis}^M = \frac{1}{2} \frac{2\pi}{3} (\mu_n - \mu_p)^2 \frac{e^2}{\hbar c} \left(\frac{\hbar}{Mc}\right)^2 (k^2 + \alpha^2) k^{-1} I^2. \tag{11.31}$$

The main differences are related to the fact that the γ-ray momentum is $\hbar\omega/c$ and its energy $\hbar\omega = \varepsilon + k^2$. There is also a factor $\frac{1}{2}$ in front of the equation. In the case of capture, the γ-rays were emitted with definite polarisation from the $m=0$ state at angle θ. In the case of photodisintegration the γ-rays are unpolarised and the probability that there is a polarised γ-ray such that M1 absorption can take place is just $\frac{1}{2}$.

The integral I has been evaluated before. However now we cannot neglect k^2 compared to α^2. The expression for the cross section is thus

$$\sigma_{\text{dis}}^{M} = \frac{2\pi e^2}{3\hbar c}\left(\frac{\hbar}{Mc}\right)^2 (\mu_{\text{n}}-\mu_{\text{p}})^2 \frac{k\alpha(1-\alpha a_{\text{ss}})^2}{(k^2+\alpha^2)(1+k^2 a_{\text{ss}}^2)}. \tag{11.32}$$

Near threshold the cross section varies as k, the square root of the excess energy $E_\gamma - \varepsilon$. It reaches a maximum at $k^2 = a_{\text{ss}}^{-1}$, that is, at an energy corresponding to the position of the unbound singlet state, which will be approximately $2.23+0.07$ MeV. Thereafter it will decrease as k^{-1} and then, as k^2 exceeds α^2, as k^{-3}. Since we are dealing with an S→S transition, the angular distribution will be isotropic. The peak value of the photomagnetic cross section is about 0.5 mb.

The photoelectric part of the disintegration cross section will involve a $^3\text{S}_1 \to {}^3\text{P}$ transition. No spin flip is necessary as in the magnetic case but the orbital angular momentum changes by one unit. The general form of the matrix element is

$$Q_{10}(\text{i},\text{f}) = e\left(\frac{3}{4\pi}\right)^{\frac{1}{2}} \sum_k \int \phi_{\text{f}}^* z_k \phi_{\text{i}}\, d\tau. \tag{11.33}$$

Since there is only one proton, there is only one term in the sum and z_k is the coordinate of the proton relative to the neutron since the z-component of the dipole moment is just $\frac{1}{2}ez$. Note that there is no spin term in the integral, only a distance term. We again represent ϕ_{i} by the ground state of the deuteron.

$$\phi_{\text{i}} = \frac{1}{(4\pi)^{\frac{1}{2}}}\frac{u_{0t}(r)}{r}\chi_{1,0}$$

while we can write $\phi_{\text{f}} = \exp(\text{i}\boldsymbol{k}\cdot\boldsymbol{r})$ neglecting any interaction between proton and neutron in the P state which we know to be small. Substituting and taking the z-direction to be the direction of $\boldsymbol{k}$

$$Q_{10} = \frac{\sqrt{3}}{8\pi}e\int z e^{-\text{i}kz}\frac{u_{0t}(r)}{r}\,d\tau \tag{11.34}$$

or integrating over angles

$$Q_{10} = -\text{i}\frac{\sqrt{3}}{2}\frac{e}{k}\int_0^\infty u_{0t}(r)u_{1t}(r)r\,dr \tag{11.35}$$

where $u_{1t}(r)$ is the radial wavefunction for the triplet p-wave component of the plane wave, $\exp \text{i}kz$,

$$u_{1t}(r) = \frac{\sin kr}{kr} - \cos kr.$$

And the cross section is then

$$\sigma^{E}_{\text{dis}} = \frac{\pi e^2}{3\hbar c} \frac{(k^2+\alpha^2)}{k} \int_0^{\infty} u_{0t}(r)u_{1t}(r)r\,\mathrm{d}r. \tag{11.36}$$

Again using the zero range approximation the integral becomes

$$(2\alpha)^{\frac{1}{2}} \frac{2k^2}{(k^2+\alpha^2)^2}.$$

Finally the cross section is

$$\sigma^{E}_{\text{dis}} = \frac{16\pi e^2}{3\hbar c} \frac{4\alpha k^3}{(k^2+\alpha^2)^3}. \tag{11.37}$$

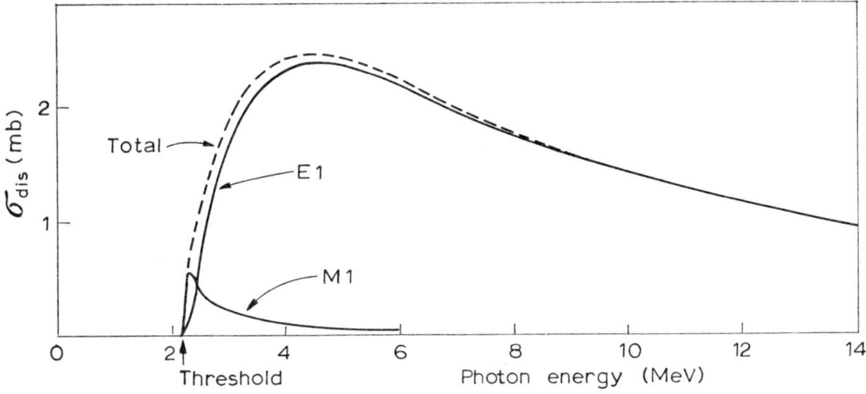

R. D. Evans, The Atomic Nucleus (McGraw-Hill, New York 1955) p. 335

Fig. 11.5.   The deuterium photodisintegration cross section

Near threshold the cross section rises slowly as k^3, that is, as $(E_y-\varepsilon)^3$. It reaches a maximum for $k=\alpha$, that is, for $E_y=2\varepsilon=4.46$ MeV. Thereafter it decreases again, cf. fig. 11.5. The angular distribution will be that for a dipole distribution, $\sigma(\theta)\propto \sin^2 \theta$, where θ is the angle between photon and proton directions. The maximum cross section is about 2.4 mb, about five times larger than the maximum photomagnetic cross section. We can estimate the order of magnitude expected for the photoelectric cross section. It should be just the 'area' of the deuteron $\pi/4\alpha^2$ times the fine structure constant $e^2/\hbar c=\frac{1}{137}$, which is the coupling constant for the electromagnetic field. This yields an estimate of 2.25 mb close to the maximum value observed.

7   The tensor force

There are still some details of the n–p system which we need to clarify and which
are not described in the simplified theory so far used. We justified our use of simple
central forces in our calculation by noting that the magnetic moment was close
to the sum of free proton and free neutron moments,

$$\mu_\mathrm{d} - \mu_\mathrm{p} - \mu_\mathrm{n} = -(0.022089 \pm 0.00012) \text{ n.m}$$

and that the electric quadrupole moment was very small, $Q_\mathrm{d} = +2.82 \times 10^{-27}$ cm^2.
However it is clear that these numbers are not zero and we now discuss their
implication. The fact that we observe a small positive quadrupole moment indicates
that there is a slight tendency for the probability density in the deuteron to be
extended along the direction of the spin vector. In pictorial terms, there is additional
binding introduced if the two nucleon spins are in line compared to them being
side by side. This tendency to deviate from the spherical state means that a small
admixture of orbital angular momentum different from zero is present. To preserve
the observed total spin of 1^+ this can only be a $^3\mathrm{D}_1$ state. This extra orbital motion
will then contribute to the magnetic moment in addition to that contributed by
the intrinsic particle moments. It is found that an admixture of about 4 % of $^3\mathrm{D}_1$
state with 96 % $^3\mathrm{S}_1$ state satisfactorily accounts for both the magnetic moment
and quadrupole moment.

The form of the additional attractive force is just that expected for the interaction of two magnetic dipoles, but it is easy to show that this electromagnetic
effect is not strong enough to account for the effect. It is a new constituent of the
nuclear force, a tensor interaction which depends on the angles between the spin
vectors of neutron and proton and the radius vector separating them.

Obviously the tensor force must exist at low energies to explain the deuteron
ground state properties. On the other hand it will not affect the singlet state as there
is no preferred spin direction ($I=0$). The singlet state really will have spherical
symmetry.

The spin–orbit force which we have met in the shell model also does not affect
the singlet state. However it must also vanish for zero energy and so cannot
affect the deuteron structure. We may expect it to show up however in higher
energy scattering experiments and we shall see that it does so.

8   The p–p and n–n systems

The remaining two-nucleon systems can only be studied through scattering experiments since no bound states of the di-neutron or di-proton are observed. The

proton–proton scattering can be studied with high precision. However the interpretation is complicated by the scattering due to the coulomb force between the protons which is in addition to the nuclear scattering and also interferes with it. In addition corrections are necessary as identical particles are being scattered.

The differential cross section for proton–proton scattering can be written

$$\frac{\mathrm{d}\sigma(\theta)}{\mathrm{d}\Omega} = \frac{e^2}{2Mv}\left[\frac{1}{\sin^4\frac{1}{2}\theta} + \frac{1}{\cos^4\frac{1}{2}\theta} - \frac{\cos(\eta\ln\tan^2\frac{1}{2}\theta)}{\sin^2\frac{1}{2}\theta\cos\frac{1}{2}\theta}\right.$$

$$\left. - \frac{2}{\eta}\sin\delta_0\left\{\frac{\cos(\delta_0+\eta\ln\sin^2\frac{1}{2}\theta)}{\sin^2\frac{1}{2}\theta} + \frac{\cos(\delta_0+\eta\ln\cos^2\frac{1}{2}\theta)}{\cos^2\frac{1}{2}\theta}\right\} + \frac{4}{\eta^2}\sin^2\delta_0\right] \quad (11.38)$$

where $\eta = e^2/\hbar v$, $M = \frac{1}{2}M_{\text{proton}}$ is the reduced mass and v is the proton velocity.

The first term is just the classical Rutherford coulomb scattering. The next two terms are the quantum mechanical correction for the scattering of identical particles and for the interference between scattered and recoil particles. The first three terms describing coulomb scattering of identical particles is called the Mott scattering. The observed p–p scattering to 300 keV is well described by Mott scattering alone. In other words no nuclear effect can be seen. Above this energy more scattering is seen than is predicted by the Mott formula and is attributed to nuclear effects (fig. 11.6). Up to 10 MeV this can be accounted for by an s-wave interaction between the protons. The fourth term is thus the interference between the nuclear and coulomb scattering and the last term is the nuclear scattering itself. Only δ_0 is involved as we assume $l=0$ and only the singlet interaction is possible as the particles are identical. You note that from the interference term between nuclear and coulomb scattering we can infer the sign of the nuclear phase shift relative to the sign of the coulomb phase shift. The interference is noted to be destructive hence the signs are opposite. The coulomb phase shift is negative corresponding to a repulsive force while the nuclear phase shift is positive corresponding to an attractive force. Again the effective range theory can be used to relate a scattering length to the nuclear s-wave phase shift δ_0. The formulation is as follows:

$$\frac{\pi\cot\delta_0}{(\exp 2\pi\eta - 1)} + h(\eta) = \frac{\hbar^2}{M_\mathrm{p}e^2}\left(-\frac{1}{a_{\mathrm{ss}}} + \tfrac{1}{2}r_{0\mathrm{s}}k^2\right) \quad (11.39)$$

where $h(\eta)$ is a calculable function of $\eta = e^2/\hbar v$. The denominator of the first term is part of the coulomb penetration. We can thus derive values for scattering length and effective range for the p–p system.

$$a_{\mathrm{ss}} = -17 \pm 0.6 \times 10^{-13} \text{ cm,}$$

$$r_{0\mathrm{s}} = 2.65 \pm 0.08 \times 10^{-13} \text{ cm.}$$

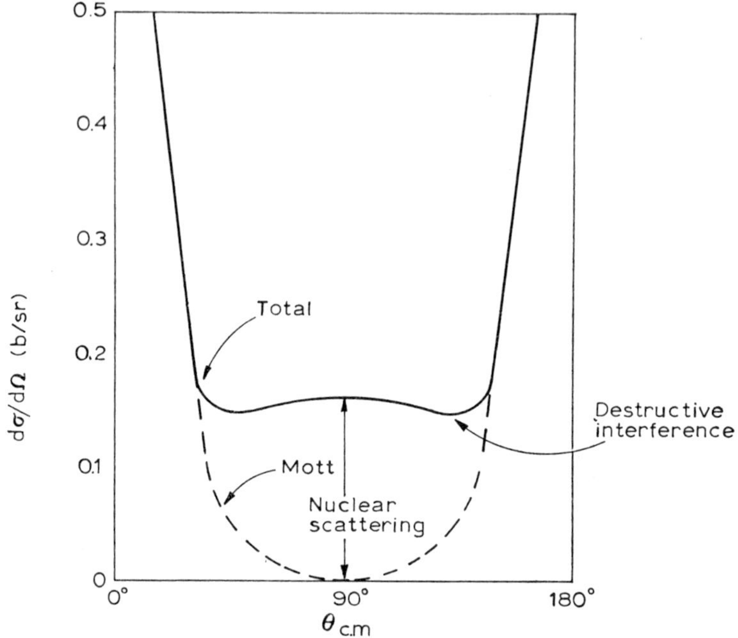

R. D. Evans, The Atomic Nucleus (McGraw-Hill, New York 1955) p. 340

Fig. 11.6.   Angular distribution of proton–proton scattering at 2.5 MeV

The scattering length is again negative indicating that the di-proton is unbound and its magnitude suggests that it is unbound by a slightly larger amount than is the singlet deuteron. The singlet effective range is very similar for the n–p and p–p systems.

We cannot directly study the other two-nucleon system, the di-neutron as we do not have neutron targets. Charge independence of the force would lead us to guess that like the singlet p–p and n–p systems it is also unbound. We have to extract the scattering length from final state interaction experiments such as $D+n \rightarrow n+n+p$ or other reactions with two neutrons in the final state. The results are somewhat discrepant at present but a value close to that for the p–p singlet system is likely correct.

9   *High energy nucleon–nucleon scattering*

We have found from our study of two-nucleon systems at low energy that the systems can be rather well described by a central attractive potential whose shape does not matter very much. With the use of just two parameters, the scattering length and effective range, the main features of deuteron binding and low energy scattering could be precisely characterised. In addition the rather small effect of a tensor force could be inferred from the moments of the deuteron ground state. To find out more about the nucleon–nucleon force we have to go to higher energy scattering experiments.

For energies above 10–20 MeV higher angular momenta with $l > 0$ are expected to become important and effective range theory applicable only for s-wave interaction is no longer valid. Further details of the potential are certain to show up but the number of parameters, that is, the number of phase shifts will also increase. This means that the information to be extracted from a given experiment is more limited. Above about 300 MeV inelastic meson production becomes possible.

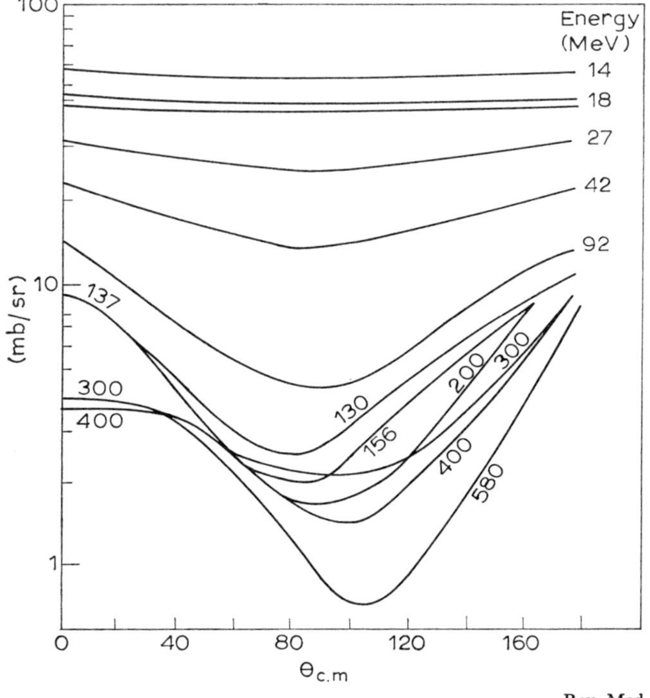

Rev. Mod. Phys. **30** (1958) 370

Fig. 11.7.   Neutron–proton scattering angular distributions at high energies

For n–p scattering, which is closely isotropic in the centre of mass system up to about 14 MeV, we see a change above this energy to an angular distribution with forward and backward peaks and a minimum at 90° (fig. 11.7). The interaction potential between neutron and proton is of the order of 30 MeV. Accordingly for energies much greater than this we should expect less deviation produced in the path of the neutron. That is we should expect a forward peaking of the scattered neutrons which we indeed observe. But we also observe a backward peak of comparable magnitude and this is direct evidence for a charge exchange interaction. In this case the incident neutron becomes a proton which goes forward and the target proton becomes a neutron which goes backward (fig. 11.8). The fact that

Fig. 11.8.   Direct and exchange scattering

experimentally we see comparable probabilities for the two interactions with and without exchange, indicates that ordinary forces of non-exchange character and charge exchange forces are comparable in strength for the n–p system.

The p–p scattering cross section in contrast with the n–p cross section remains almost isotropic up to 400 MeV, as shown in fig. 11.9. Here we get only $T = 1$ inter-

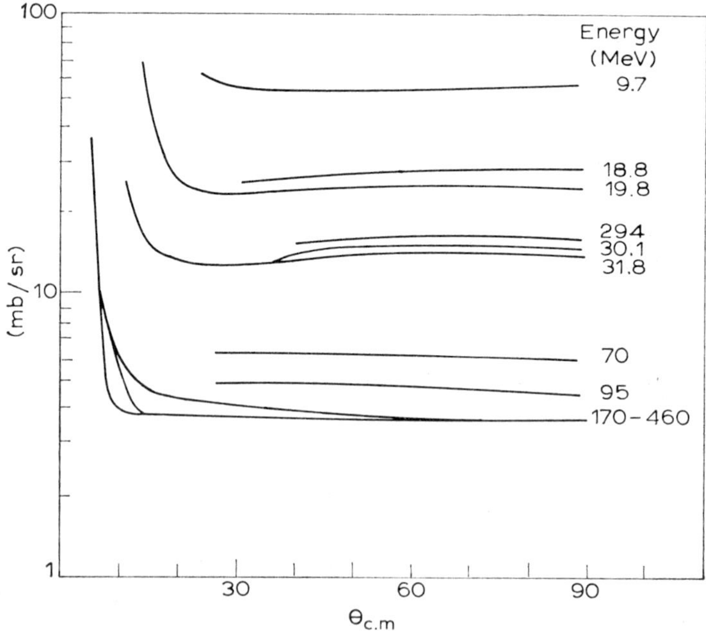

Rev. Mod. Phys. **30** (1958) 370

Fig. 11.9.   Proton–proton scattering angular distributions at high energies

actions as the particles are identical while for the n–p case we get $T=0$ and $T=1$. The isotropy above about 10–20 MeV cannot be the result of s-wave interactions alone. If this were the case we should expect a differential cross section of about λ^2. At 400 MeV this is about 2 mb/ster while a value of about 3.6 mb/ster is observed. The higher l-waves present combine to give isotropy. For the $T=1$ interaction the Pauli principle allows interaction only in the antisymmetric states ^{1}S, ^{3}P, ^{1}D, ^{3}F etc. Measurement of the proton polarisations at 315 MeV indicates the presence of a spin–orbit interaction. Since this acts only for triplet states the ^{3}P and possibly the ^{3}F states are certainly involved.

As we mentioned above when more than $l=0$ waves are involved, the number of parameters, phase shifts and mixing parameters, increases rapidly. If l_{max} is the highest value of l to be considered we require $5l_{max}+1$ parameters for n–p scattering and $\frac{1}{2}(5l_{max}+3)$ parameters for p–p scattering. Thus in addition to the total cross section and angular distribution we must do many other sorts of experiments involving polarisation and depolarisation measurements. A search is then made for a set of parameters which are consistent with all available experiments. Typically several sets of parameters can be found which describe the data. One clear point which emerges however is that the ^{1}S$_0$ phase shift which is positive at low energies indicating an attractive potential changes sign as the energy is increased and at 315 MeV is negative. This indicates that as the interaction radius decreases with increased energy the attractive potential becomes a repulsive potential. That is, there is a repulsive core in the potential surface.

10   Saturation of the nuclear force

A study of the binding energy of more complex nuclei shows up further properties of the nuclear force. If we measure the size of nuclei we find that the radius is proportional to A, that is that the nuclear density remains constant. However when we measure the binding energies of nuclei we find that it is proportional approximately to A, that is each nucleon is bound by the same amount no matter how many nucleons are present. These facts lead us to conclude that the nuclear force *saturates*, that is that the number of nucleons with which a given nucleon interacts strongly is strictly limited.

For example in the Fermi gas model we showed (ch. 8) that a nucleus in which all pairs interact would collapse to a radius of the order of the range of the force. Hence all pairs cannot be interacting. This shows up clearly as we add particles to the deuteron and examine the binding energy per interacting pair, as in table 11.3. Thus binding occurs between 2, 3 and 4 particles, but the fifth particle is not bound at all to the other four and successive particles are bound much more weakly.

TABLE 11.3

Nucleus	Binding energy (MeV)	Binding energy/$\frac{1}{2}A(A-1)$ (MeV)
^{2}D	2.23	2.23
^{3}He	7.6	2.53
^{4}He	28.0	4.67
^{5}He	−0.95	−0.095
^{6}Li	2.77	0.185
^{7}Li	2.46	0.115

There are two possible characteristics of the force which can lead to the observed situation. One possibility is that at very short distances the attractive nuclear force becomes repulsive (fig. 11.10), then at $A=4$ it may be that this nucleus is as tightly packed as the repulsive core will allow. Thereafter other nucleons will be further away and less bound. If the attractive well depth is about 30 MeV per nucleon, then the shorter range repulsive potential must be much larger, perhaps 300 MeV or more. We have seen that high energy nucleon–nucleon scattering indeed shows the existence of such a repulsive core. However in addition to a repulsive core there

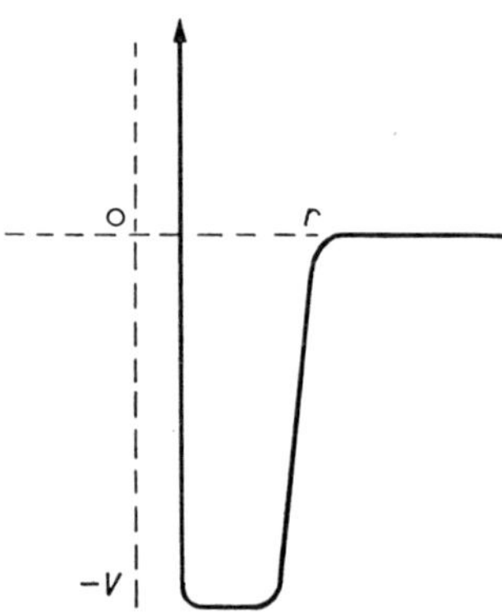

Fig. 11.10.   Potential well with repulsive core

may be other features of the force which can also lead to saturation. For example, in the binding of hydrogen atoms into molecules we find that two hydrogen atoms are bound into H_2 but a third hydrogen atom is not bound. This again is saturation and is due to a force of *exchange* type which corresponds to the two electrons jointly occupying the same spatial orbit with their spins opposed as required by the Pauli principle. A third electron cannot occupy this state by the Pauli rules and so a third atom is not bound.

In nuclei again the Pauli principle limits the number of like particles in a given spatial state to two with spins opposed. We have two kinds of nucleons, neutron and proton. Thus four particles, two paired protons and two paired neutrons, fill a given spatial state (fig. 11.11). Further particles must fill different spatial states and hence be less bound. For the lightest nuclei which we considered in

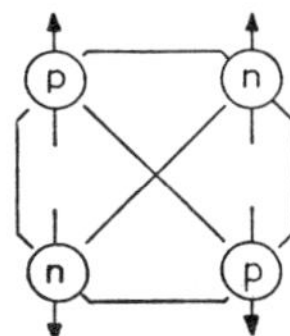

Fig. 11.11.   Nucleon bonds in ⁴He

table 11.3 the s-orbital state is filled at ^{4}He. An exchange interaction also gives attraction only for symmetric wavefunctions. The next nucleon in ^{5}He is in a p-state which is antisymmetric and the exchange force is repulsive. We already have seen that we require an exchange component of the n–p force to explain the backward peaking in the high energy n–p scattering.

We now consider the possible components of the nuclear force, ignoring tensor and $l \cdot s$ components. For charge independence of the force we assume these act between all types of nucleons, equally. Consider a singlet n–p pair.

Part of the force may be of non-exchange or ordinary type. No exchange takes place (fig. 11.12a). This is called a Wigner force.

Part of the force may involve exchange of charge or isospin (fig. 11.12b). This is called a Heisenberg exchange force and it can be seen to be equivalent to a simultaneous exchange of coordinate and spin.

Part of the force may involve exchange of coordinate or spatial state (fig. 11.12c).

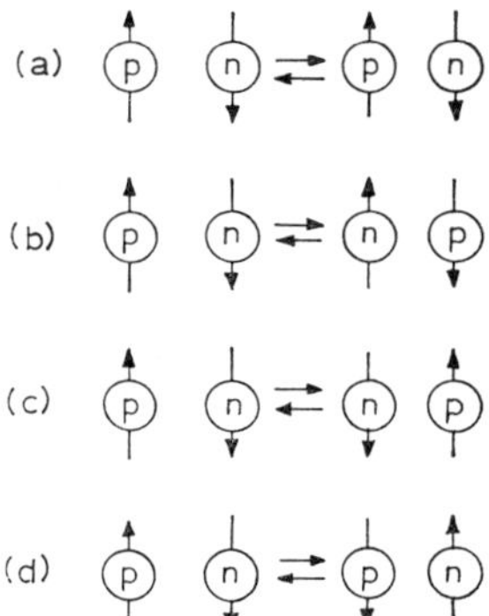

Fig. 11.12. The four types of nucleon-nucleon force, Wigner, Heisenberg, Majorana and Bartlett

This is called a Majorana exchange force and it can be seen to be equivalent to a simultaneous exchange of charge and spin.

Finally we may consider the possibility of exchange of spin (fig. 11.12d). This is called a Bartlett exchange force and is equivalent to a simultaneous exchange of charge and coordinates.

More formally we can express these as operators which act upon the wavefunction of the pair of particles $\psi(r_1\sigma_1;r_2\sigma_2)$, a function of the coordinate r and the spin σ of the particles. The operators associated with each type of force give the following results

$$P_W\psi(r_1\sigma_1;r_2\sigma_2) = \psi(r_1\sigma_1;r_2\sigma_2) \qquad \text{Wigner}$$
$$P_H\psi(r_1\sigma_1;r_2\sigma_2) = \psi(r_2\sigma_2;r_1\sigma_1) \qquad \text{Heisenberg}$$
$$P_M\psi(r_1\sigma_1;r_2\sigma_2) = \psi(r_2\sigma_1;r_1\sigma_2) \qquad \text{Majorana}$$
$$P_B\psi(r_1\sigma_1;r_2\sigma_2) = \psi(r_1\sigma_2;r_2\sigma_1) \qquad \text{Bartlett}$$

The potential energy for such a pair will be simply

$$\int \psi^* V P \psi \, \mathrm{d}\tau \tag{11.40}$$

where V is the radial wavefunction. The exchange will be either a symmetric or antisymmetric one.

$$P\psi = \pm\psi, \qquad P = \pm 1.$$

For exchange of spatial state or coordinates, even l-values are symmetric and odd l-values antisymmetric hence $P_M=(-1)^l$. For exchange of spin, triplet states are obviously symmetric and singlet states antisymmetric, hence $P_B=(-1)^{S+1}$. Similarly $P_H=(-1)^{l+S+1}$ or $(-1)^{T+1}$ and P_W must always be $+1$. Table 11.4 shows the values of P. We note that Heisenberg and Bartlett forces can account for the spin dependent force in the deuteron in which the triplet state $^3S_1$ is bound

TABLE 11.4

Force		Symmetry			
Spatial:		Even l		Odd l	
Spin:		Triplet	Singlet	Triplet	Singlet
Wigner		+1	+1	+1	+1
Heisenberg		+1	−1	−1	+1
Majorana		+1	+1	−1	−1
Bartlett		+1	−1	+1	−1

while the singlet $^1S_0$ is not. The saturation of the force which is observed cannot be accounted for by these types of force. The Bartlett force gives no change as we go from a symmetric spatial state to an antisymmetric one. In ^{4}He we have six internucleon bonds. Two are pure singlet interaction, two are pure triplet interaction and two are mixed singlet and triplet in equal amounts. Thus there is an equal number of singlet and triplet interactions and the Heisenberg force also is unchanged as we go to a different spatial state. The Majorana force is thus the only exchange force which can lead to saturation. For an s-state all pairs are attractive while for a p-state the force is repulsive.

If we choose then a mixture of these forces such that saturation occurs and the triplet–singlet splitting in the deuteron is accounted for, we find $V_M > 4V_W$ and $V_B \approx V_H < \frac{1}{10} V_M$. But you recall that in the high energy scattering of neutrons by protons comparable strengths of ordinary and exchange interactions were called for. Serber was able to fit the observations assuming a force in which $V_M = V_W$ and $V_B = V_H \approx 0$. It is clear that this force is zero for odd states where $P_W = +1$ and $P_M = -1$. It was a symmetric force to give the observed symmetrical angular distribution. However with this force which does not act in odd states we no longer get saturation of the force.

However saturation could also be produced by the repulsive core of the potential which we know to be present. Thus a mixture of approximately equal parts of Wigner and Majorana forces (the Serber mixture) together with a repulsive core to give saturation and sufficient Heisenberg and Bartlett forces ($\approx 10\,\%$) to give the spin dependence of the force, is in agreement with all the facts. The Majorana force is responsible also for the main part of the symmetry energy term in the semi-empirical binding energy formula.

11   Properties of the nuclear force

Another important feature of the nuclear force which we can infer from the properties of complex nuclei is its charge independence. We have already discussed the evidence from mirror and self-conjugate nuclei in ch. 8. To summarise then the nuclear force has the following properties:
1) An attractive well of range about 2×10^{-13} cm and depth 30–50 MeV/nucleon combined with a repulsive core of range about 0.4×10^{-13} cm and magnitude at least 300 MeV.
2) It is charge symmetric, i.e. n–n bonds = p–p bonds and also charge independent, n–n bonds = n–p bonds = p–p bonds.
3) It contains exchange components as well as non-exchange components. Spatial as well as spin symmetry control the exchange parts.

4) There is a non-central tensor component of the force.
5) There is a velocity dependent spin–orbit interaction.
Obviously the force is quite complex in character.

12   *Meson theory of nuclear forces*

In electrodynamics an elegant connection is made between the coulomb force, the electromagnetic field and the electromagnetic radiation. We now wish to follow this analogy, and see whether the strong force can be similarly described. In electromagnetic theory the scalar electric potential V obeys the wave equation

$$\nabla^2 V - \frac{1}{c^2}\frac{\partial^2 V}{\partial t^2} = 0. \tag{11.41}$$

For a static field we have the Laplace equation for the field in free space,

$$\nabla^2 V = 0. \tag{11.42}$$

The wave equation just described the propagation of electromagnetic waves in free space. Any change in the potential is transmitted from one point to another at velocity c. The wavelength and frequency are thus related $\lambda v = c$. In the quantum theory of the electromagnetic field we assume that the energy of the radiation is carried as discrete quanta or photons. Since these quanta travel at the velocity of light they have no rest mass and their energy is $E = hv$ and momentum is $p = h/\lambda = hv/c$. These are just the relations confirmed for the photon in the Compton effect. The energy and momentum of the photon are thus related $E^2 - p^2 c^2 = 0$. The potential with which this field is associated is long range varying inversely as r.

The nuclear potential however has a much shorter range. Yukawa wrote the wave equation as follows

$$\nabla^2 U - \frac{1}{c^2}\frac{\partial^2 U}{\partial t^2} - K^2 U = 0 \tag{11.43}$$

where U is the nuclear force potential and K is a constant. The solution gives a relation between wavelength and frequency,

$$-\frac{1}{\lambda^2} + \frac{v^2}{c^2} - \frac{K^2}{4\pi^2} = 0 \tag{11.44}$$

and between energy and momentum of a field quantum

$$E^2 - p^2 c^2 - M^2 c^2 = 0 \tag{11.45}$$

where $M = \hbar K/c$. This is just the relativistic energy–momentum relation for a *particle of mass M.* The static case of the Yukawa wave equation is

$$\nabla^2 U - K^2 U = 0 \tag{11.46}$$

which leads to a potential of the form

$$u = g\mathrm{e}^{-Kr}/r = g\mathrm{e}^{-Mcr/\hbar}/r \tag{11.47}$$

where g is a constant playing the same role as the charge in the electromagnetic case. It assumes a source of the field at the origin. We can choose a value for M, the mass of the field particle, such that the potential has a range corresponding to that observed for the nuclear force. The mass turns out to be about 300 times the electron mass. The particle has been identified as the π-meson or pion.

We thus assume that the force between two nucleons can be considered to arise from the transfer of virtual mesons between the two nucleons. The coulomb field can be similarly described as the continuous transfer of virtual photons between the two charges. For the nuclear force we do not have free mesons involved since we should have to supply an extra energy of the order of the rest energy of the pion, ≈ 140 MeV, and the production of mesons at zero energy is forbidden in classical physics, without violating energy conservation. However in quantum mechanics the energy $\Delta E = M_\pi c^2$ may be borrowed provided it is returned again in a time specified by the uncertainty principle, $\Delta E \cdot \Delta t \leqq \hbar$. The violation of energy conservation will be *unobservable*. This is what is meant by a transfer of *virtual* mesons. In a time Δt the meson can travel at most a distance $c\Delta t$ and this distance thus puts an upper limit on the range of the nuclear force, a:

$$a \leqq c\Delta t \leqq \frac{c\hbar}{\Delta E} = \frac{c\hbar}{Mc^2} = \frac{\hbar}{Mc} = \frac{1}{K}. \tag{11.48}$$

The force range is thus related to the Compton wavelength of the field quantum.

Originally Heisenberg suggested that electrons might be virtually exchanged. The range would then be 300 times greater but a more serious difficulty was the realisation that the electron–nucleon coupling is extremely weak, far too weak for the electron to be connected with the strong nuclear force field. In 1935 Yukawa predicted the mass of the expected field particle as we have done above from the nuclear force range. Two years later, in 1937, Anderson and Neddermeyer discovered a particle in the cosmic radiation which had a mass of 207 electron masses. For a time this was thought to be the Yukawa meson but it was soon found that the extreme penetrating power of this particle through great thicknesses of lead precluded it having any stronger nuclear interaction than the electron. The new particles were called μ-mesons or more suitably muons since they are not mesons in the Yukawa sense. It was not until ten years later, in 1947, that C. F. Powell discovered slightly heavier particles in the cosmic radiation at high altitudes.

Roughly speaking if a particle could penetrate the atmosphere it was unlikely to be the strongly interacting particle. The meson discovered by Powell was the π-meson or pion. With a mass 273 times the electron it is somewhat closer to Yukawa's prediction.

The positively and negatively charged pions so discovered are thus held to be field quanta of the nuclear force field. Kemmer pointed out that though the charged mesons could give us forces between neutron and proton, the charge independence of nuclear forces required that there be neutral pions as well, to account for the n–n and p–p forces. The neutral pion was subsequently found with a mass slightly less (264 electron masses) than that of the charged mesons.

Fundamental Particles:

Properties

So far we have referred to the fundamental particles listed in table 12.1. We now consider the experimental data from which we derive the properties of the newest additions to the list, the pion and muon.

TABLE 12.1

Name	Charge states		Mass (MeV)	J^P	Isospin T	T_3	Decay
Nucleons	proton	$+$	938	$\tfrac{1}{2}^+$	$\tfrac{1}{2}$	$+\tfrac{1}{2}$	stable
	neutron	0				$-\tfrac{1}{2}$	$p+e^-+\bar{\nu}$ (10^3 sec)
Mesons	pion	$\pm$	139.6				
		0	135				
Leptons	muon	$\pm$	105.7				
	electron	$\pm$	0.511	$\tfrac{1}{2}$			
	neutrino	0	0	$\tfrac{1}{2}$			
Photon		0	0	1^-			

1   Pions

Pions are produced when nucleons are bombarded by other nucleons or photons with an energy such that the rest mass of the pion, 140 MeV, is supplied in the centre of mass system. The target nucleons may or may not be a part of a nucleus. The reactions are

$$p+p \rightarrow p+n+\pi^+ - 140 \text{ MeV}$$
$$\rightarrow p+p+\pi^0 - 135 \text{ MeV}$$
$$p+n \rightarrow p+p+\pi^- - 140 \text{ MeV}$$
$$\rightarrow p+n+\pi^0 - 135 \text{ MeV}.$$

The threshold energy in the laboratory system is given by

$$E_{\text{th}} = |Q| \left(1 + \frac{m_1}{m_2} + \frac{|Q|}{2m_2 c^2}\right)$$

where m_1 is the mass of the bombarding particle, m_2 the mass of the target particle and Q is the energy balance. Thus the threshold for π-meson production is about 290 MeV. As we go to higher energies production of two or more pions will become energetically possible and will occur. After production the charged mesons may traverse cloud chambers, bubble chambers or emulsions where their energy, ionization and orbit in a magnetic field enable us to fix their charge and mass.

The two charged mesons differ in their interaction with matter. Negative pions after having been slowed down by ionization losses can be captured by the coulomb field of a nucleus and travel in Bohr orbits round the nucleus just as electrons do. These orbits have however much smaller radii ($r = h^2/4\pi mZ^2 e^2$) than the electron orbits because of the larger mass of the pion (280 times electron mass). The π-mesic X-rays associated with transitions in these orbits are also of correspondingly larger energies (5 MeV or more). Eventually the negative pion is captured by the nucleus. In three quarters of the cases the capture results in a visible 'star', that is, one or more charged particles result. The large amount of energy released by the capture $\pi^- + p \rightarrow n + 140$ MeV heats up the nucleus so that protons, α-particles etc. can be boiled off. In the other quarter of the cases which do not result in a star only neutral particles result, that is, one or more neutrons carry off the energy.

For the positive pion the situation is different. It cannot enter a Bohr orbit when it is slowed down as it is repelled by the nuclear field. Thus it decays by a weak decay analogous to β-decay,

$$\pi^+ \rightarrow \mu^+ + \nu + 34 \text{ MeV}.$$

The lifetime for this decay is $2.55 \pm 0.03 \times 10^{-8}$ sec. The corresponding weak decay for the π^-

$$\pi^- \rightarrow \mu^- + \nu + 34 \text{ MeV}$$

can be observed for decays in flight. The lifetime is $2.55 \pm 0.19 \times 10^{-8}$ sec. The identity of the two lifetimes indicates the identity of the two pions or the invariance under charge conjugation of the charged pions.

The spin of the pion is obtained by a comparison of the two reactions

$$p+p \to D+\pi^+$$
$$\pi^+ +D \to p+p.$$

The cross sections for the two inverse reactions are related by the principle of detailed balance, which states that

$$\frac{\sigma_{pp}}{\sigma_{\pi D}} = \frac{(2s_D+1)(2s_\pi+1)}{(2s_p+1)(2s_p+1)} \frac{k_{\pi D}^2}{k_{pp}^2} = \frac{3k_{\pi D}^2}{4k_{pp}^2}(2s_\pi+1),$$

where s represents the spin of the particles involved and the k wave numbers corresponding to the energies at which the two cross sections are measured. The spin of the pion is expected to be integral since angular momentum is conserved in reactions like

$$p+p \to \pi^+ +D.$$

The cross section ratio indicates that the spin is zero.

The intrinsic parity of the pion is inferred from a study of the reaction $\pi^- +D\to n+n$. There is evidence from a study of the π-mesic X-rays that the πD system is in an s-state. If this is so the initial system has a total angular momentum $1\hbar$, since the deuteron has spin 1, the pion spin 0 and zero orbital angular momentum. The system will have the intrinsic parity of the pion, since the deuteron has positive parity and $l=0$. On the right hand side we must have total angular momentum 1. Since we have two identical particles we cannot have all of the possible states $^3S_1$, $^1P_1$, $^3P_1$, $^3D_1$, but only those which are antisymmetric by the exclusion principle. We recall the values of the eigenfunctions for space exchange $P_M=(-1)^l$ and for spin exchange $P_B=(-1)^{S+1}$. So we have

nn state	P_M	P_B	
$^3S_1$	+	+	symmetric
$^1P_1$	−	−	symmetric
$^3P_1$	−	+	antisymmetric
$^3D_1$	+	+	symmetric

and only the $^3P_1$ state is allowed. Since this has $L=1$ the parity is negative and assuming parity conservation in the reaction, then the pion parity is negative relative to the nucleon parity (taken to be positive by convention). Thus the pion has $J^P=0^-$ which says that its wavefunction is invariant under rotation of axes but changes sign under reflection of axes. Such a system is called a pseudoscalar. A common example is the helicity of a screw. A right-handed screw remains right-handed if rotated in any way but in a mirror becomes a left-handed screw.

There is no direct determination of the parity of π^+ or π^0 but it follows from charge independence that π^0 is also 0^- and from charge conjugation invariance that π^- has 0^-. The pion, having integral spin, is a *boson*.

We now consider the neutral pion which behaves rather differently from the charged pions. This is because it can decay electromagnetically into photons which the charged pions cannot do as charge would not be conserved. We find that the π^0 decays into two photons. It cannot decay into one photon as this would violate charge conjugation symmetry. We recall that charge conjugation changes a particle into its antiparticle. For a spinless particle of zero charge like the π^0, obviously this changes nothing; 'it is its own antiparticle'. Hence it is even under charge conjugation. The photon however is odd under charge conjugation (changing charges from $+$ to $-$ changes the field direction). So the minimum number of photons into which the π^0 can decay is 2. The γ-rays were studied in the reaction

$$\pi^- + p \rightarrow n + \gamma \qquad E_\gamma = 140 \text{ MeV}$$

and

$$\pi^- + p \rightarrow n + \pi^0$$
$$\pi^0 \rightarrow \gamma + \gamma \qquad E_\gamma = \tfrac{1}{2} \times 135 = 68 \text{ MeV}.$$

The two energies of γ-rays are emitted with roughly equal intensity. Obviously the *energy* with which the π^0 is emitted depends on the mass difference between π^- and π^0. Since the two γ-rays are emitted by the π^0 in flight, the 68 MeV line will be Doppler-broadened and we can estimate the π^0 velocity. From this we find $m_{\pi^-} - m_{\pi^0} = 5.3$ MeV, and we get a π^0 mass of 135 MeV.

The lifetime of the π^0 is very short being an electromagnetic decay. It has been observed by starting with a π^0 produced as the decay product of a heavier charged meson

$$K^+ \rightarrow \pi^0 + \pi^+.$$

The path length travelled before decay into $\pi^0 \rightarrow e^+ + e^- + \gamma$ (an alternative mode of decay to $\gamma + \gamma$) is measured and gives a mean life $\tau_{\pi^0} = (1.78 \pm 0.26) \times 10^{-16}$ sec. The decay is 10^8 times faster than for charged pions as it is an electromagnetic rather than a weak decay.

As the pion has three charge states it is natural to consider it an isospin triplet with $T = 1$ and $T_3 = +1, 0, -1$.

2   *The weak decay of the pion*

We now consider further the weak decay of the charged pions which result in two leptons each of spin $\tfrac{1}{2}$,

$$\pi^{\pm} \rightarrow \mu^{\pm} + \nu$$

It is easy to see that the two decay particles must be emitted in opposite directions to conserve linear momentum and with opposite spin sense to conserve angular momentum (cf. fig. 12.1). If we call the μ^- a lepton and μ^+ an antilepton (like the electrons) they are accompanied by antineutrino and neutrino. In the two decays

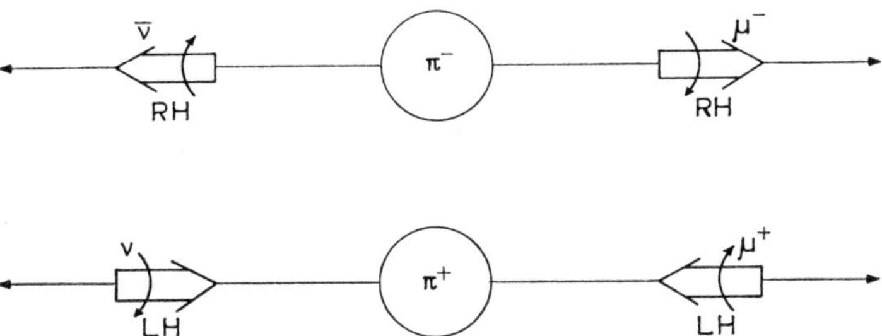

Fig. 12.1.   Pion decay

the helicities (screw senses) of the μ and v are the same. The neutrino is left-handed and the antineutrino right-handed. Now turning back to β-decay you recall that it is a consequence of parity violation that the β-particles are polarised in their direction of motion, that is they have a helicity. The degree of polarisation depends on the electron velocity and approaches 100% as $v \to c$. The experiments to measure the polarisation show that the *sense* of the polarisation indicates that for electrons the helicity is negative like a left-handed screw with p and σ opposed while for positrons the helicity is positive like a right-handed screw with p and σ aligned.

However in the pion decay we seem to have an exception. The μ^- is emitted as if it were a right-handed particle with helicity $+1$, and the μ^+ with helicity -1. This is related to our remark that the polarisation of β-particles was dependent on the velocity, which is a consequence of special relativity (ch. 2 § 6.5). Neutrinos being massless always travel at velocity c and have a definite helicity. Particles with mass have a polarisation proportional to v/c. In the pion decay the two particles must be emitted as shown to conserve linear and angular momentum. The neutrino must have a definite helicity and the helicity violation must occur for the massive particle with $v < c$.

We may now ask why the decay $\pi^+ \to e^+ + v$ should not occur? It does so but rather improbably, the branching ratio being $e_{\text{decay}}/\mu_{\text{decay}} = 1.2 \times 10^{-4}$. The reason is that the Q-value is higher and the electron is lighter and thus has a greater velocity nearer to c. Helicity violation is much less probable. The effect can be calculated and the agreement with the observed branching is good.

We summarise now the decay modes of the pions.

$$\pi^{\pm} \rightarrow \mu\nu \qquad \approx 100\% \quad \text{lifetime } 2.55 \times 10^{-8} \text{ sec}$$

$$e\nu \qquad 1.2 \times 10^{-4}$$

$$\mu\nu\gamma \qquad 1.2 \times 10^{-4}$$

$$\pi^{0}e\nu \qquad 1.1 \times 10^{-8} \text{ (discussed in ch. 14)}$$

$$\pi^{0} \rightarrow \gamma\gamma \qquad 98.8\% \quad \text{lifetime } 1.78 \times 10^{-16} \text{ sec}$$

$$\gamma e^{+}e^{-} \qquad 1.2\%$$

3   Muons

We now consider the particle resulting from the decay of the charged pion, the muon, which we have called a lepton. It was after its discovery called a μ-meson, incorrectly. It is *not* a meson according to our modern definitions. Muons are produced only via the weak interaction decays of pions and other heavier elementary particles. Again the passage of positive and negative muons through matter differs as the negative muon can be captured into a Bohr orbit of an atom and make a μ-mesic atom. Because of its small interaction with the nucleus and relatively long decay lifetime (2.2×10^{-6} sec) the μ-mesic atom is relatively long lived. The muon eventually decays either by $\mu^{-} \rightarrow e^{-} + \nu + \bar{\nu}$ (fig. 12.2) or is captured in a process analogous to electron capture (also a *weak* interaction)

$$\mu^{-} + p \rightarrow n + \nu + 106 \text{ MeV (the binding of the muon in the atomic orbit)}.$$

A few (10%) cases of visible stars are seen following μ^{-}-capture. That is, the neutrino and one or two neutrons carry away the available 106 MeV of energy in 90% of the cases rather than heating up the nucleus so that protons or α-particles are also emitted giving a star. The decay rate of the muon in an orbit is rather

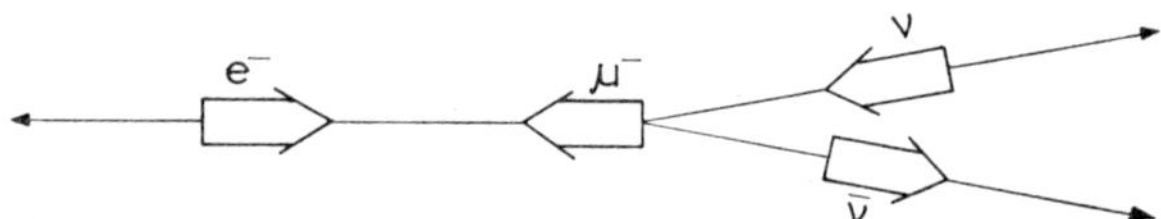

Fig. 12.2.   Muon decay

slower than for a free muon. The binding energy of the muon in a heavy atom can be as much as 10 MeV and the decay rate varies as the fifth power of the energy available. Thus

$$\left(\frac{m_{\mu}c^{2} - 10 \text{ MeV}}{m_{\mu}c^{2}}\right) = \left(\frac{96}{106}\right)^{5} = 0.6.$$

Thus the lifetime is almost twice as long as for the free decay.

Since the μ^+ cannot be captured into a Bohr orbit it always decays with the free lifetime, 2.2×10^{-6} sec.

When we measure the energy of the muons resulting from *pion decay* we observe a fixed muon energy. This is just the energy left to it after dividing with the neutrino to give momentum conservation. It is clearly a two-particle decay. In the case of *muon decay* we observe one electron produced but its energy spectrum extends from zero up to the maximum available just as in β-decay. Thus it is a three-body decay and we deduce that *two* neutrinos must accompany the electron.

$$\mu^\pm \rightarrow e^\pm + \nu_1 + \nu_2.$$

They cannot be *identical* neutrinos since if we consider the case where the electron is emitted in one direction with half of the energy and the two neutrinos are emitted in the opposite direction each with one quarter of the energy, the Pauli principle forbids this for identical neutrinos. Since we do observe electrons with half the energy the neutrinos must be non-identical. The two neutrinos clearly must have opposite helicity if the electron has the same spin as the μ^-. So they are certainly ν and $\bar{\nu}$. However there is a more subtle difference between them.

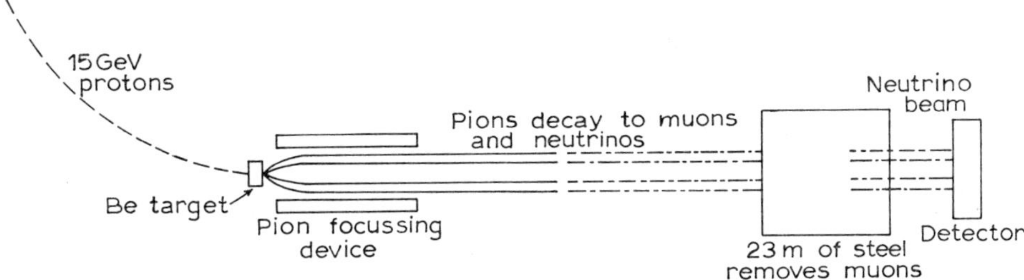

Fig. 12.3.   Arrangement for producing a neutrino beam

Experiments over the past few years at Brookhaven and at CERN have been concerned with neutrino-induced reactions. A neutrino beam is produced as shown in fig. 12.3. We might expect two reactions in the detectors (bubble and spark chambers),

$$\nu + N \rightarrow \mu + N$$

and

$$\nu + N \rightarrow e + N$$

(N is a nucleon) in other words, the inverse reactions to μ-capture and electron capture. The electron production is favoured by statistical factors. In fact only events corresponding to the first reaction are seen. We conclude that neutrinos accompanying muons in pion decay are basically different from neutrinos accompanying electrons. Furthermore the absence of two other decays

$$\mu^{\pm} \rightarrow e^{\pm} + e^{+} + e^{-}$$

and

$$\mu^{-} + N \rightarrow e^{-} + N \text{ (near a nucleus)}$$

indicates that two separate conservation laws apply: 1) μ and v_{μ} must be conserved (conservation of meptons), 2) e and v_{e} must be conserved (conservation of eptons). Together they give the conservation of leptons.

Thus the two neutrinos in muon decay are intrinsically different in addition to their opposite helicities. One is a v_{μ} to go with the muon and the other is a v_{e} to go with the electron

$$\mu^{+} \rightarrow e^{+} + v_{e} + \bar{v}_{\mu}$$
$$\mu^{-} \rightarrow e^{-} + \bar{v}_{e} + v_{\mu}.$$

If we work out the coupling constant for the weak interaction involved in muon decay we find

$$g_{\mu} = 1.43 \times 10^{-49} \text{ erg cm}^{3}.$$

This is rather close to the coupling constant derived for β-decay

$$g_{\beta} = 1.40 \times 10^{-49} \text{ erg cm}^{3}.$$

They are so nearly the same that it is logical to assume that the forces involved are identical – that there is a Universal Fermi Interaction. This assumption is often expressed in the form of a triangle of interactions called the Puppi triangle (fig. 12.4). The Universal Fermi Interaction assumption says all three sides are the

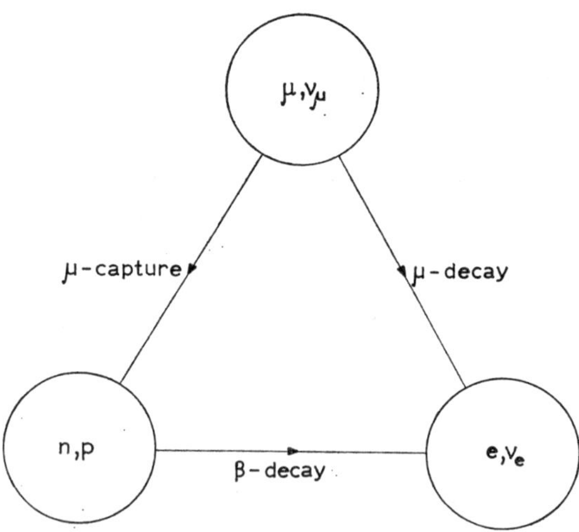

Fig. 12.4.   The Puppi triangle

same. We do not observe μ-decay of the nucleons since the energy difference between proton and neutron is too small, but the phenomenon of μ-capture like electron capture is well known.

At each vertex of the Puppi triangle there is a separate conservation law: 1) Conservation of baryons. The number of baryons minus the number of antibaryons is a constant. 2) Conservation of meptons. The number of meptons minus the number of antimeptons is a constant. 3) Conservation of eptons. The number of eptons minus the number of antieptons is a constant.

We return to these ideas in ch. 14.

4   The magnetic moment of the muon

Classically the electron is expected to have a magnetic moment $e\hbar/2m_ec$, the Bohr magneton. In fact it differs from this value by a small but significant amount. The gyromagnetic ratio $g=\mu/s$ where μ is measured in Bohr magnetons, $s=\frac{1}{2}$

$$g_e = 2 \times 1.001160 \quad \left(\mu_e = 1.001160 \frac{e\hbar}{2m_ec}\right).$$

The so-called anomalous magnetic moment is attributed to the virtual emission and reabsorption of photons in the electric field of the electron.

The proton is expected classically to have a magnetic moment $e\hbar/2M_Nc$ where now M_N is the proton mass. This is called the nuclear magneton. The neutron is expected to have a magnetic moment of zero. In fact the magnetic moments of proton and neutron are

$$\mu_p = 2.7927 \text{ n.m}, \qquad g = 2 \times 2.7927,$$
$$\mu_n = -1.91316 \text{ n.m.}$$

These anomalous nuclear magnetic moments are attributed to the virtual emission and reabsorption of pions in the strong field of the nucleon. The extra magnetic moment is due to currents in the cloud of charged pions.

We now can ask if the muon has an anomalous magnetic moment? If it does it might indicate that some new type of field is involved. If it is identical to the electron only the anomalous moment due to the electromagnetic field will be seen. In this case

$$g_\mu(\text{theor}) = 2 \times 1.00116 \quad \left(\mu_\mu = 1.00116 \frac{e\hbar}{2m_\mu c}\right).$$

A measurement performed at CERN* consisted of sending a beam of longitudi-

---

* Charpack, Farley, Garwin, Muller, Sens and Zichichi, Phys. Letters **1** (1962) 16.

nally polarised muons (from pion decay) into a magnetic field of a value such that the frequency of rotation is just equal to the spin precession frequency if $g_\mu = 2$. After 1500 turns the resultant polarisation of the muons was checked by observing the decay asymmetry. The result was $g_\mu(\text{obs}) = 2 \times (1.001145 \pm 0.000022)$. Thus to high accuracy the muon has no anomalous moment different from the electron. There is no cloud of virtual field particles of some new kind surrounding the muon.

5   *The interaction of muons and electrons*

Positive muons can form with electrons a neutral quasi-atomic system called muonium (fig. 12.5), analogous to positronium. The ground state has two possible spin values $J = 0, 1$. In a magnetic field the precession frequency of muonium can

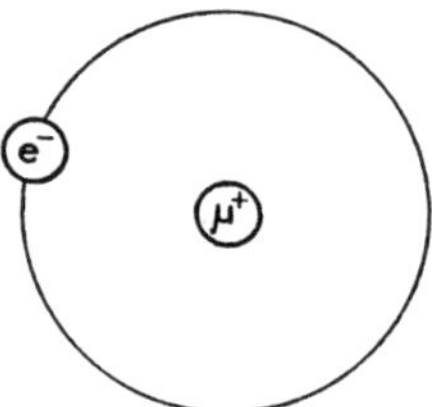

Fig. 12.5.   Muonium

be measured and we can derive a value for the hyperfine splitting of the ground state. If there is no anomaly in the muon–electron binding (purely electromagnetic) we expect the splitting to be

$$\Delta v(\text{theor}) = 4463.16 \pm 0.10 \text{ MHz}.$$

The observed* value is

$$\Delta v(\text{obs}) = 4463.16 \pm 0.06 \text{ MHz}.$$

Again no anomalous effect is detected.

We can ask whether as in positronium we can get annihilation of an electron and muon into two γ-rays. It would not violate any electromagnetic law, but it does not happen as it would violate *separate* conservation of meptons and eptons. Experimentally it happens less than 2.5×10^{-7} times as often as for positronium. Presumably we could get $\mu^+ - \mu^-$ annihilation but it has not been observed.

* V. W. Hughes, Proc. Rochester Conference, 1962.

6   *What is the muon?*

We summarise our knowledge of the muon accompanied by its own kind of neutrino.

1) It has a mass (rest energy) 207 times that of the electron.

2) It is conserved, along with its neutrinos, separately from the electron and its neutrinos, implying some as yet unrevealed quantum number.

3) It has no anomalous magnetic moment different from the electron.

4) It binds by the electromagnetic force to an electron of opposite charge exactly as an electron would.

5) Like the electron it participates only in the electromagnetic and weak interactions. It has no strong force interaction.

Thus it looks more and more like a *heavy* electron. The origin of the mass difference is a major unsolved problem, though the reason for the zero mass of both uncharged neutrinos is another possibly related problem. Turning the problem around we might understand the zero rest mass of the neutrinos since we have no neutral weak force currents and no electric charge. But electromagnetic and weak force effects cannot account satisfactorily even for the electron rest mass let alone for the larger muon rest mass.

7   *The K-meson*

The K-meson (kaon) was first observed by Rochester and Butler in emulsions exposed to cosmic rays at high altitudes. The production process looked as in fig. 12.6.

The V-shaped track consists of the decay products of a neutral particle which does not produce a track. In all cases where a K-meson was produced it was

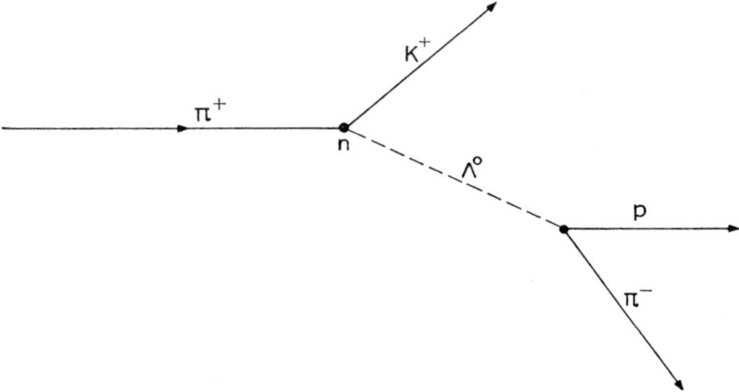

Fig. 12.6.   Associated production

accompanied by another heavy particle with a mass greater than a nucleon as it can decay into nucleon and pion. This is the phenomenon of associated production. Whenever one strange new particle is produced it is accompanied by another strange new particle. To account for this we have to assume a new property for the particles, or rather we say there is a new quantum number to be conserved which for previous fundamental particles was zero. This property is called *strangeness*. We say the kaon has $S = +1$ and the new heavier neutral particle has $S = -1$. Since pion and neutron both have $S=0$ strangeness is conserved.

$$\pi^+ + n \rightarrow K^+ + \Lambda^0.$$
$$S = 0 \quad 0 \quad +1 \quad -1$$

The heavier particle was named Λ^0 doubtless from the appearance of its decay particle tracks.

We observe all three charge states for kaons, K^+, K^-, and K^0. The charged kaons have a mass of 494 MeV and the neutral kaon has a mass of 498 MeV. The lifetime of the charged kaons is 1.22×10^{-8} sec.

Two different decay modes for particles of about 1000 electron masses were observed at an early stage. One type decayed into three pions and was called the τ-meson, the other decayed into two pions and was called the θ-meson.

$$\tau \rightarrow \pi + \pi + \pi$$
$$\theta \rightarrow \pi + \pi.$$

Since we know the pion has spin and parity 0^- and since the angular distribution of the decay products indicated zero orbital angular momentum we conclude that the τ-meson has $J^P = 0^-$ and the θ-meson has $J^P = 0^+$. As measurements improved in accuracy the masses and lifetimes of the τ- and θ-mesons became closer and closer to being identical. This puzzle led Lee and Yang* to suggest that in fact τ and θ were one and the same particle which had two alternative modes of decay in which however parity was *not* conserved as we had assumed in assigning spins to τ and θ. This was a radical suggestion but when Lee and Yang surveyed the evidence for parity conservation they discovered 1) that there was strong evidence that parity *is* conserved in strong nuclear interactions, 2) that there was no real evidence that parity is conserved in weak interactions like β-decay. They suggested that if parity is violated in the τ–θ decay then it might be violated in β-decay. This initiated the experiments of Wu et al. which showed that parity is also violated in β-decay (ch. 2 § 6.5).

Many decay modes of the charged K-meson are now recognised:

* T. D. Lee and C. N. Yang, Phys. Rev. **108** (1957) 1611.

$$\begin{aligned}
K^{\pm} &\to \mu^{\pm}\nu & &+388 \text{ MeV} \quad 63.1\,\% \\
&\to \pi^{\pm}\pi^{0} & &+219 \text{ MeV} \quad 21.5\,\% \quad \theta\text{-mode} \\
&\to \pi^{\pm}\pi^{+}\pi^{-} & &+\ 75 \text{ MeV} \quad\ 5.5\,\% \\
&\to \pi^{\pm}\pi^{0}\pi^{0} & &+\ 84 \text{ MeV} \quad\ 1.7\,\% \\
&\to \pi^{0}\mu^{\pm}\nu & &+253 \text{ MeV} \quad 3.4\,\% \\
&\to \pi^{0}e^{\pm}\nu & &+358 \text{ MeV} \quad 4.8\,\%
\end{aligned}$$

(the $\tau^{+}\pi^{-}$ and $\pi^{\pm}\pi^{0}\pi^{0}$ lines are bracketed together as τ-mode)

These are all weak decays but note that the τ- and θ-modes do not involve leptons. Also in these cases strangeness is violated as it changes by one unit.

The spin and parity of the K-meson turns out to be 0^{-} like the pion. The isospin however turns out to be $\tfrac{1}{2}$. Thus K^{+}, K^{-} must be antiparticles and K^{+}, K^{0} constitute an isospin doublet like proton and neutron. We also must assume K^{0} has an antiparticle $\overline{K}^{0}$ like the antineutron. The antikaons K^{-} and $\overline{K}^{0}$ must have strangeness -1 opposite to that for kaons if we assume baryon conservation in associated production.

The two neutral kaons K^{0} and $\overline{K}^{0}$ must differ only in strangeness as they have zero spin and zero charge. But in weak interactions we concluded for the charged K-decay that strangeness is violated. Hence if we produce some K^{0} particles and wait for a while we may expect to see a mixture of K^{0} and $\overline{K}^{0}$ result.

We have already referred in ch. 2 to the fact that in β-decay charge conjugation and parity are separately violated but that the product of the two operations leaves the system invariant. Thus for the weak decay of K^{0} and $\overline{K}^{0}$ CP is conserved, and we get two mixtures, two eigenstates of CP as follows

$$|K_{1}^{0}\rangle = \frac{1}{\sqrt{2}}\{|K^{0}\rangle+|\overline{K}^{0}\rangle\} \qquad CP \text{ even}$$

$$|K_{2}^{0}\rangle = \frac{1}{\sqrt{2}}\{|K^{0}\rangle-|\overline{K}^{0}\rangle\} \qquad CP \text{ odd.}$$

The pion is even under charge conjugation and odd parity hence it is odd under CP. Thus two pions are even and three pions are odd and we have decays of the mixtures

$$\begin{aligned}
K_{1}^{0} &\to 2\pi, & \tau &= 0.9 \times 10^{-10} \text{ sec} \\
K_{2}^{0} &\to 3\pi, & \tau &= 5.8 \times 10^{-8} \text{ sec.}
\end{aligned}$$

The longer lifetime of the K_{2}^{0}-decay reflects the more complicated decay mode. The fact that the lifetime differs means that in a beam of K^{0}-particles which can be considered as a 50–50 mixture of K_{1}^{0} and K_{2}^{0}, the K_{1}^{0}-particles will decay away quickly and after a few nanoseconds the beam will be pure K_{2}^{0} with about half the original intensity. This was observed by Fowler et al.* In the Berkeley proton

* W. B. Fowler, R. L. Lander and W. M. Powell, Phys. Rev. **113** (1959) 918.

synchrotron the 6.2 GeV internal proton beam struck a Be target

$$p + n \rightarrow p + p + \pi^-$$

producing π^--mesons. These entered a momentum analyser which selected 1.25 GeV/c pions. The analyser consisted of two quadrupoles and a bending magnet. The focussed beam of pions struck an aluminium target to produce reactions like

$$\pi^- + p \rightarrow K^0 + \Lambda^0.$$

The neutral particles travelled a distance of 9.3 feet during which a magnetic field swept out all charged particles. The neutral beam then entered a propane bubble chamber. During the traversal of the 9.3 feet all of the K_1^0 ($\tau = 10^{-10}$ sec) and the Λ^0 ($\tau = 2.51 \times 10^{-10}$ sec) will have decayed. So the beam entering the bubble chamber was pure K_2^0 since it has the longer lifetime. It contains some $\overline{K}^0$-particles with $S = -1$. So reactions such as

$$\overline{K}^0 + p \rightarrow \Lambda^0 + \pi^+$$
$$S = {-1} \quad 0 \quad -1 \quad 0$$

are seen, thus proving that starting out with K^0-particles ($S = +1$) we now have $\overline{K}^0$-particles since we see Λ^0 decaying into $p + \pi^-$ associated with a π^+. There may also be elastic scattering of the K^0 in the bubble chamber. This is a strong nuclear interaction, CP is conserved and the K_2^0-mixture will be upset so that we again have some K_1^0. So we see K_1^0 decays into 2π as well as the predominant K_2^0 decays into 3π.

In 1964 however very careful experiments were done by Christenson et al.* in which strong interactions such as scattering were minimised. They concluded that in a very few cases the K_2^0 beam decayed to *two* pions. The branching ratio was $2\pi/3\pi = (4.4 \pm 0.2) \times 10^{-3}$. This could be interpreted as an impurity in the CP eigenstate, in other words CP was not *exactly* conserved. If this is so then we may infer that time reversal invariance is also violated. An unusual feature is the rather small breakdown of the invariance. In contrast the C- and P-violation in β-decay is complete, that is the states are about 50–50 mixtures of the two parities for example. The small CP-violation is another unsolved problem of fundamental particle physics.

The decay modes of the neutral kaon may be summarised

$$K_1^0 \rightarrow \pi^+\pi^- \qquad 69.3\,\% \qquad \tau = 0.881 \times 10^{-10} \text{ sec}$$
$$ \rightarrow \pi^0\pi^0 \qquad 30.7\,\%$$

---

* J. H. Christenson, J. W. Cronin, V. L. Fitch and R. Turlay, Phys. Rev. Letters **13** (1964) 138.

$$K_2^0 \rightarrow \pi^0\pi^0\pi^0 \qquad 23.5\% \qquad \tau = 5.77 \times 10^{-8} \text{ sec}$$
$$\rightarrow \pi^+\pi^-\pi^0 \qquad 11.5\%$$
$$\rightarrow \pi\mu\nu \qquad 27.5\%$$
$$\rightarrow \pi e\nu \qquad 37.4\%$$
$$\rightarrow \pi^+\pi^- \qquad 0.15\%$$

8   The lambda hyperon

We now turn to some fundamental particles heavier than the nucleon which are normally referred to as hyperons. Together with the nucleon we refer to them as baryons. The strongly interacting particles lighter than the nucleon have been called mesons. We shall see that all the baryons have half-integral intrinsic spin, that is, are fermions whilst the mesons are bosons with integral spin. Some mesons heavier than the nucleon are now known, and can be recognised as they have integral spin and have baryon number 0. The whole class of mesons and baryons, that is strongly interacting particles are referred to as hadrons, to distinguish them from photons and leptons which do not have strong interactions.

We noted that a hyperon, the Λ^0, was associated with the K-meson in associated production. Its normal mode of decay is into a nucleon and a pion

$$\Lambda^0 \rightarrow \text{p} + \pi^- + 40 \text{ MeV}$$
$$\rightarrow \text{n} + \pi^0$$

with a lifetime of 2.51×10^{-10} sec characteristic of a weak decay. We do not see the track of a Λ^0 in a visual chamber but identify it by the characteristic V-shape of its charged decay product tracks. We note that the Λ^0 cannot interact strongly with nucleons by reactions like

$$\Lambda^0 + \text{p} \rightarrow \text{n} + \text{p}$$
$$S = -1 \quad 0 \quad \quad 0 \quad 0$$

as this violates strangeness. The mass of the Λ^0 is 1115.4 MeV. It has only one charge state, neutral, so it must have isospin 0. Its spin and parity are the same as for the nucleon, $\frac{1}{2}^+$. The branching for the two decay modes is

$$\Lambda^0 \rightarrow \text{p}\pi^- \qquad 66.3\%$$
$$\rightarrow \text{n}\pi^0 \qquad 33.7\%.$$

We note that $|\Delta S| = 1$ for this weak decay like the K-meson decay and that parity is violated. Furthermore we note that $|\Delta T_3| = \frac{1}{2}$. It has also been suggested that $|\Delta T|$ also is $\frac{1}{2}$ where T is the total isospin. We note that in the Λ^0 decay we can have two possible isospin values for the final state.

$$\Lambda^0 \to p + \pi^- \qquad \Lambda^0 \to n + \pi^0.$$

$$
\begin{array}{c|ccc|ccc}
T_3 & 0 & +\tfrac{1}{2} & -1 & 0 & -\tfrac{1}{2} & 0 \\
T & 0 & \tfrac{1}{2} & 1 & 0 & \tfrac{1}{2} & 1 \\
& & \underbrace{}_{\tfrac{1}{2}\ \text{or}\ \tfrac{3}{2}} & & & \underbrace{}_{\tfrac{1}{2}\ \text{or}\ \tfrac{3}{2}} &
\end{array}
$$

The probability ratio for the two possibilities will be simply the ratio of the Clebsch–Gordan coefficients squared which fix the probability of combining the two vectors in the two ways. These are written $(T'T''T_3'T_3''|T'''T_3''')$. For final isospin $\tfrac{3}{2}$ the ratio

$$\frac{p\pi^-}{n\pi^0} = \frac{(\tfrac{1}{2}\,1\,\tfrac{1}{2}\,-1|\tfrac{3}{2}\,-\tfrac{1}{2})^2}{(\tfrac{1}{2}\,1\,-\tfrac{1}{2}\,0|\tfrac{3}{2}\,-\tfrac{1}{2})^2} = \frac{(\sqrt{\tfrac{1}{3}})^2}{(\sqrt{\tfrac{2}{3}})^2} = \tfrac{1}{2}.$$

For final isospin $\tfrac{1}{2}$ the ratio

$$\frac{p\pi^-}{n\pi^0} = \frac{(\tfrac{1}{2}\,1\,\tfrac{1}{2}\,-1|\tfrac{1}{2}\,-\tfrac{1}{2})^2}{(\tfrac{1}{2}\,1\,-\tfrac{1}{2}\,0|\tfrac{1}{2}\,-\tfrac{1}{2})^2} = \frac{(\sqrt{\tfrac{2}{3}})^2}{(\sqrt{\tfrac{1}{3}})^2} = \tfrac{2}{1}.$$

We note that the experimental branching ratio is

$$\frac{p\pi^-}{n\pi^0} = \frac{66.3}{33.7} \approx 2.$$

Thus the final isospin appears to be $\tfrac{1}{2}$ not $\tfrac{3}{2}$ and $|\Delta T| = \tfrac{1}{2}$. This seems to be true for all strange particle decays where it can be checked.

9   *The sigma hyperon*

If we bombard nucleons with pions at somewhat higher energy than that required to produce Λ^0 (threshold 760 MeV) we see another hyperon produced above a threshold energy of 880 MeV. This is called the Σ-hyperon.

$$\pi^- + n \to \Sigma^- + K^0.$$

It is again produced along with a kaon so it must have $S = -1$. We observe it with three charge states Σ^+, Σ^0, Σ^-. It is a fermion with spin and parity $\tfrac{1}{2}^+$. The masses of the three charge states are

$$
\begin{array}{ll}
\Sigma^+ & 1189.4 \text{ MeV} \\
\Sigma^0 & 1192.3 \text{ MeV} \\
\Sigma^- & 1197.2 \text{ MeV}
\end{array}
$$

As the masses of Σ^+ and Σ^- are not identical we conclude that they are not anti-particles and that the isospin $T = 1$. There should then be three antiparticles, $\overline{\Sigma}^+$

with negative charge, $\overline{\Sigma}{}^0$ neutral and $\overline{\Sigma}{}^-$ with positive charge and these have been observed. The decay lifetimes of the charged sigmas are

$$\begin{aligned}
\Sigma^+ &\to p\pi^0 & 53\,\% & & \tau &= 0.810\times10^{-10}\ \text{sec} \\
&\to n\pi^+ & 47\,\% & & & \\
\Sigma^- &\to n\pi^- & 100\,\% & & \tau &= 1.65\times10^{-10}\ \text{sec.}
\end{aligned}$$

The decay of the Σ^+ is twice as fast since it has two decay channels available to it whilst Σ^- has only one.

The neutral sigma is now heavier than another particle of strangeness -1. So it can decay by a strangeness- and T_3-conserving interaction namely by electromagnetic or strong interaction. However it is only 77 MeV heavier than the Λ^0 so it has not enough energy to make a pion ($mc^2 = 139$ MeV). Hence it decays by an electromagnetic process

$$\Sigma^0 \to \Lambda^0 + \gamma$$

with a lifetime similar to the π^0, $\tau < 1.0\times10^{-14}$ sec. Note that isospin, T, is violated in this decay as the electromagnetic field is not charge independent!

10   *The xi hyperon*

If we now bombard nucleons with considerably higher energy pions above a threshold energy of 2.23 GeV we observe another particle in reactions such as

$$\pi + p \to \Xi^- + K^+ + K^0.$$

This particle is produced in association with *two* kaons of strangeness $+1$. So it must have strangeness -2. It occurs in two charge states Ξ^- and Ξ^0 and has isospin $\tfrac{1}{2}$. Its spin and parity are $\tfrac{1}{2}^+$. The masses of the two particles are Ξ^- 1320.8 MeV and Ξ^0 1314.3 MeV. The production threshold is higher since we have to produce *two* kaons. The decays are

$$\begin{aligned}
\Xi^- &\to \Lambda^0\pi^- & 100\,\% & & \tau &= 1.75\times10^{-10}\ \text{sec} \\
\Xi^0 &\to \Lambda^0\pi^0 & 100\,\% & & \tau &= 3.05\times10^{-10}\ \text{sec.}
\end{aligned}$$

The lifetimes indicate that these are weak decays and $|\Delta S| = 1$, $|\Delta T| = \tfrac{1}{2}$. Note that the Ξ^0 cannot decay by an electromagnetic decay to Λ^0 or Σ^0 as this would violate T_3 which must be conserved in electromagnetic processes.

11   *Charge conservation*

Let us now review some of the quantum numbers we have been using:

Conserved in Interaction?

Quantum number	Strong	Electromagnetic	Weak
Isospin T	yes	no	no (for non-leptonic decays $\|\Delta T\|=\tfrac{1}{2}$)
Isospin projection T_3	yes	yes	no (for non-leptonic decays $\|\Delta T_3\|=\tfrac{1}{2}$)
Baryon number B ($+1$, baryons -1, antibaryons 0, mesons)	yes	yes	yes
Charge Q	yes	yes	yes
Strangeness S (0, pions, nucleons $+1$, K-mesons -1, Λ, Σ -2, Ξ)	yes	yes	no (for non-leptonic decays $\|\Delta S\|=1$)

Our definition of T_3 for a nucleus was $\tfrac{1}{2}(Z-N)=\tfrac{1}{2}(2Z-A)$ or $Q-\tfrac{1}{2}B$ where there are A baryons. Thus the charge is $Q = T_3+\tfrac{1}{2}B$. Hence we have

$$
\begin{aligned}
\text{for pions} \quad \tfrac{1}{2}B=0 \qquad Q=T_3 &= +1 \quad \text{for} \quad \pi^+ \\
&= 0 \quad \text{for} \quad \pi^0 \\
&= -1 \quad \text{for} \quad \pi^-
\end{aligned}
$$

$$
\begin{aligned}
\text{for nucleons} \quad \tfrac{1}{2}B=\tfrac{1}{2} \qquad & Q=T_3+\tfrac{1}{2} \\
\text{for proton} \qquad & T_3=+\tfrac{1}{2} \qquad Q=1 \\
\text{neutron} \qquad & T_3=-\tfrac{1}{2} \qquad Q=0
\end{aligned}
$$

$$
\begin{aligned}
\text{for antinucleons} \quad \tfrac{1}{2}B=-\tfrac{1}{2} \qquad & Q=T_3-\tfrac{1}{2} \\
\text{for antiproton} \qquad & T_3=-\tfrac{1}{2} \qquad Q=-1 \\
\text{antineutron} \qquad & T_3=+\tfrac{1}{2} \qquad Q=0
\end{aligned}
$$

But now for strange particles this scheme does not work. Consider the Λ^0 with $B=1$ and $T_3=0$. By our rule above $Q=\tfrac{1}{2}$ which is not correct. The charge must be connected with strangeness also

$$
Q = T_3+\tfrac{1}{2}B+\tfrac{1}{2}S.
$$

Since B and S go together it is usual to lump these two together into a new quantum number Y, called the hypercharge. This then has values $+1$ for nucleons and kaons, 0 for Λ, Σ and pions, and -1 for Ξ. Since the Y operation is equivalent to an axis inversion in isospin space it might have been more suggestive to call it isoparity. Our charge formula is now $Q = T_3 + \tfrac{1}{2}Y$. Y is also twice the average charge of a charge multiplet. Then we can assign quantum numbers as in table 12.2.

TABLE 12.2

Particle	T_3	B	S	Q	Y	Antiparticle	T_3	B	S	Q	Y
Λ^0	0	1	-1	0	0	$\overline{\Lambda}^0$	0	-1	$+1$	0	0
Σ^+	$+1$	1	-1	1	0	$\overline{\Sigma}^+$	-1	-1	$+1$	-1	0
Σ^0	0	1	-1	0	0	$\overline{\Sigma}^0$	0	-1	$+1$	0	0
Σ^-	-1	1	-1	-1	0	$\overline{\Sigma}^-$	$+1$	-1	$+1$	$+1$	0
Ξ^-	$-\tfrac{1}{2}$	1	-2	-1	-1	$\overline{\Xi}^-$	$+\tfrac{1}{2}$	-1	$+2$	$+1$	$+1$
Ξ^0	$+\tfrac{1}{2}$	1	-2	0	-1	$\overline{\Xi}^0$	$-\tfrac{1}{2}$	-1	$+2$	0	$+1$
n	$-\tfrac{1}{2}$	1	0	0	$+1$	$\overline{\text{n}}$	$+\tfrac{1}{2}$	-1	0	0	-1
p	$+\tfrac{1}{2}$	1	0	$+1$	$+1$	$\overline{\text{p}}$	$-\tfrac{1}{2}$	-1	0	-1	-1

12 *Pion–nucleon resonances*

We now consider the interactions of the various particles as a function of energy. If we look at the total cross section for pions bombarding nucleons we see the curve of fig. 12.7. The curve for $\pi^- + \text{n}$ resembles that for $\pi^+ + \text{p}$ and the curve for $\pi^+ + \text{n}$ resembles that for $\pi^- + \text{p}$. We see resonances in the cross section and just as in low energy nuclear physics we interpret them as states in the (πp) system with lifetimes long compared to the transit time of the pion past the proton ($\approx 4 \times 10^{-24}$ sec). The width of the 200 MeV resonance is about 120 MeV which corresponds to a lifetime of 5.5×10^{-24} sec. We are thus observing short-lived excited states of the nucleon. Let us examine the isospin involved. The cross section for forming a compound state of isospin T''', T'''_3 from initial particles of isospins T', T'_3 and T'', T''_3 respectively is just proportional to the square of the Clebsch–Gordan coefficient $(T'\ T''\ T'_3\ T''_3 | T'''\ T'''_3)$. Thus for $\pi^+ + \text{p}$ forming a state of $T = \tfrac{3}{2}$

$$\sigma_{\tfrac{3}{2}} \propto (1\ \tfrac{1}{2}\ 1\ \tfrac{1}{2} | \tfrac{3}{2}\ \tfrac{3}{2})^2 = 1$$

and since $T'''_3 = \tfrac{3}{2}$ we cannot form a state of $T = \tfrac{1}{2}$ [since $(1\ \tfrac{1}{2}\ 1\ \tfrac{1}{2} | \tfrac{1}{2}\ \tfrac{3}{2})^2 = 0$]. Similarly for $\pi^- + \text{p}$ forming a state of $T = \tfrac{3}{2}$

$$\sigma_{\tfrac{3}{2}} \propto (1\ \tfrac{1}{2}\ -1\ \tfrac{1}{2} | \tfrac{3}{2}\ -\tfrac{1}{2})^2 = \tfrac{1}{3}$$

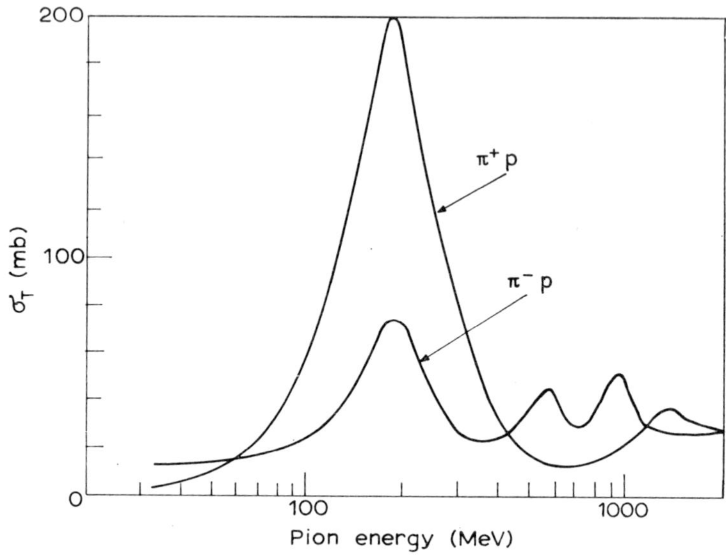

S. de Benedetti, Nuclear Interactions (Wiley, New York 1964) p. 455

Fig. 12.7.   Pion–nucleon cross section

and for forming a state of $T=\tfrac{1}{2}$

$$\sigma_{\frac{1}{2}} \propto (1\,\tfrac{1}{2}\,-1\,\tfrac{1}{2}|\tfrac{1}{2}\,-\tfrac{1}{2})^2 = \tfrac{2}{3}.$$

Now since we see the states at 200 MeV and 1400 MeV in both the π^+ and π^- reactions we conclude that they are at least partly $T=\tfrac{3}{2}$. The fact that at the 200 MeV resonance we observe a cross section ratio

$$\frac{\sigma(\pi^+\mathrm{p})}{\sigma(\pi^-\mathrm{p})} = \frac{200}{70} \approx 3$$

indicates that this state is pure $T=\tfrac{3}{2}$. Since this state has $T=\tfrac{3}{2}$ the charge multiplicity must be 4. These are (designating the intermediate state as Δ)

$$\pi^- + \mathrm{n} \to \Delta^-$$
$$\pi^- + \mathrm{p} \to \Delta^0$$
$$\pi^+ + \mathrm{n} \to \Delta^+$$
$$\pi^+ + \mathrm{p} \to \Delta^{++}.$$

So in this case we have a state with charge 2.

From our scattering theory of ch. 7 § 4 the maximum elastic scattering cross section was $4\pi\lambdabar^2 g$ where $\lambdabar$ is the pion wavelength and

$$g = \frac{2J+1}{(2I+1)(2s+1)}$$

where J is the compound state spin, I ($=\frac{1}{2}$ here) is the target spin and s ($=0$ here) is the bombarding particle spin. Then $\sigma_{max}=\frac{1}{2}(2J+1)\pi\lambda^2$. The peak cross section in $\pi^+ + p$ is close to $8\pi\lambda^2$ at the 200 MeV resonances indicating an angular momentum $J=\frac{3}{2}$. The angular momentum cannot be zero since at low energies the cross section is energy independent for s-interaction. Hence it must be in a p-state as indicated by the sharp rise from low momenta. If the scattering is in a p-state the parity of the resonance must be positive since pion and nucleon have opposite intrinsic parity. The angular distribution of the scattering has the form $1+3\cos^2\phi$ also indicating p-wave scattering. This state is sometimes designated $N^*_{\frac{3}{2}}$, the subscript indicating isospin. We prefer the alternative designation $\varDelta$ where this means the excited nucleon ($B=1$, $S=0$) with isospin $\frac{3}{2}$. The state is also labelled by its mass in MeV, thus $\varDelta(1236)$ since its mass is 1236 MeV.

The low energy pion–nucleon scattering is thus dominated by this resonance and we can treat the problem in the framework of the effective range theory in exactly the way the low energy nucleon–nucleon scattering was treated. We can then try to relate the scattering length to the nuclear force problem. We treat this problem in more detail in ch. 15 § 6.

Careful work has shown that the 600 and 900 MeV resonances seen in the $\pi^- + p$ total cross section have isospin $\frac{1}{2}$ and spin parity $\frac{3}{2}^-$ and $\frac{5}{2}^+$ respectively that is, they are $d_{\frac{3}{2}}$ and $f_{\frac{5}{2}}$ resonances in the pion–nucleon system. We label them N(1518) and N(1688) respectively. The 1400 MeV resonance seen in $\pi^+ + p$ has isospin $\frac{3}{2}$ and spin-parity $\frac{7}{2}^+$, an $f_{\frac{7}{2}}$ resonance labelled $\varDelta(1924)$. We label the isospin $\frac{1}{2}$ states N meaning an excited nucleon ($B=1$, $S=0$) with isospin $\frac{1}{2}$ like the nucleon [N(938)]. The energy widths of these resonances are as follows

Resonance	Width
$\varDelta(1236)$	120 MeV
N(1518)	120 MeV
N(1688)	100 MeV
$\varDelta(1924)$	200 MeV

We now must consider the semantic problem, when do we call a fundamental entity a particle and when do we call it a resonance? The reason we can observe the state $\varDelta(1238)$ as a resonance is simply that its mass and quantum numbers are such that it can decay by a *strong* interaction in a time like 10^{-24} sec. The Σ^0 cannot decay by a strong interaction to $n+\pi^0$ as strangeness would be violated or to $\varLambda^0+\pi^0$ as there is insufficient energy, so it decays by an electromagnetic interaction in a time like 10^{-14} sec. The $\varLambda^0$ cannot decay by a *strong* interaction to $n+\pi^0$ as this would violate strangeness and it cannot decay by an electromagnetic transition as this would also violate strangeness, so it decays by a weak interaction in a time like 10^{-10} sec. The neutron also decays to the proton by a weak interaction in a time like 10^{-3} sec since it is heavier. Thus the *only* truly stable state is the proton (lifetime $>6 \times 10^{27}$ y).

The situation is exactly comparable to the situation in nuclear physics for a nucleus like ^{7}Be whose decay scheme is shown in fig. 12.8. The state at 4.55 MeV excitation can energetically decay by a strong interaction to ^{3}He$+^4$He and does

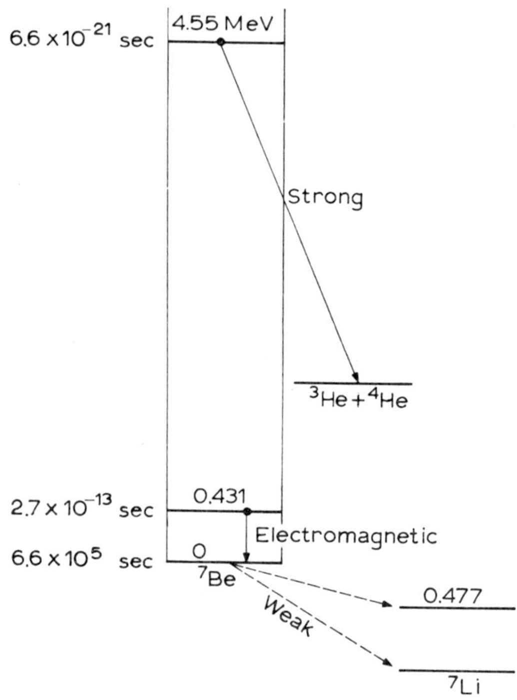

Fig. 12.8.   Decay modes of a nucleus, ^{7}Be

so with $\tau = 6.6 \times 10^{-21}$ sec. The state at 0.431 MeV cannot decay by a strong interaction so it decays by electromagnetic radiation to the ground state with $\tau = 2.7 \times 10^{-13}$ sec. The ground state cannot decay by a strong or electromagnetic interaction and conserve charge, so it decays by a weak interaction (K capture) with $\tau = 6.6 \times 10^5$ sec. These are lifetimes for three ^{7}Be 'particles'! Thus the distinction between particles like N(938), Λ(1115), Σ(1189) and Ξ(1320) and resonances like Δ(1236) or N(1518) is simply one of stability and it is more reasonable to call them all fundamental particles having different decay modes and lifetimes.

13   Meson resonances

If we bombard nucleons with pions of energy greater than about 171 MeV, in addition to elastic scattering we may also see an inelastic process

$$\pi + N \rightarrow N + \pi + \pi - 140 \text{ MeV.}$$

This is a three-body reaction and we can analyse the results in different ways. For example we could study the mass spectrum of coincident pions and nucleons or we could study the mass spectrum of coincident pions. If it is a completely prompt three-particle break-up as we indicate above we will see a mass distribution extending from zero to the maximum available which is controlled only by the momentum space available to the third unobserved particle. On the other hand if the reaction proceeds sequentially through intermediate relatively long lived combinations of particles such as

$$\pi + N \rightarrow (N\pi) + \pi$$
$$(N\pi) \rightarrow N + \pi$$

or

$$\pi + N \rightarrow N + (\pi\pi)$$
$$(\pi\pi) \rightarrow \pi + \pi$$

we will see a peak in the mass spectrum at the mass corresponding to the mass of the 'resonance' and with a width corresponding to the lifetime of the intermediate state.

In fact such final state interactions are seen. The (πN) states correspond exactly to the ones found in the $\pi + N$ scattering. The $(\pi\pi)$ states which are also seen are new since we cannot do pion–pion scattering as we have no pion targets. The situation is similar for neutrons. We cannot do neutron–neutron scattering but must study the neutron–neutron interaction as a final state interaction in such reactions as $n + p \rightarrow n + n + p$.

We can search for $(\pi\pi)$ resonances in four reactions

$$\pi^+ + p \rightarrow p + \pi^+ + \pi^0$$
$$\rightarrow n + \pi^+ + \pi^+$$
$$\pi^- + p \rightarrow p + \pi^- + \pi^0$$
$$\rightarrow n + \pi^- + \pi^+.$$

The solid-line 'background' curve of fig. 12.9 is simply the distribution expected for simultaneous break-up. The resonance superimposed in three cases is a new particle called the ϱ-meson. It is a meson as it is made up of two other mesons. Its mass is 765 MeV. It obviously decays into two pions. Its width is 124 MeV. It exists in three charge states, $+$, $-$ and neutral so the isospin is 1. A study of the

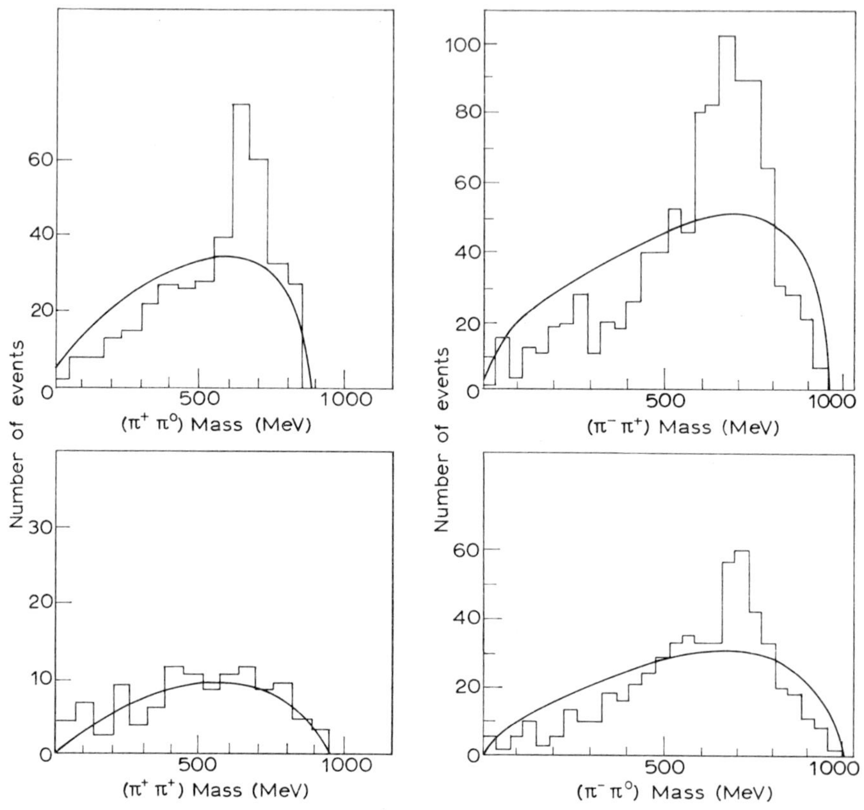

S. de Benedetti, Nuclear Interactions (Wiley, New York 1964) p. 506, 507

Fig. 12.9.   Evidence for the pion resonance

angular distribution of the two pions indicates that $J=1$. Since a two-pion system has zero spin and positive parity, the two pions in this case must be in a p-state with negative parity. Thus $J^P=1^-$.

Other resonances can be observed in reactions like

$$\pi^+ + D \rightarrow p + p + (\text{resonance})$$
$$(\text{resonance}) \rightarrow \pi^+ + \pi^- + \pi^0.$$

All charged particles are observed and their momentum measured. The energies and directions of the other particles completely determine the energy and direction of the π^0. We find two resonances in this case, one of mass 783 MeV and the other of mass 549 MeV. By other experiments we determine that they only occur in the neutral charge state hence we infer $T=0$ in both cases.

The heavier meson is called the ω-meson. It has a width of 12 MeV for decay into 3 pions; it is much longer lived than the ϱ-meson because of the more complicated decay. Angular distribution measurements on the decay products show that it has $J^P = 1^-$. It belongs to a new class of meson as it has different isospin from the pion.

The lighter meson is called the η-meson. It has a decay width of less than 10 keV. In only 27 % of the cases does it decay into two charged pions. Its decay modes are as follows

$$
\begin{aligned}
\eta &\to \gamma + \gamma & 31\,\% \\
&\to \gamma + \gamma + \pi^0 & 21\,\% \\
&\to \pi^0 + \pi^0 + \pi^0 & 21\,\% \\
&\to \pi^+ + \pi^- + \pi^0 & 22\,\% \\
&\to \pi^+ + \pi^- + \gamma & 5\,\%
\end{aligned}
$$

The 2γ decay mode tells us that the spin of η must be $J=0$, 2 or more. There is no anisotropy observed in the decay products, hence $J=0$. As we do not observe a 2π decay we conclude the η has odd parity.

Thus the η has the same spin and parity as the π-meson, $J^P = 0^-$. We now consider the problem of how the relatively weak electromagnetic decay mode can compete with the strong decay modes. The lifetime as indicated by the width is greater than 6×10^{-20} sec, rather slower than other strong interaction decays. Let us consider two operators, 1) charge conjugation C which changes a particle into its antiparticle, 2) rotation in isospin space $R_2(\pi)$, about the 2-axis say, which reverses the sign of T_3 and hence of the charge, the charge symmetry operator. We note that although particle wavefunctions may not be invariant under C or under $R_2(\pi)$ separately, they may be invariant under the product of these two operations. We call this operator G or G-parity. For the pion G is odd [C even, $R_2(\pi)$ odd], hence for three pions G is also odd. If we now assume that G is even for η [C even, $R_2(\pi)$ even], the decay into 3 pions violates a selection rule if G must be conserved. Decay into two photons is allowed by this rule since for the photon C is odd and $R_2(\pi)$ is not a good quantum number (no charge symmetry). Hence the photon decay can compete successfully with the G-violating 3π decay. Why is 3π decay observed at all? It is thought that it must also come about from a virtual electromagnetic transition.

By similar methods some 15 more mesons have been identified. Two of special interest are

$$\Phi, \qquad \text{mass } 1020 \text{ MeV}, \qquad T = 0, \qquad J^P = 1^-$$

and

$$X^0, \qquad \text{mass } 959 \text{ MeV}, \qquad T = 0, \qquad J^P = 0^-$$

which we shall be discussing later. Several excited states of the K-meson are also seen. Of special interest is K*(891), $T = \tfrac{1}{2}$, $J^P = 1^-$.

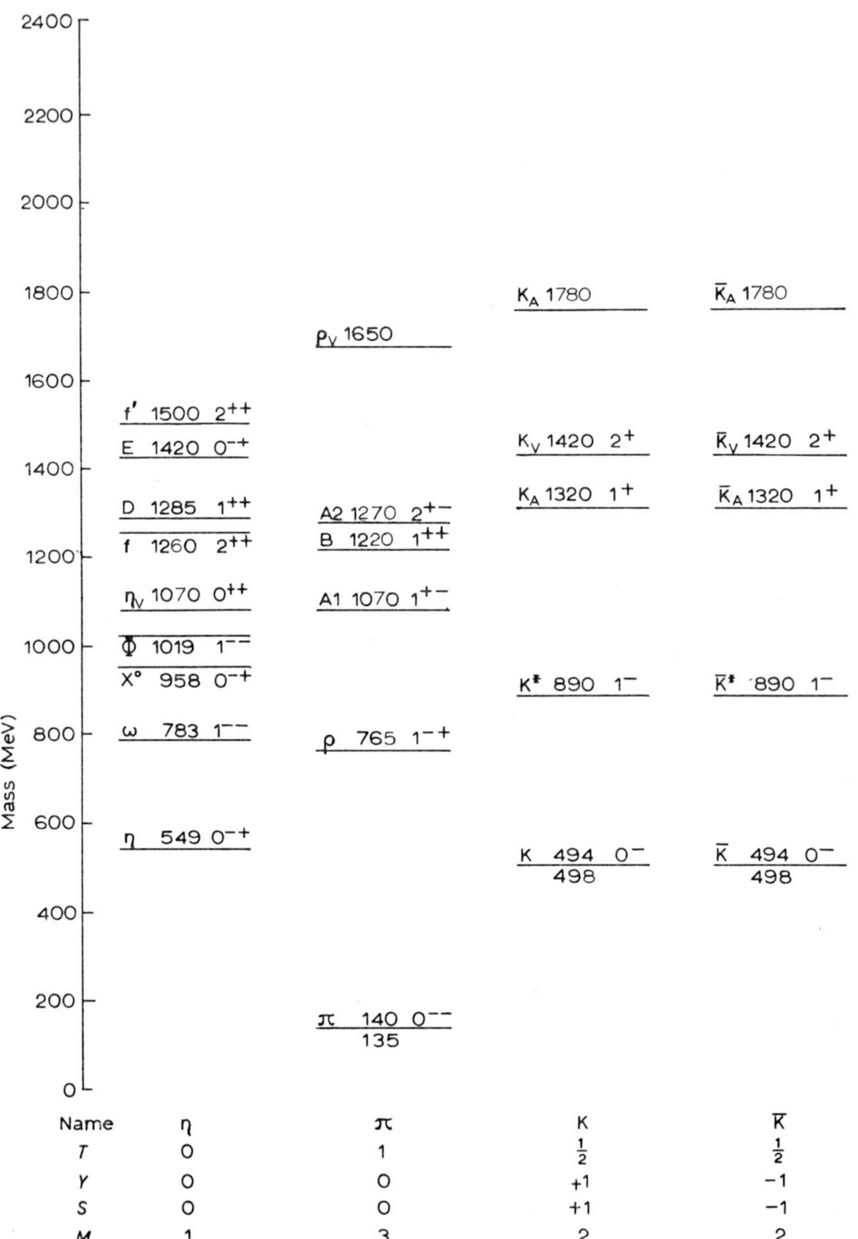

Fig. 12.10.   Mesons

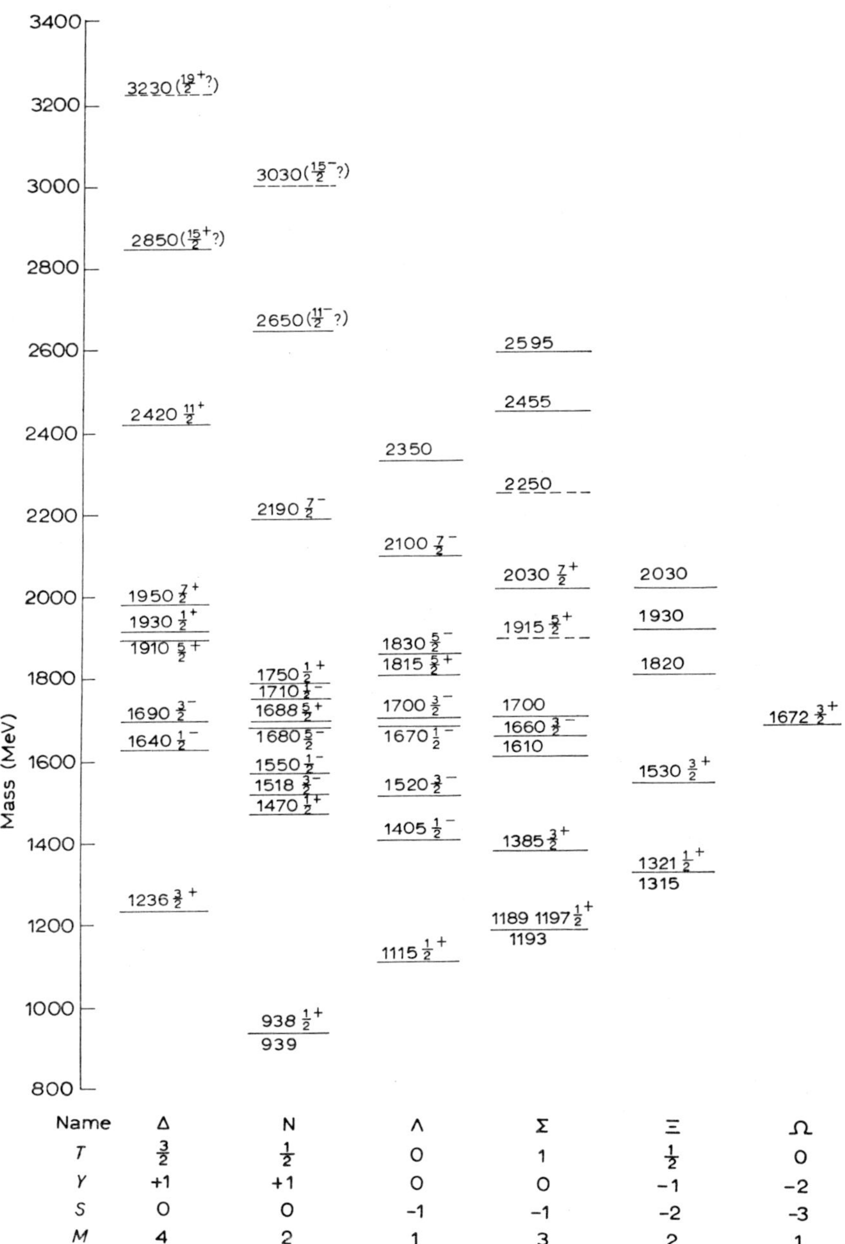

Fig. 12.11.   Baryons

14   More baryons

We have already discussed excited baryon states in the scattering of pions by nucleons. We see evidence for these as we have noted also in final state interactions and many more species are found. For example Curtis et al.* studied the reaction $\pi^- + p \rightarrow K^0 + R^0$ where R^0 denotes some neutral baryon state. The mass spectrum for R^0 showed three peaks, one at $7.99 m_\pi$ (1115 MeV) corresponding to a Λ^0, a second at $8.55 m_\pi$ (1189 MeV) corresponding to Σ^0; a third at $9.92 m_\pi$ (1383 MeV) corresponds to a new particle. It must have strangeness -1 like Σ^0 and Λ^0. It turns out to have isospin 1 and $J^P = \frac{3}{2}^+$. Thus it is an excited state of the Σ.

By similar methods many baryon states have been identified the total list being at present about 28.

We can display as level diagrams the particles currently identified. Fig. 12.10 shows the mesons while fig. 12.11 shows the baryons. A third figure identical to the latter would show the antibaryons. The grand total of fundamental particles thus stands at 76 if we count the antibaryons or at 176 if we count all possible charge states as well. In the meson table the antikaon levels have the same masses and J^P values as the kaons (by the *CPT* rule).

15   Antinucleons

We have mentioned antinucleons from time to time. We should now discuss their properties in more detail. From our general ideas about antiparticles we expect that the antiproton will have identical mass to the proton, one unit of *negative* charge, isospin $T = \frac{1}{2}$, $T_3 = -\frac{1}{2}$, spin $\frac{1}{2}$, parity opposite to that of the proton i.e. negative according to our convention, the spin and magnetic moment oppositely directed so that the magnetic moment is -2.793 n.m. The antineutron will have identical mass to the neutron, will be neutral with isospin $T = \frac{1}{2}$, $T_3 = +\frac{1}{2}$, spin $\frac{1}{2}$, parity negative and magnetic moment reversed, that is, $+1.913$ n.m. Like their antiparticles the antiproton will be stable while the antineutron will try to decay to the antiproton. However in this case there is not sufficient energy difference to create a positron (energy available 0.78 MeV, positron requires at least 1.02 MeV). So decay would depend on there being an electron close enough to be captured.

Of course the ultimate fate of antinucleons is to annihilate with their corresponding nucleon since our world is well stocked with the latter. Because of the strength of the force we see very many field particles, in other words multiple meson production is observed.

* L. J. Curtis, C. T. Coffin, D. I. Meyer and K. M. Terwilliger, Phys. Rev. **132** (1963) 1771.

In order to produce antinucleons we must supply enough energy to produce a nucleon–antinucleon *pair*. Thus we must supply twice the nucleon rest mass (2×939 MeV $= 1.98$ GeV) in the centre of mass system.

The corresponding laboratory threshold energy is about 5.6 GeV for a reaction like $p + p \rightarrow p + p + p + \bar{p}$. The Berkeley proton synchrotron with a design energy of 6 GeV was the first accelerator able to produce antinucleons. Antiprotons were identified by the Berkeley group* as illustrated in fig. 12.12. A time-of-flight measurement between scintillation counters S1 and S2 was performed. The time

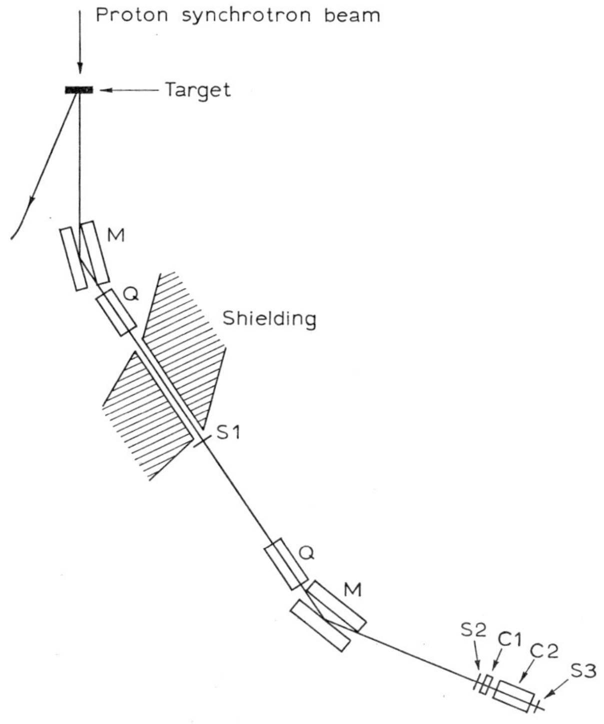

Phys. Rev. **100** (1956) 947

Fig. 12.12.   The detection of antiprotons. M, analysing magnets; Q, quadrupole focussing magnets; S1, S2, S3, scintillation counters; and C1, C2, Cerenkov counters

over the 12 metre distance is 40 nsec for π^- and 51 nsec for $\bar{p}$, C1 is a threshold Cerenkov counter to detect only charged particles with $v/c > 0.79$. The $\bar{p}$ of specified momentum have $v/c \approx 0.765$ so they do *not* trigger C1 while π^- have $v/c > 0.79$ and do so. C2 is a differential Cerenkov counter which detects particles with v/c

* O. Chamberlain, E. Segre, C. Weigand and T. Ypsilantis, Phys. Rev. **100** (1956) 947.

between 0.75 and 0.78. Finally S3 is a scintillation counter to show that $\bar{p}$ passed through C2. So a coincidence S1, delay, S2 C2, S3 not C1 specified an antiproton of momentum 1.19 GeV/c emerging from S3. The certified beam of antiprotons (some 300–600/hour in this experiment) then passed into an emulsion or bubble chamber where annihilation would take place.

$$\bar{p}+p\rightarrow(5 \text{ to } 10 \text{ pions})+ \text{ a few nucleons}+4\,\% \text{ KK}.$$

The pions are likely the primary product but some are reabsorbed in the same nucleus to produce various reactions.

Antineutrons were identified in a related experiment.* They were produced by a charge exchange collision of an antiproton $\bar{p}+p\rightarrow\bar{n}+n$. We start with the antiprotons emerging from S3 (fig. 12.13). LS is a large liquid scintillator and C3 is a

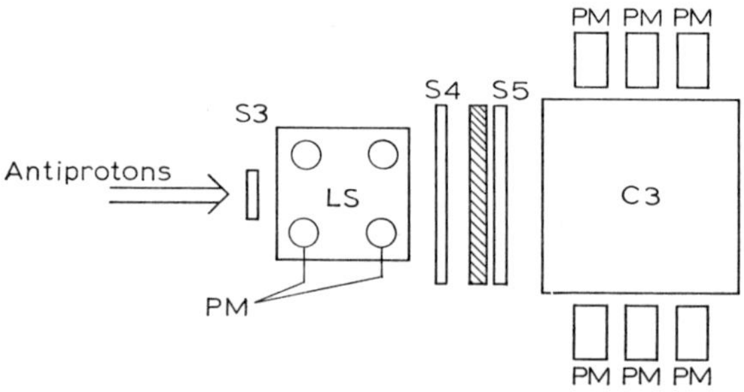

Phys. Rev. **104** (1956) 1193

Fig. 12.13.   The detection of antineutrons. S3, S4, S5, scintillation counters; LS, liquid scintillator; C3, Cerenkov counter; PM, photomultipliers

lead glass Cerenkov detector. The antiproton enters LS and either annihilates giving a large pulse corresponding to a fraction of 1.88 GeV or it exchanges charge with a neutron to give $\bar{n}$ which gives a pulse of <100 MeV. The $\bar{n}$ must now pass S4 and S5 without counting. S4 will detect charged particles from annihilation escaping from LS. The lead block will convert high energy γ-rays from the annihilations to charged particles which cause S5 to count. Finally the antineutron will enter C3 and annihilate. Annihilation pulses in C3 not accompanied by S4, S5 or large pulses in LS correspond to antineutrons.

---

* O. Chamberlain et al., Phys. Rev. **104** (1956) 1193.

13

Fundamental Particles:

Classification

1 Introduction

Many attempts have been made to understand the order underlying this wealth of mesons and baryons. One can start by grouping the particles according to various quantum numbers and energy. For example in figs. 12.10 and 12.11 we have grouped particles according to values of Y and T and shown the various energy states with different J^P as a level sequence. Alternatively we could have grouped states of the same J^P and displayed them as a function of Y and T. We seek to understand how the rest energies change with quantum number.

2 Regge recurrences

This approach looks for dynamic effects which can cause particles with the same Y and T to have masses which depend on the J^P value. In other words within a given system the quantisation of angular momentum may result in a series of states of different total energy and J but otherwise similar properties. An analogy would be the rotational states in nuclei and molecules wherein the quantised rotational angular momentum gives rise to levels whose energy depends upon J only. For baryons, Regge suggested that states differing by two units of angular momentum should occur approximately equally spaced in rest mass. That is, states with J, $J+2$, $J+4$ etc. should be equally spaced. In some cases this seems to be roughly satisfied. For example, the states of Δ with masses $1236(\tfrac{3}{2}^+)$, $1920(\tfrac{7}{2}^+)$,

391

$2420(\frac{11}{2}{}^{+})$, $2850(\frac{15}{2}{}^{+})$ and $3230(\frac{19}{2})^{+}$ show mass differences of 665, 500, 430 and 380 MeV. The states of N with masses $1525(\frac{3}{2}{}^{-})$, $2190(\frac{7}{2}{}^{-})$, $2650(\frac{11}{2}{}^{-})$ and $3030(\frac{15}{2}{}^{-})$ show mass differences of 665, 460 and 380 MeV. A few other Regge recurrences may be identified. The Regge theory also makes predictions about baryon and pion scattering but here the theory does not work so well. Thus the exact significance of the recurrences is still in some doubt.

3 Symmetry classification: baryons

Another approach is at present more successful. In this case we group together particles with the same spin and parity, that is, the same 'mechanical' properties but different values of the Y and T quantum numbers and seek approximate degeneracies indicating approximate symmetry. Since the masses are not exactly the same we assume the multiplet is split by some interaction into different substates labelled with different values of Y and T. The symmetry is broken to give the differing observed masses.

The first obvious set we consider in this way are the lowest mass baryons with $J^{P}=\frac{1}{2}{}^{+}$. They do not differ in mass very much ($\pm 20\%$). There are 8 such particles with various values of Y and T_{3} and we can display them as shown in fig. 13.1. Without having to make detailed assumptions about the symmetry or the

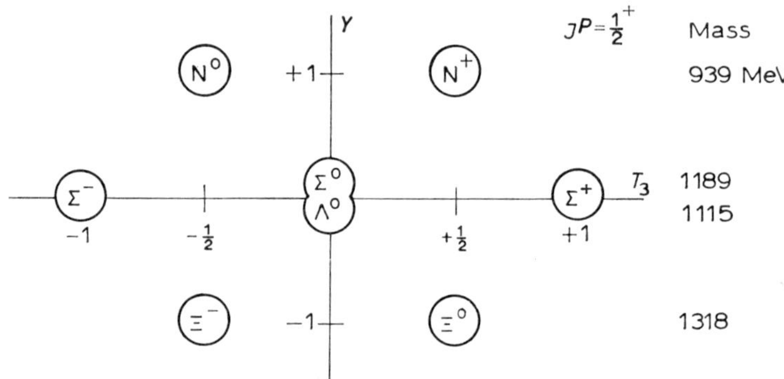

Fig. 13.1.   The baryon octet for $J^{P}=\frac{1}{2}{}^{+}$

force which breaks it, a relation between the masses of the octet can be derived as follows

$$\tfrac{1}{2}M_{\mathrm{N}}+\tfrac{1}{2}M_{\Xi} = \tfrac{3}{4}M_{\Lambda}+\tfrac{1}{4}M_{\Sigma}.$$

Substituting MeV values

$$470+ 660 = 836+298$$
$$1130 = 1134.$$

The agreement is remarkable. We ignore in this the small differences in mass between different charge states having the same value of Y. These are well explained as electromagnetic effects resulting from the different charges.

This is the only complete octet that we can identify at present amongst the known baryons.* Portions of octets with $J^P = \frac{1}{2}^-$ (N1570, Λ1670), with $J^P = \frac{3}{2}^-$ (N1525, Λ1700, Σ1660), with $J^P = \frac{5}{2}^-$ (N1670, Σ1770) and with $J^P = \frac{5}{2}^+$ (N1688, Λ1820) may have been identified. The states Λ1405, Λ1520 and Λ2100 are thought to be singlet states with $J^P = \frac{1}{2}^-$, $\frac{3}{2}^-$ and $\frac{7}{2}^-$ respectively.

We have not yet considered the lowest Δ state with mass 1236 and $J^P = \frac{3}{2}^+$. As this has $T = \frac{3}{2}$, it is a charge quartet and cannot be included in an octet. The lowest multiplet which could include it would be a ten-fold decuplet. In this case the masses in the broken decuplet are expected to be equally spaced (cf. fig. 13.2).

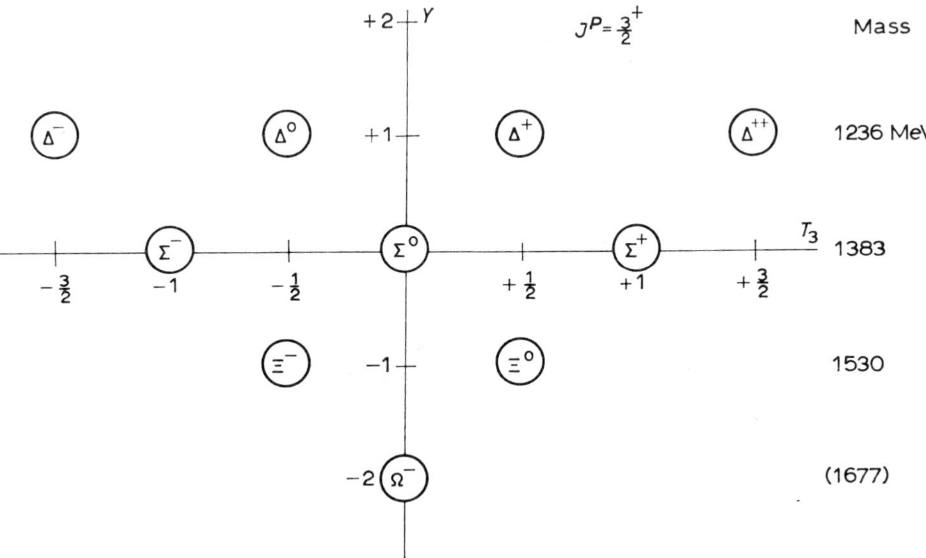

Fig. 13.2.   The baryon decuplet for $J^P = \frac{3}{2}^+$

At the time this classification of $J^P = \frac{3}{2}^+$ states was proposed, the Ω^- particle had not been seen. Its properties could be predicted in detail. It should have mass 1677 MeV, $J^P = \frac{3}{2}^+$, $T = 0$, $T_3 = 0$, $Y = -2$. Since it is a baryon $B = +1$, hence $S = -3$ and it will have one charge state of charge $Q = T_3 + \frac{1}{2}Y = -1$. A search was made for this particle and it was found with mass 1674 ± 3 MeV, a brilliant success for this classification scheme.

* R. D. Tripp et al., Nucl. Phys. **B3** (1967) 10.

A possible reaction for producing the Ω^- is

$$
\begin{array}{cccccc}
 & \mathrm{K}^- + \mathrm{p} \to & \Omega^- & + \mathrm{K}^+ & + \mathrm{K}^0 & -1.235\ \mathrm{GeV} \\
T_3 & -\tfrac{1}{2} \quad +\tfrac{1}{2} & 0 & +\tfrac{1}{2} & -\tfrac{1}{2} \\
Y & -1 \quad +1 & -2 & +1 & -1 \\
J^P & 0^- \quad \tfrac{1}{2}^+ & \tfrac{3}{2}^+ & 0^- & 0^- \\
l & \quad 0 & & 1 &
\end{array}
$$

The Ω^- decays as follows with a lifetime 1.5×10^{-10} sec:

$$
\begin{aligned}
\Omega^- &\to \Xi^- + \pi^0 \\
&\to \Xi^0 + \pi^- \\
&\to \Lambda^0 + \mathrm{K}^-.
\end{aligned}
$$

No other decuplets are seen amongst the baryons so far. The Regge recurrence states with $J^P = \tfrac{7}{2}^+$, $\Delta(1920)$ and $\Sigma(2035)$ have a mass difference (135 MeV) close to that observed in the $\tfrac{3}{2}^+$ decuplet. To complete the set a Ξ at mass 2170 and a Ω at 2305 would be required. These would require higher energies and intensities than are available with present accelerators.

4   Symmetry classification: mesons

No Regge recurrences have been positively identified amongst the mesons where they are expected to be equally spaced in mass squared. A pion recurrence with $J^P = 2^-$ would be expected at ≈ 1700 MeV. The resonance $\pi(1640)$ may be a candidate.

The symmetry classification has been very successful however. The same mass formulae are expected to hold for octets etc., except that now the mass squared should be used.

The mesons with $J^P = 0^-$ are $\pi(138)$, $\mathrm{K}(496)$, $\mathrm{K}(496)$, $\eta(549)$ and $\mathrm{X}^0(958)$, and these comprise altogether 9 mesons (fig. 13.3). The η and the X^0 both have $T_3 = 0$, $Y = 0$. The interpretation is that one is the member of the octet and the other is a singlet which, we shall see, we expect theoretically. Since the two have identical quantum numbers they will mix causing the two to be pushed apart in mass. This means the η will become lighter and the X^0 heavier.

The mass predicted for the η by the mass formula

$$
\tfrac{1}{2}M_\mathrm{K}^2 + \tfrac{1}{2}M_\mathrm{\bar K}^2 = \tfrac{3}{4}M_\eta^2 + \tfrac{1}{4}M_\pi^2
$$

is 565 MeV. We conclude that mixing with the X^0 caused the η mass to shift down by 16 MeV. The X^0 was presumably shifted up to mass 958 from mass 942 MeV.

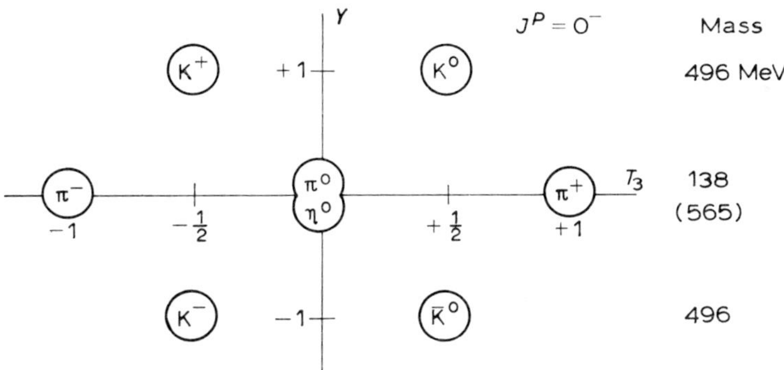

Fig. 13.3.   The meson octet for $J^P = 0^-$

Similarly we find there are nine $J^P = 1^-$ mesons, K*(890) and its antiparticles, the $\varrho(760)$ and two $T_3 = 0$, $Y = 0$ mesons: $\omega(783)$ and $\phi(1019)$, cf. fig. 13.4. Here the predicted position of the member of the octet is 930 MeV. In this case the unperturbed position of the singlet was at 873 MeV, and the mixing is rather complete.

A third nonet of $J^P = 2^+$ is also recognised composed of K(1420), A2(1300), f(1250) and f'(1500). The predicted position of the octet $T_3 = 0$, $Y = 0$ meson is 1440 MeV. The unperturbed position of the singlet would then be 1313 MeV and the mixing is again rather strong.

No other complete multiplets of mesons are seen although members of nonets of $J^P = 0^+$ and 1^+ may have been recognised. Thus this scheme has also been very successful in classifying the mesons.

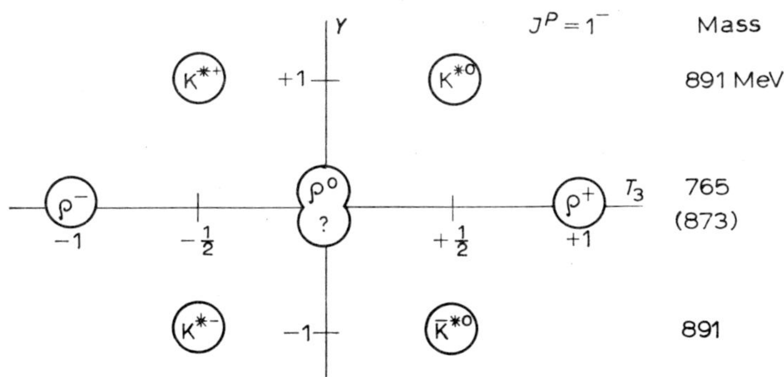

Fig. 13.4.   The meson octet for $J^P = 1^-$

5   *Conservation laws, invariance and symmetry groups*

It seems clear from the foregoing that multiplets are seen amongst the fundamental particles, each multiplet having a given value of spin and parity and the individual members having masses consistent with a symmetry breaking situation. We now want to discover what the existence of such multiplets can tell us about the basic physics involved.

It will be desirable to discuss the concepts of symmetry invariance which the multiplets imply and the conservation laws which may be inferred. Along the way we shall discuss some other examples from nuclear physics where such ideas are fruitful.

To study the symmetry of any physical system we perform certain operations upon it and study the change or lack of change in the system as a result. That is, we study the invariance of the properties of a system by employing a set of operations which change the coordinates or other parameters of the system. For example we can perform rotations about the three space axes of a solid object. A sphere will retain its sphericity under such a set of operations. The operators act upon the three Euler angles. The set of such operators designed to perform a symmetry operation is called a symmetry group and the operators are called symmetry group operators. In our example the operators are members of the rotation group. We could also study the mirror reflection of our object. In this case the operator changes the coordinates (x, y, z) to $(-x, -y, -z)$ and is the parity operation. Parity is conserved if the object is symmetric under reflection. Thus with each symmetry operation goes a conservation law. So we consider next the various conservation laws we have been assuming. Violation of a conservation law implies non-symmetry under the corresponding operation.

1) *Conservation of energy.* Classically this is expressed as $dE/dt = 0$; the total energy of a system is constant with time. Quantum mechanically, the time derivative of an operator $\mathcal{O}$ is

$$\frac{d\mathcal{O}}{dt} = \frac{1}{i\hbar}(\mathcal{O}\mathcal{H} - \mathcal{H}\mathcal{O}) + \frac{\partial\mathcal{O}}{\partial t}$$

where $\mathcal{H}$ is the Hamiltonian operator. But now E corresponds to $i\hbar\partial/\partial t$. Hence

$$\frac{dE}{dt} = \frac{\partial}{\partial t}\mathcal{H} - \mathcal{H}\frac{\partial}{\partial t} = 0.$$

The Hamiltonian of an isolated system must not contain the time variable explicitly. The laws of motion or the interaction do not depend on the time at which they are measured. The laws of physics are invariant under translations along the time axis in space–time. The system is symmetric under translations in time.

2) *Conservation of momentum.* Classically this is expressed as $dp_i/dt=0$ where $p_i=p_x, p_y, p_z$. Quantum mechanically

$$\frac{1}{i\hbar}(p_i \mathcal{H} - \mathcal{H} p_i) + \frac{\partial p_i}{\partial t} = 0.$$

But p_i corresponds to $-i\hbar\partial/\partial x_i$ where $x_i = x, y, z$. Hence

$$\frac{\partial}{\partial x_i} \mathcal{H} - \mathcal{H} \frac{\partial}{\partial x_i} = 0.$$

The Hamiltonian of a free particle must not contain the position coordinates explicitly. The laws of motion or the interaction do not depend on where they were measured. The laws of physics are invariant under translation in space, that is, space is homogeneous.

3) *Angular momentum.* Orbital angular momentum is defined classically as $L = r \times p$ or $L_k = r_i p_j - r_j p_i$ where i, j, k take values 1, 2, 3. Quantum mechanically we define

$$L = -i\hbar r \times \nabla$$

or

$$L_k = i\hbar \left(x_i \frac{\partial}{\partial x_j} - x_j \frac{\partial}{\partial x_i} \right).$$

If we use polar coordinates (r, ϕ) with the axis in the z-direction, $L_z = -i\hbar\partial/\partial\phi$ and the conservation law is

$$\left(\frac{\partial}{\partial \phi} \mathcal{H} - \mathcal{H} \frac{\partial}{\partial \phi} \right) = 0.$$

The Hamiltonian is invariant under rotation or the laws of motion do not depend on the orientation in space. Space is isotropic.

However orbital angular momentum is only conserved in *central* potentials, for example for particles with no intrinsic spin or if the wavefunction of the particle is a scalar (one-valued) function of position. However our particle may itself possess intrinsic spin, that is, there is an internal axis whose orientation we must consider. Then to preserve isotropy of space it is the *total* angular momentum which is conserved. In this case orbital angular momentum may not be separately conserved. The precession of the spin axis must be compensated by a change in orbital angular momentum in order to conserve total angular momentum. That is $J_z = L_z + S_z$ and $J = L + S$.

4) *Parity.* This property defines the symmetry of the system under inversion of the space axes. The parity operator P_r inverts the direction of the three coordinate axes $P_r(x, y, z) = (-x, -y, -z)$. Note that a reflection about *one* axis $P_z(x, y, z) =$

$(x, y, -z)$ is equivalent to inversion and rotation $P_z = P_r R_z(\pi)$ where $R_z(\pi)$ means a rotation of π about the z-axis.

The conservation law is $(P_r \mathscr{H} - \mathscr{H} P_r) = 0$. The energy does not depend on the right- or left-handedness of the system. As in the case of angular momentum a particle may possess an intrinsic parity which need not be even. It is the total parity of a system which is conserved, that is the product of intrinsic parity and the parity associated with the orbital angular momentum of the particle (even for even l-values, odd for odd l), or $P_r = P_r(\text{i})P_r(\text{o})$ where $P_r(\text{i})$ inverts the internal axis of the particle, $P_r(\text{o})$ inverts the spatial axes x, y, z, and P_r, the total parity operator, inverts *both* sets of axes.

The previous three symmetry principles, the conservation of energy, momentum and angular momentum, are universally observed in all types of interaction. In the case of parity we have seen that it is only conserved in strong and electromagnetic interactions and is definitely not conserved in weak interactions.

5) *Charge conjugation.* This is an operation which changes particles into their antiparticles. Normally a wavefunction does not specify the nature of the particles, only their linear and angular momenta and their positions. However, in general the wavefunction is a complex function and it is found that the complex conjugate of the wavefunction has opposite charge and current densities. It specifies an antiparticle. The charge conjugation operator tests the symmetry between particle and antiparticle.

For example an electromagnetic field must change sign if all the electrons are changed to positrons. Thus we say a photon is odd under C, the charge conjugation operator. Since π^0 decays only into two photons we infer that π^0 is even under C if we assume this conservation law.

The strong nuclear interaction is invariant under C. The electromagnetic interaction is certainly invariant for changing leptons to antileptons as mentioned above. The weak interaction is not invariant under C. This can be inferred from the magnitude of the asymmetry observed in the β-decay of aligned ^{60}Co. This asymmetry resulting from interference between parity conserving and parity non-conserving parts of the interaction does not occur if C is invariant. However the two modes of K^+ indicate parity violation directly and indicate nothing about C invariance.

6) *CP invariance.* If both C and P are violated in the weak interaction then we can assume the product CP operation *does* lead to invariance. However in the decay of a strange particle K^0 we get an apparent small violation of CP.

7) *CPT invariance.* It is a general result of field theory that any interaction is invariant under the product of charge conjugation, parity and time reversal operations. A proof of this is related to the identity of rest masses and lifetimes of particles and antiparticles. That is, a particle in a certain state of motion and an

antiparticle in the time reversed state are indistinguishable, that is, they have the same total energy, same total momentum, and hence the same rest mass.

8) *Baryon number and lepton number.* In general for a field we must specify that the difference between the number of sources and the number of sinks is invariant. This is referred to as the gauge invariance of the field. For the strong field we note that baryons appear and disappear in pairs of particle–antiparticle. That is, the number of baryons is conserved. We introduced baryon number $B = +1$ for baryons and -1 for antibaryons and demand that it be conserved. The total net number of baryons in our universe thus is a constant.

We have also seen that sources and sinks of the weak field are similarly conserved via conservation of lepton number. In fact we noted the individual conservation of eptons and meptons which of course need not imply two types of field but only a hitherto undetected new quantum number which distinguishes muons from electrons.

9) *Charge conservation.* As we found symmetry principles and conservation laws in ordinary space–time we deduce similar principles in isospin space. Charge conservation obviously corresponds to gauge invariance of isospace. Counting charges as $+1$ and -1 such charges appear and disappear in pairs so our world remains electrically neutral.

10) *Charge symmetry.* This operation, $R_2(\pi)$, corresponds either to a rotation about the 2-axis in isospace or to a reflection about the equatorial plane. It reverses T_3, changing a neutron into a proton for example.

11) *Hypercharge.* Invariance under the operation of inversion in isospace corresponds to conservation of $B+S$, baryon number and strangeness. We called this Y, the hypercharge. A better name might have been isoparity.

12) *T_3 conservation.* This corresponds to the operation of a cylindrical rotation in isospace. Invariance under this operation together with inversion in isospace corresponds, we have seen, to charge conservation. $Q = I_3 + \frac{1}{2}Y$.

13) *T conservation.* This corresponds to invariance under *all* rotations in isospace. We have seen that T is only conserved in the charge independent strong interactions.

These and some other symmetries are listed in table 13.1 together with conservation or violation observed in various interactions. This obviously does not exhaust the symmetry invariances we can study under even more complex combinations of operations in space time and isospin space. We examine some of these later using the mathematical theory of groups to predict the effects of groups of operations.

TABLE 13.1

Conservation (+) and violation (—) of properties in three fields

Property	Strong	Electromagnetic	Weak
Energy	+	+	+
Momentum	+	+	+
Total angular momentum	+	+	+
Helicity or chirality ($m=0$, $v=c$)	+	+	+
Parity	+	+	—
Charge conjugation	+	+	—
Time reversal	+	+	?
CP	+	+	?
CPT	+	+	+
Baryon number	+	+	+
Lepton number	?	+	+
Charge	+	+	+
Strangeness	+	+	—
Hypercharge	+	+	—
T_3	+	+	—
Charge symmetry, $R_2(\pi)$	+	—	
Isospin T	+	—	—
G-parity, $R_2(\pi)C$	+	—	

6   *Multiplets and broken symmetry*

Let us review the concept of multiplets. They occur as degenerate energy states of a system when some specified set of operations can change some quantum numbers but leave a certain quantum number unchanged. In the absence of an external force each of the states has the same energy. For example an atomic state is labelled by the L and S of the appropriate electronic orbit as ^{3}D. However if we look at the corresponding spectral line we find that it is actually split into three lines due to the effect of the $\boldsymbol{L} \cdot \boldsymbol{S}$ force on the electron. This force is small compared to the total binding hence the splitting is small compared to the next electronic state. The $\boldsymbol{L} \cdot \boldsymbol{S}$ force breaks the symmetry so that we see the separate fine structure components. L and S no longer serve to label the components of the multiplet. Instead $\boldsymbol{J} = \boldsymbol{L} + \boldsymbol{S}$ gives us the three possible J-values ^{3}D$_1$, ^{3}D$_2$, ^{3}D$_3$. This is an example of a multiplet of the rotation group or we can call it the group SU(2). The 2 implies that it has *two* operators, one a generator J_3 and the other an invariant J^2 corresponding to the quantum numbers m and j. The multiplicity or dimension of the multiplet is in this case simply $(2j+1)$. In the case above the states of a given J-value could be split up further by an external magnetic field into the magnetic substates. These would be $3+5+7=15=(2L+1)(2S+1)$ lines altogether.

We can use group theory shorthand to indicate the multiplets expected from SU(2) or the rotation group. One particle of $j=\frac{1}{2}$, $m=\pm\frac{1}{2}$ has dimension or multiplicity [2]. If we combine two such particles we find multiplets as follows

$$[2] \otimes [2] = [3] \oplus [1].$$

This is just the rule of vector addition. Two spin $\frac{1}{2}$ particles can produce $j=1$, $m=+1, 0, -1$ and $j=0$, $m=0$, a triplet and a singlet. For three particles $[2]\otimes[2]\otimes[2]=[2]\oplus[4]$, a doublet and a quartet. And so on.

A multiplet is strictly speaking the degenerate set of substates. It is only split up into its constituent states or the symmetry is only broken through some new force which gives each substate a slightly different energy. This new force may be an external force as the magnetic field in the Zeeman splitting or it may be a force not considered in the first approximation whose strength is much less than the force we have assumed in describing the multiplet. The splitting will therefore be small compared to the spacing of the multiplets themselves.

In this way the $l \cdot s$ force in nuclei splits the l-labelled multiplets of the harmonic oscillator potential into j-labelled components of the multiplet which corresponded to the observed shell model states. Similarly the coulomb force splits the degenerate isospin multiplets. So we say the multiplets are the result of a certain symmetry under certain symmetry operations, in the above mentioned cases core rotations in ordinary space or isospin space. The operations constitute a group, in this case SU(2) or the rotation group. *Conversely* we can say that the observation of split multiplet structure implies that a certain symmetry existed which in principle we can guess at from the multiplicity of the multiplet and that a symmetry breaking force exists which tends partially to break this symmetry. For example the observation of multiplets of dimension $2j+1$ and eigenvalues $j(j+1)$ leads us at once to infer the rotation group or SU(2) as a description of the operations involved. Similarly the $2T+1$ states of isospin multiplets imply the same group and we must imagine now a rotation in some new space connected with the charge.

It is for this reason that the observation of groupings into octets of the fundamental particles led to consideration of the simplest group which could be involved, which is SU(3), the next more complicated group than SU(2). As SU(2) implies fundamental doublets of spin $\pm\frac{1}{2}$ or isospin $\pm\frac{1}{2}$ so SU(3) implies fundamental triplets. We discuss later what we can infer about these triplets.

First we consider another approach to a nuclear model suggested by Wigner in

1937. Wigner was one of the pioneers in applying group theory to physical problems.

7   Nuclear supermultiplets

Suppose we assume that a reasonable nuclear wavefunction can be constructed using a nuclear force which is spin and isospin independent. That is we use the ordinary or Wigner part (non-exchange) and the Majorana exchange part which involved, you recall, exchange of position only, that is, our force is only spatially dependent. We thus neglect or treat only in first approximation the spin dependent central exchange forces (Bartlett and Heisenberg), the electrostatic force, the tensor force and the spin–orbit force. We assume the deviations produced by these parts of the force will be small compared to the total binding.

The states we can produce have both the total spin S and the isospin T as good quantum numbers. So we should have at least $(2S+1)(2T+1)$ degenerate states having the same energy and substates ranging from $S \rightarrow -S$ and $T \rightarrow -T$. These correspond to rotations in ordinary space and isospin space respectively. But we can get even more degenerate states by exchanging S and T without changing energy. So in addition we get states with isospin equal to S and spin equal to T. We see therefore that we have a much larger set of degenerate states since we have two good quantum numbers and their combination. Hence we call the set of states a *supermultiplet*. It is based on a larger group, SU(4), with the states labelled by 3 quantum numbers rather than one as for ordinary spin multiplets.

The symmetry properties may be described by a partition of the total number of nucleons A into four integers

$$f_1 + f_2 + f_3 + f_4 = A$$

where the four integers refer to the number of neutrons with spin up, neutrons with spin down, protons with spin up and protons with spin down respectively.

Fig. 13.5 shows an example of neutrons and protons filling up nuclear states. An alternative partition would be in terms of the number of states with 1 nucleon (λ_1), the number with two (λ_2), the number with three (λ_3) and the number with four (λ_4). The symmetry may be represented in convenient form by a Young tableau shown in fig. 13.5 for the same case. Each row represents a given state of the particle and each column represents the occupation of a different energy level. Note that in this case the symmetry depends only on the unfilled energy levels. Thus λ_4 is not needed to specify the symmetry and the two left hand columns of the Young tableau may be omitted. In general for an SU(n) group, $n-1$ numbers specify the symmetry. The three numbers generally used to specify nuclear super-

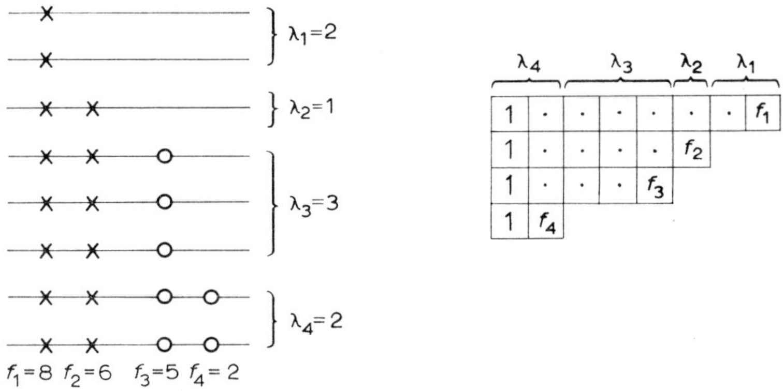

Fig. 13.5.   Partitions of nucleon types and the corresponding Young tableau

multiplets are

$$P' = \tfrac{1}{2}(f_1+f_2-f_3-f_4) = \tfrac{1}{2}(N-Z)$$
$$P'' = \tfrac{1}{2}(f_1-f_2+f_3-f_4) = \tfrac{1}{2}(\text{spins up} - \text{spins down})$$
$$P''' = \tfrac{1}{2}(f_1-f_2-f_3+f_4) = \tfrac{1}{2}[(\text{neutrons up}+\text{protons down})-(\text{neutrons down}$$
$$+\text{protons up})].$$

Obviously one of these numbers is just the projection of the isospin T_3, another is the z-component of the spin S_z and the third is associated with the difference of the z-component of total neutron and total proton spins. We may call the latter D_z. We order the P', P'', P''' such that $P' \geq P'' \geq P'''$ consistent with the definitions above.

We can now assign supermultiplet identification to the ground states of nuclei

1) for even–even nuclei $(|T_3|, 0, 0)$
2) for odd-A nuclei $(|T_3|, \tfrac{1}{2}, \tfrac{1}{2})$
3) for odd–odd nuclei with $T_3 > 0$ $(T_3, 1, 0)$
4) for odd–odd nuclei with $T_3 = 0$ $(1, 0, 0)$.

In general the supermultiplet having the lowest values of P', P'', P''' corresponds to the greatest symmetry for spatial exchange and will be the most tightly bound (lie lowest). For example, the $T_3 = 0$ ground states of even–even $A = 4n$ nuclei with supermultiplet $(0, 0, 0)$ are most stable, the odd-A mirror nuclei are next with $(\tfrac{1}{2}, \tfrac{1}{2}, \tfrac{1}{2})$. The $4n+2$ even–even nuclei and odd–odd nuclei with $(1, 0, 0)$ should be next and so on.

In fact this part of the binding energy which depends on the spatial symmetry is just the symmetry energy term in the semi-empirical binding energy formula. This was introduced as a disruptive term to indicate departure from complete symmetry, that is, the (0, 0, 0) state, and was written $a_a(A-2Z)^2/A$, which is just proportional to T_3^2/A, the ratio of the symmetry energy to the volume binding energy which is proportional to A.

Now we examine the breaking of this higher symmetry by the forces which we assumed to be zero. In particular the coulomb force is charge dependent and gives us isospin multiplets and the spin dependent part of the nuclear force (Bartlett and Heisenberg exchange forces) gives us broken spin multiplets.

For example the supermultiplet (1, 0, 0) is split by the coulomb force into an isospin triplet with $T_3 = +1, 0, -1$ (fig. 13.6). The spin dependent force further splits the $T_3=0$ state into two states with $S_z=1$ and $S_z=0$, the $S_z=1$ triplet state lying lower, being more tightly bound.

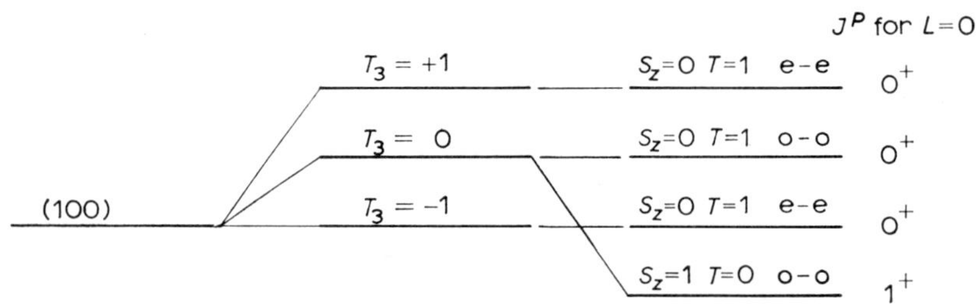

Fig. 13.6.   The supermultiplet (100) split first by the coulomb force and then by the spin dependent force

This spin dependent splitting is just the effect we studied in the deuteron which puts the $T=0$, $J^P=1^+$ ground state 2.23 MeV lower than the $T=1$, $J^P=0^+$ singlet excited state (fig. 13.7). In the two-particle system there is no splitting of the nn and np states by the coulomb force. We have stable odd–odd nuclei of the $A=4n+2$ type up until the splitting due to the coulomb force exceeds that due to the spin dependent force. This happens between mass 14 and 18. So ^{14}N is the heaviest stable odd–odd nucleus and ^{18}F is unstable as now the $T_3=-1$, $S_z=0$, $T=1$ ^{18}O nucleus is more stable.

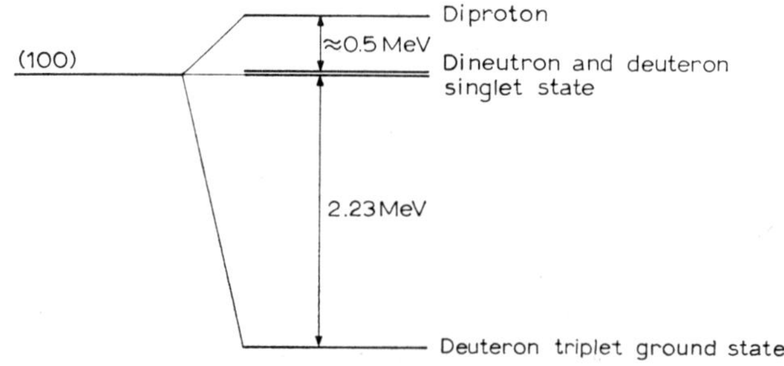

Fig. 13.7.   The two-nucleon system

The supermultiplet splitting for the nuclei of mass 6, 10, 14 and 18 of the $A = 4n + 2$ type with supermultiplet (1, 0, 0) is shown in fig. 13.8a,b,c,d resp.

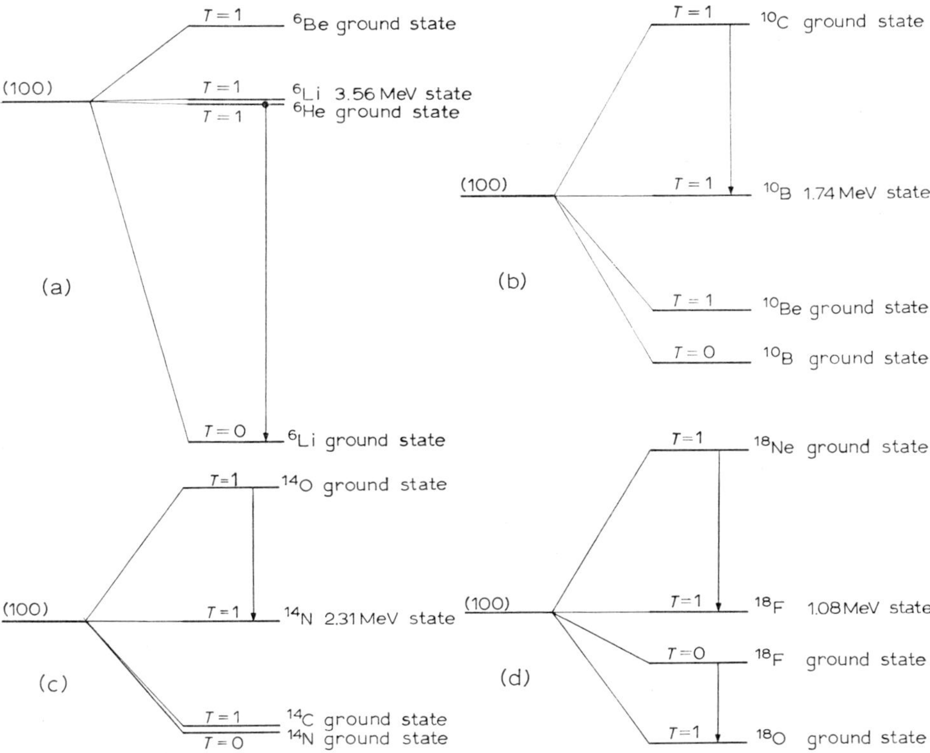

Fig. 13.8.   a, b, c, d) The 6, 10, 14 and 18-nucleon system resp.

When the coulomb force becomes comparable to or stronger than the symmetry energy the multiplets are completely mixed up and the supermultiplet structure is no longer a useful model. However we can still see some effects of the super-multiplet symmetry. Strictly speaking any transition between supermultiplets is forbidden and transitions are only allowed within the same supermultiplet. Thus when the weak interaction causes β-decay transitions between nuclear states we should find allowed transitions between members of the same supermultiplet and forbidden transitions between numbers of different supermultiplets. In the case of mirror nuclei, whose ground states are members of the $(\frac{1}{2}, \frac{1}{2}, \frac{1}{2})$ super-multiplet, we indeed observe *superallowed* β-transitions whose $\log ft$ values are appreciably smaller (3–3.5) than for ordinary allowed transitions ($\log ft = 4.0$–5.0). The latter are in this sense 'super-forbidden'. In the (1, 0, 0) supermultiplet we have the $A = 4n + 2$ nuclei and between the $S_z = 0$, $T = 1$ state of the even–even nucleus and the $S_z=1$, $T=0$ or the $S_z=0$, $T=1$ states of the odd–odd nucleus, we also observe superallowed transitions. These are indicated in fig. 13.8.

To the extent that the model works it implies that T is a good quantum number and also S. The latter is difficult to check directly since we do not measure it directly but rather the total spin $I = L+S$. We also have noted that L–S coupling was experimentally less valid, for nuclei heavier than about mass 20, than the j–j coupling scheme on which the shell model was based. In other words the spin–orbit force also seriously breaks the supermultiplet symmetry. This can be seen already for the $^{14}\text{C} \rightarrow ^{14}\text{N}$ decay which should be superallowed but in fact has a large $\log ft$ (9.03). This is because already the configurations for the two nuclei are quite different so that other selection rules are operating.

8   *Fundamental particle multiplets*

Returning now to the problem of fundamental particle classification, we seemed to be observing multiplets of particles with the same value of spin and parity – octets and singlets for the mesons and octets and decuplets for the baryons. The multiplets were split into substates of different mass labelled by hypercharge Y and isospin projection T_3. We seek a group of operations which could give rise to such multiplets.

Let us compare the strong force field and the electromagnetic field. If we remove a source of the electromagnetic field (a charge) by giving a high momentum to an electron, the source-free field which is left behind gives rise to one or two field quanta, photons. On the other hand we have seen that if we give a high momentum to a source of the strong field such as a baryon, the source-free field left behind gives rise to many field quanta, that is, pions and other mesons. As many pions will be produced as are energetically possible because the sources are now strongly coupled to the field in contrast to the coupling of the charge to the electromagnetic field.

Two light quanta could form a positron–electron pair bound together into positronium $\gamma+\gamma \rightarrow \text{e}^+ +\text{e}^-$. Conversely the bound electron and positron could virtually annihilate and form photons. The sources disappear and leave a source-free field $\text{e}^+ +\text{e}^- \rightarrow \gamma+\gamma$.

The probability of the reaction going in one direction or the other depends on the relative distances involved for the two conditions. The Bohr radius for positronium is *large* compared to the radius for annihilation. Thus positronium exists in the (e^+e^-) condition predominately and we do not see 'excited states' of the photon at least at present energies.

For nucleonium, the compound state of a baryon and an antibaryon, the Bohr radius is very much smaller because of the larger masses. It is now comparable to the annihilation radius and the condition of field quanta is very probable, that is, it is a meson for a large fraction of the time even at present energies. We thus can observe many field quanta and it is attractive to consider whether the mesons can be considered as different states of the baryon–antibaryon system. Consider first the mesons we can make by combining a nucleon with an antinucleon. Recall that nucleons and antinucleons are assigned the quantum numbers given in table 13.2. Then the states of $N\bar{N}$ are as listed in table 13.3.

TABLE 13.2

	T_3	J^P	B	S
p	$+\frac{1}{2}$	$\frac{1}{2}^+$	1	0
$\bar{\text{p}}$	$-\frac{1}{2}$	$\frac{1}{2}^-$	-1	0
n	$-\frac{1}{2}$	$\frac{1}{2}^+$	1	0
$\bar{\text{n}}$	$+\frac{1}{2}$	$\frac{1}{2}^-$	-1	0

TABLE 13.3

L	T-spin	J-spin	T	J^P	Nucleonium state	Meson
0	↑↑	↑↓	1	0^-	$^{31}S_0$	π
	↑↑	↑↑	1	1^-	$^{33}S_1$	ρ
	↑↓	↑↓	0	0^-	$^{11}S_0$	η
	↑↓	↑↑	0	1^-	$^{13}S_1$	ω

TABLE 13.4

L	T-spin	J-spin	T	J^P	S	Baryonium state	Meson
0	↑0	↑↓	$\frac{1}{2}$	0^-	$+1$	$^{21}S_0$	K
	↑0	↑↑	$\frac{1}{2}$	1^-	$+1$	$^{23}S_0$	K*(891)

For mesons with strangeness it is natural to consider combinations of nucleons with the lightest strange antibaryon of $T=0$, the $\bar{\Lambda}$, which has $J^P=\frac{1}{2}^-$, $B=-1$, $S=+1$. Then the states of $N\bar{\Lambda}$ are as listed in table 13.4. The antikaons are similar combinations of $\bar{N}\Lambda$. So with this model we account for the two meson octets. The octet with 2^+ would be produced with $L=1$ and so on. The Sakata model, as this is generally called, might be used also to generate baryons by combining a meson and a baryon to get $B=1$. We already considered the $\Delta(1238)$ particle to be a πN resonance. So in terms of our fundamental baryons we consider combinations like $N\overline{N}N$, $N\overline{N}\Lambda$, $N\overline{\Lambda}N$, $N\overline{\Lambda}\Lambda$, $\Lambda\overline{N}\Lambda$ etc. To get positive parity for the baryons we must consider the third baryon and the meson to be in a relative p-state with $L=1$. Thus

$N\overline{N}N$ gives us nucleons and Δ-particles with $J^P=\frac{1}{2}^+$ and $\frac{3}{2}^+$, i.e. combining π, ϱ, η, ω with nucleons. (We do not see nucleons of $\frac{3}{2}^+$ or Δ of $\frac{1}{2}^+$.)

$N\overline{N}\Lambda$ gives us Λ and Σ of spins $\frac{1}{2}^+$, $\frac{3}{2}^+$, $\frac{5}{2}^+$, i.e. combining π, ϱ, η, ω with a Λ.

$\Lambda\overline{N}\Lambda$ gives us the Ξ-particle, i.e. $\overline{K}+\Lambda$.

$N\overline{\Lambda}\Lambda$ gives us nucleons again, which is disturbing.

$N\overline{\Lambda}N$ gives us Λ- and Σ-particles with $B=+1$, $S=+1$ and $Y=2$, which we clearly do not observe.

The property of the Sakata model of predicting unobserved baryons has cast serious doubt on the idea that the fundamental particles could be considered as compounds of themselves.

9   The quarks

What more fundamental entities can then be invented whose groupings will account for the multiplets of particles observed? We do not see such objects experimentally so we have to deduce their properties from their compounds. We require a triplet of them to give the observed multiplets, two of them must be an isospin doublet like the nucleons and one at least must possess strangeness. These hypothetical particles are called quarks and have the quantum numbers listed in table 13.5.

TABLE 13.5

Quark	I	T_3	S	B	Y	Q
α	$\frac{1}{2}$	$\frac{1}{2}$	0	$\frac{1}{3}$	$\frac{1}{3}$	$\frac{2}{3}$
β	$\frac{1}{2}$	$-\frac{1}{2}$	0	$\frac{1}{3}$	$\frac{1}{3}$	$-\frac{1}{3}$
γ	0	0	-1	$\frac{1}{3}$	$-\frac{2}{3}$	$-\frac{1}{3}$

The assignment of baryon number $\frac{1}{3}$ to the quarks is the trick to enable us to construct baryons with $B=1$ out of three quarks instead of two particles and an antiparticle as in the Sakata model. The electric charges come out to be fractions of the electronic charge which is a strange feature and should make quarks easy to identify. It is surprising if they do exist that this feature has not shown up experimentally in the past.

The mesons are again formed from quark–antiquark combinations. In the group theory shorthand for SU(3) if we combine a triplet with an antitriplet

$$[3] \otimes \overline{[3]} = [1] \oplus [8]$$

we obtain an octet and a singlet as observed. Forming the baryons by a combination of three quarks we obtain

$$[3] \otimes [3] \otimes [3] = [10] \oplus [8] \oplus [8] \oplus [1]$$

which includes the observed octet and decuplet. In contrast the Sakata model gives

$$[3] \otimes [3] \otimes \overline{[3]} = [15] \oplus [6] \oplus [3] \oplus [3]$$

multiplets which are *not* observed.

So far we have considered the multiplets separately labelled by spin and parity. However we could consider a higher symmetry by taking account of the quark spin, just as the isospin symmetry of SU(2) and the spin symmetry of SU(2) were combined to give supermultiplets of SU(4). The 0^- meson octet + singlet suggests a quark–antiquark combination with spins opposed and $l=0$. The 1^- mesons would have the quark–antiquark spins aligned with $l=0$. The $\frac{1}{2}^+$ baryon octet would have the quark spins ↑↑↓ and the $\frac{3}{2}^+$ baryon decuplet would have ↑↑↑ both also with $l=0$. The larger supermultiplets labelled by orbital angular momentum would thus arise from a larger group, SU(6).

Then for the mesons we have

$$[6] \otimes [6] = [35] \oplus [1]$$

i.e. an octet of spin 0, multiplicity $(2j+1) \times 8 = 8$
an octet of spin 1, multiplicity $(2j+1) \times 8 = 24$
a singlet of spin 1, multiplicity $(2j+1) \times 1 = \underline{3}$
$$35$$
a singlet of spin 0, multiplicity $(2j+1) \times 1 = \underline{1}$
$$36$$

For the baryons we have again with $l=0$

$$[6] \otimes [6] \otimes [6] = [56] \oplus [70] \oplus [70] \oplus [20]$$

i.e. an octet of spin $\frac{1}{2}$, multiplicity $(2j+1) \times 8 = 16$
a decuplet of spin $\frac{3}{2}$, multiplicity $(2j+1) \times 10 = \underline{40}$
$$56$$

corresponding to the first of the expected multiplets.

Introducing now higher l-values, for the mesons $l=1$ would imply octets and singlets with $J^P=0^+$, 1^+ and 2^+, and $l=2$ would imply multiplets with $J^P=1^-$, 2^-, 3^-. A nonet of $J^P=2^+$ is well established. Some members of 0^+, 1^+ and 2^- nonets may have been seen. For the baryons octets of $J^P=\frac{1}{2}^-$, $\frac{3}{2}^-$, $\frac{5}{2}^-$ ($l=1$), an octet of $J^P=\frac{5}{2}^+$ ($l=2$), and a decuplet of $J^P=\frac{3}{2}^+$ ($l=2$) may have been partially established.

The validity of the SU(3) and SU(6) does not depend only on the arguments used above. Group theory can make many predictions for matrix elements, transition probabilities, moments etc. in such multiplets and in those cases where these can be checked the classification is extremely well confirmed.

We should now consider the symmetry breaking forces which cause the degeneracy to be removed for the various members of the multiplets. Clearly to split the basic octets into members with different masses requires a strong force. A force of comparable strength splits the multiplets of different spin. We get more binding with quark spins antiparallel; that is, paired off similar to the nucleon pairing force. Finally there must be a super-strong quark binding force to give the SU(6) supermultiplets.

The quark–antiquark spin S_q is indicated in fig. 13.9. The multiplicity of the broken SU(3) multiplets is also indicated. The further small scale breaking of the multiplets by the coulomb force is indicated on the right. We see that two kinds of strong force are indicated, a main quark binding force giving the largest multiplets and a less strong symmetry breaking force which contains a spin dependent part

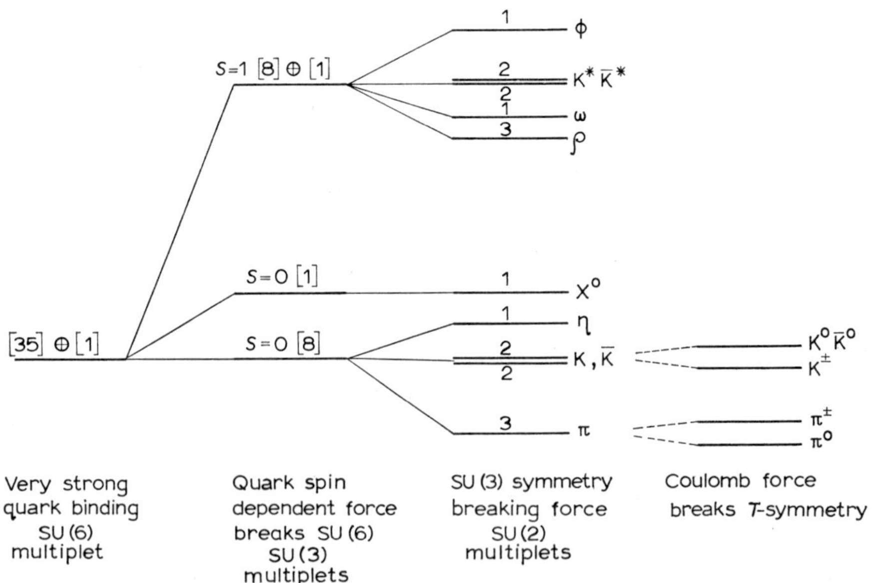

Fig. 13.9.   The quark supermultiplet for mesons

and a Y-dependent part. It may be that the range of the super-strong force is very short, which would imply a large mass for the quarks and this might be why they have not yet been observed. This however would imply a very large ratio of binding energy to rest mass for the fundamental particles. It may also be that the field theoretical ideas of the quark model are not a unique description of the structures we see. Until quarks are observed this question will remain an open one. Many careful experiments designed to detect particles of fractional charge have so far been completely unsuccessful.

14

The Weak Interaction

1   Introduction

We have referred many times to the weak interaction in discussing nuclear β-decay, muon decay and the decay of fundamental particles. In β-decay and muon decay leptons were always involved. In fundamental particle decays both leptonic and non-leptonic decays are observed. We wish now to see if all these effects can be unified into a consistent model of the weak interaction.

2   Beta-decay

In β-decay the weak force acts to produce a transition between initial and final states such that a neutron, for example, is changed to a proton.

$$n \rightarrow p + \beta^- + \bar{\nu}.$$

This reaction can also be written

$$n + \nu \rightarrow p + \beta^-.$$

Thus the interaction must contain an operator to change a neutron into a proton. We write the operator τ_+ which destroys a neutron and creates a proton. It is just the operator T_3. We need also an operator to destroy a neutrino and create an electron. By analogy we call this τ_+^L. We require also a dependence on distance for the interaction. This is known to be a very short distance so we use a delta function $\delta(r - r_L)$ where r and r_L are positions of nucleon and lepton. Finally there may be other operators acting upon nucleon and lepton whose nature we can only predict from experiment. We call these $\mathscr{O}$ and $\mathscr{O}_L$. The Hamiltonian is

$$H_\beta = (\mathscr{O}\mathscr{O}_L \tau_+ \tau_+^L + \mathscr{O}^\dagger \mathscr{O}_L^\dagger \tau_- \tau_-^L)\delta(r - r_L).$$

The second term is just the Hermitian conjugate of the first and this just represents the other sort of β-decay namely

$$p + \beta^- \rightarrow n + \nu \qquad \text{(electron capture)}$$

or equivalently

$$p \rightarrow n + \beta^+ + \nu \qquad \text{(positron decay)}.$$

The main problem of β-decay studies for the 25 years following Fermi's formulation of β-decay theory in 1934 was the determination of the form of the operators $\mathscr{O}$ and $\mathscr{O}_L$. We can restrict their possible form if we make the following assumptions: 1) the theory must be relativistically invariant; this must certainly be true as we are dealing with light and indeed mas less particles, 2) the operators do not contain a momentum dependence. The transition probabilities of course do contain a momentum dependence but we assume the interaction strength itself does not. There is now considerable experimental evidence that this assumption is correct. Theories containing a momentum dependence give incorrect answers.

H_β must be independent of the coordinate system used. On the other hand it may depend on the handedness of the axes since the neutrino being massless has a definite helicity or handedness. A scalar is a quantity which is invariant under any change in axes, while a pseudoscalar is a quantity which changes sign on reflection but does not change otherwise. So we assume that H_β is a linear combination of a scalar part and a pseudoscalar part.

Nucleon and lepton wavefunctions have the form of four-component spinors, the components corresponding to particle spin up, particle spin down, antiparticle spin up, antiparticle spin down. Two such wavefunctions ϕ, the lepton wavefunction and ψ, the nucleon wavefunction may be multiplied together with any four by four matrix M to give a number $(\phi^* M \psi)$. All possible four by four matrices can be generated by multiplication or addition of a set of five independent standard matrices. These are called the Dirac matrices. We find that we can construct the required scalar–pseudoscalar mixture by such operations of addition and multiplication. However only five different ways are possible. These may be named after the form of the *nuclear* matrix element,

1)   scalar,
2)   pseudoscalar,
3)   vector,
4)   axial vector or pseudovector,
5)   tensor.

The selection rules derived from these various forms for the interaction are

for vector and scalar	$\Delta J=0,$	no parity change
for axial vector and tensor	$\Delta J=0,\,1,$	no parity change (not $0\to0$)
for pseudoscalar	$\Delta J=0,$	parity change.

You note that the V or S selection rules correspond to Fermi allowed transitions, the A or T selection rules correspond to Gamow–Teller allowed transitions and the P selection rule corresponds to a forbidden transition.

We can also predict the electron–neutrino correlations, the electron polarisations and the spectral shapes expected for each of these types of interaction. These may be compared with experiment and a choice made between the different types. If we assume that the interaction is a weighted sum of all the possibilities and label the various coupling constants with the helicity assumed for the neutrino, then we can evaluate the various coupling constants from experiment. For example C_{AL} is the coupling constant for the term arising from an axial-vector interaction with left-handed neutrinos. The experimental results lead to the following conclusions.

(i) Electron polarization requires $C_{\mathrm{SL}}=C_{\mathrm{VR}}=C_{\mathrm{TL}}=C_{\mathrm{AR}}=0$.

(ii) Electron–neutrino correlations require for pure GT transitions which could involve C_{TR} and C_{AL}, $C_{\mathrm{TR}}=0$ while for pure Fermi transitions which could involve C_{VL} and C_{SR}, $C_{\mathrm{SR}}=0$.

(iii) Shapes of forbidden spectra indicate that $C_{\mathrm{P}}=0$.

So we conclude that the β-interaction contains only vector and axial-vector terms and that the neutrinos are left-handed as we already had concluded. The V interaction shows up in pure form in pure Fermi allowed transitions ($J=0\to J=0$, no) while the A interaction shows up in pure form in pure GT allowed transitions ($\Delta J=1$, no). By measurement of the ft-value in these two cases we can derive absolute values of the coupling constants for the two interactions. This is not completely straightforward as knowledge of the wavefunctions of the nuclear states is also required. However the best values are as follows:

$$g_{\mathrm{A}} = -(1.18\pm0.02)\,g_{\mathrm{V}}$$
$$g_{\mathrm{V}} = 1.4029\pm0.0022\times10^{-49}\ \mathrm{erg\ cm^{3}}.$$

3 Muon decay and conserved vector coupling

We can now apply our knowledge of the form of the interaction and its strength derived from nuclear β-decay to the next simplest case, that of muon decay.

$$\mu^- \to e^- + \nu_\mu + \bar{\nu}_e$$
$$\mu^+ \to e^+ + \bar{\nu}_\mu + \nu_e.$$

To account for the half-life, spectral shape and electron polarisation we require $g_A = -g_V$ exactly and

$$g_V = 1.4350 \pm 0.0011 \times 10^{-49} \text{ erg cm}^3.$$

Now the fact that the vector coupling constants for muon decay and for β-decay are identical to a few per cent is very surprising for the following reason. We are comparing the two transitions

$$n \to p + e^- + \bar{\nu}$$
$$\mu^- \to e^- + \nu + \bar{\nu}.$$

We know from the anomalous magnetic moments of neutron and proton that we must consider nucleons to be continuously surrounded by a cloud of virtual mesons. In other words the neutron is $p + \pi^-$ for an appreciable part of the time while the proton is $n + \pi^+$ also for part of the time. But in our analysis of β-decay we considered only the decay of a bare nucleon to a bare nucleon and have ignored the effects of the pions which in reality 'dress up' the nucleon. So we should expect that the observed coupling constant should be appropriately reduced to take account of this. On the other hand we have seen experimentally that the muon has no anomalous magnetic moment different from the electron. Both have a small amount due to virtual photons arising from their charge. But the essentially bare muon behaves identically to the 'dressed up' nucleon so far as the vector interaction is concerned. Note however that the axial vector coupling constants are not the same but differ by 18 %.

There is an analogous situation to this, however. A bare electron behaves essentially as a point electric charge while the dressed up proton with its cloud of charged mesons has a measurable charge distribution. And yet the total charge of a proton is identical to that of an electron. That is, it is the same as we would expect for a bare proton. All the interactions are arranged so that the equality between the physically observed charge and the bare proton charge is not disturbed. We say that the vector electrostatic potential couples to a *conserved* charge current. On the other hand the magnetic effect of the pions does not cancel out and we get an anomalous magnetic moment in addition to the magnetic moment due to the charge of the bare proton. We say therefore that the axial vector magnetic potential cannot lead to a conserved axial vector current.

Accordingly Gell-Mann and Feynman* by analogy suggested that in the weak interaction also the vector part of the interaction was coupled to a conserved

* M. Gell-Mann and R. P. Feynman, Phys. Rev. **109** (1958) 193.

vector current of something analogous to charge carried by the pions. Such conservation could then lead to the remarkable identity of the coupling constants for nucleon and muon decay. We assume that the 18 % discrepancy in g_A is due to the pion cloud and corresponds to 'weak magnetism'.

Much experimental work has been devoted to the measurement of the two coupling constants with high precision in order to determine whether the few percent discrepancy is real or not. It does appear to be real and we shall consider later a theoretical reason for this.

4   A unified weak interaction theory

We now consider whether β-decay, muon decay and fundamental particle decay can all be put on a common basis. A useful approach is based on the following assumptions.

A) We assume that the decays are generated in proportion to the product of two currents, that is the Lagrangian for the weak interaction has the form

$$L = \tfrac{1}{2}\sqrt{2}\,g j_\alpha j_\alpha^\dagger$$

where g is the coupling constant. The current j_α contains several terms

$$j_\alpha = j_\alpha^l + j_\alpha^h + j_\alpha^{hs}$$

where j_α^l is the leptonic current responsible for transformations

$$e^- \to \nu_e, \qquad e^+ \to \bar{\nu}_e, \qquad \mu^- \to \nu_\mu, \qquad \mu^+ \to \bar{\nu}_\mu;$$

j_α^h is the current of strongly interacting particles (hadrons) with no change in strangeness, responsible for $(n \to p)$ etc.;
j_α^{hs} is the current of hadrons with change in strangeness such as is responsible for $(\Lambda \to p)$ etc.

The first current j_α^l deals with bare leptons – no quantum number changes are involved except charge; $\Delta Q = 1$. The second current j_α^h deals with nucleons and pions, the latter considered as nucleon–antinucleon pairs. The quantum number changes are $\Delta Q = 1$, $\Delta S = 0$, $\Delta T = 0$, 1. The third current j_α^{hs} deals with strange particle decays where $\Delta S = 1$, $\Delta Q = 1$, $\Delta T = \tfrac{1}{2}$.

Combinations of these currents then give us all observed types of weak decay. For example
1) $j_\alpha^h j_\alpha^{hs\dagger}$ corresponds to $\Xi^- \to \Sigma^0 + \pi^-$; the current $j_\alpha^{hs\dagger}$ destroys a Ξ^- and creates a Σ^0, the current j_α^h creates a N and destroys a N and since $\pi \equiv N\bar{N}$, corresponds to creation of a pion.

2) $j_\alpha^l j_\alpha^{hs\dagger}$ corresponds to $\Lambda \to p + e^- + \bar{\nu}$; here $j_\alpha^{hs\dagger}$ destroys a Λ and creates a p, j_α^l creates an electron and destroys a neutrino (equivalent to creating an anti-neutrino).

3) $j_\alpha^l j_\alpha^{h\dagger}$ corresponds to $n \to p + e^- + \bar{\nu}$ (β-decay) or $\pi \to \mu + \nu_\mu$ (pion decay).

4) $j_\alpha^l j_\alpha^{l\dagger}$ corresponds to $\mu^- \to e^- + \bar{\nu}_e + \nu_\mu$ (muon decay).

B) The second assumption is that of a universal Fermi interaction. This corresponds to the use of a single coupling constant g for all the types of current. This is suggested by the virtual equality of the vector coupling constant for muon and β-decay.

C) The third assumption is the conserved vector current (CVC) hypothesis of Gell-Mann and Feynman.

The weak decays of the strange particles have a coupling constant only about 10 % that for the weak decays of non-strange hadrons. This might be thought to be a contradiction of the universal Fermi interaction hypothesis. However it is at present assumed that this is not a basic feature of the weak interaction but is an 'external' effect due to the strong interaction. For example Cabibbo* noted that the quantum numbers carried by the j^h current are just those of the charged pion, $Q = \pm 1$, $S = 0$, $T = 1$, while those carried by the j^{hs} current are just those of the charged kaon $Q = \pm 1$, $S = \pm 1$, $T = \frac{1}{2}$. All members of the pseudoscalar octet of mesons might then correspond to currents except that since no neutral *leptonic* current is seen we can see no effects due to them. For example $K^0 \to \mu^+ + \mu^-$ is not observed. If the currents are related thus to a broken SU(3) symmetry we should only expect a common coupling constant for strict SU(3) with no symmetry breaking, that is, if K, π, η all have the same mass. Since we *do* have symmetry breaking of the octet we should expect a relation between the coupling constants for j^h and j^{hs} of the following form:

$$j_\alpha = (\tfrac{1}{2}\sqrt{2})(g j^l + g \cos\theta\, j^h + g \sin\theta\, j^{hs}).$$

Then comparing decay rates of strange (K) and non-strange (π) particles:

$$\frac{K^+ \to \pi^0 e^+ \nu}{\pi^+ \to \pi^0 e^+ \nu} \propto \tan^2\theta_V \approx \theta_V = 0.26 \text{ radian}$$

and

$$\frac{K^+ \to \mu^+ \nu}{\pi^+ \to \mu^+ \nu} \propto \tan^2\theta_A \approx \theta_A = 0.258 \text{ radian}$$

* N. Cabibbo, Phys. Rev. Letters **10** (1963) 531.

where θ_V and θ_A refer to vector and axial vector coupling. This agreement for the two couplings lends weight to Cabibbo's suggestion that the value of θ depends only on the SU(3) symmetry breaking, a strong interaction effect. We can also infer from this argument that the vector coupling constant for β-decay should equal the vector coupling constant for μ-decay multiplied by $\cos\theta$. Using $\theta = 0.26$ rad, $\cos\theta = 0.966$. In other words g_μ is greater by 3.4 %. We observe a discrepancy in this direction of 2.2 %. It is not clear at present if the residual discrepancy of 1.2 % is real or not.

5   Some predictions of the theory

There are several interesting consequences of the theory outlined above.

1) There may be self-currents like $j_\alpha^h j_\alpha^{h\dagger}$. This corresponds to a weak interaction between nucleons. This very small part of the nucleon–nucleon force should violate parity and this should show up in nuclear interactions. Attempts have been made to detect this effect by studying the polarisation of γ-rays from decaying nuclear states. The parity violating part of the wavefunction of the decaying state is expected to be a few parts in 10^7. This would lead to an admixture of E1 radiation with predominant (M1, E2) radiation, for example, in the decay between two states of the same parity. The expected discrepancy in the polarisation is less than a few tenths of a percent and the experimental results to date are inconclusive.

2) Another prediction of the theory is the occurrence of the decay $\pi^+ \to \pi^0 + e^+ + \nu$ since in the theory we assume that pions carry weak 'charge' just as nucleons do. Because of the small energy available (≈ 4.6 MeV) the branching ratio is very small. The prediction is

$$\frac{\pi^+ \to \pi^0 e^+ \nu}{\pi^+ \to \mu^+ \nu} = (1.07 \pm 0.02) \times 10^{-8}.$$

This decay has been searched for and found in laboratories in the U.S., Russia and at CERN. The experiment is very difficult since the sought-for event occurs only once in 100 million π^+ decays. The events could be tagged by the fourfold coincidence of two 70 MeV γ-rays from the π^0 decay and two 0.511 MeV γ-rays from the β^+ annihilation. The CERN result for the branching ratio is $(1.15 \pm 0.22) \times 10^{-8}$, in excellent agreement with theory.

3) A further prediction of the theory is weak magnetism. In electromagnetism although the charge currents are conserved for nucleons surrounded by their pion cloud and the electric charge is not affected by the pion cloud, we know that the magnetic moment is affected by the cloud and we observe an anomalous nucleon magnetic moment.

In the same way the weak vector currents seem to be conserved so that the weak charge is unaffected by the pion cloud but it may be that magnetic-like effects of the weak charge on the pions are not cancelled out. So we might expect an anomalous weak magnetism. An elegant experiment to check this was suggested by Gell-Mann, in which the effect could be predicted in direct analogy to measured effects of the anomalous magnetic moment. The nucleus ^{12}B, the 15.11 MeV $T=1$ state of ^{12}C and ^{12}N form an isospin triplet; ^{12}B and ^{12}N decay to the ^{12}C ground state by weak decays while the $T=1$ state decays to the ^{12}C ground state by an electromagnetic M1 decay (fig. 14.1). The anomalous magnetic moment plays a

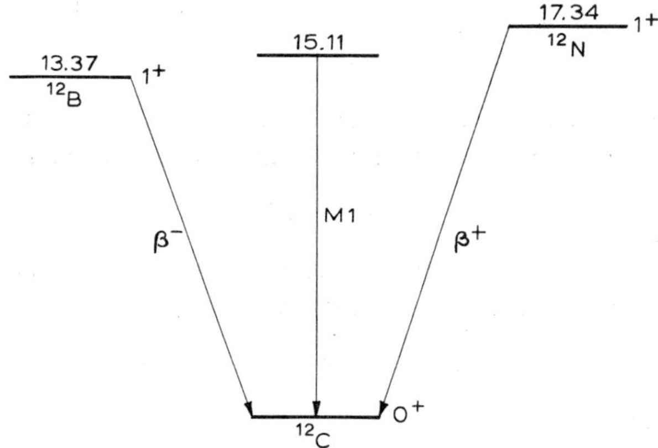

Fig. 14.1.   The mass 12 energy level diagram

basic role in the strength of the M1 transition. By analogy we can then calculate an energy dependent term in the β-decays which similarly depends on the weak magnetism. This correction term to the allowed shape has the form $1+\frac{8}{3}aE$ for the β^--transition and $1-\frac{8}{3}aE$ for the β^+-transition. The prediction based on the M1 transition strength is

$$a = (0.55 \pm 0.12)\,\%.$$

The measurements* show that

$$a(^{12}\text{B}) = (0.55 \pm 0.10)\,\%$$
$$a(^{12}\text{N}) = -(0.52 \pm 0.06)\,\%.$$

The agreement with the conserved vector current theory is excellent.

* Y. K. Lee, L. W. Mo and C. S. Wu, Phys. Rev. Letters **10** (1963) 253.

6　The weak interaction field quantum

We could attempt to describe the weak field in terms of virtual exchange of a field particle instead of in terms of currents. In electrodynamics the photon field particle formalism leads to the same result as the charge current formalism. Since the weak interaction is a vector–axial vector interaction the field particle should have spin 1 and undefined parity. Since the range of the weak force is known to be short, 10^{-14} cm or less, the field particle must be quite massive. This so-called 'intermediate vector boson' or 'weakon' will exist in charged versions W^+, W^- but will have no neutral member since we see no neutral weak currents. It must be at least heavier than the K-meson, otherwise we should expect to see

$$K^{\pm} \rightarrow W^{\pm} + \gamma.$$

A more decisive test is possible using the beams of neutrinos described in connexion with the distinguishing of ν_μ from ν_e. Reactions of this beam such as

$$\nu_\mu + N \rightarrow N + \mu^- + W^+$$

may occur if the neutrino energy is sufficient to supply the muon and W rest masses. The W^+ would then decay either via

$$W^+ \rightarrow \mu^+ + \nu_\mu \quad \text{or} \quad W^+ \rightarrow e^+ + \nu_e.$$

Since the lifetime of the W^+ is expected to be $\leq 10^{-17}$ sec we look for associated μ^-, μ^+ pairs or μ^-, e^+ pairs. No clear evidence for such pairs is seen and it is concluded that the W has a mass in excess of 3 GeV. Higher energy neutrino beams will be required to settle the question of the existence of a weak interaction field particle.

15

Miscellaneous Topics

1 Abundance of the elements and nuclear synthesis

1.1 Introduction. When the cosmic abundances of the stable elements are examined, it is clear that there is no correlation with their chemical properties but rather with their nuclear properties. The element distribution which we measure must be closely related to the mechanisms by which the elements were formed and we therefore infer that the elements were formed by nuclear processes.

When we study the *isotopic ratios* in the earth's crust, in meteorites and in the sun and stars, they are found to be closely the same. This implies the common nuclear reaction origin since chemical effects cannot vary these ratios appreciably.

The composition of the earth's crust is as follows:

46.6 % Oxygen	(99.7 % ^{16}O	double closed shell)
27.7 % Silicon	(92.3 % ^{28}Si	double closed subshell)
5.0 % Iron	(91.7 % ^{56}Fe	even–even)
3.6 % Calcium	(97.0 % ^{40}Ca	double closed shell)
2.1 % Magnesium	(78.6 % ^{24}Mg	even–even)
8.1 % Aluminium	(100 % ^{27}Al)
2.8 % Sodium	(100 % ^{23}Na)
2.6 % Potassium	(93.1 % ^{39}K)
<1 % all others totalling 1.7 %.		

Thus more than 86 % of the earth's crust consists of even–even nuclei and more than half of double closed shell nuclei. We have seen that these are the most stable types of *nuclei,* again indicating the nuclear origin of the distribution. It reflects the distribution of the stable nuclides of which

$$165 \text{ are even–even nuclei}$$
$$55 \text{ are even } Z, \text{ odd } N$$
$$50 \text{ are odd } Z, \text{ even } N$$
$$4 \text{ are odd–odd.}$$

However, it is the cosmic abundance which is more interesting since the abundance on the earth is dependent on other factors. For example most of the hydrogen and helium has escaped from the gravitational field of the earth. The cosmic abundance is arrived at by combining many different types of data.

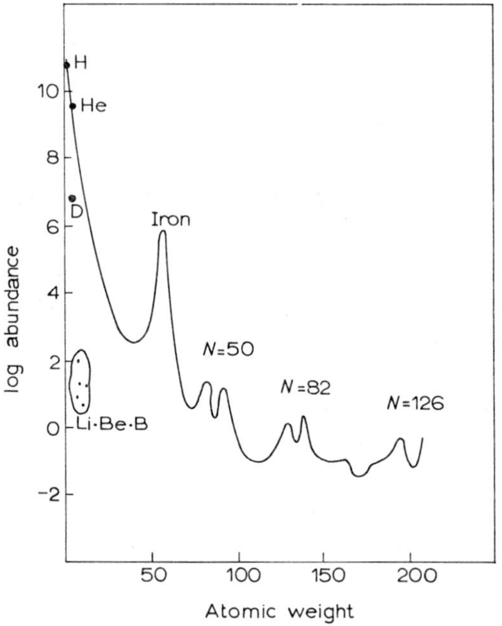

Rev. Mod. Phys. **29** (1957) 547

Fig. 15.1.   The abundance of the elements in the universe

In fig. 15.1 is plotted the relative abundance on a log scale vs A. The main features of the curve are as follows:
1) The curve is very complicated indicating that many element synthesis mechanisms are likely to be involved.
2) There is a steep exponential decrease up to $A=100$.
3) After $A=100$ the slope is quite small.
4) The nuclei D, Li, Be, B are unusually rare.

5) The peaks at low A at the nuclei which are multiples of the α-particle (^{4}He, ^{12}C, ^{16}O, ^{20}Ne, ^{24}Mg etc).

6) The prominent peak at mass 56, the iron peak.

7) The double peaks at $A = 80$ and 90; 130 and 138; 196 and 208, close to the occurrence of magic neutron numbers 50, 82 and 126.

Our theory of nuclear synthesis must then explain each of these features.

1.2 Cosmological theories. We consider first the various theories of the orgin of the universe itself. The age of the rocks on earth can be dated by assaying for ^{238}U ($t_{\frac{1}{2}} = 4.5 \times 10^{10}$ y) or for ^{40}K ($t_{\frac{1}{2}} = 1.3 \times 10^9$ y) and assuming that the ^{206}Pb in the uranium ore resulted from the uranium decay and that the isotope ^{40}K originally had a similar abundance to ^{39}K and ^{41}K. These yield 2.6×10^9 y for the age of the oldest rocks and about 4.5×10^9 y for the age of meteorites. Hence the latter figure is taken as the probable age of the solar system. Galaxies in outer space exhibit a red-shift in their spectra which if interpreted as a Doppler effect gives us their velocity. They are all flying away from our galaxy at tremendous velocities which increase with distance. If we calculate backwards they would have started at a point in space some 10^{10} y ago. The time scale is thus taken to be

$$\text{Age of Universe } 10\text{--}20 \times 10^9 \text{ y}$$
$$\text{Age of Stars } 5\text{--}10 \times 10^9 \text{ y}$$
$$\text{Formation of Solar System } 4.5 \times 10^9 \text{ y}$$

The cosmological theories proposed are as follows:

1) *Equilibrium theory*. This assumes an equilibrium condition at the beginning. With 8 MeV/nucleon the heavy elements have binding energies of ≈ 2000 MeV. This implies extremely high temperatures, in excess of 8×10^9 °K and densities of $\approx 10^7$ g/cm^3. The abundance would fall exponentially with binding energy and mass number throughout the periodic table. This would give too low abundances for the heavy elements by a factor of 10^{50}. Hence this theory is quite unsatisfactory.

2) *'Big Bang' theory*. This assumes a primordial fireball in which element synthesis and expansion began during a short period of time. This was first proposed by Alpher, Bethe and Gamow. It starts from a localised region of space containing mostly neutrons, say 10^{17}/cm^3. The initial conditions are a density of 10^{-7} g/cm^3, a temperature of $\approx 10^9$ °K (≈ 100 keV) and an initial rate of expansion as is observed at present via the galactic red-shift. In a few minutes some neutrons decay to protons, protons capture neutrons to form deuterons and the heavier nuclides are then built up by successive (n, γ) reactions followed by β-decay to stable nuclides. In the course of an hour or so the process is finished due to β-decay of uncaptured neutrons and reduction of density and temperature. The assumption is that today we see the products of this primordial cataclysm with a mean density of 10^{-30} g/cm^3 still expanding.

This theory gives the correct overall shape for the abundance curve since the (n, γ) cross sections increase up to $A \approx 100$ and then level off at an approximately constant value for $A > 100$ with anomalies at the magic numbers. However it fails to account for many details of the abundance curve such as the double peaks near the magic numbers, α-nuclei peaks and the iron peak.

Recently, possible support for this theory has been reported. An unexpectedly large intensity of background radiation of wavelength 7.4 cm and 3.2 cm has been observed from space. This could be interpreted as the tail of a black-body distribution of radiation of temperature about 3 °K. It is suggested that this resulted from the primordial cosmic fireball postulated in the 'big bang' theory.

3) *Continuous creation.* This theory due to Bondi, Hoyle and Gold was popular a few years ago and sought to avoid the abnormal initial condition described above. It assumed that new matter was created in the empty space between galaxies and this new matter then coalesced into new stars and galaxies. Thus as expansion proceeds the density of galaxies remains roughly constant. However such violation of conservation of mass and energy, even though infinitesimally slow on a laboratory scale, is not very satisfactory and the new support for the 'big bang' theory has led to the virtual abandonment of this theory.

Thus we note that the 'big bang' theory is at present the most successful cosmological theory. However it fails to explain many features of element abundance. Many of these features may be explained in terms of *long term* synthesis mechanisms. Burbidge, Burbidge, Fowler and Hoyle have outlined a detailed theory of element synthesis which was originally proposed in connection with the 'continuous creation' theory. However it is so successful that it must be largely correct. It can be merged with the 'big bang' theory if we assume that element synthesis is closely connected with stellar evolution and stellar energy sources. Thus the galaxies must contain element synthesising 'factories'. We must therefore consider the sources of stellar energy.

1.3 Stellar energy sources. The sun is a rather modest size star emitting energy at a rate of 4×10^{33} erg/g · sec, and this rate of energy emission has been going on for at least 10^9 years. Chemical reactions could not maintain this rate of energy production since for example complete combustion of a carbon sun would be complete in about a few thousand years. Gravitational energy is also insufficient by about a factor of 100 for an object of the sun's mass. So we are left with nuclear energy as a source.

We know that nuclear reactions can lead to energy release of a few MeV per particle so we could have an energy source of about 10^{19} erg/g. At a rate of energy release of 2 erg/g · sec such reactions could supply energy for 10^{11} years so they are indeed adequate. What then are the possible reactions? The composition of the sun is about 90 % hydrogen and helium in about equal quantities. Hence heavy element reactions such as fission are ruled out. We must turn to the so-called fusion reactions in which lighter nuclei combine to form heavier nuclei.

Suppose that we can invent reactions wherein four protons combine to form a helium nucleus. Then the energy balance is

$$4M_{\text{H}} - M_{\text{He}} = 4 \times 1.007825 - 4.002603 = 0.028697$$
$$= 26.7 \text{ MeV} = 4.3 \times 10^{-5} \text{ erg.}$$

Thus we get 10^{-5} erg per proton used and since 1 g contains $\approx 10^{23}$ protons we have 2×10^{18} erg/g available, again satisfactory. The rate of the reaction will depend critically on the temperature. At room temperature the mean thermal energy is about $\frac{1}{40}$ eV but at solar temperatures of 10–20 million degrees the mean energy is about 1 keV which is adequate for thermonuclear reactions to proceed at a reasonable rate.

There are two reaction cycles for 'burning' hydrogen.

1) Proton–proton chain

$$^1\text{H} + {}^1\text{H} \rightarrow {}^2\text{H} + \beta^+ + \nu + 0.42 \text{ MeV} \tag{1}$$
$$^1\text{H} + {}^2\text{H} \rightarrow {}^3\text{He} + \gamma + 5.5 \text{ MeV} \tag{2}$$
$$^3\text{He} + {}^3\text{He} \rightarrow {}^4\text{He} + 2{}^1\text{H} + 12.8 \text{ MeV.} \tag{3}$$

Two of the reactions (1) and (2) must occur for each reaction (3). The net effect is

$$4{}^1\text{H} \rightarrow {}^4\text{He} + 2\beta^+ + 2\gamma + 2\nu + 24.7 \text{ MeV.}$$

The neutrinos escape leaving 24.3 MeV. At the central solar temperature the mean reaction times for the reactions are (1) 14×10^9 y, (2) 6 sec, (3) 9×10^5 y.

The first reaction has never been observed in the laboratory. These lifetimes tell us that the hydrogen content of the sun will be reduced to $1/e$ of its initial value in a time about five times longer than the age so far (5×10^9 y).

Variations in the course of the proton–proton reaction could be

$$^1\text{H} + {}^1\text{H} \rightarrow {}^2\text{H} + \beta^+ + \nu \tag{1}$$
$$^1\text{H} + {}^2\text{H} \rightarrow {}^3\text{He} + \gamma \tag{2}$$
$$^3\text{He} + {}^4\text{He} \rightarrow {}^7\text{Be} + \gamma \tag{4}$$
$$^7\text{Be} + \beta^- \rightarrow {}^7\text{Li} + \nu + \gamma \tag{5}$$
$$^7\text{Li} + {}^1\text{H} \rightarrow {}^4\text{He} + {}^4\text{He} \tag{6}$$

or

$$^7\text{Be} + {}^1\text{H} \rightarrow {}^8\text{B} + \gamma \tag{5a}$$
$$^8\text{B} \rightarrow {}^8\text{Be} + \beta^+ + \nu \tag{6a}$$
$$^8\text{Be} \rightarrow {}^4\text{He} + {}^4\text{He} \tag{7a}$$

These cycles yield approximately the same energy but require ^{4}He to be already present for reaction (4). Also the temperature would have to be greater as higher coulomb barriers are involved. Thus these may play a role at a later stage in stellar history after a portion of the hydrogen is converted to helium. They also illustrate how heavier element synthesis begins to take place.

2) Another set of reactions was proposed by Bethe; it is the so-called ^{12}C cycle.

Reaction	*Energy release (MeV)*	*Neutrino energy lost (MeV)*	*Reaction time*
$^{12}\text{C}+{}^1\text{H} \rightarrow {}^{13}\text{N}+\gamma$ $\quad+$ 1.95			10^7 y
$^{13}\text{N} \rightarrow {}^{13}\text{C}+\beta^+ +\nu+$ 2.22		0.72	7 min
$^{13}\text{C}+{}^1\text{H} \rightarrow {}^{14}\text{N}+\gamma$ $\quad+$ 7.54			2×10^6 y
$^{14}\text{N}+{}^1\text{H} \rightarrow {}^{15}\text{O}+\gamma$ $\quad+$ 7.35			3×10^8 y
$^{15}\text{O} \rightarrow {}^{15}\text{N}+\beta^+ +\nu+$ 2.71		0.98	2 min
$^{15}\text{N}+{}^1\text{H} \rightarrow {}^{12}\text{C}+{}^4\text{He}$ $\quad+$ 4.96			10^5 y
Totals	26.73	1.7	

Note that the carbon which now must be present is not destroyed in the cycle but acts as a catalyst for burning four protons to helium. The net energy gain is 25.03 MeV.

These have been examples of hydrogen burning. In many stars the hydrogen is almost exhausted and the main constituent of the star is helium. In such stars gravitational attraction will have taken place, the density will have increased and the central temperature will be higher than that of the sun, perhaps 2×10^8 °K. Then helium burning is possible by a series of α-capture reactions.

$$^4\text{He}+{}^4\text{He} \rightarrow {}^8\text{Be} \quad -0.095 \text{ MeV}$$
$$^4\text{He}+{}^8\text{Be} \rightarrow {}^{12}\text{C}+\gamma+7.4 \quad \text{MeV}$$
$$^{12}\text{C}+{}^4\text{He} \rightarrow {}^{16}\text{O}+\gamma+7 \quad \text{MeV}$$

and so on up the table until the mean energy is insufficient to penetrate the coulomb barrier with sufficient probability. At still higher temperatures, further captures in heavier elements can proceed.

1.4 Element synthesis. The processes of element synthesis are then assumed to be as follows:

1) Hydrogen burning produces He from H and may also synthesize carbon and nitrogen to some extent.

2) Helium burning produces ^{12}C, ^{16}O, ^{20}Ne and other α-particle nuclei and the peaks in the abundance curve reflect this.

3) After a sufficient amount of C, O, Ne have been produced, the temperature will increase to perhaps 10^9 °K so that photo-reactions such as $^{20}\mathrm{Ne}(\gamma, \alpha)^{16}\mathrm{O}$ ($Q = -4.75$ MeV) can proceed and produce α-particles with high enough energy to produce capture in heavier nuclei. For example: $^{20}\mathrm{Ne} + \alpha \rightarrow {}^{24}\mathrm{Mg}$ etc. This is called the *α-process*.

4) As the temperature rises further a great variety of reactions becomes possible and an equilibrium condition for the abundance at a given temperature is reached. This condition can be predicted by an elaborate calculation and F. Hoyle has shown that the elements in the iron peak are preferentially formed. This is called the *e-process*.

5) Neutron capture is still the dominant process for forming the elements above $A = 25$ or so. There are however two situations in which this can occur. In the first case with moderate neutron density the capture process is slow compared to the β-decays which follow. Since nuclei with a magic number of neutrons have a small capture cross section, an excess abundance will be observed for those β-decay daughters which are magic. Neutron numbers 50, 82 and 126 account for the abundance peaks at $A = 90$, 138 and 208. This process is dominant from $A = 23$–46 and above the iron peak from $A = 63$–209. It is called the *s-process*. The neutrons are assumed to result from (α, n) reactions with the light elements.

In certain conditions however as in a stellar explosion the neutron density may be very high and neutron capture is much faster than the β-decay. Then a magic nucleus *before* decay will show less capture and its daughter will show a higher abundance. This produces a second set of peaks at $A = 80$, 130 and 194. In general this process synthesizes a different series of elements. It is called the *r-process*.

6) Finally we need one or more special processes to account for the anomalously small abundance of D, Li, Be and B. These are rare since they are rather easily burned up in stellar interiors. Thus their production can only take place in cool exterior regions of low density. This is referred to as the *x-process*.

The very heaviest elements beyond $Z = 82$ probably arise in supernovae explosions. The many fast neutrons resulting from various fusion reactions will give rise to successive neutron captures via the *r*-process producing elements up to the limit of the periodic table. Exactly this effect has been observed in high altitude thermonuclear weapon tests. Another confirmation comes from the decay of the light intensity following a supernova outburst. There is historical data on this from the one observed in 1054 by the Chinese (now called the Crab Nebula), the one observed by Tycho Brahe in 1572, the one observed by Kepler in 1604 and one observed in 1938. Other remnants are found still decaying whose original outburst was not recorded. All observations agree on a half-period of 55 days. This has been followed for the 1938 event over a diminution of a factor of 10^5. This agrees remarkably well with the only short half-life for spontaneous fission amongst heavy nuclides near the stability line, namely that for ^{254}Cf.

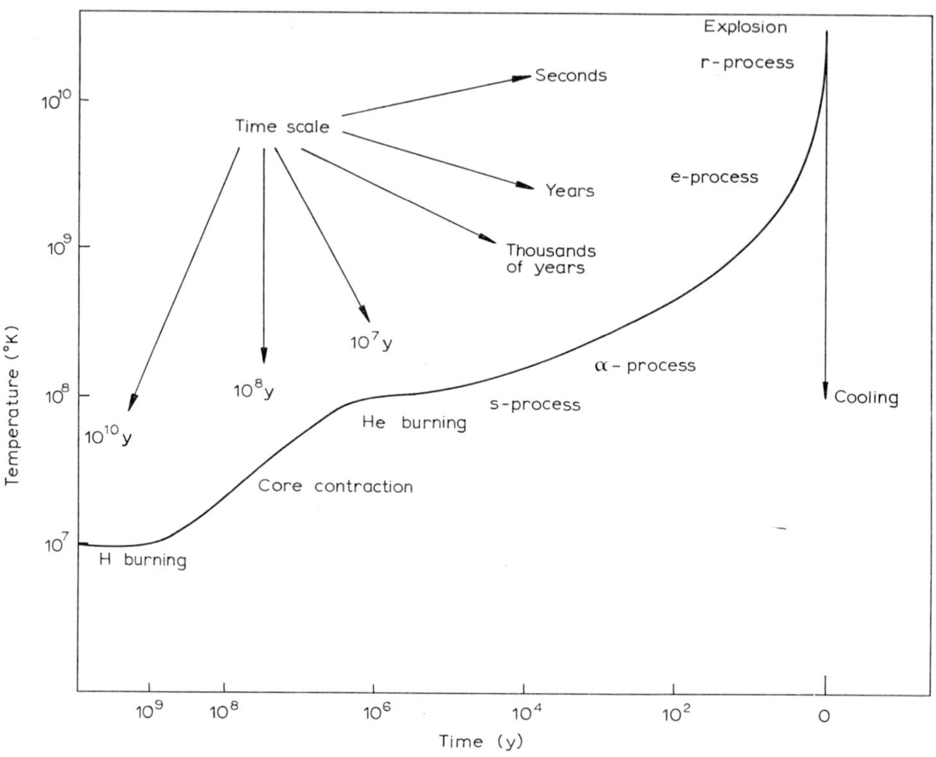

Rev. Mod. Phys. **29** (1957) 547

Fig. 15.2.   Schematic illustration of stellar evolution

Stellar evolution, somewhat idealised, then proceeds as shown in fig. 15.2. The time scales for the various nucleo-synthesis processses are

Hydrogen burning	10^6–10^9 y
Helium burning	10^7–10^8 y
s-process	in the red giant stage 10^5 y
α-process	10^2–10^4 years after branching to red giant stage
e-process	in explosive stage, seconds or minutes
r-process	100–1000 sec

In terms of temperature the processes proceed as follows:

a) Condensation of interstellar matter

b) Contraction

c) 10^6 °K;   D destroyed via D+D→He
d) 3×10^6 °K;   Li destroyed via Li+p→He
e) 4×10^6 °K;   Be destroyed via Be+p→He
f) 7×10^6 °K;   B destroyed via B+p→He
g) 8×10^6 °K;   proton–proton burning via ^{3}He→^{4}He
h) 10^7 °K;   proton–proton burning via ^{7}Be–^{7}Li→He
i) 1.7×10^7 °K;   hydrogen burning proceeds in a star with 1–1.5 times the solar mass for a long time (main sequence star).
j) $(1.7–3) \times 10^7$ °K; hydrogen burning in larger stars via carbon cycle.
k) 3×10^7 °K;   stars of a few solar masses. Oxygen and fluorine destroyed.
l) $(4–6) \times 10^7$ °K;   hydrogen becoming depleted at the centre but is still burning in the shell, the core contracts, the shell expands and we get a *red giant*.
m) 10^8 °K;   Star now 50 % depleted of hydrogen; helium burning now dominates. Mass ($\approx$1.3 solar masses) is now lost either (i) as a continuous red giant or (ii) suddenly in a nova or super-nova outburst.
n) Eventually all nuclear sources are exhausted. The mass is now about 0.6 solar mass. The remaining heat energy is then radiated over a long time scale. The star is a *white dwarf*.

It is obvious therefore that the origin of the elements is linked closely to the various possible stages of stellar evolution, which in turn are controlled by a variety of energy producing mechanisms.

2   *Cosmic rays*

The radiation from space which is continually bombarding the earth's atmosphere contains particles with energies far in excess of anything likely to be achieved with accelerators on earth. The debris of the high energy nuclear reactions occurring at the top of the atmosphere showers down upon us and contains examples of many of the fundamental particles we have discussed in chapter 12. Many of them were first observed in cosmic rays at high altitudes and many of the techniques of high energy physics were first developed for studying cosmic rays. The origin of cosmic rays is likely related to the ideas of stellar evolution and element synthesis we have just discussed.

Cosmic rays were discovered as early as 1900 by C. T. R. Wilson and about ten years later Hess showed their extraterrestrial origin by observing an increase in intensity with altitude when electroscopes were flown in balloons.

Several effects are due to the influence of the earth's magnetic field. The intensity of the radiation varies with magnetic latitude. Since the primary intensity varies inversely with momentum, this can be interpreted as an effect of the earth's magnetic field which fixes a minimum momentum for particles reaching the earth at a given latitude. Particles with lower momentum are bent away and miss the earth (fig. 15.3).

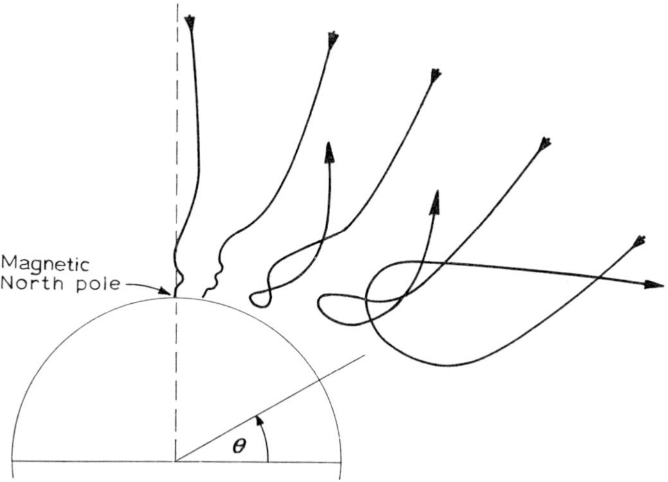

Fig. 15.3. The latitude effect for cosmic rays

Roughly the minimum momentum is given by $p_{\min} \approx 14.8 \cos^4 \theta$ GeV/c where θ is the magnetic latitude. It is clear then that a large fraction of the primaries are charged. With the aid of artificial satellites and sounding rockets, regions of trapped particles, both electrons and protons, have been found to surround the earth. These are the Van Allen belts. The charged particles spiral back and forth from pole to pole along the line of force, leaking out from time to time at the poles or when magnetic storms upset the field. One of the mechanisms supplying these belts is through β-decay of the neutrons produced in high energy cosmic ray reactions at the top of the atmosphere. The resulting proton and electron may then be trapped. However other sources must also play a role such as solar protons and electrons and the solar wind. The detailed mechanisms for supplying the various belts are not understood.

Many of the lower energy primary protons of energy less than 10–100 MeV originate in the sun since they show intensity variations associated with solar activity. Decreases in numbers of particles with energy less than 1 GeV are associated with maxima in sun-spot activity. These so-called Forbush decreases are apparently due to clouds of magnetic plasma filling the solar system and deflecting particles from the earth. Over shorter periods however increases in the number of particles with energies up to 10 GeV are observed to coincide with solar flares which must enable particles to be accelerated toward the earth.

Higher energy primary cosmic rays are also mostly protons and are not associated with the sun as they show no time variations, and are distributed isotropically in space. That the average intensity of cosmic rays has been constant for a long time can be inferred from a study of radio carbon dating. The nucleus ^{14}C with a half-life of 5600 years is produced in the atmosphere by cosmic ray neutrons through the reaction ^{14}N (n, p)^{14}C. A plant or animal while it is alive is in equilibrium with the atmospheric ^{14}C content through its biological processes. When the plant or animal dies, no further ^{14}C intake occurs and the ^{14}C content then decays with the 5600 year half-life. Thus by detecting the amount of ^{14}C remaining in something originally alive we can date the object (wood, leather etc.) if we assume the cosmic ray production rate of ^{14}C has remained unchanged over the period. When the dates can be checked by other methods, geological or archaeological, they indicate no appreciable change in cosmic ray intensity in the last 25 000 years.

The energy spectrum of the proton primaries extends up to at least 10^{20} eV and above 10^{10} eV follows a law $N(E)=\text{const }(E+5.3)^{-1.75}$, where E is in GeV. Not all the primaries are protons however and the primaries are distributed as follows: H 77 %, He 20 %, (Li, Be, B) 0.7 %, (C, N, O) 1.7 %, $(Z>10)$ 0.6 %. Nuclei as heavy as Fe have been identified as primaries. Less than 1 % of the primaries can be electrons or γ-rays. The primary distribution follows roughly the cosmic elemental abundance curve, but there are some significant differences. The Li, Be, B group are much more abundant relative to neighbouring elements, the elements with $Z\geqq 10$ are several times more abundant than in the cosmic abundance and C is several times more abundant in cosmic rays. This points to cosmic rays originating in a particular astrophysical set of conditions differing from the more general element synthesis conditions.

The energy density due to cosmic rays is about equal to the radiant energy of the stars ≈ 0.5 eV/cm^3 of space. This tells us that they cannot be isotropically distributed throughout *all* space as their total energy would then almost equal the rest mass of the universe, which is unreasonable. Thus it must be assumed that the bulk of the cosmic rays are confined within our own galaxy by the galactic magnetic field. From the magnitude of this field we can calculate that up to 10^{16}–10^{17} eV particles would be trapped in the galaxy. Since energies greater than this have been observed, at least some of the cosmic rays must originate outside the galaxy.

A mechanism for acceleration to such high energies has been suggested by Fermi. He assumed that local regions of high magnetic field could accelerate particles to high energy. From the distribution of elements in the cosmic rays we can deduce that the particles originate in the relatively cool halo region around supernovae or large stars. They must be cool to give the high abundance of Li, Be and B. The large carbon abundance also would indicate large stars where helium burning or carbon-cycle hydrogen burning is dominating. It must be pointed out, however, that no completely self-consistent picture of cosmic ray origins is yet available.

At sea level, we observe two general types of radiation. The *penetrating* component is composed mostly of muons resulting from decay of charged pions produced at the top of the atmosphere, together with some neutrons and a very few primaries. The *soft component* is made up of electrons, positrons and photons. These may sometimes occur in bursts containing many thousands of approximately coincident particles. These showers (fig. 15.4) are initiated mainly by the π^0 mesons produced

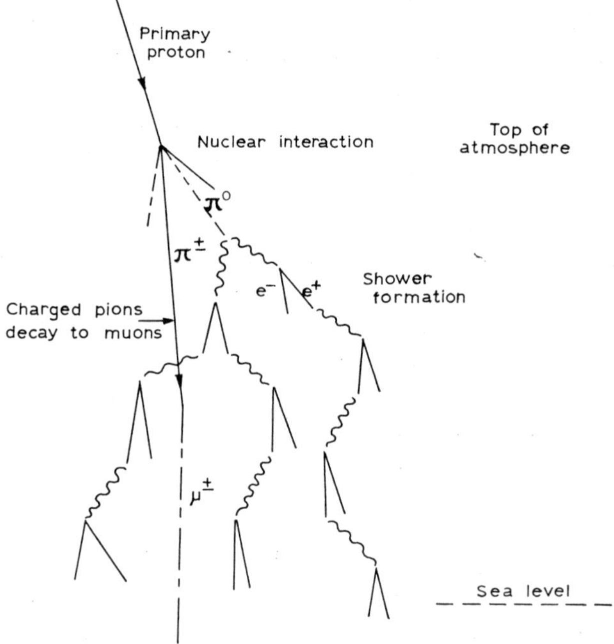

Fig. 15.4.   The origin of cosmic ray showers

in the high energy reactions which decay in flight ($\tau = 10^{-16}$ sec) to produce two γ-rays of high energy since the kinetic energy of the π^0 as well as its rest mass is shared between the photons. On the average after about 300 metres the γ-ray creates an electron–positron pair. These in turn create more γ-rays by bremsstrahlung and so we get a cascade process. The original energy of the π^0 is divided amongst the constituents until they finally strike the earth. The higher the energy of the initiating primary the larger will be the shower and the larger area it will cover on the earth. For example a shower initiated by a primary of 10^{14} eV energy will contain about 25–50 particles per square metre. The showers also give flashes of light resulting from Cerenkov radiation produced in the atmosphere and these are also observed.

Since we can relate shower size and intensity to primary energy, the observation of showers is almost the only way to study the higher energy regions of the primary spectrum. Several mechanisms might cause deviations from the power law for intensity vs energy in the ultra-high energy region. If it is assumed that the cosmic rays all originate in our own galaxy a cut-off at about 10^{17} eV might be expected. In addition interactions with the interstellar black body radiation recently observed would cause particles of energy $> 10^{18}$ eV to be less frequent than predicted by the power law. The recent observation of a 10^{20} eV shower is thus so extremely improbable on this basis that we may conclude that for particles of energy greater than 10^{17} eV, which must originate from outside our galaxy, the decrease with energy is not so rapid. The rates however are so low in this region that it will take a long time for these questions to be settled.

3 Controlled thermonuclear processes

For some time the idea of using the stellar energy production idea to produce energy in the laboratory has been studied. Most of the actual stellar energy reactions are much too slow. However some of the light element reactions do result in very favourable energy release.

$$\begin{aligned}
{}^2\text{H} + {}^2\text{H} &\rightarrow {}^3\text{H} + {}^1\text{H} + 4 \ \text{MeV} \\
&\rightarrow {}^3\text{He} + \text{n} + 3.3 \ \text{MeV} \\
{}^2\text{H} + {}^3\text{H} &\rightarrow {}^4\text{He} + \text{n} + 17.6 \ \text{MeV} \\
{}^2\text{H} + {}^3\text{He} &\rightarrow {}^4\text{He} + {}^1\text{H} + 18.3 \ \text{MeV}.
\end{aligned}$$

If all of this could be made to go we might get 7.2 MeV per deuteron burned of which the neutrons carry off 2.76 MeV. Although the energy release is much less than that in fission (≈ 200 MeV) the energy yield per unit mass is rather greater.

To penetrate the coulomb barrier with sufficient probability we must have about 100 keV energy. If we merely start with an accelerated beam we do not get an energy gain as most of the energy is used up in ionizing the target material. Hence we need to have the target material already ionized.

Also we know the most probable process is not one of the above but is $\text{D} + \text{D} \rightarrow \text{D} + \text{D}$, elastic scattering. So the target and projectile must be in a container in which many successive collisions can take place so that the reactions have a chance of taking place.

In fig. 15.5 we plot the cross section for the reaction vs energy in this low energy region. Since $E \propto kT$, 1 eV corresponds to a temperature of 1.16×10^4 °K and 10 keV corresponds to a temperature of 10^8 °K. At such high temperatures all atoms

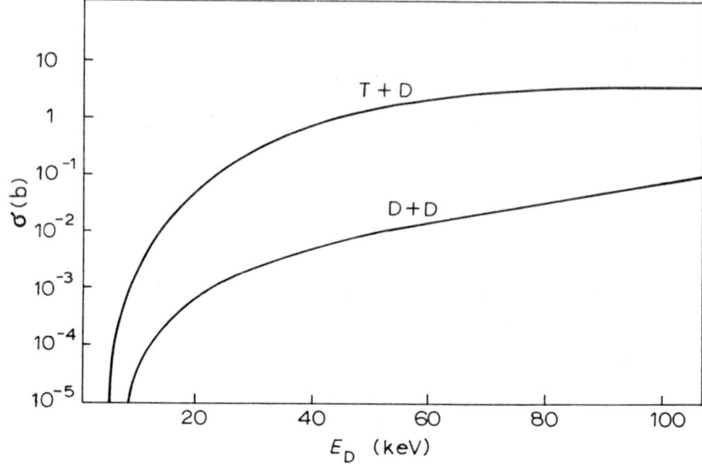

Fig. 15.5.   Low energy form of the fusion reactions

are stripped of all electrons. Completely ionized gas is called a plasma consisting of nuclei and electrons in rapid motion. No material wall can contain such a high temperature plasma. In the sun and other stars they are contained by gravitational pressure. In the laboratory we try to contain them in a magnetic field, if the plasma pressure is not too high. Hence the particle density is limited to about 10^{14}–10^{18} particles/cm^3.

We hope then to reach temperatures and containment times long enough to get energy generation in excess of the various energy losses – by neutrons escaping, by escape of plasma from confinement and by radiation losses from the plasma. This critical temperature is about 4×10^8 °K for D–D and about a tenth of this for D–T.

The problems that prevent this from being achieved so far are connected with the confinement of the plasma, which tends to become unstable in most magnetic field configurations that have been tried.

4   Nuclear photodisintegration

4.1   Introduction. When photons interact with nuclei, certain specific experimental features show up and it is convenient to consider photo-reactions separately from other nuclear reactions. For orientation let us consider the total cross section as a function of photon energy for a hypothetical fairly heavy nucleus (cf. fig. 15.6).

At the lowest energies the γ-ray can only shake the whole nucleus giving elastic nuclear Thomson scattering. This is an electric dipole effect with differential cross section

$$\frac{d\sigma}{d\Omega} = \frac{Z^2 e^2}{AMc^2}\left(\frac{1+\cos^2\theta}{2}\right).$$

At higher energies excited states of the target nucleus can be excited and inelastic as well as elastic scattering is possible above the first excited state of the nucleus. Above E_T, the photodisintegration threshold which corresponds to the lowest particle separation energy in the target, the photon can eject a proton, neutron or

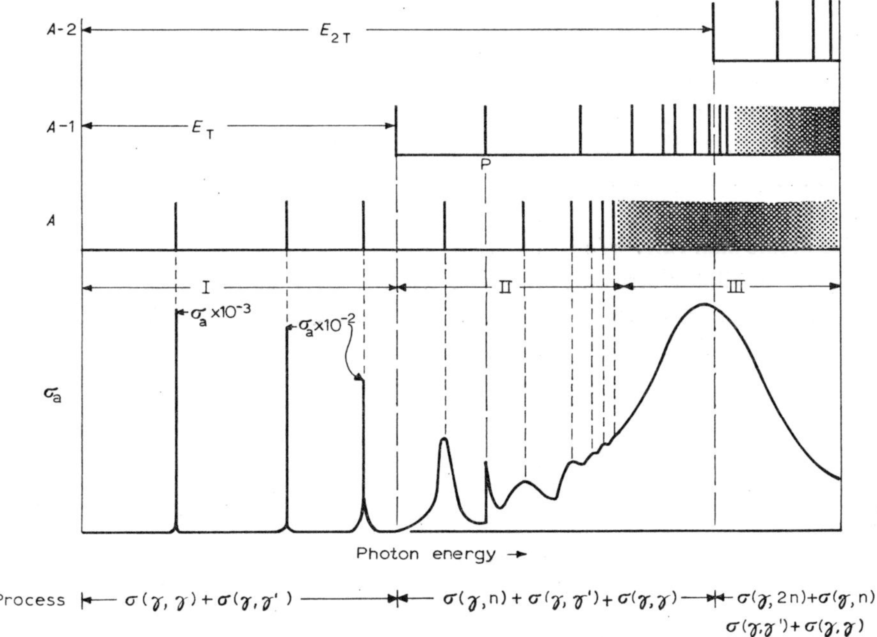

P. M. Endt and P. B. Smith (eds.), Nuclear Reactions 2 (North-Holland, Amsterdam 1962) p. 115

Fig. 15.6.   Photon interactions with nuclei as a function of energy

α-particle from the nucleus. The shape of the cross section just above the threshold will depend on how close a level of the target nucleus (the compound nucleus in this case) is to the threshold energy. Structure may also be seen when a particular level of the product nucleus is closely similar in structure to the ground state of the target less one nucleon. That is, the product nucleus level is a parent of the target nucleus ground state. The cross section will show an increase when a transition to this product nucleus level becomes energetically possible. At higher energies the levels broaden and overlap, except in the lighter nuclei, and the cross section becomes smooth. However it continues to increase and somewhere between threshold and 30 MeV a very large peak in the cross section occurs. It is roughly of dispersion shape and has a width of 3–8 MeV. It can be represented by

$$\sigma(E) = \sigma_0 \frac{E^2\Gamma^2}{(E_0^2 - E^2)^2 + E^2\Gamma^2}. \tag{15.1}$$

The angular distribution of the product particles indicates that this 'giant photo-nuclear resonance' is due mainly to electric dipole absorption. Of course it is actually made up of many individual states of the excited target nucleus ranging from 5–10 in a light nucleus to many millions in a heavy nucleus. These share the probability of electric dipole absorption with the distribution seen as the giant resonance. Above the giant resonance region the cross section falls fairly slowly until the threshold for meson production is reached at about 140 MeV when the cross section rises again. The most striking feature of the photodisintegration cross section is the giant dipole absorption resonance and we now consider this in more detail.

4.2   The dipole sum rule. We can start by considering the maximum possible nuclear dipole oscillator strength when we assume that all the Z protons and N neutrons in the nucleus participate. For atoms, the Thomas–Reiche–Kuhn sum rule states that the total dipole oscillator strength summed over discrete atomic levels and those in the continuum is simply equal to Z, the number of charges in the system. For nuclei, this has to be modified to take account of the relative motion between a nucleon and the recoiling nucleus and also to take account of the fact that the nuclear force, not just the coulomb force, operates between the nucleons.

The oscillator strength of the kth level located at energy E_k is defined as

$$f_k = \frac{2ME_k}{\hbar^2}\left|\int \psi_k^* z\psi_0\, d\tau\right|$$

$$= \frac{2ME_k}{\hbar^2}|z_{k0}|^2 \tag{15.2}$$

where M is the nucleon mass, and z_{k0} is the matrix element connecting the ground state ψ_0 with the kth excited state ψ_k via the dipole operator, z. The integrated absorption cross section is then

$$\int \sigma_k(E)dE = \frac{2\pi^2 e^2\hbar}{Mc}\, q_k^2 f_k \tag{15.3}$$

where q_k is the effective charge of the nucleon making the transition in units of e. Finally integrating over all states k connected by an electric dipole transition with the ground state,

$$\int \sigma(E)dE = \frac{2\pi^2 e^2\hbar}{Mc}\sum_k q_k^2 f_k. \tag{15.4}$$

We must now evaluate the effective charges of the nucleons. Suppose a proton of charge e is given a displacement r_i relative to the centre of the mass. Then there is a recoil of charge $(Z-1)e$ and mass $(A-1)$ which is displaced $R_i = -r_i/(A-1)$ from the centre of mass. The induced dipole is

$$
\begin{aligned}
D_i &= e[r_i + (Z-1)R_i] \\
&= er_i \left[1 - \frac{Z-1}{A-1} \right] \\
&= er_i \left[\frac{A-Z}{A-1} \right] \\
&= er_i \frac{N}{A-1}.
\end{aligned}
\tag{15.5}
$$

Now expressing r_i in terms of the position of the proton with respect to the recoiling nucleus,

$$
\begin{aligned}
\rho_i &= r_i - R_i \\
&= r_i \left(\frac{A-1+1}{A-1} \right) \\
&= r_i \frac{A}{A-1}
\end{aligned}
\tag{15.6}
$$

and

$$
D_i = e \frac{N}{A-1} \frac{A-1}{A} \rho_i = \frac{N}{A} e\rho_i.
\tag{15.7}
$$

Hence the effective charge of the proton is Ne/A.

Treating the neutron similarly, the recoil of charge Z, mass $A-1$ is displaced $R_i = -r_j/(A-1)$ where r_j is the neutron displacement.

$$
\begin{aligned}
D_i &= -\frac{Z}{A-1} er_i \\
&= -\frac{Z}{A} e\rho_j.
\end{aligned}
$$

Thus the effective charge of the neutron is $-Ze/A$. In general for other electric multipoles of order L

$$
q(\text{proton}) = \left(\frac{A-1}{A} \right)^L + (-)^L \left(\frac{Z-1}{A^L} \right)
$$

$$
q(\text{neutron}) = (-1)^L \frac{Z}{A^L}.
$$

Substituting the appropriate values of q into the dipole absorption integral we get

$$\int \sigma(E)\mathrm{d}E = \frac{2\pi^2 e^2 \hbar}{Mc}\left[\sum_{i=1}^{Z}(N/A)^2 f_i + \sum_{j=1}^{N}(-Z/A)^2 f_j\right]$$

$$= \frac{2\pi^2 e^2 \hbar}{Mc}\frac{NZ}{A}\beta \tag{15.8}$$

where β is a correction term to allow for enhancement when nuclear exchange forces are included. For an equal mixture of ordinary and exchange forces $\beta = 1.4$. For no exchange forces $\beta = 1$. This integral has been measured for many elements up to about 25–30 MeV photon energy. When the integral is plotted in units of $0.06NZ/A$ (corresponding to $\beta = 1$) vs Z we obtain fig. 15.7.

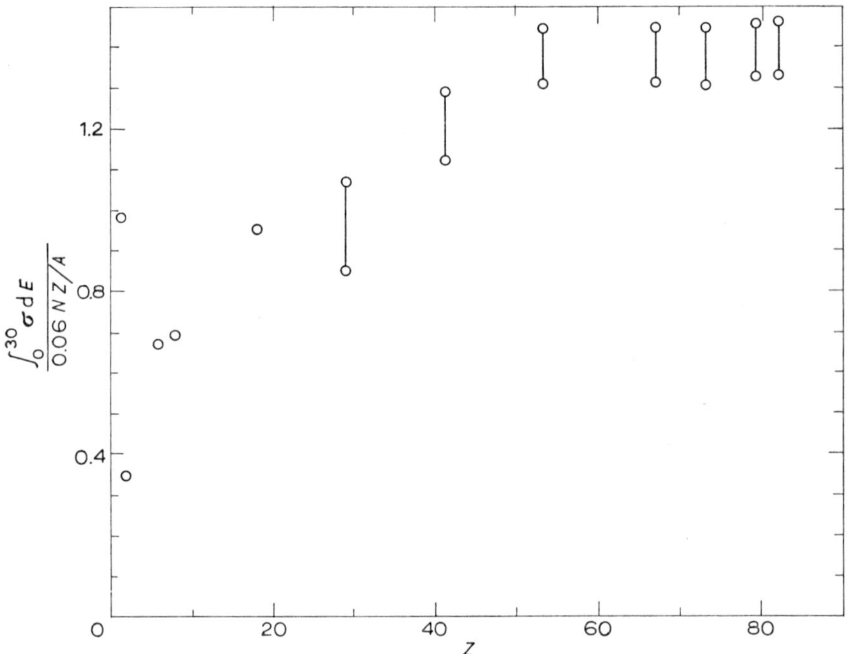

P. M. Endt and P. B. Smith (eds.), Nuclear Reactions **2** (North-Holland, Amsterdam 1962) p. 152

Fig. 15.7.   The integral of the giant resonance as a function of Z. The bars represent the uncertainties resulting from extrapolation of the absorption cross section of the heavier nuclei

The theoretical sum rule is exhausted for $\beta = 1.4$ for $Z > 50$. The approximately equal mixture of ordinary and exchange forces implied by high energy n–p scattering is thus confirmed. The fact that the observed integral exceeds the theoretical value indicates that our neglect of absorption of multipoles higher than dipole may be subject to question. The falling off below $Z = 50$ is attributed to the fact that the experimental integration was not carried to high enough energy to include all the dipole strength.

4.3 Energy and width of the dipole resonance. A similar simplified approach may be used to estimate the energy at which the giant resonance peaks. The polarisability, the induced dipole moment per unit field strength, is related to this energy by

$$\varepsilon_0 = \frac{\hbar c}{2\pi^2}\,\frac{1}{E_0^2}\int \sigma(E)\mathrm{d}E. \tag{15.9}$$

Substituting the value for the dipole integral

$$E_0 = \left[\frac{e^2\hbar^2}{M}\,\frac{NZ}{A}\,\beta\,\frac{1}{\varepsilon_0}\right]^{\frac{1}{4}}. \tag{15.10}$$

Migdal* has calculated the polarisability ε_0, assuming two interpenetrating neutron and proton fluids and using the liquid drop model,

$$\varepsilon_0 = \frac{e^2 R^2 A}{40 a_{\mathrm{a}}} \tag{15.11}$$

where a_{a} is the coefficient for the symmetry energy term in the semi-empirical mass formula. From this we find $E_0 \approx 82 A^{-\frac{1}{3}}$ MeV. If we plot experimental values of E_0 as $E_0 A^{\frac{1}{3}}$ vs A (fig. 15.8) we find good agreement for $A > 100$. Below this the

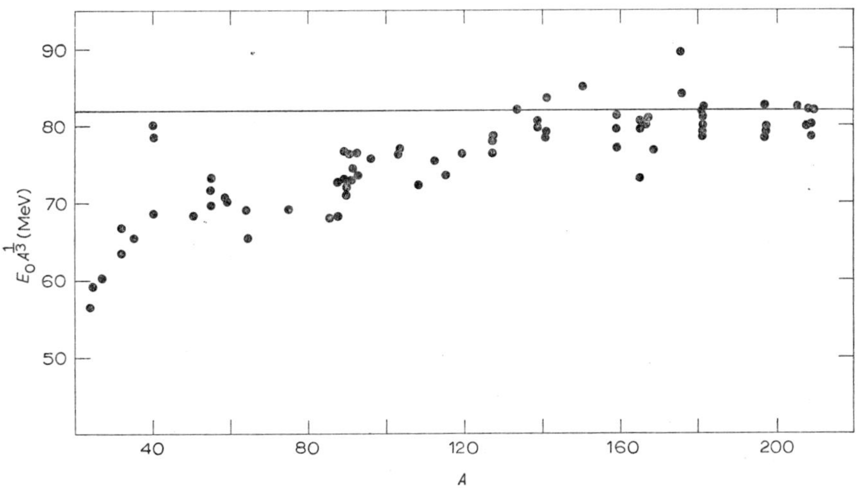

P. M. Endt and P. B. Smith (eds.), Nuclear Reactions 2 (North-Holland, Amsterdam 1962) p. 164

Fig. 15.8.   The energy of the giant resonance as a function of A

experimental E_0 is lower than the theoretical value which may again indicate that the true centre of gravity of the giant resonance should lie higher than that estimated.

* A. Migdal, J. Phys. USSR **8** (1944) 331.

The giant resonance widths lie between 3 and 8 MeV. The narrower widths are found near closed shells. The largest widths are found in the deformed nucleus regions. This is understood to be due to the superposition of two giant resonances corresponding to oscillations along the two short axes and the long axis of an axially symmetric deformed nucleus. The former oscillation will have higher energy and be twice the intensity of the latter. The two separate peaks have been identified in a few cases e.g. in Ho and Er (fig. 15.9).

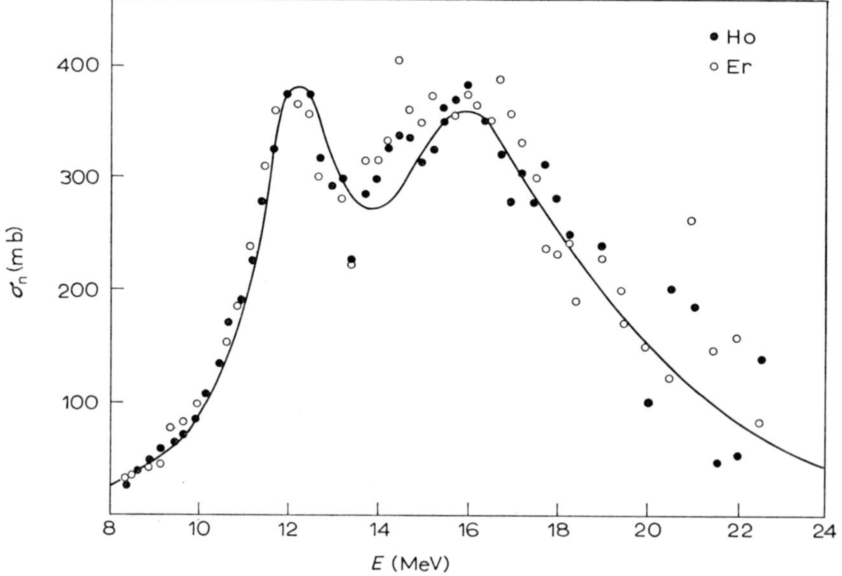

P. M. Endt and P. B. Smith (eds.), Nuclear Reactions **2** (North-Holland, Amsterdam 1962) p. 169

Fig. 15.9.   The split giant resonance for deformed nuclei

4.4 The dipole resonance in light nuclei. The photodisintegration of the deuteron has already been discussed. The cross section peaks at 4.4 MeV but is otherwise featureless. For helium the peak occurs at 27.5 MeV and the cross section curve is again featureless. This lack of structure is expected since no excited states of the nucleon or of the mass three system are known. For heavier nuclei such as Be, C etc. structure can be seen. Many details of the dipole absorption can be inferred from a study of the inverse reaction (p, γ_0). By identifying the ground state electric dipole transitions following proton capture in the product nucleus, the variation of the dipole absorption cross section with energy can be studied with good energy resolution. Most of the photo-reaction studies have been carried out using a bremsstrahlung source. To derive the cross section as a function of energy requires the

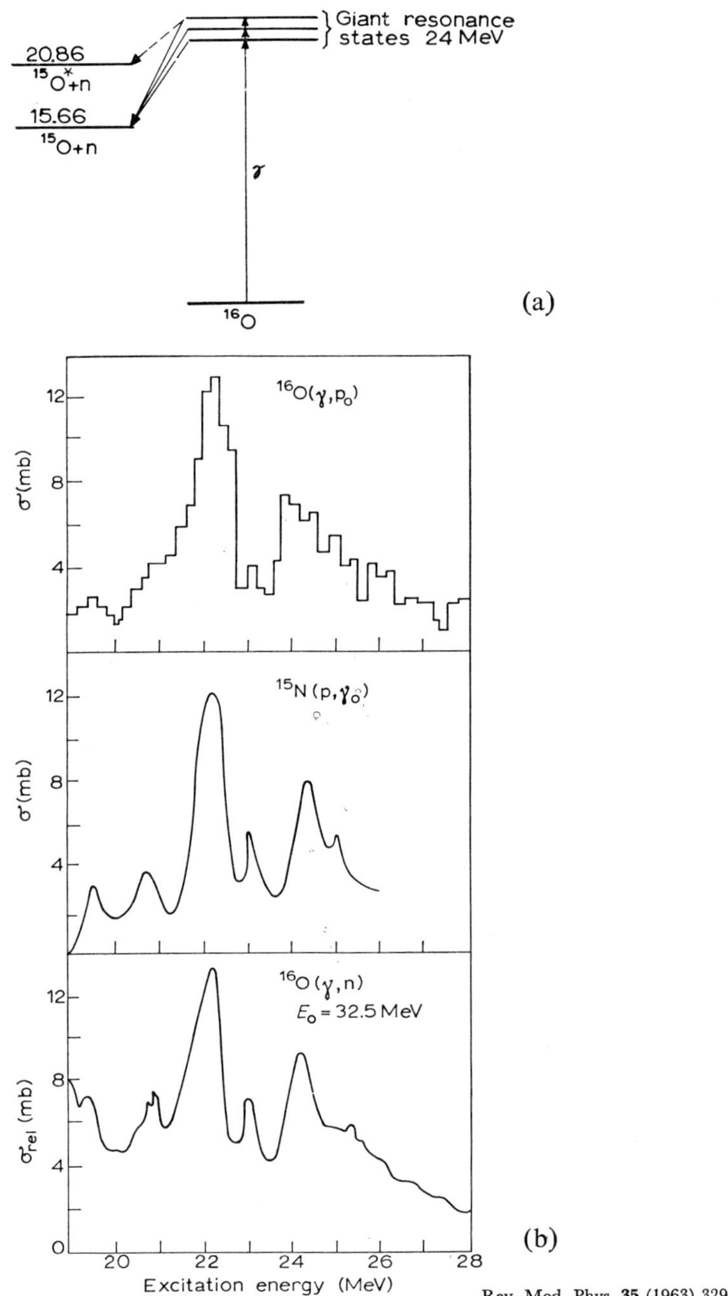

Rev. Mod. Phys. **35** (1963) 329

Fig. 15.10.   a) States of the giant resonance observed by detecting the neutrons from the γ–n reaction. b) ^{16}O(γ, p), ^{15}N(p, γ) and ^{16}O(γ, n) excitation functions

bremsstrahlung spectrum to be unfolded from the observed curve of yield vs maximum bremsstrahlung energy. This introduces uncertainty into the yield and poor energy definition. Even with photon sources resulting from positron annihilation in flight or by the tagged photon technique, resolution of better than a few hundred kilovolts is difficult to achieve. A better method is to use a pulsed source of bremsstrahlung from typically an electron linear accelerator and measuring the resulting neutron spectrum by time of flight. Over the neutron energy range $E_\gamma(\max) - S_n > E_n > E_\gamma(\max) - S_n - E_1$ where E_1 is the energy of the first excited state of the product nucleus (^{15}O in the case shown in fig. 15.10a*) and $E_\gamma(\max)$ is the maximum bremsstrahlung energy, only ground state neutrons are seen and their energies will correspond to the excitation of the component states of the giant resonance less S_n. Lower energy neutrons may correspond to either ground or excited state neutron groups. These can be sorted out to some extent by changing the maximum bremsstrahlung energy.

This same region can also be studied by the inverse reaction ^{15}N(p, γ_0)^{16}O. We expect that at such high energy the effect of the coulomb barrier on the proton widths would be not important so that the yield curve for the (p, γ) and (γ, n) reactions should be essentially identical. This is found to be the case, as shown in fig. 15.10b.

4.5 Microscopic theory of the dipole resonance. We must now consider how this giant dipole absorption resonance, which we have so far described in terms of a featureless liquid drop, can be explained on the independent particle model of nuclear structure. Wilkinson** has developed a general description

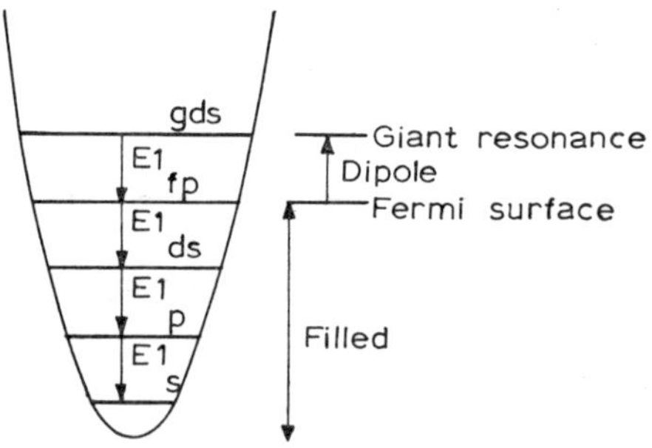

Fig. 15.11.   The single particle model of the giant resonance

along these lines. In the simple harmonic oscillator independent particle model, which was the starting point for the shell model, we had evenly spaced levels ordered according to the orbital angular momentum l (fig. 15.11). Transitions

* F. W. K. Firk, Nucl. Phys. **52** (1964) 437.
** D. H. Wilkinson, Ann. Rev. Nucl. Sci. **9** (1959) 1.

between these states would be electric dipole as we require. A certain number of the lower states would be filled and the next excited single particle state one oscillator spacing above would decay by E1, to the level at the top of the Fermi sea.

However we know that the spin–orbit force splits up these states and mixes states of the different oscillator levels. How can we get a concentration of states of the correct spin and parity which combine with the ground state via dipole transitions? Wilkinson's argument is as follows. Because of the Pauli principle only those states near the top of the Fermi sea are available to combine with unbound states. This will be subshells of a major l-shell of definite parity. The effect of using a finite oscillator is partially to restore the clustering of states around the evenly spaced harmonic oscillator states. Finally the dipole transition picks out states differing by a single quantum jump and thus harmonic oscillator states are favoured by the size of the matrix elements. Detailed calculations along these lines show that the main independent particle transitions take place from the closed shells that lie within an oscillator spacing of the Fermi surface. These transitions exhaust the dipole sum and the energies agree well with the giant resonance positions.

For example Elliott and Flowers* and Brown et al.** have used shell model wavefunctions to calculate the energies and strengths of the five E1 transitions in ^{16}O connecting the closed 1p shell to the next excited 2s1d shell. In an even–even nucleus with 0^+ ground state the excited states must have $J^P = 1^-$, $T = 1$ and in ^{16}O are located at 13.1, 17.3, 20.4, 22.6, and 25.2 MeV excitation. The two highest states contain almost all the dipole strength and constitute the giant resonance in ^{16}O. This structure is clearly seen in the photoneutron spectrum for neutrons leading to the ground state of ^{15}O and are also identified in the inverse reaction ^{15}N(p, γ_0)^{16}O. The energies and strengths observed agree excellently with those calculated. Similar comparisons have been made for nuclei up to Ca.

5 *Resonant absorption of γ-rays*

Since electromagnetic radiation interacts predominantly with the atomic electrons in matter in several ways, the direct observation of purely nuclear absorption is difficult as it is masked by the atomic and field effects. In photodisintegration we pick out the nuclear effect by observing the specific nuclear particles or radioactivity produced resulting from excitation of an unbound state. In a few cases we are able to observe absorption between bound states where the decay of the excited state leads to an isomeric level whose decay can then be observed (fig. 15.12). In

* J. P. Elliott and B. H. Flowers, Proc. Roy. Soc. London **A242** (1957) 57.
** G. E. Brown, Castillejo, and Evans, Nucl. Phys. **22** (1961) 1.

this way the resonant energy, E_R, is picked out of a continuous spectrum of γ-rays.

We now wish to consider the resonant absorption in more detail. The cross section for absorption is $\sigma_\gamma = 4\pi\lambdabar^2 g\Gamma_\gamma/\Gamma$ from our general reaction formula. For radiation of energy 500 keV and for $\Gamma_\gamma \approx \Gamma$, the cross section is about 10^4 barns.

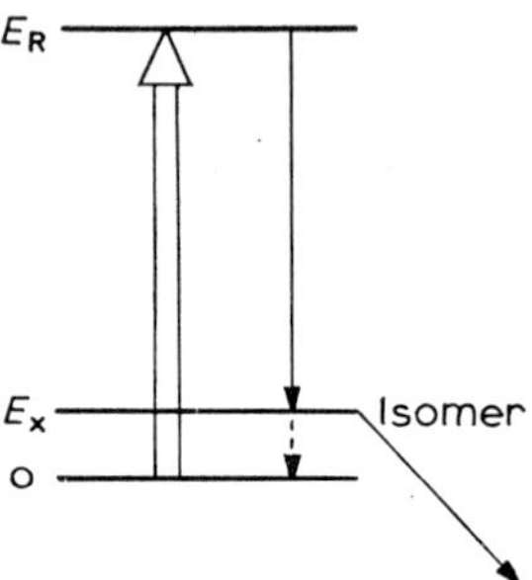

Fig. 15.12.   Resonant absorption leading to an isomeric state

The *electronic* cross section for absorption of similar radiation in a light element is about 10^2 barns. So it would seem that observation of the nuclear absorption should be easy. But this is not the case since for the above cross section formulation to be applicable we must have γ-ray energy resolution of the order of the radiative width, a small fraction of an electron volt. We do not have monochromatic γ-radiation available except at a few energies but use a bremsstrahlung continuous spectrum and the cross section we observe is averaged over the energy interval ΔE, accepted by the detector. This effective cross section is then $\bar{\sigma}_\gamma = 2\pi^2\lambdabar^2 g\Gamma_\gamma/\Delta E$. Since Γ_γ is less than 1 eV and ΔE is typically 1 MeV, $\Gamma_\gamma/\Delta E < 10^{-4}$ and our observed cross section is much reduced. The differential cross section for the electronic scattering is strongly peaked forward as we have seen so we can observe nuclear absorption at a backward angle. In this way absorption lines have been observed in a bremsstrahlung spectrum at 145° from boron and carbon and radiative widths of the order of 0.5 eV deduced from the cross sections.

Instead of exciting the level using a white spectrum, it would be simpler to use radiation of exactly the right energy which we can derive from decay of the same level as that we wish to excite (fig. 15.13). The cross section would be an average over the line shape and the effective cross section $\bar{\sigma}_\gamma = 2\pi\lambdabar^2 g\Gamma_\gamma/\Gamma$. In the case of bound levels only radiative processes are possible and the γ-ray width Γ_γ plus the width for internal conversion equals the total width.

$$\bar{\sigma}_\gamma = 2\pi\lambdabar^2 g\,\frac{1}{1+\alpha} = \pi\lambdabar^2\,\frac{2J'+1}{2J+1}\,\frac{1}{1+\alpha}$$

where J', J are spins of the excited and ground states. This ideal experiment we have described is however not straightforward, because of recoil and Doppler shifts in the transition energy.

In the emission of resonance radiation of energy E_γ from a stationary source atom of mass M, a momentum $p_\gamma = E_\gamma/c$ must be given the emitting atom. It recoils with energy $\frac{1}{2}E_\gamma^2/Mc^2$ and the emitted radiation falls short of the transition energy by this amount. The source radiation then has energy

$$E_s \approx E_\gamma - \frac{p_\gamma^2}{2M} = E_\gamma - \frac{E_\gamma^{\,2}}{2Mc^2}.$$

In the absorption process the energy required for resonance is

$$E_\alpha \approx E_\gamma + \frac{E_\gamma^2}{2Mc^2}$$

since the absorbing atom also takes up momentum. Therefore the emission and absorption energies are shifted by $2R = E_\gamma^2/Mc^2$. For $A = 50$ and $E = 1$ MeV the recoil energy is about 10 eV. We recall that $\Gamma \approx 10^{-2}$ eV and it is clear that no resonant absorption could take place without arranging to supply the recoil energy.

The second effect is the thermal motion of the source and absorber atoms which broadens the lines by the Doppler effect. This Doppler width is

$$\Delta = p_\gamma (2kT/M)^{\frac{1}{2}} = \frac{E_\gamma}{c}(2kT/M)^{\frac{1}{2}}$$

and has the effect of increasing the overlap between the radiation energy spread and the resonance width.

We note the relative size of the effects in the case of atomic radiation resonance in Na and the cases of nuclear resonance in two cases.

	λ (cm)	E_γ (eV)	Γ (eV)	2Δ at 300 °C (eV)	$2R$ (eV)
Na atom	5.89×10^{-5}	2.11	4.5×10^{-8}	6.6×10^{-6}	2.1×10^{-10}
^{198}Hg nucleus	3×10^{-10}	4.12×10^5	2.1×10^{-5}	0.4	0.9
^{57}Fe nucleus	8.6×10^{-9}	1.44×10^4	4.6×10^{-9}	0.02	0.004

We note that in the atomic case $R \ll \Delta$ and resonant absorption is easily observed while in the nuclear case $R \approx \Delta$ and the absorption is extremely weak or absent.

The recoil and Doppler effects make the size of the absorption very sensitive to Γ and thus if it can be observed it is a sensitive method for the measurement of Γ and hence of the decay lifetime of the level. Several methods have been used to restore the recoil energy loss. The source may be mounted on the tip of a high

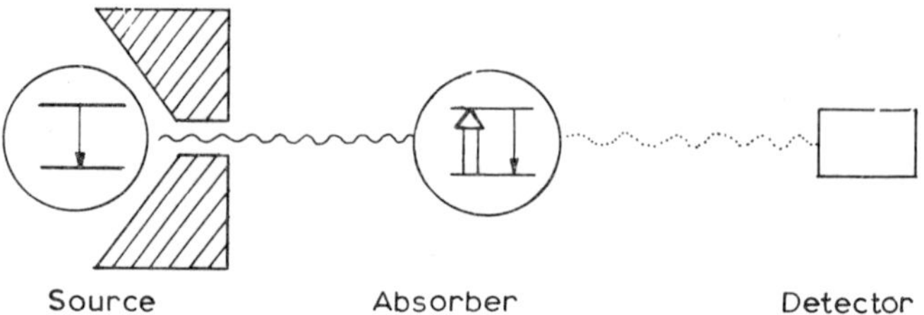

Fig. 15.13.   Resonant absorption of a γ-ray

speed rotor and the Doppler effect used to increase the frequency of the emitted radiation, that is its energy. The velocity required is given by $E_\gamma v/c = E_\gamma^2/Mc^2$. Thus for mass 200 and $E_\gamma = 500$ keV the velocity required is 8.2×10^4 cm/sec. In some cases the source may be heated to temperatures in excess of 1000 °C. The Doppler broadening is increased so that the recoil energy shift is partially compensated. Finally in other cases a preceding nuclear transition or reaction may supply the decaying atom with the required velocity so that again the Doppler shift compensates the recoil shift.

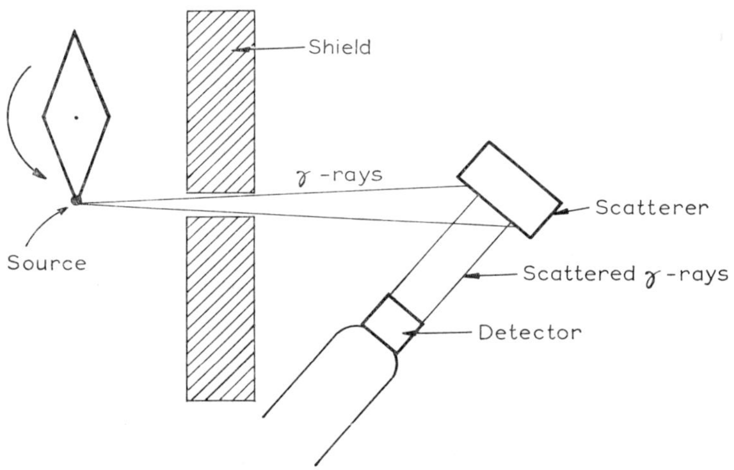

Proc. Phys. Soc. **64** (1951) 76

Fig. 15.14a.   Resonant scattering: experimental set-up

Nuclear resonance was first demonstrated by Moon and his collaborators* at Birmingham in 1951. The decay of ^{198}Au was studied.

$$^{198}\text{Au} \;\rightarrow\; ^{198}\text{Hg*} + \beta^- + \bar{\nu}$$
$$^{198}\text{Hg*} \;\rightarrow\; ^{198}\text{Hg} + \gamma + 412 \text{ keV}.$$

* P. B. Moon, Proc. Phys. Soc. London **64** (1951) 76.

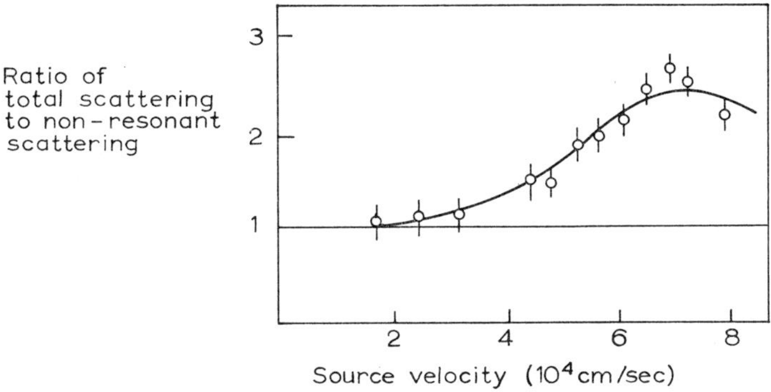

Proc. Phys. Soc. 64 (1951) 76

Fig. 15.14b.   Resonant scattering: the result

The source was mounted on the tip of a high speed rotor and resonance was looked for in the radiation scattered from a Hg sample (fig. 15.14a). When the Doppler shift introduced at a particular rotor speed just compensates the recoil energy loss of about 0.8 eV, resonance scattering is observed. A width for the 0.412 MeV level of ^{198}Hg of $(2.1\pm0.4)\times10^{-5}$ eV was derived. This corresponds to 3.1×10^{-11} sec with an error of about 5×10^{-12} sec, just beyond the limit of present electronic techniques.

The method of compensation by heating the source has been applied* to the 161 keV state of ^{117}Sn. A 0.5 mCi source of ^{117}Sn was heated in an oven to 700 °C. An increase in the resonance scattering of only 1 % was observed as the temperature was changed from 27 °C to 700 °C. However this allowed a determination of the mean lifetime of the state with an error of about 40 %.

The method of compensation using reaction kinetics has been applied also by Metzger and his group** to many cases. A level is excited by inelastic scattering or another reaction. The de-excitation γ-rays are then resonantly scattered from a sample of the same material. An example (Metzger 1961) is the 1.61 MeV level in ^{25}Mg which is the $\frac{7}{2}$ member of the $K=\frac{5}{2}$ rotational band based on the ground state. The level is excited by the ^{25}Mg(p, p')^{25}Mg* reaction (fig. 15.15). The ^{25}Mg* travels forward along with the centre of mass. As the γ-ray is emitted it is thus Doppler shifted up in energy which compensates the recoil loss. Resonant scattering is observed and then the self-absorption of the resonant radiation is also measured. The analysis must also include corrections for the angular correlations at each stage. In this case a mean life of $(2.5\pm0.5)\times10^{-14}$ sec was derived. This corresponds to a $B(E2)$ for the transition which is 53 times the single particle

* F. Metzger and W. B. Todd, Phys. Rev. **95** (1953) 627.
** F. Metzger, Prog. Nucl. Phys. **7** (1959) 53.

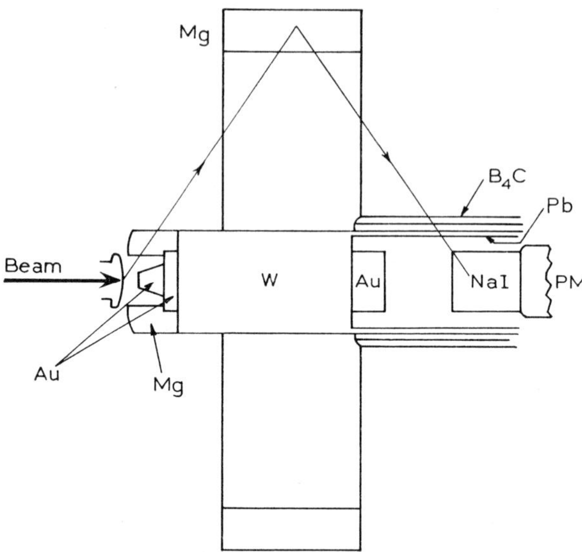

Phys. Rev. **123** (1961) 1386

Fig. 15.15.   Resonant scattering with the recoil compensated by the reaction kinematics

(Weisskopf) unit demonstrating the collective enhancement of this transition.

In certain special cases the recoil energy loss can be essentially eliminated. We have assumed so far that the target nuclei in a solid, liquid or gas were all equally free to move as free particles and take up the full recoil energy $E_\gamma^2/2Mc^2$. But suppose now that the target atom is chemically bound into a solid lattice and that the γ-ray energy is low enough so that the recoil atom is not freed from its chemical bond. Then the recoil is taken up not by the target atom alone but by all the neighbouring atoms as well, in fact by the whole crystal. The mass M in the recoil energy formula then becomes enormous and the recoil energy loss is very small. Under these conditions the radiation emitted by such a crystal contains an extremely sharp line of natural width, Γ_γ, and unshifted frequency. The same effect takes place upon absorption. In 1958 Mössbauer* observed the recoilless emission in ^{191}Ir in which he cooled both source and absorber to reduce thermal Doppler broadening. He was able to measure the line width Γ by Doppler shifting the source by a small amount (fig. 15.16, 15.17). Velocities of a few cm per second correspond to the natural line breadth. Compare the velocity of 10^4 cm per second required to compensate the full recoil in the Moon experiments. This phenomenon of absorption of momentum by a solid lattice without appreciable absorption of

* R. L. Mössbauer, Naturwiss. **45** (1958) 538.

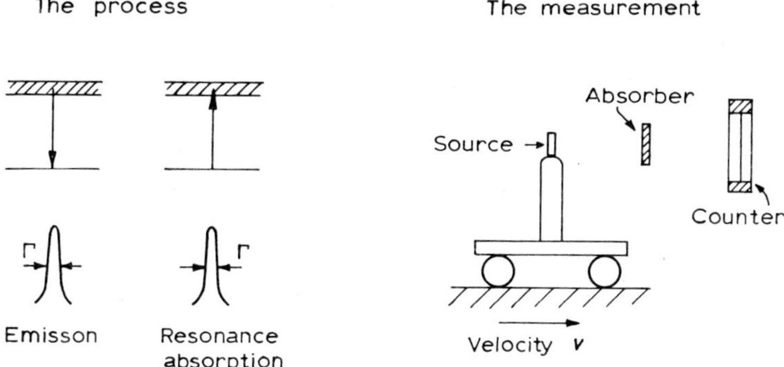

The result

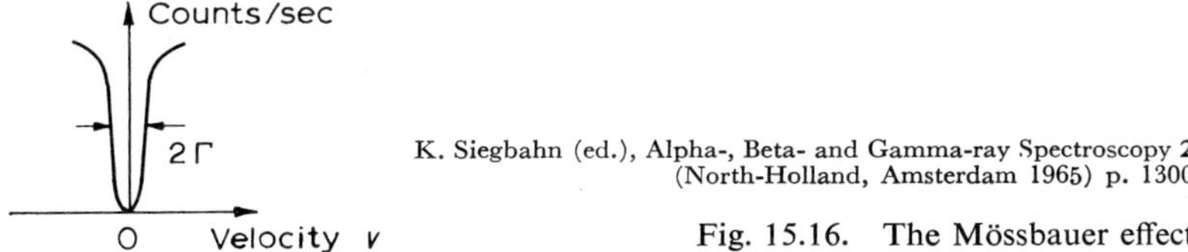

K. Siegbahn (ed.), Alpha-, Beta- and Gamma-ray Spectroscopy **2**
(North-Holland, Amsterdam 1965) p. 1300

Fig. 15.16.   The Mössbauer effect

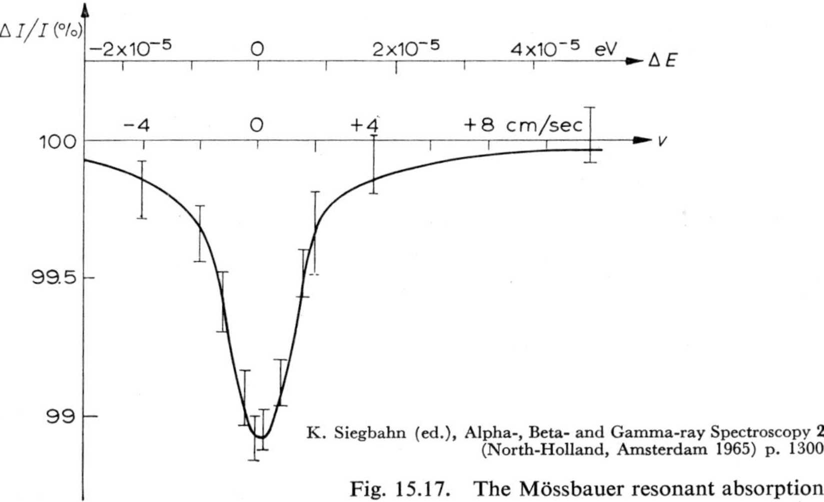

K. Siegbahn (ed.), Alpha-, Beta- and Gamma-ray Spectroscopy **2**
(North-Holland, Amsterdam 1965) p. 1300

Fig. 15.17.   The Mössbauer resonant absorption

energy is just what happens in the Bragg reflection of X-rays from crystal lattice
planes.

The strongest Mössbauer effect occurs in ^{57}Fe. In this case cooling is not required to observe recoilless emission of the 14.4 keV γ-ray. Since the conditions for recoilless emission depend on the solid state nature of the crystal lattice, it can only be observed in selected cases. The hyperfine splitting of nuclear states and the nuclear Zeeman effect can be observed for these cases. Since in the case of ^{57}Fe, $\Gamma_\gamma/E_\gamma \approx 3 \times 10^{-13}$ it can be used to observe very small shifts in frequency. It was used by Pound et al.* to check that the change in frequency of a γ-ray when it fell through a distance of 50 feet was consistent with the relativity theory prediction of the behaviour of photons in a gravitational field. The main application of the Mössbauer effect is, however, in investigation of the solid state. The details of the atomic binding in solids and of the molecular fields can be elucidated in these studies.

6   Mesons and nuclear forces

We have indicated the expected connexion between mesons and nuclear forces, but we have not really tried to make use of it. Recall our discussion of the n–p forces and the deuteron. We required

1) an ordinary central force term, $V_{\mathrm{W}}(r)$,

2) a spin dependent central force term $V_\sigma(r) \cdot \frac{1}{2}(1 + \boldsymbol{\sigma}_{\mathrm{p}} \cdot \boldsymbol{\sigma}_{\mathrm{n}}) = V_\sigma(r)P_\sigma$,

3) a non-central force term. Since our nucleons have spin $\frac{1}{2}$ we could not have moments of any kind higher than dipole. In electromagnetism the potential energy of an electric dipole $\boldsymbol{d}_1$ in the field of another dipole $\boldsymbol{d}_2$ at distance r is

$$-\frac{1}{r^3}\left[3(\boldsymbol{d}_1 \cdot \boldsymbol{r})(\boldsymbol{d}_2 \cdot \boldsymbol{r}) - \boldsymbol{d}_1 \cdot \boldsymbol{d}_2\right].$$

If we assume our nucleon non-central force has a similar form, we still do not know the radial dependence of the potential acting between the 'mesic' charges which we assume as sources of the strong field. So we replace the r^{-3} dependence by an arbitrary function, and write $V_{\mathrm{T}}(r)S_{\mathrm{pn}}$ with

$$S_{\mathrm{pn}} = 3(\boldsymbol{\sigma}_{\mathrm{p}} \cdot \boldsymbol{r})(\boldsymbol{\sigma}_{\mathrm{n}} \cdot \boldsymbol{r}) - \boldsymbol{\sigma}_{\mathrm{p}} \cdot \boldsymbol{\sigma}_{\mathrm{n}}$$

the tensor force operator. So our interaction is

$$V_{\mathrm{W}}(r) + V_\sigma(r)P_\sigma + V_{\mathrm{T}}(r)S_{\mathrm{pn}}.$$

We now proceed to try to compute the potential resulting from a field of pions. We assume a static meson cloud, i.e. the rest energy is just compensated by the binding energy and we need to add $m_\pi c^2$ to make mesons appear. The cloud

* R. V. Pound and G. A. Rebka, Phys. Rev. Letters **3** (1959) 439.

extends a distance like the de Broglie wavelength of the meson $\hbar/m_\pi c$. The mesons must be in p states to preserve the positive parity of the nucleon, since the meson has odd parity. If we assume the bare nucleon is a point we get

$$U = f^2 \tau_1 \cdot \tau_2 \left\{ \left(\frac{3}{r^2} + \frac{3}{r} + 1 \right) S_{12} + \sigma_1 \cdot \sigma_2 \right\} \frac{e^{-r}}{3r} - \tfrac{4}{3}\pi f^2 (\tau_1 \cdot \tau_2)(\sigma_1 \cdot \sigma_2)\delta(r).$$

This is in units such that $m_\pi = c = \hbar = 1$. The factor f^2 is a coupling constant related to the pion–nucleon force as the fine structure constant $e^2/\hbar c = e^2$ in these units is related to the electromagnetic field; $(3r^{-2} + 3r^{-1} + 1)S_{12}$ is a tensor force non-central as we require; $\sigma_1 \cdot \sigma_2$ is a spin exchange force as we require; e^{-r}/r is the Yukawa potential shape giving the expected short range of the force. The last term is a term of opposite sign to the main term which is connected with the nucleon itself. If it is not a point but has a small extension it is a core force; a repulsive core since the main term is attractive.

This is all non-relativistic and assumes that the coupling is weak, i.e. higher order terms and meson–meson interactions are neglected. We want now to compare this with the force, for example, in the singlet S state of the deuteron. $\tau_1 \cdot \tau_2$ is -3, an isotriplet, $\sigma_1 \cdot \sigma_2$ is 1, a spin singlet; hence $(\tau_1 \cdot \tau_2)(\sigma_1 \cdot \sigma_2) = -3$ and the force is attractive in the first term. The δ-function repulsion does not affect our solution of the radial wavefunction. Also the tensor force does not act in singlet states ($S_{12} = 0$). So the potential reduces to

$$U_{\text{singlet}} = -f^2 e^{-r}/r.$$

Now we know that the singlet state is just unbound. For a Yukawa well we have the condition that

$$f^2 r_0^2 = (0.6 M_N)^{-1}$$

for zero binding where r_0 is the de Broglie wavelength of the pion (1 in our units) and M_N is the nuclear mass ($6.7 m_\pi$ in our units). This just corresponds to $Kr_0 = \tfrac{1}{2}\pi$. From this we find

$$f^2 = \frac{1}{0.6 \times 1 \times 6.7} = \tfrac{1}{4}.$$

So although f^2 is less than 1 it is considerably greater than $e^2 = \tfrac{1}{137}$ for electromagnetic theory and our assumption of weak coupling is clearly wrong and likely our value of f^2 is then also in error.

It will obviously be better to derive f^2 directly from a study of π–N scattering and we shall do so presently.

Now let us turn to the triplet state of the deuteron, the bound ground state. Obviously now we must take the tensor force term into account. This gives added attraction and can account for the binding of the deuteron. But now if we try to solve the Schrödinger equation we are in real trouble since the term in $1/r^3$ diverges much too strongly at the origin and gives us an infinitely small deuteron, with infinite binding energy. But, of course, it is just at small distance corresponding to high momentum transfers that we expect our theory to be wrong as it was non-relativistic and also multi-meson transfers will become more probable at small distances.

So we have to resort to a theoretician's trick to retain the agreement we have so far. We resort to a cut-off of our theory at a given radius and ignore what goes on inside or if you wish we specify a maximum momentum transfer above which we ignore any effects. This is equivalent to assuming the bare nucleon is spread out over a finite volume – our repulsive core again. You could say that we use a large repulsive core to counterbalance the large attractive effect of the divergent function. Feynman calls this 'hiding the dirt under the rug'.

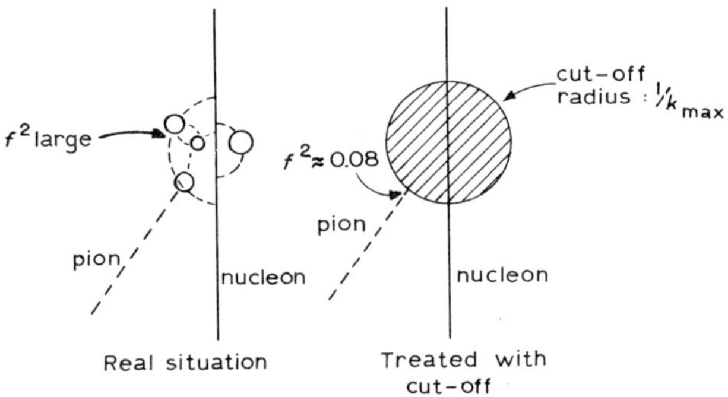

S. de Benedetti, Nuclear Interactions (Wiley, New York 1964) p. 452

Fig. 15.18.   Illustration of the cut-off radius

Thus the pion–nucleon strong interaction which actually deals with complex nucleon–pion–nucleon–antinucleon pairs is treated at *low* energies as in fig. 15.18. In other words we deal only with the peripheral collisions at large distances and low momentum transfers. And then again we can hope that a perturbation treatment will apply to the nucleon–nucleon potential.

So we approach the π–N scattering which we have seen is p-wave, in the spirit of zero range approximation. We have seen that, for p-wave,

$$k^{2l+1} \cot \delta_l = k^3 \cot \delta_l$$

is a constant. For π-scattering

$$\frac{f^2 k^2}{3\omega} \lambda_\alpha = \frac{1}{k \cot \delta_\alpha - ik} \qquad \text{for} \quad \omega \approx 0$$

where λ_α is a matrix element depending on the phase shift involved, δ_α, and ω is the total energy, i.e. (rest mass + kinetic energy). The term ik is negligible for low energies and

$$k \cot \delta_\alpha = \frac{3\omega}{f^2 k^2} \frac{1}{\lambda_\alpha} \qquad (\omega \approx 0)$$

$$\frac{k^3 \cot \delta_\alpha}{\omega} = \frac{3}{\lambda_\alpha} \frac{1}{f^2} = \text{const.}$$

which is close to the zero range approximation expression.

In this approach we have not yet had to introduce a cut-off or a form factor for the nucleon. The slowly varying factor $1/\omega$ on the left hand side reduces to the

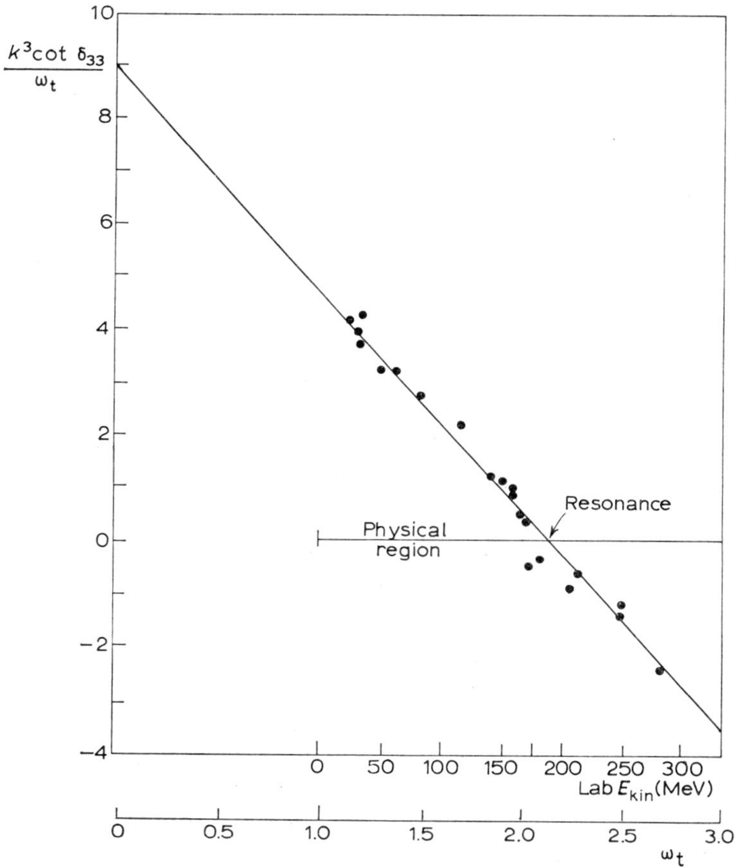

S. de Benedetti, Nuclear Interactions (Wiley, New York 1964) p. 473

Fig. 15.19. The extrapolation from the pion–nucleon resonance to the bound pion

constant rest mass for low energy, i.e. $\omega=0$ corresponds to pions bound into the meson cloud with rest mass equalling the binding energy $-E$.

Now we know that the lowest resonance in the π–N scattering is the $T=\tfrac{3}{2}$, $J=\tfrac{3}{2}$ resonance at 200 MeV. So we might try $\delta_{33}=(0.234k^3)$radian where k is in units of $m_\pi c$. For the 33 resonance $\Delta(1938)$, $\lambda_{33}=4$. Thus $(k^3\cot\delta_{33})/\omega=\tfrac{3}{4}f^{-2}$ whence $f^2\approx\tfrac{3}{4}\times0.235=0.17$. You recall that we estimated 0.25 from the singlet scattering length. But this is a bit unfair – we assumed ω was zero, i.e. zero energy. Yet we take the value of δ_{33} from data at positive energy. So we in fact find we need to modify the zero range approximation to allow for variation in ω.

Let us replace ω by a new variable $\omega_t=\omega+k^2/2M$. Then we plot the experimental values of $(k^3\cot\delta_{33})/\omega_t$ vs ω_t (fig. 15.19). The straight line thus gives

$$\frac{k^3\cot\delta_{33}}{\omega_t}\approx 9-4.1\omega_t$$

which corresponds to $\approx\tfrac{3}{4}f^{-2}-R_{33}\omega_t$ i.e. the constant term corresponds to the zero range approximation and the term in ω_t is an effective range correction. We find $f^2=0.08$ and $R_{33}=4.1$. R_{33} has the dimension of a momentum and it corresponds to a *cut-off* momentum $k_{max}=6$, i.e. about 1 GeV/c.

Thus our analysis of the π–N scattering leads to two parameters similar to the strength and cut-off we derived before. The result of this computation gives a potential shape for the triplet n–p force as shown in fig. 15.20.

We now go back to the nucleon–nucleon problem, for the s state. You recall that for one-pion exchange with no cut-off we got $f^2=0.25$, quite different from our above value. So we have to introduce the cut-off idea into this problem. Suppose we take the quadrupole moment. This is obviously affected only by the 'outside' of the nucleons since it is the radial integral of the nucleon wavefunction *multiplied* by r^2. The triplet effective range is sensitive to the inside of the nucleon, since it is the radial integral of the *difference* of the asymptotic and true value of wavefunctions.

Thus we try to compute the quadrupole moment using as parameters the triplet effective range which we know from measurement and the coupling constant f^2. We find f^2 lies between 0.07 and 0.09 in good agreement with the value from π–N scattering. We can also try to compute the anomalous magnetic moments of the nucleons in the same spirit. Using $f^2=0.08$ and the usual cut-off we get $\mu_{anomalous}=1.76\tau_3$ compared to $\mu_a=+1.79$ for protons ($\tau_3=1$) and $\mu_a=-1.91$ for neutrons ($\tau_3=-1$). Again the agreement is quite good.

We cannot of course use this static approximation $\omega\approx0$ to describe the spin–orbit part of the force. This is certainly a velocity dependent force and we have

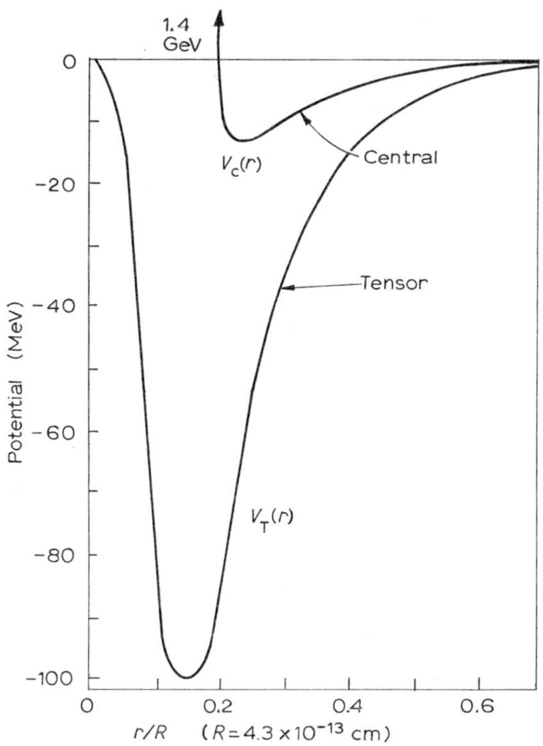

S. de Benedetti, Nuclear Interactions (Wiley, New York 1964) p. 55

Fig. 15.20.   Potential for the triplet n–p force derived from pion–nucleon data

assumed zero velocity. But there is only the mathematical difficulty of dealing with the strong coupling that prohibits its inclusion.

Attempts have been made to study the effect of the known meson resonances ϱ and ω on the multipion exchange processes and some consistent results have been obtained but the mathematical problems still remain.

So far, so good, but of course we are left with the unsatisfactory feature of the cut-off which limits the knowledge we can derive about the internal details of a real nucleon and its force field. We were forced to introduce the cut-off so that we could use a mathematical technique we could handle. An important problem of current theoretical research is to devise mathematical techniques which can handle the real strong coupling situation. Some progress has been made but we are still a long way from the final answer.

7 *Electromagnetic structure of nucleons*

We recall that high energy electron scattering from nuclei enabled us to measure the charge radii of the nuclei. The effect of smearing out the charge over a finite volume was to give some destructive interference between electron waves scattered from different parts and so give a net reduction in cross section at certain angles. This effect is normally expressed as a *form factor* which equals 1 for a point charge distribution and is smaller than 1 for an extended distribution. This factor multiplies the differential cross section.

We now want to study these effects with respect to the nucleons themselves. Let us consider first the neutron–electron interactions. The neutron has a magnetic moment which we interpret as due to currents in the virtual meson cloud. We can distinguish several possible effects.

1) The interaction of the magnetic moment of the neutron and the magnetic moment of the electron.

2) The interaction of the magnetic moment of the neutron and the electrostatic field of the electron.

3) The possibility of a non-zero electric charge distribution for the neutron which is sometimes (proton $+ \pi^-$).

The first effect is studied by scattering neutrons from magnetised iron in which the electron spins and moments are prealigned. We can make the usual arrangement of polariser and analyser, and by this technique the neutron magnetic moment is measured. A study of the intensity of the magnetic scattering can yield form factors which indicate the spatial distribution of the electron magnetic moments doing the scattering. This is of great importance for studying the solid state and is now a basic experimental tool. In the study of other materials antiferromagnetism was discovered by this technique. Completely polarised beams of neutrons can be produced by reflection from magnetised mirrors, i.e. only one of the neutron spin states is reflected.

The second effect due to the interaction of the neutron moment with the electric field of the electron results in a net attractive force between neutron and electron. The theoretical value assuming a square well is 4.08 keV. A non-magnetic scatterer is used so that the scattering through effect (1) cancels out. Slow neutrons are used and one technique is to look for departures from spherical symmetry in the scattering. This experiment indicated a small effect corresponding to a potential of 4100 eV $\pm 10\,\%$. The energy dependence of the scattering can also be studied to look for deviations from the expected dependence. This gives a value of 4165 ± 265 eV. The fact that these values are so close to that expected for the magnetic moment–electric field effect is taken as evidence that no non-zero electric charge distribution is detectable.

We now turn to the results of scattering of electrons by nucleons. Since the nucleon size that we are attempting to probe is very small we must use electrons of high energy. Already at 400 MeV ($\lambda_e = 0.5 \times 10^{-13}$cm) we observe a significant deviation from the scattering expected from a point charge. Note that the differential cross section is reduced by about a factor of 10 at 140° for an incident energy of 550 MeV (fig. 15.21).

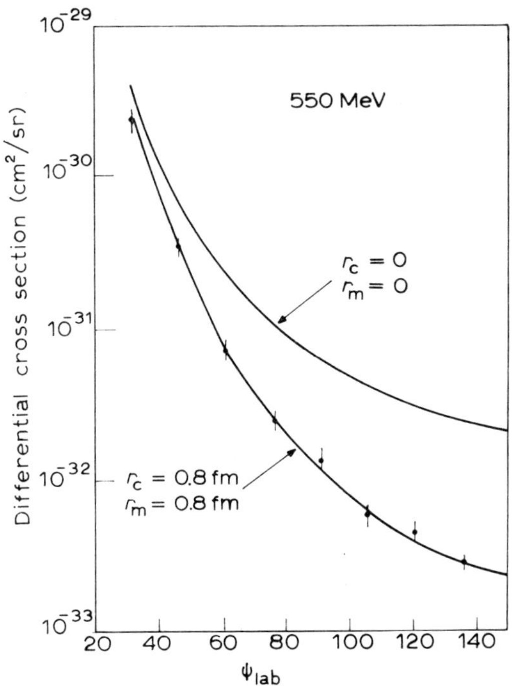

Fig. 15.21.   Electron scattering cross-section. The upper curve is for a point nucleon

The two radii refer to two distinct effects. r_c is the charge radius and the magnetic moment is that associated with a spinning electric charge, i.e. for the bare nucleon. This magnetic moment is 1 n.m for the proton and 0 for the neutron and is often called the Dirac moment. r_m is the radius associated with the anomalous part of the magnetic moment, i.e. due to the meson cloud. This magnetic moment is 1.79 n.m for the proton and -1.91 n.m for the neutron. It is often called the Pauli moment.

Note that there is no anomalous charge effect since the vector currents are conserved so as to give zero electric charge due to the meson cloud. This is what suggested the CVC theory for the weak 'charge'.

Much work has been done at higher electron energies, ≈ 1 GeV ($\lambda_e = 0.2$ fm), and analysis of the angular distributions enables form factors to be extracted for the two effects of charge and of anomalous magnetic moment. The form factors are related to the charge distribution or moment distribution via $F = \int e^{i\mathbf{q}\cdot\mathbf{r}} g(r)\, dr$ where $g(r)$ is the distribution function and q is the momentum transfer. When we plot the distributions so derived* from scattering of 1 GeV electrons, we have the curves of fig. 15.22. We can derive similar quantities for the neutron via electron–deuteron scattering. These are also shown in fig. 15.22.

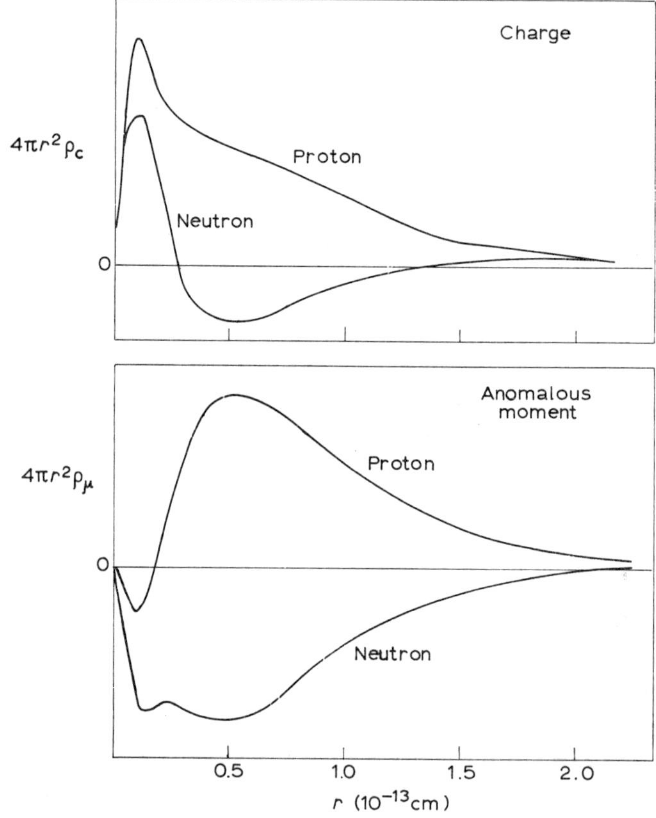

S. de Benedetti, Nuclear Interactions (Wiley, New York 1964) p. 188

Fig. 15.22. Form factors for the nucleons

The anomalous moment effect extends somewhat further out than the charge effect. The neutron charge distribution radius is zero hence the failure to detect a charge distribution by neutron scattering.

So with the aid of electromagnetic probes we have a photograph, if you like,

* R. Littauer, H. Schopper and R. R. Wilson, Phys. Rev. Letters **7** (1961) 141, 144.

taken in electromagnetic colour of the complex structure of nucleon+meson cloud which is a physical strongly interacting particle.

In all this of course we assume two things.
1) The electron can be considered a real point charge.
2) The laws of electrodynamics hold for these very short distances and high energies.

If either of these assumptions is wrong our form factors would of course be wrong – overestimated for the nucleon. It is probable that the electron does have a radius due to its cloud of virtual photons but this must be smaller than the nucleon radius in roughly the ratio of the force constants, i.e. two orders of magnitude. The measured anomalous magnetic moment of the electron confirms this. It is only 0.1 % while for the proton the anomalous moment is 180 %.

Very recently the laws of electrodynamics have been checked in experiments using electron–electron scattering at several GeV and are known to hold down to less than 10^{-14} cm. So at least down to this distance our results should be satisfactory.

Now what light do these observations throw on our meson theories. The first point to note is that the observed r.m.s. radius of 0.9×10^{-13} cm is significantly less than the pion Compton wavelength 1.4×10^{-13} cm. It can also be shown that two or more virtual pions must be involved to couple to the electromagnetic field. (π^0 is even under charge conjugation while the photon is odd.) Thus, in 1959, before any meson resonances were known, in order to explain the form factors deduced, it was concluded that a two-pion resonance of mass ≈ 600–700 MeV had to be postulated. This of course agrees with the ϱ-resonance subsequently found of mass 765 and decay $\varrho \rightarrow 2\pi$.

As we go to higher and higher momentum transfers and probe deeper and deeper into the nucleus it seems that the influence of higher mass multipion resonances will be seen.

Remember we are merely plotting *electromagnetic* effects of the meson cloud. These effects extend only to 0.9×10^{-13} cm. Pion exchange is possible presumably out to the pion Compton wavelength of 1.4×10^{-13} cm and this is more properly the range of the *strong* force. We measure in the electromagnetic experiments the two-pion exchange force range.

8   Nuclear forces and nuclear structure

You recall that we discussed nuclear structure and various model descriptions of it and then later discussed nuclear forces mostly in terms of the nucleon–nucleon forces. However the feed-back from our ideas of the nuclear force to its operation in complex nuclei has been at best rather vague. There is a good reason for this and we want now to examine this in a little more detail.

For example we found in the shell model that we could consider single particles to act as if they were essentially free and also in the complex potential model we derived the mean free path in nuclear matter which was remarkably large. And yet when we talk of nuclear forces we say they are extremely *strong* which should exclude any possibility for a single particle picture. Of course in the shell model we assume a common central potential to keep the particle in the confines of the nuclear volume. But suppose we consider a very large nucleus, then motions in the central region will be negligibly affected by the boundary conditions. In such uniform nuclear matter the approximation of free motion with constant momentum corresponds to the shell model approximation for real nuclei.

Suppose we consider the collision of two nucleons. As they approach their motions are affected by the force – they scatter. If we analyse the scattering in nuclear matter we derive the result that the motions are essentially those of a free particle except when they approach to within about 1.0 fm. On the other hand we know the average separation of nucleons in nuclear matter is considerably greater than this, viz. ≈ 1.66 fm. This merely confirms the fact that two-particle correlations are not as important as we might have expected. The distance of 1.0 fm outside which the wavefunction is essentially that of free particles is called the *healing distance*.

The small healing distance compared to the average internucleon distance also makes it improbable that a third particle will be within the healing distance of the first two. Hence we need not consider that clusters of more than two particles will be of any importance.

For the collision of two nucleons in vacuo, the healing distance is infinite, i.e. the scattered wave is phase shifted with respect to free particle motion forever. Inside nuclear matter the situation is very different. In vacuo when two nucleons collide the momentum in the final state may be divided in many different ways. In the nucleus the numbers of possible final states is greatly reduced since many of the available states are already occupied and the exclusion principle does not allow further occupation. We have to choose our nuclear force in a very special way to get a short healing distance.

1) We require a repulsive core to prevent collapse of the nucleus.

2) We require the repulsive core to be small enough to give a short healing distance, yet large enough to give a reasonably low nuclear density.

3) We require an attractive force to give binding yet weak enough to maintain low density and reasonable nucleon spacing.

4) The attractive force must be of such a shape, strength and range, that the main momenta arising from it are less than the momentum of the Fermi surface.

It is clear that the behaviour of a particle will depend on its momentum. That is, we have a *momentum dependent potential*. This can be defined as the difference between the total energy of the particle and its kinetic energy $p^2/2M$. In infinite nuclear matter the properties of the matter are the same at all points in space and the energy of the particle does *not* depend on position but may depend on momentum. In ordinary space such a potential is called *non-local*. The energy of the particle at point r depends not only upon r but also on the wavefunction at other points r' near r. This is equivalent to describing the correlations in nuclear matter. The presence of a particle at point r influences the possibility of finding another particle at point r' which in turn affects the energy of the particle at r. Of course the particles are not localised because of the uncertainty principle and so this is not inconsistent with overall constant density of nuclear matter.

Suppose we represent the wavefunction as plane waves of constant momentum

$$\psi = \exp\left(\mathrm{i}\boldsymbol{p}\cdot\boldsymbol{r}/\hbar\right).$$

Then the energy is

$$E = \frac{p^2}{2M} + \int \mathrm{d}r'\, V(r,r')\exp\left(\frac{\mathrm{i}\boldsymbol{p}\cdot(\boldsymbol{r}'-\boldsymbol{r})}{\hbar}\right).$$

If $V(r,r')$ depends only on the coordinate $r'-r$, the integral is not a function of r but only of p. It is just the Fourier transform of $V(r'-r)$. Then $E=(p^2/2M)+V(p)$ where $V(p)$ is the Fourier transform. We can split up $V(p)$ and express it as a quadratic in p for momentum values up to the Fermi surface momentum. Then

$$V(p) = V_0 + bp^2$$

and

$$E = \frac{p^2}{2M} + V_0 + bp^2.$$

We could collect terms in p^2 and write

$$E = \frac{p^2}{2M^*} + V_0$$

where $(M^*)^{-1} = M^{-1} + 2b$. M^* is termed the effective mass as the last equation is just that for a free particle of mass M^*.

The asymptotic forms of the wavefunctions are solutions of Schrödinger's equation containing $V(p)$ and thus the energy–momentum relation is not that of free particles. We express the change by use of an effective mass M^*.

The extremely complex calculation of the characteristics of nuclear matter using such a momentum dependent non-local potential has been carried out for infinite nuclear matter by Brueckner and his group. The nuclear binding is calculated and compared with the volume energy term of the semi-empirical mass formula. The result is -15.5 MeV with the radius of the central core of the nucleus ≈ 1.07 fm which can be compared to the r_0 parameter deduced from electron scattering.

Of course the effective mass parameter b is also momentum dependent

$$\frac{1}{M^*} = \frac{1}{M} + \left(\frac{1}{p}\frac{\mathrm{d}V}{\mathrm{d}p}\right)_{p_F}.$$

So for $p \gg p_F$ (the Fermi surface momentum) $M^* \to M$, the real mass.

If we turn now to a finite nucleus we expect the local part of the potential to be position dependent as in the shell model potential while the non-local part is not too different from that applicable in the infinite nucleus core. The main difficulty is properly to take account of the density variation near the edge which is of course quite rapid and we get a variation in the effective potential from that in uniform nuclear matter at the centre to that for particles in free space at the edge.

The spin–orbit part of the shell model potential can also be included in this formulation and gives splittings in general agreement with those found.

Self-consistency calculations have been done for three nuclei, ^{16}O, ^{40}Ca and ^{90}Zr, by Brueckner's group. In order to make the calculations possible even for the largest computers several simplifying assumptions had to be made. The binding energies per nucleon came out rather too small by about 3 MeV (5 MeV instead of 8). However binding energy differences for single particle states came out exactly as expected in the shell model with about 1 MeV accuracy.

These successes have indicated that the general picture of nuclear matter is not too far from the truth.

There are of course other particle correlations which may occur but which we have neglected in the belief that they will not affect the *binding energy*, that their effect is negligible compared to the single-pair type of correlation we have been discussing. However in cases where we are studying very fine details of the binding they may show up in a rather more striking way.

For example, we believe there is a correlation of particles of the same type whose spins are opposed. This is the *pairing correlation* which you recall had the effect of binding the *ground state only* of even–even nuclei by an additional amount of the order of 1 or 2 MeV. This is still small compared to the *total binding* of about $8A$ MeV that is, 160 MeV for a light nucleus and 1600 MeV for a heavy nucleus.

You recall this could be interpreted as introducing a fuzziness of order of half the gap energy in the Fermi surface.

Another small effect is the polarising effect of a few outer nucleons on the core leading to a non-spherical nucleus with hence a non-spherical effective potential.

These have the effect of long range nucleus – wide particle correlations and give rise to the collective effects. Again since in considering excited states we are looking at differences of a MeV or so in many hundreds of MeV binding energy, we have been justified in ignoring it in connexion with the nuclear matter problem. However in calculating nuclear spectra we have to consider all these small effects and usually prefer to treat them by the use of particular models.

Further Reading

Chapter 1

Mass:

Everling, F., König, L. A., Mattauch, J. H. E., and Wapstra, A. H., Nucl. Phys. **18** (1960) 529

Duckworth, H. E., (ed.), Proc. Internat. Conf. Nuclidic Masses (Toronto University Press, Toronto 1960)

Mass spectrometers:

Bainbridge, K. T., *in* E. Segre (ed.), Experimental Nuclear Physics **1** (Wiley, New York 1953)

Mayne, K. I., Rept. Progr. Phys. **15** (1952) 24

Duckworth, H. E., Mass Spectroscopy (Cambridge University Press, Cambridge 1958)

Radius:

Rev. Mod. Phys. **30** (1958) 412–569

Electron scattering:

Hofstadter, R., Nuclear and Nucleon Structure (Benjamin, New York 1963)

Muonic atoms:

West, D., Rept. Progr. Phys. **21** (1958) 271

Primakoff, H., Rev. Mod. Phys. **31** (1959) 802
Feinberg, G., and Lederman, L. M., Ann. Rev. Nucl. Sci. **13** (1963) 431

Isotope shift:
Brix, P., and Kopfermann, H., Rev. Mod. Phys. **30** (1958) 517
Mack, J. E., and Arroe, H., Ann. Rev. Nucl. Sci. **6** (1956) 117

Mirror nuclei:
Kofoed-Hansen, O., Rev. Mod. Phys. **30** (1958) 449

Alpha scattering and decay:
Rasmussen, J. O., Rev. Mod. Phys. **30** (1958) 424

Chapter 2

Alpha decay:
Rasmussen, J. O., *in* K. Siegbahn (ed.), Alpha-, Beta- and Gamma-ray Spectro-
scopy **1** (North-Holland Publ. Co., Amsterdam 1965) ch. XI
Hanna, G. C., *in* E. Segre (ed.), Experimental Nuclear Physics **3** (Wiley, New
York 1959)

Beta decay:
Reines, F., Ann. Rev. Nucl. Sci. **10** (1960) 1
Lipkin, H. J., Beta Decay for Pedestrians (North-Holland Publ. Co., Amsterdam
1962)
Konopinski, E. J., Ann. Rev. Nucl. Sci. **9** (1959) 99
Siegbahn, K., (ed.), Alpha-, Beta- and Gamma-ray Spectroscopy **2** (North-Hol-
land Publ. Co., Amsterdam 1965) chs. XXIII, XXIV

Log ft calculation:
Moszkowski, S. A., Phys. Rev. **82** (1951) 35

Gamma decay:
Deutsch, M., and Kofoed-Hansen, O., *in* E. Segre (ed.), Experimental Nuclear
Physics **3** (Wiley, New York 1959)
Rose, M. E., Internal Conversion Coefficients (North-Holland Publ. Co., Am-
sterdam 1959)
Rose, M. E., Multipole Fields (Wiley, New York 1955)
Siegbahn, K., (ed.), Alpha-, Beta- and Gamma-ray Spectroscopy **2** (North-Hol-
land Publ. Co., Amsterdam 1965) chs. XV, XVI

Chapter 3

Siegbahn, K., (ed.), Alpha-, Beta- and Gamma-ray Spectroscopy **1** (North-Holland Publ. Co., Amsterdam 1965) chs. I, II

Evans, R. D., The Atomic Nucleus (McGraw-Hill, New York 1955) chs. 18–25

Bethe, H. A., and Askin, J., *in* E. Segre (ed.), Experimental Nuclear Physics **1** (Wiley, New York 1953)

Bohr, N., Dan. Kgl. Vid. Selsk. Mat.-Fys. Medd. **18** (1948) 8

Jelley, J. V., Čerenkov Radiation (Pergamon Press, London 1958)

Wapstra, A. H., Nijgh, G. J., and van Lieshout, R., Nuclear Spectroscopy Tables (North-Holland Publ. Co., Amsterdam 1959)

Chapter 4

Siegbahn, K., (ed.), Alpha-, Beta- and Gamma-ray Spectroscopy **1** (North-Holland Publ. Co., Amsterdam 1965) chs. III–VIII

Flügge, S., (ed.), Encyclopedia of Physics **45** (Springer-Verlag, Berlin 1958)

Yuan, L., and Wu, C. S., (ed.), Methods of Experimental Physics (Academic Press, New York 1963)

Segre, E., (ed.), Experimental Nuclear Physics (Wiley, New York 1959)

Fretter, W. B., Ann. Rev. Nucl. Sci. **5** (1955)

Korff, S. A., Electron and Nuclear Counters (Van Nostrand, New York 1955)

Marshall, J., Ann. Rev. Nucl. Sci. **4** (1954)

Wilkinson, D. H., Ionization Chambers and Counters (Cambridge University Press, Cambridge 1950)

Dearnaley, G., and Northrop, D. C., Semiconductor Counters (Spon, London 1963)

Snowdon, M., The Diffusion Cloud Chamber, *in* Progr. Nucl. Phys. **3** (1953) 1

Bradner, H., Bubble Chambers, *in* Ann. Rev. Nucl. Sci. **10** (1960) 109

Barkas, W., Nuclear Research Emulsions (Academic Press, New York 1963)

Jelley, J. V., Čerenkov Radiation (Pergamon Press, London 1958)

Ritson, D. M., Techniques of High Energy Physics (Wiley-Interscience, New York 1961)

Chase, R. L., Nuclear Pulse Spectrometry (McGraw-Hill, New York 1961)

Chapter 5

Livingston, M. S., and Blewett, J. P., Particle Accelerators (McGraw-Hill, New York 1962)

McMillan, E. M., Particle Accelerators, *in* E. Segre (ed.), Experimental Nuclear Physics **3** (Wiley, New York 1959)

Tandems:
Van de Graaff, R. J., Nucl. Instr. Meth. **8** (1960) 195

Ion sources:
Moak, C. D., Reese, H., Jr., and Good, W. M., Nucleonics **9** (1951) 18
Moak, C. D., Banta, H. E., Thurston, J. N., Johnson, J., and King, R. F., Rev. Sci. Instr. **30** (1959) 694

Chapter 7

Tobocman, W., Theory of Direct Nuclear Reactions (Oxford University Press, Oxford 1961)
Weinberg, A., and Wigner, E., Physical Theory of Neutron Chain Reactors (Chicago University Press, Chicago 1958)

Neutron cross sections:
Brookhaven National Laboratory Report BNL-325 (1965)

Optical model:
Brown, G. E., Rev. Mod. Phys. **31** (1959) 893

R-matrix theory:
Lane, A. M., and Thomas, R. G., Rev. Mod. Phys. **30** (1958) 257

Nuclear fission:
Halpern, I., Ann. Rev. Nucl. Sci. **9** (1959) 245

Chapter 8

Mayer, M. G., and Jensen, J. H. D., Elementary Theory of Nuclear Shell Structure (Wiley, New York 1955)
Mayer, M. G., Jensen, J. H. D., and Kurath, D., *in* K. Siegbahn (ed.), Alpha-, Beta- and Gamma-ray Spectroscopy **1** (North-Holland Publ. Co., Amsterdam 1965) ch. IX
Talmi, I., and Unna, I., Ann. Rev. Nucl. Sci. **10** (1960) 353

de Shalit, A., and Talmi, I., Nuclear Shell Theory (Academic Press, New York 1963)

Chapter 9

Nathan, O., and Nilsson, S. G., *in* K. Siegbahn (ed.), Alpha-, Beta- and Gamma-ray Spectroscopy **1** (North-Holland Publ. Co., Amsterdam 1965) ch. X
Marshalek, Persson, and Sheline, Rev. Mod. Phys. **35** (1963) 108
Kisslinger, L. S., and Sorensen, R. A., Rev. Mod. Phys. **35** (1963) 853

Chapter 10

Roberts, L. D., and Dabbs, J. W. T., Ann. Rev. Nucl. Sci. **11** (1961) 175
Blin-Stoyle, R. J., and Grace, M. A., *in* S. Flügge (ed.), Encyclopedia of Physics **42** (Springer-Verlag, Berlin 1957) p. 555
Daniels, J. M., and Goldemberg, J., Rept. Progr. Phys. **25** (1962) 1
Fagg, L. W., and Hanna, S. S., Rev. Mod. Phys. **31** (1959) 711
Siegbahn, K., (ed.), Alpha-, Beta- and Gamma-ray Spectroscopy **2** (North-Holland Publ. Co., Amsterdam 1965) ch. XIX

Triple correlations:
Smith, P. B., *in* P. M. Endt and P. B. Smith (ed.), Nuclear Reactions **2** (North-Holland Publ. Co., Amsterdam 1962)

Chapter 11

Adair, R. K., Rev. Mod. Phys. **22** (1950) 249
Blatt, J. M., and Weisskopf, V. F., Theoretical Nuclear Physics (Wiley, New York 1952)
Preston, M. A., Physics of the Nucleus (Addison-Wesley, Reading, Mass. 1962)
de Benedetti, S., Nuclear Interactions (Wiley, New York 1964)

Chapter 12

Gell-Mann, M., and Rosenfeld, A. H., Ann. Rev. Nucl. Sci. **7** (1957) 407
Nishyima, K., Fundamental Particles (Benjamin, New York 1963)

Williams, S. C., Introduction to Elementary Particles (Academic Press, New York 1962)

Jackson, J. D., Physics of Elementary Particles (Princeton University Press, Princeton 1958)

de Benedetti, S., Nuclear Interactions (Wiley, New York 1964)

List of particles:
Rosenfeld, Barbaro-Galtieri, Podolsky, Price, Soding, Wohl, Roos and Willis, Rev. Mod. Phys. **39** (1967) 1

Pionic resonances:
Puppi, G., Ann. Rev. Nucl. Sci. **13** (1963) 287

Muons:
Feenberg, G., and Lederman, L. M., Ann. Rev. Nucl. Sci. **13** (1963) 431

Chapter 13

Yang, C. N., Elementary Particles (Princeton University Press, Princeton 1962)

Ford, K. W., The World of Elementary Particles (Blaisdell Publ. Co., New York 1963)

Gell-Mann, M., and Ne'eman, Y., The Eightfold Way (Benjamin, New York 1964)

Lipkin, H., Lie Groups for Pedestrians (North-Holland Publ. Co., Amsterdam 1965)

McVoy, K., Rev. Mod. Phys. **37** (1965) 84

Pais, A., Rev. Mod. Phys. **38** (1966) 215

Chew, G. F., Gell-Mann, M., and Rosenfeld, A. H., Scientific American (February 1964)

Weisskopf, V. F., and Matthews, P. T., *in* E. H. S. Burhop (ed.), High Energy Physics **1** (Academic Press, New York 1967)

Chapter 14

Nataf, R., *in* Mme P. Gugenberger (ed.), Comptes Rendus du Congrès International de Physique Nucléaire (CNRS, Paris 1964)

Wu, C. S., Rev. Mod. Phys. **36** (1964) 618

Preston, M. A., Physics of the Nucleus (Addison-Wesley, Reading, Mass. 1962)

Siegbahn, K., (ed.), Alpha-, Beta- and Gamma-ray Spectroscopy **2** (North-Holland Publ. Co., Amsterdam 1965) chs. XXII, XXIII

Chapter 15

Section 1:

Suess, H. E., and Urey, H. C., Rev. Mod. Phys. **28** (1956) 53

Burbidge, E. M., Burbidge, G. R., Fowler, W. A., and Hoyle, F., Rev. Mod. Phys. **29** (1957) 547

Cameron, A. G. W., Astrophys. J. **130** (1959) 429

Bashkin, S., *in* Aller and McLaughlin (ed.), Origin of the Chemical Elements in Stellar Structure (Chicago University Press, Chicago 1965)

Section 2:

Galbraith, W., Extensive Air Showers (Butterworth, London 1958)

Section 3:

Glasstone, S., Controlled Thermonuclear Reactions (Van Nostrand, New York 1960)

Section 4:

Fuller, E. G., and Hayward, E., *in* P. M. Endt and P. B. Smith (ed.), Nuclear Reactions **2** (North-Holland Publ. Co., Amsterdam 1962)

Section 5:

Metzger, F., Progr. Nucl. Phys. **7** (1959) 53

Section 7:

Thouless, D. J., Quantum Mechanics of Many Body Systems (Academic Press, New York 1961)

Day, B. D., Rev. Mod. Phys. **39** (1967) 719

Problems

Chapter 1

1 ^{62}Zn is heavier than ^{62}Cu by 1.68 MeV and decays to it by both positron emission and electron capture. Calculate the maximum neutrino energy in the positron decay and the neutrino energy in the electron capture.

2 The Q-value for the reaction ^{12}C(d, α)^{10}B is -1.342 MeV for α-particles leading to the ground state of ^{10}B and -2.059 MeV for α-particles leading to the first excited state. Find the mass of ^{10}B and the difference in mass between the ground and first excited states.

3 Use the semi-empirical mass formula to calculate the neutron separation energy for $^{51}_{21}$Sc.

4 Use mass tables to calculate the energy release in the reaction ^{42}Ca(d, p)^{43}Ca.

5 Suppose you have an accelerator which can produce 4 MeV beams of hydrogen and helium ions. Using mass tables show which reactions can be used to study the nucleus ^{19}O.

6 If matter were compressed so that the nuclei essentially touched (as is nearly the case in neutron stars) estimate the increase in density.

7 The atomic masses of nuclei with $A = 64$ are as follows

^{64}Ni 63.94813 ^{64}Zn 63.94932 ^{64}Cu 63.94994 ^{64}Ga 63.95710.

Which of these nuclei are stable? Which are radioactive and how would they decay? What are the β-spectra end points and what neutrino velocities are expected?

471

Chapter 2

1 Calculate the volume of helium produced per month by 1 curie of radium which is free of daughter products.

2 In the decay of ^{11}Be the β-spectrum indicates two end points of 11.48 ± 0.15 MeV and 9.34 ± 0.15 MeV and a γ-ray of energy 2.12 ± 0.10 MeV. Show how these fit into a decay scheme and if $M - A$ for ^{11}B is 8.668 ± 0.001 MeV, calculate the mass of ^{11}Be in mass units.

3 A radioactive nucleus $_z$A shows the following properties in its decay to nucleus $_{z-1}$B: 1) positrons whose spectrum shows two end points at 1.0 and 0.6 MeV. Each part has the allowed shape. 2) Gamma-rays of energy 0.4, 0.8 and 1.2 MeV. 3) Characteristic X-rays of atom B. 4) Coincidence measurements show the 0.6 MeV positron spectrum to be in coincidence with the 0.4 MeV γ-ray but no other $\beta^+ - \gamma$ coincidences are detected. 5) Measurements on internal conversion lines show that both the 0.4 and 1.2 MeV γ-rays are pure quadrupole radiations.

If the nucleus $_{z-1}$B is an even–even nucleus construct a decay scheme showing excited state energies and probable spin assignments, justifying each conclusion from the experimental results.

4 ^{160}Tb decays by β-emission to states of ^{160}Dy. Two of the γ-rays resulting are associated with an internal conversion line spectrum as follows: 32.9, 78.0, 84.7, 86.1, 143.0 keV. The binding energies for K, L, M and N shell electrons in dysprosium are 53.4, 8.6, 1.9, and 0.4 keV respectively. Find the energies of the γ-ray transitions, identifying each conversion line with the transition and the shell from which it arises.

Chapter 3

1 Alpha-particles from the reaction ^{19}F(p, α)^{16}O ($Q = 8.119$ MeV) are observed at 90° to the proton beam. Elastically scattered protons are also observed in the same detector both from the fluorine and from the calcium with which it is combined in the target. If the energy of the proton beam is 3 MeV calculate the energies of the different particle groups seen.

2 Gamma-rays of 1.022 MeV pass through a 0.1 g/cm^2 sheet of lead. What is the probability of 0.511 MeV γ-rays emerging from the lead.

3 What are the relative numbers of photons in the same energy interval below the maximum energy emitted by electrons of 0.1, 1, 10 and 100 MeV striking the lead sheet of problem 2.

Chapter 4

1 A hydrogen filled ionization chamber is designed to detect fast neutrons. If it has a sensitive volume of 1 litre and is filled to a pressure of 10 atmospheres, what is the minimum neutron flux that can be detected if the detecting equipment can measure currents down to 10^{-14} amperes. Take the hydrogen scattering cross section to be 4.5 barns and assume that half the neutron energy is lost in each collision.

2 Design a Cerenkov detector using glass of refractive index 1.6 which will distinguish between mesons of rest mass 300 times that of the electron having momentum 100 MeV/c and electrons of the same momentum.

3 A 4π β-counting equipment has a paralysis time of 10 μsec and can be considered to be 100% efficient. An irradiated sample is removed from a reactor at 0 h and counted at the following times:

Time	Counts per second
1 h	6948
2 h	5960
3 h	5184
10 h	2567
20 h	1245
30 h	631

If the counter background is 10 cps, find the half-lives involved in the decay and the initial activities of the components.

Chapter 5

1 The target in an experiment consists of a 1 cm^2 plate of aluminium 0.5 mm thick. If the plate is thermally isolated how long will it stand up to a uniform beam of

$10\ \mu A$ of protons at 5 MeV? What will happen if the aluminium is made 0.01 mm thick? (Melting point of Al, 659 °C; specific heat, 0.25 cal/g.)

2 A fixed frequency cyclotron has a magnetic field of 10 kilogauss. The pole diameter is 100 cm. If a target is installed 13 cm in from the edge of the field what energies of 1) protons, 2) deuterons, 3) α-particles will strike it when the cyclotron is arranged to accelerate these different ions?

3 A particular ion source is 1% efficient in producing ions. If it produces 100 A of protons for 1000 hours, how much hydrogen gas will be required.

Chapter 6

1 Devise an experiment to detect shadow scattering directly assuming that a detector 1 mm square is available and outline the difficulties which have to be overcome.

2 Prepare graphs of proton and neutron penetrabilities for ^{12}C and ^{89}Y for orbital angular momenta 0, 1, 2, and 3.

3 Estimate the neutron energies for which a) p-wave, b) d-wave neutron resonances might just be seen for neutron bombardment of a) argon, b) silver. Assume that the total cross section at these resonances must be at least a hundredth of the cross section for neighbouring resonances.

Chapter 7

1 Using the simple ideas of § 14.5 compute the critical radius of graphite-moderated spheres containing ^{233}U, ^{235}U and ^{239}Pu. Assume $f = 0.88$, $\varepsilon = 1.02$ and $p = 0.89$ as in the natural uranium example quoted.

2 Estimate the position of the first maxima in the angular distributions for the following reactions at 10 MeV:

$$^{16}O(^{3}He, \alpha)^{15}O, \qquad ^{12}C(d, p)^{13}C, \qquad ^{10}B(n, d)^{9}Be.$$

3 A resonance in the reaction $^{14}N + n$ at 1.12 MeV is observed to have a width of 13 keV. Assuming that the resonance is described approximately by a single particle interaction what are the possible angular momenta of the state.

4   The 2200 m/sec neutron absorption cross section of gold is 98.8 barns. The first neutron resonance seen in gold is at 4.9 eV. It has total width 0.140 eV and radiation width 0.124 eV. What fraction of the thermal cross section can be accounted for by this resonance?

Chapter 8

1   The lowest states of two light nuclei are listed as follows:

$^{13}_{6}N_7$		$^{17}_{8}O_9$	
E_{ex} (MeV)	J^P	E_{ex} (MeV)	J^P
0	$\frac{1}{2}^-$	0	$\frac{5}{2}^+$
2.37	$\frac{1}{2}^+$	0.871	$\frac{1}{2}^+$
3.51	$\frac{3}{2}^-$	3.058	$\frac{1}{2}^-$
3.56	$\frac{5}{2}^+$	3.846	$\frac{7}{2}^-$
		4.555	$\frac{3}{2}^-$

Outline the shell model interpretation of each of these states and comment on the energy spacing observed.

2   Calculate the positions of the lowest $T=\frac{3}{2}$ state in each of the odd-A nuclei from ^{7}Li to ^{23}Na. How many of these have been observed to date?

3   The first two excited states of ^{19}F at 110 and 197 keV have lifetimes of 0.7 and 60 nsec respectively. What conclusions can be drawn about the multipolarity of the γ-radiations to the ground state (no other decay modes are observed).

Chapter 9

1   The lowest states of ^{20}Ne are ground 0^+, 1.63 MeV 2^+, and 4.25 MeV 4^+. Compare the moment of inertia indicated by the spectrum with the moment for a rigid body and comment.

2   The lowest states of ^{21}Ne are ground $\frac{3}{2}^+$, 0.35 MeV $\frac{5}{2}^+$, and 1.74 MeV $\frac{7}{2}^+$. Compare this spectrum with the theoretical rotational spectrum derived from the collective model and the spectrum of ^{20}Ne (preceding problem). Suggest possible reasons for the discrepancies found.

3 The lowest states of ^{18}O are ground 0^+, 1.98 MeV 2^+, and a triplet near 3.75 MeV with 4^+, 0^+ and 2^+. Calculate the theoretical vibrational spectrum predicted by the phonon model (ch. 9 § 3.2) and comment on the discrepancies observed.

Chapter 10

1 The angular distribution of γ-rays to the ground state from a reaction has the form $W(\theta) = A + B\cos^2\theta + C\cos^4\theta$. If the target is an even–even nucleus and the bombarding particle is an α-particle, what can be concluded about the resonant state and the γ-ray multipolarity?

2 Design an experiment to establish the parity of the resonant state of Problem 1.

3 In the experiment on circular polarisation of ^{60}Co and ^{22}Na γ-rays find the values of P_3 if particles of 250 keV *only* are detected. Interpret these results in terms of the helicities involved.

Chapter 11

1 Some of the mesons discovered since the pion may play a role in nuclear forces. What range would the force have mediated by a) the η-meson, mass 549 MeV, b) the K-meson, mass 496 MeV?

2 Prepare a graph of the energy of the singlet state of the deuteron vs the ratio of the ortho-to-parahydrogen neutron cross sections.

3 Estimate how much the diproton system is unbound using the quoted values of the scattering length. Remember that protons also carry electric charge.

Chapter 12

1 The reaction $\pi^- + d \to 2n + \pi^0$ is not observed for pions at rest. Why?

2 Calculate the threshold photon energy required for photo-production of 1) π^0, 2) K^0-mesons. Assume a liquid hydrogen target is used.

3 The Λ^0 particle can sometimes be bound into a nucleus in place of a neutron.

Assuming the Λ-nucleon force is of an exchange type what particle is likely virtually exchanged and what is the expected range of the force?

4 The following K^+ decay modes have not been seen (branching less than 1%). Can you suggest why?

$$K^+ \rightarrow \pi^+ + \gamma$$
$$K^+ \rightarrow e^+ + \gamma.$$

5 The meson A1(1090) decays predominantly to $\pi + \varrho$. From this decay what can we conclude about the isospin, ordinary spin and parity of this state.

Chapter 13

1 Calculate mass values or relations between the masses of the missing members of the various negative parity octets mentioned in § 3. Show how the quark model might describe these particles.

2 Plot the positions of the known baryons on a graph of J^P vs rest mass and connect the Regge recurrences. List the positions and properties of the new baryons predicted up to a mass of 4 GeV.

3 What isospin states can be involved in the reactions

$$\pi^- + p \rightarrow \Lambda^0 + K^0$$
$$\pi^- + d \rightarrow \Lambda^0 + K^0 + n$$
$$\pi^+ + p \rightarrow \Sigma^+ + K^+$$
$$K^- + d \rightarrow \Lambda^0 + p + \pi^-.$$

Chapter 15

1 Show why the giant photonuclear resonance for $N = Z$ nuclei must have $T = 1$.

2 Calculate the temperature a source of ^{57}Co would have to be raised to, so that the recoil of the ^{57}Fe nucleus was compensated upon emission of the 14.4 keV γ-ray and resonance scattering can be observed. Assume that the cobalt is not in crystalline form.

Subject Index

Solutions and Hints

Chapter 1

1 0.458 MeV and 1.471 MeV.

2 10.01294 m.u. and 0.717 MeV.

3 7.56 MeV.

4 5.703 MeV.

5 $^{18}O(d, p)$ and $^{17}O(t, p)$.

6 About a factor 10^{12}.

7 Stable are: ^{64}Ni and ^{64}Zn.
^{64}Ga is β^+-active, end point 6.23 MeV,
^{64}Cu is β^+-active, end point 0.664 MeV
and β^--active, end point 0.578 MeV.
All neutrino velocities are c.

Chapter 2

1 3.6×10^{-6} litre.

2 See references mentioned in § 6.

3 After a procedure analogous to solution 2, the solution is straightforward.

Chapter 7

1 See formulae in § 14.4.

2 Estimate the transferred angular momentum from the energy of the particle and the radius of the nucleus. Use fig. 7.13b.

3 See eq. (7.9) ff.

4 All data are available to calculate from the Breit–Wigner level formula the resonance contribution to the cross section at a neutron energy equivalent to 2200 m/sec.

Chapter 8

1 $^{13}_{6}\mathrm{N}_7$

J^P	E (level) (MeV)	Possible configurations for protons:	neutrons:
$\frac{5}{2}^+$	3.56	$(1s\frac{1}{2})^2(1p\frac{3}{2})^4$	$(1s\frac{1}{2})^2(1p\frac{3}{2})^4(1d\frac{5}{2})^1$
$\frac{3}{2}^-$	3.51	$(1s\frac{1}{2})^2(1p\frac{3}{2})^4$	$(1s\frac{1}{2})^2(1p\frac{3}{2})^2(2s\frac{1}{2})^2$
$\frac{1}{2}^+$	2.37	$(1s\frac{1}{2})^2(1p\frac{3}{2})^4$	$(1s\frac{1}{2})^2(1p\frac{3}{2})^4(2s\frac{1}{2})^1$
$\frac{1}{2}^-$	0	$(1s\frac{1}{2})^2(1p\frac{3}{2})^4$	$(1s\frac{1}{2})^2(1p\frac{3}{2})^4(1p\frac{1}{2})^1$

In all three excited states the odd neutron is lift up to the next shell, which causes the high excitation energy. In the $\frac{3}{2}^-$ state even two neutrons are lift up to the next shell. The excitation energy of this state should be twice the excitation energy of the $\frac{1}{2}^+$ state (4.74 MeV), the difference may be caused by e.g. differences in pairing energy, l and s depending contributions to the energy, configuration mixing.

$^{17}_{8}O_9$	J^P	E (level) (MeV)	Possible configurations for protons:	neutrons:
	$\tfrac{3}{2}^-$	4.551	$(1s\tfrac{1}{2})^2(1p\tfrac{3}{2})^4(1p\tfrac{1}{2})^2$	$(1s\tfrac{1}{2})^2(1p\tfrac{3}{2})^3(1p\tfrac{1}{2})^2(1d\tfrac{5}{2})^2$
	$\tfrac{7}{2}^-$	3.846	$(1s\tfrac{1}{2})^2(1p\tfrac{3}{2})^4(1p\tfrac{1}{2})^2$	$(1s\tfrac{1}{2})^2(1p\tfrac{3}{2})^4(1p\tfrac{1}{2})^2(1f\tfrac{7}{2})^1$
	$\tfrac{1}{2}^-$	3.058	$(1s\tfrac{1}{2})^2(1p\tfrac{3}{2})^4(1p\tfrac{1}{2})^2$	$(1s\tfrac{1}{2})^2(1p\tfrac{3}{2})^4(1p\tfrac{1}{2})^1(1d\tfrac{5}{2})^2$
	$\tfrac{1}{2}^+$	0.871	$(1s\tfrac{1}{2})^2(1p\tfrac{3}{2})^4(1p\tfrac{1}{2})^2$	$(1s\tfrac{1}{2})^2(1p\tfrac{3}{2})^4(1p\tfrac{1}{2})^2(2s\tfrac{1}{2})^1$
	$\tfrac{5}{2}^-$	0	$(1s\tfrac{1}{2})^2(1p\tfrac{3}{2})^4(1p\tfrac{1}{2})^2$	$(1s\tfrac{1}{2})^2(1p\tfrac{3}{2})^4(1p\tfrac{1}{2})^2(1d\tfrac{5}{2})^1$

Again the high excitation energies of the 2nd, 3rd and 4th excited state are caused by lifting up the odd neutron to the next shell (see above).

In both cases the order and spacings of the single particle levels can be deduced from the data.

2 With the aid of the formula given in section 6.1 the positions can be calculated. Example: The lowest $T = \tfrac{3}{2}$ state in ^{13}C should have an excitation energy of about 16.3 MeV; $(M_n - M_p)c^2 = 0.8$ MeV and the value for the constant $3e^2/5R$ in formula (8.11) is then deduced from the known difference in ground state energy of ^{13}C and ^{13}N (3.2 MeV).

3 0.110 MeV transition: E1; 0.197 MeV transition: E2.

Chapter 9

1 From the spectrum it follows that $h^2/2\mathscr{I}_{\text{exp}} = 0.24$ MeV which has to be compared with the value deduced from the formula $\mathscr{I}_{\text{rig}} = \tfrac{2}{5}AMR_0^2$. $\mathscr{I}_{\text{exp}}$ appears to be smaller than $\mathscr{I}_{\text{rig}}$. Clearly a strictly independent particle model fails in predicting $\mathscr{I}$.

2 From the excitation spectrum it follows that $h^2/2\mathscr{I}_{\text{exp}} = 0.11$ MeV. Conclusion: $\mathscr{I}_{\text{exp}}(^{20}\text{Ne}) < \mathscr{I}_{\text{exp}}(^{21}\text{Ne})$. Possible reasons for the discrepancy may be the neglected rotation–particle coupling and the change in the deformation of the nucleus by adding one nucleon.

3 From the one-phonon state at 1.98 MeV a 0^+, 2^+, 4^+ triplet is predicted by the pure vibration model at twice the energy, i.e., at 3.96 MeV. The experimentally observed splitting of the triplet and the difference of 0.21 MeV in the observed and calculated energy may be caused by anharmonic terms in the Hamiltonian, the assumption on the nuclear matter or the neglect of particle–vibration coupling and particle excitations.

Chapter 10

1 The angular momentum of the resonant state equals the orbital angular momentum of the α-particle, l_α, as does the angular momentum of the γ-ray. If $C \neq 0$, then l_α should be at least 2.

2 Measurement of the polarisation of the γ-ray by Compton scattering (see section 6.1).

3 $P_3(^{22}\text{Na}) = +0.3 \cos \theta$; $P_3(^{60}\text{Co}) = -0.3 \cos \theta$. The opposite values of P_3 indicate that the helicities of the neutrino and the anti-neutrino are opposite.

Chapter 11

1 If the force mediated by the π-meson has the range a_0, the forces mediated by the η-meson and the K-meson would have ranges $0.25a_0$ and $0.30a_0$, respectively.

2 From eq. (11.23).

3 440 keV.

Chapter 12

1 From the arguments mentioned on page 363 it follows that not both parity and angular momentum can be conserved.

2 Consider this production in the momentum zero system. Then the formula of page 362 follows with $m_1 = 0$ and $Q = 135$ and 498 MeV, respectively.

3 Since the strong interaction is strangeness conserving the particle will have strangeness 1 and is probably a K-meson. Then the range of the force is $h(2\pi m_\text{K}c)^{-1}$ = 0.4 fm.

4 $K^+ \rightarrow \pi^+ + \gamma$ is not observed because of angular momentum conservation. In the zero momentum frame the angular momentum component along the direction of motion of the γ is zero for the orbital angular momenta and ± 1 for the spin of the photon.

$K^+ \rightarrow e^+ + \gamma$ is not possible since $e^+ + \gamma$ has half-integral angular momentum. Also lepton number is not conserved.

5   Assuming that it is not a weak decay T_3 is conserved and therefore T must be integral. If $\pi + \varrho$ has $l = 0$ then J^P is 1^+.

Chapter 13

1   See text.

2   See text.

3   Consider the isospin coupling Clebsch–Gordan coefficients for both initial and final states. The deuteron has $T = 1$.

Chapter 15

1   In this case the dipole operator (15.7) may be written as $D_i = e\varrho_i\tau_i$ where τ_i is the isotopic spin operator. The operator $\sum D_i$ transforms as $T = 1$ under rotations in isovector space. The ground state of the $N = Z$ nuclei has $T = 0$; therefore the excited state must have $T = 1$.

2   Hint: The key to the solution is the equality of the Doppler broadening of the line, and the recoil energy (pp. 445–447).

2   Beta-branches from ^{11}Be to ground state and second excited state at 2.12 MeV in ^{11}B. Mass ^{11}Be: $M{-}A = 0.0216 \pm 0.0001$ m.u.

3   The $Z-1$ nucleus has 0^+ ground state, 2^+ excited state at 0.4 MeV, 2^+ excited state at 1.2 MeV, all three are fed from spin 1 $_zA$ nucleus.

4   86.5 K, L, M and N line and, very probably, 196.4 K line.

Chapter 3

1   2.70 and 2.85 MeV protons, 8.75 MeV α-particles.

2 3.1×10^{-8} per keV energy interval.

3 (0.6×1.1); (1.5×1.2); (10.5×2.1); (100.5×2.8).

Chapter 4

1   3.0 MeV/E_n neutrons per cm^2 per sec.

2   Partial answer: The angles of emission are 51.5° for electrons and 41.5° for mesons.

3   3 h and 40 h.

Chapter 5

1   It will start melting after 1.8 sec. The 0.01 mm foil will last longer: the absorption loss decreases faster for thinner foils than the weight.

2 $E_p = 36.0$ MeV, $E_d = 18.3$ MeV, $E_\alpha = 36.7$ MeV.

3 8.2×10^6 litre or 370 kg.

Chapter 6

1   Specify the bombarding particles and their energies after choosing a nucleus. Estimate experimental conditions from the expected angular distribution.